Y0-DEU-535

WITHDRAWN
Stafford Library
Columbia College
1001 Rogers Street
Columbia, MO 65216

Earth Science: Physics and Chemistry of the Earth

Earth Science: Physics and Chemistry of the Earth

Volume 1

EDITOR

Joseph L. Spradley, Ph.D.
Wheaton College, Illinois

Salem Press
A Division of EBSCO Publishing
Ipswich, Massachusetts   Hackensack, New Jersey

Stafford Library
Columbia College
1001 Rogers Street
Columbia, MO 65216

Cover Photo: New Zealand - North Island - Rotorua area Wai-O-Tapu (Sacred Waters)
© Barry Lewis/In Pictures/Corbis

Copyright © 2012, by Salem Press, A Division of EBSCO Publishing, Inc.
All rights reserved. No part of this work may be used or reproduced in any manner whatsoever or transmitted in any form or by any means, electronic or mechanical, including photocopy, recording, or any information storage and retrieval system, without written permission from the copyright owner. For permissions requests, contact permissions@ebscohost.com.

The paper used in these volumes conforms to the American National Standard for Permanence of Paper for Printed Library Materials, Z39.48-1992 (R1997).

Library of Congress Cataloging-in-Publication Data

Earth science : physics and chemistry of the earth / editor, Joseph L. Spradley.
    p. cm.
  Includes bibliographical references and index.
  ISBN 978-1-58765-989-8 (set) – ISBN 978-1-58765-973-7 (set (1 of 4)) – ISBN 978-1-58765-974-4 (vol. 1) – ISBN 978-1-58765-975-1 (vol. 2) 1. Earth sciences. I. Spradley, Joseph L.
  QE26.3.E27 2012
  550–dc23
                    2012004322

First Printing

PRINTED IN THE UNTED STATES OF AMERICA

CONTENTS

Publisher's Note · · · · · · · · · · · · · · · · · · · vii
Introduction · ix
Contributors · xi
Common Units of Measure · · · · · · · · · · · xiii
Complete List of Contents · · · · · · · · · · · · xix
Category List of Contents · · · · · · · · · · · · xxi

Asteroid Impact Craters · · · · · · · · · · · · · · · 1
Biogeochemistry · 5
Carbon Sequestration · · · · · · · · · · · · · · · · 9
Climate Change: Causes · · · · · · · · · · · · · · 14
Continental Drift · 19
Creep · 24
Cross-Borehole Seismology · · · · · · · · · · · · 29
Deep-Earth Drilling Projects · · · · · · · · · · 35
Deep-Focus Earthquakes · · · · · · · · · · · · · 41
Discontinuities · 47
Earth-Moon Interactions · · · · · · · · · · · · · 53
Earthquake Distribution · · · · · · · · · · · · · 57
Earthquake Engineering · · · · · · · · · · · · · 63
Earthquake Hazards · · · · · · · · · · · · · · · · 69
Earthquake Locating · · · · · · · · · · · · · · · · 75
Earthquake Magnitudes and Intensities · · · · · · · 80
Earthquake Prediction · · · · · · · · · · · · · · · 87
Earthquakes · 93
Earth's Age · 98
Earth's Core ·102
Earth's Differentiation · · · · · · · · · · · · · · ·107
Earth's Interior Structure · · · · · · · · · · · · ·112
Earth's Lithosphere · · · · · · · · · · · · · · · · ·117
Earth's Magnetic Field · · · · · · · · · · · · · · ·122
Earth's Mantle ·127
Earth's Oldest Rocks · · · · · · · · · · · · · · · ·131
Earth Tides ·137

Elastic Waves ·141
Electron Microprobes · · · · · · · · · · · · · · ·146
Electron Microscopy · · · · · · · · · · · · · · · ·150
Elemental Distribution · · · · · · · · · · · · · ·155
Engineering Geophysics · · · · · · · · · · · · ·161
Environmental Chemistry · · · · · · · · · · · ·167
Experimental Petrology · · · · · · · · · · · · · ·173
Experimental Rock Deformation · · · · · · ·178
Faults: Normal ·183
Faults: Strike-Slip · · · · · · · · · · · · · · · · · ·189
Faults: Thrust ·194
Faults: Transform · · · · · · · · · · · · · · · · · ·199
Fission Track Dating · · · · · · · · · · · · · · · ·204
Fluid Inclusions ·209
Freshwater Chemistry · · · · · · · · · · · · · · ·215
Geobiomagnetism · · · · · · · · · · · · · · · · · ·221
Geochemical Cycle · · · · · · · · · · · · · · · · ·226
Geodetic Remote Sensing Satellites · · · · · · · ·233
Geodynamics ·238
The Geoid ·243
Geologic and Topographic Maps · · · · · · ·248
Geothermometry and Geobarometry · · · · · · ·252
Glaciation and Azolla Event · · · · · · · · · · ·258
Gravity Anomalies · · · · · · · · · · · · · · · · · ·262
Heat Sources and Heat Flow · · · · · · · · · ·267
Importance of the Moon for Earth Life · · · · · · ·273
Infrared Spectra · · · · · · · · · · · · · · · · · · ·278
Isostasy ·283
Isotope Geochemistry · · · · · · · · · · · · · · ·288
Isotopic Fractionation · · · · · · · · · · · · · · ·292
Jupiter's Effect on Earth · · · · · · · · · · · · ·297
Lithospheric Plates · · · · · · · · · · · · · · · · ·302
Lunar Origin Theories · · · · · · · · · · · · · · ·306

v

PUBLISHER'S NOTE

Salem Press's *Physics and Chemistry of the Earth* provides a two-volume introduction to the major topics of study in the earth's physical processes and structures. These volumes provide a comprehensive revision and update to an earlier edition, with the same title, published by Salem Press in 2001. The essays in this collection cover a wide range of subject areas, including plate tectonics, geochemical processes, geochronology, and seismology. The editor, Joseph L. Spradley, Ph.D., has reviewed each article for scientific authority, as well as ensured each article's currency for new developments and events significant in the field. Designed for high school and college students and their teachers, these volumes provide hundreds of expertly written essays supplemented by illustrations, charts, and useful reference materials, resulting in a comprehensive overview of each topic. Librarians and general readers alike will also turn to this reference work for both foundational information and current developments.

Each essay topic begins with helpful reference information, including a summary statement that explains its significance in the study of the earth and its processes. *Principal Terms* define key elements or concepts related to the subject and *Background and History* details important contextual information in the topic. The text itself is organized following informative subheadings that guide readers to areas of particular interest. An annotated *Bibliography* closes each essay, referring the reader to external sources for further study that are of use to both students and nonspecialists. Finally, a list of *Cross-References* directs the reader to other subject-related essays within the title. At the end of every volume, several appendices are designed to assist in the retrieval of information, including a *Glossary* that defines key terms contained in each set; a list of forty *Common Earth Minerals* with formulas and descriptions; *Earth Data* listing fundamental facts about the earth relative to surface and size, mass, density, and gravity, structure, and sun and moon relationships; and a *Timeline of Geophysical History* that lists the major names and findings that shape the field of study.

Salem Press's *Physics and Chemistry of the Earth* is part of a series of earth science books that includes *Earth's Surface and History, Earth Materials and Resources*, and *Earth's Weather, Water, and Atmosphere*.

Many hands went into the creation of this work. Special mention must be made of its editor, Joseph Spradley, who played a principal role in shaping the reference work and its contents. Thanks are also due to the many academicians and professionals who worked to communicate their expert understanding of earth science to the general reader; a list of these individuals and their affiliations appears at the beginning of the volume. The contributions of all are gratefully acknowledged.

INTRODUCTION

Physics and Chemistry of the Earth offers a survey of the planet's structure from crust to core, as well as an introduction to the major developments in areas of study relative to the earth's composition, climate, and mineral resources. Earth's ability to support a diverse array of life is the serendipitous result of many complex and interrelated processes. A major theme of *Physics and Chemistry of the Earth* is the interconnected nature of these processes and structures, as well as their relation to life-sustaining features, which make Earth unique among all known planets. Connections between the core, mantle, and crust of the earth are evident in the differentiation processes that formed the earth and made it a habitable environment through the shaping effects of plate tectonics, volcanism, and mountain building. Geochemical processes added to these shaping influences in forming the atmosphere and hydrosphere that govern climate and other environmental features. Gravitational interactions also reveal how closely the earth is connected to its neighbors in space as demonstrated by asteroid impacts, lunar influences, and Milankovitch cycles. Earth's magnetism also reveals these interconnections, extending from its liquid-core source to its solar-wind interactions that protect life on Earth. The rare Earth hypothesis dramatizes these connections by showing how closely they all relate to Earth life.

Several new essays on the earth's structure describe in detail how the planet's differentiation began when it was in the molten state, with heavier metals sinking to the center and the lower-density materials floating to the top. This resulted in a solid-metal core surrounded by a liquid-metal core, a plastic-flowing mantle, and a solid rocky crust at the surface as it cooled. In the cooling process, the crust was thin enough to crack into large crustal plates, and the mantle was warm enough from radioactive decay to provide convection currents to drive the crustal plates and produce volcanoes, mountains, and continents in a process called plate tectonics. This process is the result of a delicately balanced internal structure and thin crust that appears to be unique to the earth among all known planets.

Essays on geological processes show how plate tectonics contributed to continental drift and volcanism, with its associated outgassing that helped produce the atmosphere and form the oceans by condensation. The formation of continents ensured that the earth would not be completely covered with water, and continuing plate tectonic activity brings essential minerals to the surface and helps to control long-term climate patterns by recycling carbon dioxide. The study of plate tectonics has helped us to understand earthquakes and may lead to better methods of prediction. Earth is the only known planet that supports all these interconnected processes and their associated results that appear to be prerequisites for life.

Gravitational interactions connect Earth with other astronomical bodies, and several new essays describe the results of these interactions and their effects on Earth. One result that has become apparent from satellite photographs is a long history of asteroid collisions, which have produced craters of all sizes over the surface of the earth. Most of these have been largely hidden by weathering, but it has become clear recently that the larger of these collisions have played a role in mass extinctions at various times in the past. Such an event appears to have helped to end the dinosaur age some 65 million years ago and led to the dominance of mammals, and others may have ultimately contributed to increasing the diversification of animal life and the emergence of humans. The large mass and gravity of Jupiter has helped to deflect and absorb many asteroids and comets before they could reach the earth and cause more frequent mass extinctions. Recent evidence suggests that the moon was formed from a glancing collision of a Mars-size object, helping to remove greenhouse gases, increase the size of the iron core and rotation rate, thin the surface crust enough to permit plate tectonics, and provide other beneficial results. Gravitational interactions also have had long-term effects on the earth's climate, including the Milankovitch cycles that appear to have contributed to the ice-age glaciations that have occurred over the last several million years.

Earth's magnetism plays an important role for life on Earth, as discussed in some of the new essays. The earth's large liquid-iron core and rapid rotation rate, apparently resulting from the giant impact that formed the moon, produce a magnetic field about a hundred times stronger than that of any other rocky planet. This magnetic field is strong enough to deflect most of the solar wind and other high-energy cosmic rays into the polar regions, causing the northern lights. Recent evidence from satellite probes has shown that the solar wind strips atmospheric gases such as water vapor from Mars and Venus, which lack the magnetic protection that Earth has. Without this protection, Earth would have been exposed to high-energy particles, preventing the emergence of life. Earth's magnetism also undergoes occasional reversals, which have helped to reveal the mechanisms of plate tectonics and have probably accelerated genetic mutations.

All of the essays in *Physics and Chemistry of the Earth* reveal the interconnectedness of Earth structures and processes and describe how closely they are related to the existence of life on Earth. A special note of thanks is due to all the professionals and academicians who have shared their expertise in the earth sciences in a form accessible to the general reader. A list of their names and affiliations is included.

Joseph L. Spradley, PhD, Wheaton College, Illinois

CONTRIBUTORS

Arthur L. Alt
College of Great Falls

Valentine J. Ansfield
University of South Dakota

Richard W. Arnseth
Science Applications International

Michael P. Auerbach
Marblehead, MA

Alvin K. Benson
Brigham Young University

Elizabeth K. Berner
Yale University

David M. Best
Northern Arizona University

Rachel Leah Blumenthal
Somerville, MA

Joseph I. Brownstein
Atlanta, GA

James A. Burbank, Jr.
Western Oklahoma State College

Byron D. Cannon
University of Utah

Roger V. Carlson
Jet Propulsion Laboratory, California Institute of Technology

Robert S. Carmichael
University of Iowa

Habte Giorgis Churnet
University of Tennessee at Chattanooga

Robert L. Cullers
Kansas State University

E. Julius Dasch
National Aeronautics and Space Administration

Ronald W. Davis
Western Michigan University

James A. Dockal
University of North Carolina at Wilmington

Dean A. Dunn
University of Southern Mississippi

Steven I. Dutch
University of Wisconsin–Green Bay

George J. Flynn
State University of New York, Plattsburgh

Robert G. Font
Strategic Petroleum, Inc.

A. Kem Fronabarger
College of Charleston

Daniel G. Graetzer
University of Washington, Seattle

Hans G. Graetzer
South Dakota State University

Kristina Grifantini
Cambridge, MA

Gina Hagler
Washington, DC

Edward C. Hansen
Hope College

William Hoffman
Independent scholar

Earl G. Hoover
American Institute of Professional Geologists

Ruth H. Howes
Ball State University

Micah L. Issitt
Independent scholar

Pamela Jansma
Jet Propulsion Laboratory

Pamela R. Justice
Collin County Community College

Gary G. Lash
State University of New York College, Fredonia

M. Lee
Independent scholar

Gary R. Lowell
Southeast Missouri State University

David N. Lumsden
Memphis State University

Paul Madden
Hardin-Simmons University

David W. Maguire
C. S. Mott Community College

Mehrdad Mahdyiar
Leighton and Associates

Glen S. Mattioli
University of California, Berkeley

Otto H. Muller
Alfred University

Edward B. Nuhfer
University of Wisconsin–Platteville

Steven C. Okulewicz
City University of New York, Hunter College

Donald F. Palmer
Kent State University

Donald R. Prothero
Occidental College

Donald F. Reaser
University of Texas at Arlington

Jeffrey C. Reid
North Carolina Geological Survey

Richard M. Renneboog
Independent Scholar

Mariana Rhoades
St. John Fisher College

Raymond U. Roberts
Pacific Enterprises Oil and Gas Company

David M. Schlom
California State University, Chico

Kenneth J. Schoon
Indiana University Northwest

John F. Shroder, Jr.
University of Nebraska at Omaha

Elizabeth Shugart
Chattanooga State University

Stephen J. Shulik
Clarion University of Pennsylvania

R. Baird Shuman
University of Illinois at Urbana-Champaign

xi

Contributors

Joseph L. Spradley
Wheaton College

David Stewart
Southeast Missouri State University

Leslie V. Tischauser
Prairie State College

D. D. Trent
Citrus College

Ian Williams
University of Wisconsin–River Falls

Shawn V. Wilson
Independent scholar

Grant R. Woodwell
Mary Washington College

COMMON UNITS OF MEASURE

Notes: Common prefixes for metric units—which may apply in more cases than shown below—include giga- (1 billion times the unit), mega- (one million times), kilo- (1,000 times), hecto- (100 times), deka- (10 times), deci- (0.1 times, or one tenth), centi- (0.01, or one hundredth), milli- (0.001, or one thousandth), and micro- (0.0001, or one millionth).

UNIT	QUANTITY	SYMBOL	EQUIVALENTS
Acre	Area	ac	43,560 square feet 4,840 square yards 0.405 hectare
Ampere	Electric current	A *or* amp	1.00016502722949 international ampere 0.1 biot *or* abampere
Angstrom	Length	Å	0.1 nanometer 0.0000001 millimeter 0.000000004 inch
Astronomical unit	Length	AU	92,955,807 miles 149,597,871 kilometers (mean Earth-sun distance)
Barn	Area	b	10^{-28} meters squared (approx. cross-sectional area of 1 uranium nucleus)
Barrel (dry, for most produce)	Volume/capacity	bbl	7,056 cubic inches; 105 dry quarts; 3.281 bushels, struck measure
Barrel (liquid)	Volume/capacity	bbl	31 to 42 gallons
British thermal unit	Energy	Btu	1055.05585262 joule
Bushel (U.S., heaped)	Volume/capacity	bsh *or* bu	2,747.715 cubic inches 1.278 bushels, struck measure
Bushel (U.S., struck measure)	Volume/capacity	bsh *or* bu	2,150.42 cubic inches 35.238 liters
Candela	Luminous intensity	cd	1.09 hefner candle
Celsius	Temperature	C	1° centigrade
Centigram	Mass/weight	cg	0.15 grain
Centimeter	Length	cm	0.3937 inch
Centimeter, cubic	Volume/capacity	cm^3	0.061 cubic inch
Centimeter, square	Area	cm^2	0.155 square inch
Coulomb	Electric charge	C	1 ampere second

Unit	Quantity	Symbol	Equivalents
Cup	Volume/capacity	C	250 milliliters 8 fluid ounces 0.5 liquid pint
Deciliter	Volume/capacity	dl	0.21 pint
Decimeter	Length	dm	3.937 inches
Decimeter, cubic	Volume/capacity	dm^3	61.024 cubic inches
Decimeter, square	Area	dm^2	15.5 square inches
Dekaliter	Volume/capacity	dal	2.642 gallons 1.135 pecks
Dekameter	Length	dam	32.808 feet
Dram	Mass/weight	dr *or* dr avdp	0.0625 ounce 27.344 grains 1.772 grams
Electron volt	Energy	eV	$1.5185847232839 \times 10^{-22}$ Btu $1.6021917 \times 10^{-19}$ joule
Fermi	Length	fm	1 femtometer 1.0×10^{-15} meters
Foot	Length	ft *or* '	12 inches 0.3048 meter 30.48 centimeters
Foot, cubic	Volume/capacity	ft^3	0.028 cubic meter 0.0370 cubic yard 1,728 cubic inches
Foot, square	Area	ft^2	929.030 square centimeters
Gallon (U.S.)	Volume/capacity	gal	231 cubic inches 3.785 liters 0.833 British gallon 128 U.S. fluid ounces
Giga-electron volt	Energy	GeV	$1.6021917 \times 10^{-10}$ joule
Gigahertz	Frequency	GHz	—
Gill	Volume/capacity	gi	7.219 cubic inches 4 fluid ounces 0.118 liter
Grain	Mass/weight	gr	0.037 dram 0.002083 ounce 0.0648 gram

Unit	Quantity	Symbol	Equivalents
Gram	Mass/weight	g	15.432 grains 0.035 avoirdupois ounce
Hectare	Area	ha	2.471 acres
Hectoliter	Volume/capacity	hl	26.418 gallons 2.838 bushels
Hertz	Frequency	Hz	$1.08782775707767 \times 10^{-10}$ cesium atom frequency
Hour	Time	h	60 minutes 3,600 seconds
Inch	Length	in or "	2.54 centimeters
Inch, cubic	Volume/capacity	in^3	0.554 fluid ounce 4.433 fluid drams 16.387 cubic centimeters
Inch, square	Area	in^2	6.4516 square centimeters
Joule	Energy	J	$6.2414503832469 \times 10^{18}$ electron volt
Joule per kelvin	Heat capacity	J/K	$7.24311216248908 \times 10^{22}$ Boltzmann constant
Joule per second	Power	J/s	1 watt
Kelvin	Temperature	K	-272.15 degree Celsius
Kilo-electron volt	Energy	keV	$1.5185847232839 \times 10^{-19}$ joule
Kilogram	Mass/weight	kg	2.205 pounds
Kilogram per cubic meter	Mass/weight density	kg/m^3	$5.78036672001339 \times 10^{-4}$ ounces per cubic inch
Kilohertz	Frequency	kHz	—
Kiloliter	Volume/capacity	kl	—
Kilometer	Length	km	0.621 mile
Kilometer, square	Area	km^2	0.386 square mile 247.105 acres
Light-year (distance traveled by light in one Earth year)	Length/distance	lt-yr	5,878,499,814,275.88 miles 9.46×10^{12} kilometers
Liter	Volume/capacity	L	1.057 liquid quarts 0.908 dry quart 61.024 cubic inches
Mega-electron volt	Energy	MeV	—
Megahertz	Frequency	MHz	—

Common Units of Measure

Unit	Quantity	Symbol	Equivalents
Meter	Length	m	39.37 inches
Meter, cubic	Volume/capacity	m³	1.308 cubic yards
Meter per second	Velocity	m/s	2.24 miles per hour 3.60 kilometers per hour
Meter per second per second	Acceleration	m/s²	12,960.00 kilometers per hour per hour 8,052.97 miles per hour per hour
Meter, square	Area	m²	1.196 square yards 10.764 square feet
Metric. See unit name			
Microgram	Mass/weight	mcg *or* μg	0.000001 gram
Microliter	Volume/capacity	μl	0.00027 fluid ounce
Micrometer	Length	μm	0.001 millimeter 0.00003937 inch
Mile (nautical international)	Length	mi	1.852 kilometers 1.151 statute miles 0.999 U.S. nautical mile
Mile (statute or land)	Length	mi	5,280 feet 1.609 kilometers
Mile, square	Area	mi²	258.999 hectares
Milligram	Mass/weight	mg	0.015 grain
Milliliter	Volume/capacity	ml	0.271 fluid dram 16.231 minims 0.061 cubic inch
Millimeter	Length	mm	0.03937 inch
Millimeter, square	Area	mm2	0.002 square inch
Minute	Time	m	60 seconds
Mole	Amount of substance	mol	6.02×10^{23} atoms or molecules of a given substance
Nanometer	Length	nm	1,000,000 fermis 10 angstroms 0.001 micrometer 0.00000003937 inch
Newton	Force	N	0.224808943099711 pound force 0.101971621297793 kilogram force 100,000 dynes
Newton-meter	Torque	N·m	0.7375621 foot-pound

Unit	Quantity	Symbol	Equivalents
Ounce (avoirdupois)	Mass/weight	oz	28.350 grams 437.5 grains 0.911 troy or apothecaries' ounce
Ounce (troy)	Mass/weight	oz	31.103 grams 480 grains 1.097 avoirdupois ounces
Ounce (U.S., fluid or liquid)	Mass/weight	oz	1.805 cubic inch 29.574 milliliters 1.041 British fluid ounces
Parsec	Length	pc	30,856,775,876,793 kilometers 19,173,511,615,163 miles
Peck	Volume/capacity	pk	8.810 liters
Pint (dry)	Volume/capacity	pt	33.600 cubic inches 0.551 liter
Pint (liquid)	Volume/capacity	pt	28.875 cubic inches 0.473 liter
Pound (avoirdupois)	Mass/weight	lb	7,000 grains 1.215 troy or apothecaries' pounds 453.59237 grams
Pound (troy)	Mass/weight	lb	5,760 grains 0.823 avoirdupois pound 373.242 grams
Quart (British)	Volume/capacity	qt	69.354 cubic inches 1.032 U.S. dry quarts 1.201 U.S. liquid quarts
Quart (U.S., dry)	Volume/capacity	qt	67.201 cubic inches 1.101 liters 0.969 British quart
Quart (U.S., liquid)	Volume/capacity	qt	57.75 cubic inches 0.946 liter 0.833 British quart
Rod	Length	rd	5.029 meters 5.50 yards
Rod, square	Area	rd^2	25.293 square meters 30.25 square yards 0.00625 acre
Second	Time	s or sec	$\frac{1}{60}$ minute $\frac{1}{3,600}$ hour

Common Units of Measure

Unit	Quantity	Symbol	Equivalents
Tablespoon	Volume/capacity	T or tb	3 teaspoons 4 fluid drams
Teaspoon	Volume/capacity	t or tsp	0.33 tablespoon 1.33 fluid drams
Ton (gross or long)	Mass/weight	t	2,240 pounds 1.12 net tons 1.016 metric tons
Ton (metric)	Mass/weight	t	1,000 kilograms 2,204.62 pounds 0.984 gross ton 1.102 net tons
Ton (net or short)	Mass/weight	t	2,000 pounds 0.893 gross ton 0.907 metric ton
Volt	Electric potential	V	1 joule per coulomb
Watt	Power	W	1 joule per second 0.001 kilowatt $2.84345136093995 \times 10^{-4}$ ton of refrigeration
Yard	Length	yd	0.9144 meter
Yard, cubic	Volume/capacity	yd^3	0.765 cubic meter
Yard, square	Area	yd^2	0.836 square meter

COMPLETE LIST OF CONTENTS

Volume 1

Contents · v
Publisher's Note · vii
Introduction · ix
Contributors · xi
Common Units of Measure · xiii
Complete List of Contents · xix
Category List of Contents · xxi

Asteroid Impact Craters · 1
Biogeochemistry · 5
Carbon Sequestration · 9
Climate Change: Causes · 14
Continental Drift · 19
Creep · 24
Cross-Borehole Seismology · 29
Deep-Earth Drilling Projects · 35
Deep-Focus Earthquakes · 41
Discontinuities · 47
Earth-Moon Interactions · 53
Earthquake Distribution · 57
Earthquake Engineering · 63
Earthquake Hazards · 69
Earthquake Locating · 75
Earthquake Magnitudes and Intensities · 80
Earthquake Prediction · 87
Earthquakes · 93
Earth's Age · 98
Earth's Core · 102
Earth's Differentiation · 107
Earth's Interior Structure · 112
Earth's Lithosphere · 117
Earth's Magnetic Field · 122
Earth's Mantle · 127
Earth's Oldest Rocks · 131

Earth Tides · 137
Elastic Waves · 141
Electron Microprobes · 146
Electron Microscopy · 150
Elemental Distribution · 155
Engineering Geophysics · 161
Environmental Chemistry · 167
Experimental Petrology · 173
Experimental Rock Deformation · 178
Faults: Normal · 183
Faults: Strike-Slip · 189
Faults: Thrust · 194
Faults: Transform · 199
Fission Track Dating · 204
Fluid Inclusions · 209
Freshwater Chemistry · 215
Geobiomagnetism · 221
Geochemical Cycle · 226
Geodetic Remote Sensing Satellites · 233
Geodynamics · 238
The Geoid · 243
Geologic and Topographic Maps · 248
Geothermometry and Geobarometry · 252
Glaciation and Azolla Event · 258
Gravity Anomalies · 262
Heat Sources and Heat Flow · 267
Importance of the Moon for Earth Life · 273
Infrared Spectra · 278
Isostasy · 283
Isotope Geochemistry · 288
Isotopic Fractionation · 292
Jupiter's Effect on Earth · 297
Lithospheric Plates · 302
Lunar Origin Theories · 306

Volume 2

Contents · v
Common Units of Measure · vii

Complete List of Contents · xiii
Category List of Contents · xv

Magnetic Reversals	311
Magnetic Stratigraphy	317
Mantle Dynamics and Convection	322
Mass Extinction Theories	328
Mass Spectrometry	334
Metamorphism and Crustal Thickening	340
Milankovitch Hypothesis	344
Mountain Building	348
Neutron Activation Analysis	353
Notable Earthquakes	358
Nucleosynthesis	365
Ocean Drilling Program	371
Ocean-Floor Drilling Programs	377
Oxygen, Hydrogen, and Carbon Ratios	383
Petrographic Microscopes	388
Phase Changes	394
Phase Equilibria	400
Plate Motions	407
Plate Tectonics	413
Plumes and Megaplumes	419
Polar Wander	424
Potassium-Argon Dating	429
Radioactive Decay	435
Radiocarbon Dating	441
Rare Earth Hypothesis	448
Relative Dating of Strata	453
Remote-Sensing Satellites	458
Rock Magnetism	462

Rubidium-Strontium Dating	468
Samarium-Neodymium Dating	473
San Andreas Fault	478
Seismic Observatories	485
Seismic Reflection Profiling	491
Seismic Tomography	496
Seismic Wave Studies	502
Seismometers	507
Slow Earthquakes	514
Soil Liquefaction	519
Solar Wind Interactions	525
Stress and Strain	530
Subduction and Orogeny	535
Tectonic Plate Margins	541
Tsunamis and Earthquakes	547
Uranium-Thorium-Lead Dating	553
Volcanism	559
Water-Rock Interactions	564
X-ray Fluorescence	571
X-ray Powder Diffraction	576
Appendixes	581
Glossary	583
Bibliography	603
Common Earth Minerals	627
Earth Facts	629
Historical Timeline of Geophysical History	631
Subject Index	637

CATEGORY LIST OF CONTENTS

EARTH AS A MAGNET
Earth's Magnetic Field, 122
Geobiomagnetism, 221
Geodetic Remote Sensing Satellites, 233
Magnetic Reversals, 311
Magnetic Stratigraphy, 317
Polar Wander, 424
Rock Magnetism, 462
Solar Wind Interactions, 525

EARTHQUAKES
Deep-Focus Earthquakes, 41
Earthquake Distribution, 57
Earthquake Engineering, 63
Earthquake Hazards, 69
Earthquake Locating, 75
Earthquake Magnitudes and Intensities, 80
Earthquake Prediction, 87
Earthquakes, 93
Notable Earthquakes, 358
San Andreas Fault, 478
Slow Earthquakes, 514
Soil Liquefaction, 519
Tsunamis and Earthquakes, 547

EARTH'S STRUCTURE AND INTERIOR
Asteroid Impact Craters, 1
Deep-Earth Drilling Projects, 35
Deep-Focus Earthquakes, 41
Discontinuities, 47
Earth's Core, 102
Earth's Differentiation, 107
Earth's Interior Structure, 112
Earth's Lithosphere, 117
Earth's Mantle, 127
Geologic and Topographic Maps, 248
Glaciation and Azolla Event, 258
Heat Sources and Heat Flow, 267
Lithospheric Plates, 302
Mantle Dynamics and Convection, 322
Plumes and Megaplumes, 419
Rare Earth Hypothesis, 448

EXPLORING EARTH'S INTERIOR
Deep-Earth Drilling Projects, 35
Engineering Geophysics, 161

Geodynamics, 238
Ocean Drilling Program, 371
Ocean-Floor Drilling Programs, 377
Seismic Reflection Profiling, 491

GEOCHEMICAL PHENOMENA AND PROCESSES
Biogeochemistry, 5
Carbon Sequestration , 9
Climate Change: Causes, 14
Elemental Distribution, 155
Environmental Chemistry, 167
Fluid Inclusions, 209
Freshwater Chemistry, 215
Geochemical Cycle, 226
Geothermometry and Geobarometry, 252
Isotopic Fractionation, 292
Nucleosynthesis, 365
Oxygen, Hydrogen, and Carbon Ratios, 383
Phase Changes, 394
Phase Equilibria, 400
Rare Earth Hypothesis, 448
Volcanism, 559
Water-Rock Interactions, 564

GEOCHRONOLOGY AND THE AGE OF EARTH
Asteroid Impact Craters, 1
Climate Change: Causes, 14
Earth's Age, 98
Earth's Oldest Rocks, 131
Fission Track Dating, 204
Glaciation and Azolla Event, 258
Lunar Origin Theories, 306
Mass Extinction Theories, 328
Nucleosynthesis, 365
Potassium-Argon Dating, 429
Radioactive Decay, 435
Radiocarbon Dating, 441
Relative Dating of Strata, 453
Rubidium-Strontium Dating, 468
Samarium-Neodymium Dating, 473
Uranium-Thorium-Lead Dating, 553

GEODESY AND GRAVITY
Earth-Moon Interactions, 53
Earth Tides, 137
Geodetic Remote Sensing Satellites, 233

Geodesy and Gravity (continued)
Geodynamics, 238
The Geoid, 243
Gravity Anomalies, 262
Importance of the Moon for Earth Life, 273
Isostasy, 283
Jupiter's Effect on Earth, 297
Magnetic Reversals, 311
Milankovitch Hypothesis, 344
Polar Wander, 424

Plate Tectonics
Continental Drift, 19
Lithospheric Plates, 302
Metamorphism and Crustal Thickening, 340
Mountain Building, 348
Plate Motions, 407
Plate Tectonics, 413
Subduction and Orogeny, 535
Tectonic Plate Margins, 541

Seismology
Creep, 24
Cross-Borehole Seismology, 29
Deep-Focus Earthquakes, 41
Discontinuities, 47
Earthquake Distribution, 57
Earthquake Locating, 75
Earthquake Magnitudes and Intensities, 80
Elastic Waves, 141
Experimental Rock Deformation, 178

Faults: Normal, 183
Faults: Strike-Slip, 189
Faults: Thrust, 194
Faults: Transform, 199
Mountain Building, 348
San Andreas Fault, 478
Seismic Observatories, 485
Seismic Tomography, 496
Seismic Wave Studies, 502
Seismometers, 507
Stress and Strain, 530
Volcanism, 559

Techniques of Geochemistry
Cross-Borehole Seismology, 29
Earth's Age, 98
Electron Microprobes, 146
Electron Microscopy, 150
Experimental Petrology, 173
Geologic and Topographic Maps, 248
Geothermometry and Geobarometry, 252
Infrared Spectra, 278
Isotope Geochemistry, 288
Isotopic Fractionation, 292
Mass Spectrometry, 334
Neutron Activation Analysis, 353
Petrographic Microscopes, 388
Remote-Sensing Satellites, 458
Seismic Reflection Profiling, 491
X-ray Fluorescence, 571
X-ray Powder Diffraction, 576

Earth Science:
Physics and Chemistry of the Earth

A

ASTEROID IMPACT CRATERS

Impact craters are geological structures that are formed when an extraterrestrial object, most often an asteroid, hits the solid surface of a planet or a satellite. On Earth, these events have been responsible for creating dramatic land features like Arizona's Meteor Crater. In addition, some evidence links impact events with large-scale planetary effects, such as causing sudden changes in the earth's climate and triggering mass extinctions such as the one thought to have wiped out the dinosaurs.

PRINCIPAL TERMS

- **allochthonous:** rock or sediment that was not originally formed in its present location, but some distance away
- **asteroid:** a small, rocky body in orbit around the sun; the majority exist in a belt between Mars and Jupiter
- **breccia:** broken fragments of rock or mineral that have been fused in a matrix of sand or clay; often produced by impact events
- **ejecta:** the material that is thrown out of an impact crater during its formation
- **impactor:** any object, such as a meteorite, that collides with another body; the collision itself is known as an "impact event"
- **megaton:** a unit of force equivalent to the force produced by one million tons of the high-explosive TNT; used to measure the power of both impact events and nuclear weapons
- **meteorite:** a small extraterrestrial body, such as an asteroid or a comet, that has struck the surface of the earth; known as a meteor before impact and as meteoroid before it enters the earth's atmosphere
- **shatter cone:** a conical fracture in the surface of the earth, caused by an impact event and marked by distinct lines or ridges radiating outward from the apex
- **shock metamorphism:** permanent physical or chemical changes caused in rocks by a shock wave that is either generated by an impact event, or by an explosive or nuclear device
- **siderophile:** literally, "iron-loving;" refers to elements, such as platinum, palladium, osmium, and iridium, which are readily soluble in molten iron and found commonly in meteorites but extremely rare on the earth's surface
- **target rocks:** existing rocks on the surface of a planet that are smashed during a meteorite impact event
- **tektite:** a dark, glassy object, typically sphere-shaped, that is formed when molten debris flies out of an impact crater upon impact and cools in the air

SCIENTIFIC IMPORTANCE OF IMPACT EVENTS

Asteroid impact craters are formed when an extraterrestrial object hits the solid surface of a planet or a satellite. The vast majority of these objects are asteroids; a few are comets (which are made of ice and dust rather than rock). Although geologists have been identifying and studying terrestrial impact craters since the early twentieth century, the scientific community has only recently begun to explore their true geological significance in terms of understanding the Earth's past and predicting its future.

Critically, the results of space exploration have made it clear that almost every planet in our solar system has a surface that is pockmarked with signs of ancient impacts. (Other planets, including Mars, Mercury, and the moon, tend to retain many more ancient asteroid impact craters than Earth, whose surface is constantly renewing itself—burying old craters through processes such as volcanic activity, shifting tectonic plates, and erosion.) What has also become clear, largely through data gleaned from dating samples of lunar rock, is that the rate at which interplanetary bodies like asteroids smashed into planets was much higher about four billion years ago, during the formation of the solar system, than it is today. Most scientists now believe that this period of an unusually vigorous rain of rocky bodies, sometimes known as the late heavy bombardment, represented one of the most significant processes, shaping the solar system as we know it, including Earth.

For instance, one prevailing theory about the origin of the moon holds that it came into being as a result of an enormous celestial object, about the size of Mars, smashing into the still-forming Earth. This event is thought to have broken through the planet's crust, causing volatile vaporized gases, molten rock, and other debris to be ejected into space. This ejecta then clustered together and condensed, eventually forming a satellite planet that entered into orbit around Earth. It has also been suggested that meteorite impacts may have played a critical role in delivering to the proto-Earth a number of elements that were necessary for the formation of life on our planet—including both water molecules and simple organic compounds such as amino acids, proteins, and nucleotide bases.

In other words, asteroid impact craters represent physical remnants of the kinds of events that took place at the birth of our planet. Studying these structures is one way of gaining a deeper insight into Earth's complex geological history. Other asteroid researchers are engaged in the task of figuring out how to accurately predict and possibly deflect future impact events, since a large meteor on a collision course with our planet could have catastrophic effects. (It is worth noting that some popular literature and films have suggested that an asteroid hitting the earth could actually alter its mass enough to change its orbit—the shape of its path around the sun. However, the impact energy required to expel enough of the planet's mass to shift its orbit even in the most minor way is so tremendous that it would have to be generated by an asteroid that was many times larger than any scientists have ever observed; such an impactor would certainly destroy all life on the planet if it ever collided with the earth.)

FORMATION AND TYPES OF IMPACT CRATERS

Scientists have identified three main stages in the formation of an asteroid impact crater. During the first, known as compression, the impactor strikes its target, creates a small break in its surface, and delivers a shock wave that begins to flow through the impact site. This compresses the target and produces shock metamorphism effects (changes in the structure of the target rocks such as melting, vaporizing, or crushing). This stage lasts only a few microseconds, and very little material is yet being thrown up from the developing crater. During the second stage, known as excavation, the initial shock wave spreads both outward and upward from the impactor itself. This rapidly expands the size of the crater, and also shoots up a stream of vaporized and molten rock and other debris (the ejecta) that will land in the area around the crater. This period is also when the rim of the crater begins to fold over to form a lip. During the final stage, known as modification, loose or molten debris begins to collapse and fall back down into the crater; this has the potential to change the overall shape of the structure by forming shelf-like formations on the walls of the depression or even a high central peak made up of material that rebounds up from the crater floor. The entire process happens extremely quickly, especially the first two phases, which together may last only a few seconds.

Although the same three stages are seen in the formation of every asteroid impact crater, not all craters end up looking the same. Geologists generally categorize impact craters into two types: simple and complex. Simple craters typically have a bowl-shaped depression with a diameter that is about five to seven times as wide as the pit is deep. Complex craters are often shallower (the diameter of a complex crater may be up to twenty times as wide as the pit is deep), and though they may have started out with relatively steep walls, these are likely to have partially collapsed to form either a single peak or a ring of peaks within the depression. Complex craters are also generally larger in size than simple craters. Both types of craters are typically found partially filled with breccia, rock composed of broken fragments that have been cemented together.

The mass of an impactor and its velocity as it moves toward the earth are the two factors that determine its kinetic energy and, therefore, are largely responsible for the size of the crater that will be formed when it makes contact with the planet's surface. (Other slightly less critical factors that affect crater size and shape include the impactor's composition—and therefore its density—and the angle at which it strikes.) For example, if a meteorite of 30 meters (98 feet) in diameter that weighed 200,000 metric tons (about 440 million pounds) were to strike the earth at a velocity of about 30 kilometers (19 miles) per second, the kinetic energy it generated upon impact would be about 20 megatons of force, and the crater that formed would be well over a kilometer (or almost one mile) in diameter. These are approximately the forces that created Meteor Crater in Arizona some 50,000 years ago.

ARIZONA'S METEOR CRATER

Meteor Crater is an enormous pit that measures approximately 1,200 meters (almost 4,000 feet) in diameter and about 180 meters (almost 600 feet) in depth; it has an uneven rim that rises between 30 to 60 meters (100 to 200 feet) above the surrounding desert landscape. Unlike most craters formed as a result of volcanic activity, the crater floor is not found on top of a volcanic peak, but dips far below ground level. Located near the town of Winslow, Arizona, Meteor Crater is roughly bowl-shaped; however, it has four "corners" arising from tear faults in the earth's crust. These tear faults make it appear more rectangular than circular when viewed from above. Meteor Crater is also known as the Barringer Meteorite Crater, after the engineer who was the first to correctly hypothesize about its probable origin. In fact, it was the very first impact crater on the earth to be recognized for what it was.

Today, the impact event that formed Arizona's Meteor Crater is generally agreed to have been a massive extraterrestrial object smashing into the surface of the earth. The object was probably a fragment that had broken off from the asteroid belt between Mars and Jupiter about half a billion years ago and set off on a collision course with our planet.

For many years, the pit was believed to be the site of an extinct volcano, despite evidence pointing toward its extraterrestrial origin. In 1891, a mineralogist named A. E. Foote collected a large number of rock fragments from the crater. When analyzed, the rocks were found to be allochthonous: formed somewhere other than the location where they presently appear. Specifically, they were composed of nickel-iron alloys: a material that is extremely rare on the surface of the earth, but found in virtually all stony meteorites. But a series of incorrect observations and calculations led U.S. Geological Survey researcher G. K. Gilbert to discount the possibility of a falling space mass having created the pit; instead, Gilbert ended up backing the volcanic theory.

It was not until the early 1900's, when Philadelphia silver mining engineer Daniel Moreau Barringer bought the land containing the crater and conducted a set of independent drilling surveys at the site, that any serious evidence was collected to prove its origin. Among many other things, Barringer's experiments found that the ground beneath the crater contained millions of tons of silica that had been crushed to a powder, presumably by a tremendous pressure; that there were numerous spherules of iron meteorite found around the rim of the crater; and that there was no volcanic rock to be found anywhere near the site.

Despite Barringer's efforts, the final scientific confirmation that Meteor Crater was in fact formed by an ancient impact event would arrive only in the 1960's. That was when U.S. Geological Survey researchers Eugene Shoemaker, Ed Chow, and Don Milton collected samples from the site and discovered two crystallized forms of silica—coesite and stichovite—that are formed only at pressures of more than 200,000 kilograms per square meter (more than 300,000 pounds per square inch). These minerals had never been found in nature.

The research associated with the quest to clarify the true nature of Meteor Crater has had wide-ranging effects on the study of asteroid impact craters in general. The presence of coesite and stichovite, for instance, is now frequently used as a diagnostic marker of a historic impact event, along with other characteristic signs like shatter cones and scattered tektites. More than 160 confirmed impact craters have been identified across the world to date.

THE K/T EXTINCTION EVENT

The geological history of the earth has been marked by dramatic shifts in its climate and its life forms. At least one catastrophic change may have been related to an asteroid impact event that some scientists believe took place about 65 million years ago, between the time period that geologists call the Cretaceous and the Tertiary periods. This time of transition corresponds with the time at which the dinosaurs are believed to have gone extinct. Although there are conflicting theories about what actually happened during the K/T boundary, most scientists agree on a few basic facts.

One is that there was a change in the overall climate of the planet. During the Mesozoic era (an era is a unit of geologic time that is divided into periods), the climate on the earth was relatively warm and consistent. But the Cenozoic era, which followed immediately afterward, was much colder and also subject to greater fluctuations in temperature and rainfall.

In addition to these long-term changes, this era appears to have been a time of some unusual short-term weather phenomena that were unfavorable to life, such as toxic gases being emitted into the atmosphere and the falling of acid rain. These significant changes in climate are believed to have been responsible for the mass extinction that is known to have taken place at around this time. An enormous variety of organisms on land and sea, including the dinosaurs, completely disappeared from the face of the earth.

Some scientists believe that these changes in climate took place gradually, and were the result of intrinsic events on the earth's surface, such as volcanic activity and a shifting of tectonic plates that caused the oceans to recede from the land. Other scientists believe that they took place suddenly, and were the result of some extrinsic, or extraterrestrial event. The most well-accepted proposal within this school of thought holds that the event was, in fact, the collision of a large space object into Earth—in other words, an asteroid impact event. This is sometimes known as the Alvarez Hypothesis, after the University of California, Berkeley scientists Luis and Walter Alvarez who outlined it in its original form.

The Alvarez Hypothesis proposes that the widespread and catastrophic climate change that caused the extinction of the dinosaurs happened in the wake of a massive meteor, most likely an asteroid, colliding with Earth. This impact, the theory goes, resulted in the almost complete vaporization of the meteor itself—throwing up a thick cloud of dust over the planet and triggered the dramatic shifts in climate. One major piece of evidence for this theory is the fact that at many places across the world, a layer of clay containing a high level of the rare metal iridium has been found near the geological stratum, or sedimentation layer, that has been dated as having been formed during the transition between the Cretaceous and Tertiary periods. (Although iridium is rare on Earth, it is not an uncommon element in asteroids.) The same layer of sediment also contains soot, which may have been produced as a result of firestorms setting alight large swathes of forest; pieces of tektite, which may have been part of the impactor's ejecta; and quartz that showed signs of having undergone shock metamorphism. For a long time, researchers could not find a crater associated with this hypothetical impactor—but in the 1970's, a crater of plausible size was found on Mexico's Yucatán Peninsula. The crater, known as Chicxulub, is now considered to be the most likely site of the hypothesized impact event.

M. Lee

FURTHER READING

Bottke, William F., et al. "An Asteroid Breakup 160 Myr Ago as the Probable Source of the K/T Impactor." *Nature* 449 (September 2007): 48-53. A collision in the Mars-Jupiter asteroid millions of years ago, the authors argue, may have triggered both the mass extinction involving the demise of the dinosaurs and an increase in the number of impact events on Earth.

Grieve, Richard A. F. and Gordon R. Osinski. "Impact Craters on Earth." In *Encyclopedia of Solid Earth Geophysics*, edited by Harsh K. Gupta. Dordrecht, The Netherlands: Springer, 2011, 593-599. A succinct introduction to the formation, types, and significance of terrestrial impact craters.

Hodge, Paul. *Meteorite Craters and Impact Structures of the Earth*. Cambridge: Cambridge University Press, 2010. Describes each of the 139 known impact crater sites on Earth, including information about their size, age, location, and features.

Reimold, W. U. and R. L. Gibson. *Meteorite Impact!: The Danger from Space and South Africa's Mega-Impact; The Vredefort Structure*. Berlin: Springer, 2010. Focusing on a specific crater lets the authors explore geological concepts such as how scientists reconstruct details of a historic impact and how atmospheric events shaped the earth during its earliest days.

Stewart, I. S. and J. Lynch. *Earth: The Biography*. Washington, D.C.: National Geographic Society, 2007. In true National Geographic style, a spectacular book with hundreds of full-color photographs. The engaging writing stands up to the images. Chapter 1, "Impact," explores the effects of historic collisions, covers new techniques for studying collisions, and speculates on future catastrophes.

See also: Climate Change: Causes; Earth's Magnetic Field; Geologic and Topographic Maps; Mass Extinction Theories; Relative Dating of Strata.

B

BIOGEOCHEMISTRY

Biogeochemistry is the study of the inorganic chemical elements, such as carbon and nitrogen, in the biosphere, the hydrosphere, the pedosphere, the atmosphere, and the lithosphere. The focus is on the chemical cycles that are caused by or have an effect on biological activity of individual elements and their compounds at the boundary of living and nonliving systems. It is a systems discipline, closely related to systems ecology.

PRINCIPAL TERMS

- **atmosphere:** the gaseous mass or envelope surrounding a celestial body
- **biogeochemical cycle:** the cycle in which nitrogen, carbon, and other inorganic elements of the soil, atmosphere, and other parts of a region are converted into the organic substances of animals or plants and released back into the environment
- **biosphere:** the zone of the earth where life naturally occurs
- **biotic:** a product of life or living things or caused by living organisms
- **ecosystem:** a system composed of an interrelated community of animals, plants, and bacteria, together with their physical and chemical environment
- **greenhouse gases:** gases that are trapped in Earth's atmosphere and contribute to the warming of the planet
- **hydrosphere:** all the water on the surface of the earth
- **inorganic chemical elements:** composed of matter that is not animal or vegetable; not having the organized structure of living things
- **lithosphere:** the outer part of the earth; consists of crust and upper mantle
- **pedosphere:** the layer of the earth where soil is formed
- **trace element:** a chemical element, such as iron, copper, or zinc, that is essential in plant and animal nutrition, but only in minute quantities

ECOSYSTEM FEEDBACK

Biogeochemistry is a field of science that examines the interrelation of the inorganic and organic components of any given ecosystem. It encompasses a number of subdisciplines that are specific to individual ecosystems or regions, such as the lithosphere or pedosphere, in the form of cycles such as the carbon and nitrogen cycles. The practice of studying the effect of inorganic chemical cycles on the organic cycles reached prominence in the 1990's with concerns about global climate change and the growing realization that biofeedback loops were not functioning effectively at multiple levels. These failures were contributing to a warming of the planet, as well as to changes in habitat on local and global levels.

One of the works credited with spurring scientists to take a systems approach to environmental issues was the 1926 book, *The Biosphere*, by Russian scientist Vladimir Vernadsky. In his book, he laid out three spheres, each with laws governing that sphere. He also pointed out that human activity affects the spheres. With a growing appreciation for the crucial roles water and other natural cycles play in the health of an ecosystem, the next step was to understand the impact of chemical cycles in the health of those same cycles and systems.

ECOSYSTEM HEALTH

Biogeochemistry is a field of science where systems are key. This is because biogeochemical researchers focus on the health of systems. Virtually any system with a feedback mechanism for self-regulation falls under this discipline. As long as the feedback mechanism functions, the system can maintain its health and thrive. Once the feedback loop is compromised by chemical cycles at their interface, the system suffers. There is, then, reason for scientific inquiry to ascertain the extent of the damage, the implications for other systems, and the remedies that might be taken.

For example, researchers in the field of the biogeochemistry of wetlands study swamps, marshes, and floodplains. These areas provide habitats, groundwater recharge, flood control, and stabilization of the shoreline. Through biogeochemical transformations, they also improve water quality. Defining a wetland and determining the health of that wetland involves analyses of a number of factors involved in the decomposition of organic matter in wetland soils. These efforts include identifying severely limited oxygen levels, an accumulation of organic carbon, the presence of nitrogen as a major nutrient in both organic and inorganic form, and phosphorus retention. Ferric iron and manganese are also present in wetland soils, as are toxic organics, naturally occurring or synthetic chemicals that are detrimental to the wetland habitat even at low concentrations. An understanding of the interplay of these inputs gives researchers a window into the current and potential health of a wetland.

Monitoring Carbon and Nitrogen Cycles

Biogeochemistry is a discipline that is closely related to environmental concerns. Biogeochemical cycles are known by a variety of names. In the carbon cycle, carbon, an element, moves from the atmosphere to the plants through the process of photosynthesis. It is then moved to animals through the food chain, as animals eat the plants or eat other animals that have eaten the plants. Carbon is released back into the atmosphere through respiration or through the burning of fossil fuels. Bodies of water absorb some of the carbon. Some of it is trapped in the atmosphere as a greenhouse gas in the form of carbon dioxide. Too much carbon dioxide in the atmosphere is a major factor in global climate change. As a result of the carbon dioxide and other greenhouse gases, gases that trap heat in the atmosphere—the earth is becoming warmer. This causes changes in weather cycles that affect human life via phenomena such as drought, increased hurricane activity and severity, rising sea levels due to melting glacier ice and the expansion of warmer seawater, melting Arctic sea ice, rising water temperatures, negative effects on ecosystems such as coral reefs, and more acidic seawater as a result of carbon dioxide dissolving into the oceans.

Another cycle is the nitrogen cycle. On the earth, most of the nitrogen is found in the atmosphere. To be used by plants and animals, nitrogen must exist in a form that plants and animals can use. This occurs through the activity of nitrifying bacteria and the

Vladimir Vernadsky carried out pioneering work in the field of biogeochemistry. He expanded the concept of biosphere, originally defined by Eduard Suess, to the version accepted by scientists today. (RIA Novosti/Photo Researchers, Inc.)

effects of lightning strikes, which break down atmospheric nitrogen and fix it in the soil as nitrate ions. Other types of bacteria, called denitrifying bacteria, act upon excess nitrogen. As a result, the nitrogen enters the waterways and returns to the atmosphere. Too much nitrogen in the waterways and soil is of major concern as well. Too much nitrogen in lakes and bays results in an overabundance of aquatic plants. It is also a factor in ocean dead zones—areas where animals cannot survive. Excess nitrogen is also a form of air pollution and a contributor to greenhouse gases. By gaining an understanding of the role of chemical cycles in the shifting of major weather patterns, solutions may present themselves where there currently are none.

Those involved in biogeochemical research are monitoring these changes and assessing the impact they have on the systems that are vital to the health of our planet. They are doing this by keeping accurate records of things like rainfall and water temperatures, comparing those figures to historical figures, and constructing models that anticipate future events. They are also monitoring the change in glaciers, permafrost, and Arctic sea ice. An increase in the melting rates of these forms of ice is not necessarily a problem, but biogeochemists are the scientists who are in the position to make this determination.

AVENUES OF RESEARCH

Monitoring changes such as the changes in the thickness of Arctic sea ice is an excellent example of a research task in this discipline. Tracking changes in the thickness of the ice is just one part of the work. An assessment of the impact on the animals that are part of that ecosystem is also necessary. Since they are part of a cycle, the effects on them will also alter their impact on the system through changes in factors such as the birthrate. But melting Arctic sea ice is not just part of the immediate ecosystem that includes the habitat for wildlife living there. It is also a component in a larger cycle. It is part of the overall water cycle on the planet. It is also a factor in the temperature of adjacent ocean water that plays a part in other habitats, which in turn play a part in other systems.

Establishing baselines for variables such as soil health, nutrients, rainfall, the health of animals living in the habitat or ecosystem, and the quality of the air and water is another type of work done by biogeochemists. By establishing baselines in a habitat or ecosystem, researchers have a means for comparison over time. These comparisons will alert researchers to a decline or improvement in the health of the area. They are then in a position to explore the causes of these changes.

THINKING GLOBALLY AND LOCALLY

Biogeochemistry is an international discipline. Annual international conferences bring researchers from around the word together to discuss biogeochemistry and traceable elements, the biogeochemistry of wetlands, the biogeochemistry of forested areas, education, public policy, and other relevant issues. The list is long because this field of study involves cycles that stretch across continents and borders. Because of this, it is necessary to have a global perspective to appreciate activity in a given cycle. For example, climate change is not just a phenomenon affecting one continent or hemisphere. The causes and impact of climate change are global. Part of the work done by biogeochemists is work that familiarizes them with conditions in other parts of the system. It allows them to ask and answer questions such as, "What is happening to water temperatures in other hemispheres?" This information is important because a mistaken assumption will result in useless solutions and forecasts.

By the same token, some work is hyperlocal. Focusing on one wetland or one lake or the soil in one area may seem like taking too small a view, but the findings can be compared to findings from similar systems in other parts of the world because the variables in the system may prove to have characteristics in common. The hyperlocal systems may also be part of a larger system or systems, as in the case of the habitat of migrating birds that make their way from wetland to wetland along their route. Each wetland may have a specific set of characteristics, but there are shared characteristics that are essential to the success of the migratory species. By coordinating efforts, these essentials can be measured and tracked to assess the health of the habitats and the likely effect on the species that depend on them.

Scientists and researchers involved in the discipline of biogeochemistry take a systems approach. They familiarize themselves with the cycles at work in a given ecosystem and then determine which variables play roles in the health of that ecosystem. By measuring and tracking those variables along with the health of the system, they are then in a position to compare results with scientists around the world who are researching and tracking comparable systems. These comparisons may lead to collaboration on future studies or to solutions that will work on a global scale. Either way, the research addresses some of the most pressing issues of our time in a methodical manner that builds upon information already in hand while compiling new data that will be of use now and in the future.

Gina Hagler

FURTHER READING

Bashkin, V. N., and Robert W. Howarth. *Modern Biogeochemistry*. Boston: Kluwer Academic Publishers, 2002. Covers a large range of topics related to biogeochemistry, including terrestrial and aquatic ecosystems, nitrogen cycles, and the evolution of the lithosphere, atmosphere, and hydrosphere.

Biogeochemistry. Dordrecht: Martinus Nijhoff/Dr W. Junk Publishers, 1984. Internet resource. An e-journal with research on topics related to biogeochemistry, such as forest soil carbon inventories and dynamics in particular regions.

"Biogeochemistry." *Encyclopædia Britannica Online.* 2011. Web. 16 Oct. 2011. http://www.britannica.com/EBchecked/topic/65886/biogeochemistry.

Eleventh International Conference on the Biogeochemistry of Trace Elements. Information available here includes recognized authorities in the discipline as well as an overview of the research in progress on topics like arsenic and heavy metals in the soil. Web. 16 Oct. 2011. http://www.icobte2011.com.

Likens, Gene, et al. *Biogeochemistry of a Forested Ecosystem.* New York: Springer-Verlag, 1995. In-depth analysis of the biogeochemistry of a forested ecosystem; identifies components and their role in the overall system.

Reddy, Ramesh, and R. D. DeLaune. "Biogeochemical Characteristics." In *Biogeochemistry of Wetlands.* Boca Raton: Taylor & Francis, 2008. Covers wetland health and the variables that affect it.

Schlesinger, William H. *Biogeochemistry.* Amsterdam: Elsevier, 2005. Includes information on biomineralization, historic and current carbon cycles, and global oxygen, nitrogen, phosphorus, and sulfur cycles, anaerobic metabolism, the evolution of metabolism, and sedimentary hydrocarbons as biomarkers for early life.

_____. *Biogeochemistry: An Analysis of Global Change.* San Diego: Academic Press, 1997. Covers the atmosphere, lithosphere, biosphere, and hydrosphere, global water, carbon, nitrogen, phosphorus, and sulfur cycles.

See also: Carbon Sequestration; Deep-Earth Drilling Projects; Earth's Age; Earth's Oldest Rocks; Elemental Distribution; Environmental Chemistry; Freshwater Chemistry; Geobiomagentism; Geochemical Cycle; Geodynamics; Isotope Geochemistry; Lithospheric Plates; Mass Extinction Theories; Oxygen, Hydrogen, and Carbon Ratios; Radioactive Decay; Radiocarbon Dating.

C

CARBON SEQUESTRATION

With increasing levels of carbon dioxide in the atmosphere, the ability to remove and store atmospheric carbon is an important field of study. Many of the major storage areas—terrestrial, geological, and aquatic—are all interconnected and continued study is necessary to understand the impact of increased concentrations of carbon dioxide on these systems. New technologies to capture and use carbon dioxide can lead to real economical advantage if implemented correctly.

PRINCIPAL TERMS

- **acidification:** the increased presence of hydrogen or aluminum in water or soil, which decreases the water's acid/alkaline balance, or pH
- **anthropogenic:** caused by humans
- **carbon capture and storage (CCS):** the capture of carbon dioxide from industrial processes, which is stored in geological formations, biological organisms, or bodies of water
- **carbon sequestration:** the removal of carbon from the atmosphere for storage
- **carbon sink:** a reservoir of carbon that has been captured and stored for a short- or long-term period
- **cigatonnes carbon (GtC):** one billon tonnes of carbon
- **clathrate hydrates or gas hydrates:** crystal structures of gas molecules trapped or co-crystallized with water under conditions of high pressure and low temperature
- **enhanced oil recovery (EOR) method:** the use of carbon dioxide to enhance oil recovery from depleted fields
- **point source emitters:** a power or industrial plant that produces carbon dioxide during the plant's operation
- **terrestrial:** found on land

INTRODUCTION

Since the advent of the Industrial Revolution, the input of large amounts of carbon dioxide from human-derived sources into the atmosphere has increased. The annual rate of anthropogenic carbon release from energy related production was around 30 gigatonnes in 2007 and is expected to grow as more countries industrialize. Scientists claim that this large increase in carbon dioxide in the atmosphere has had a detrimental effect on the climate. Consequently, international policy has been created to mitigate carbon release, such as the Kyoto Protocol to tackle emissions of carbon dioxide and other greenhouse gases. One way to address the large amount of carbon being released into the atmosphere is to capture it, called carbon sequestration.

Carbon sequestration is the removal and storage of carbon from the atmosphere. This can be accomplished naturally, as wetlands and forests absorb carbon, or by way of anthropogenic techniques, such as power plant carbon dioxide capture and storage. A number of hurdles need to be overcome for carbon storage to be on the scale needed to compensate for the anthropogenic source production. Successful carbon storage processes are benign and inert to the environment, sufficiently long term, and commercially viable. Various natural and anthropogenic sources of carbon storage exist and are being developed, as are methods of capturing carbon dioxide at the source of production.

VEGETATION AND SOIL AS A SOURCE OF CARBON SEQUESTRATION

In the natural world, there are various reservoirs that provide both long- and short-term storage of carbon. There is a large reservoir of carbon found in plants and soil. Plant mass is estimated to hold about 600 gigatonnes carbon worldwide; soil is estimated to store slightly more than twice that amount, between 1400 and 1650 gigatonnes of carbon. A large percentage of the soil-based reserve of carbon is found in peat, which is rich in carbon-containing compounds. Terrestrial sources of carbon dioxide removal from the atmosphere account for between 1.5 and 2 gigatonnes of carbon removed per year, including crops, forests, and wetlands. The terrestrial storage can be either long- or short-term, depending on the type of plant and what occurs after the carbon is fixed into the plant matter.

Different ecosystems store carbon at different rates and it is unknown at this time what effect higher carbon dioxide levels will have on the uptake of carbon. Young forests tend to store carbon faster, but older growth forests tend to have more carbon stored, in both plant matter and soils. After a few thousand years, most reclaimed forests have equilibrated to a steady state of carbon storage where the input equals the output. With levels of carbon dioxide rising, the impact on forest growth is still under investigation. For example, studies have been conducted on the impact of higher concentrations of carbon dioxide levels on plant growth, with evidence suggesting that while initial growth is faster in most cases, carbon storage is often limited by other important nutrients such as nitrogen, phosphorus, or iron.

An important aspect of terrestrial carbon storage is to limit disturbances of previously stored carbon. Much of the present-day, terrestrial-stored carbon can be found in old growth forest (both tropical and temperate) and peat reserves found in wetlands. Limiting disturbances and increasing the size of these carbon sinks (natural or artificial reservoirs that store carbon) can increase, or at least slow down, carbon

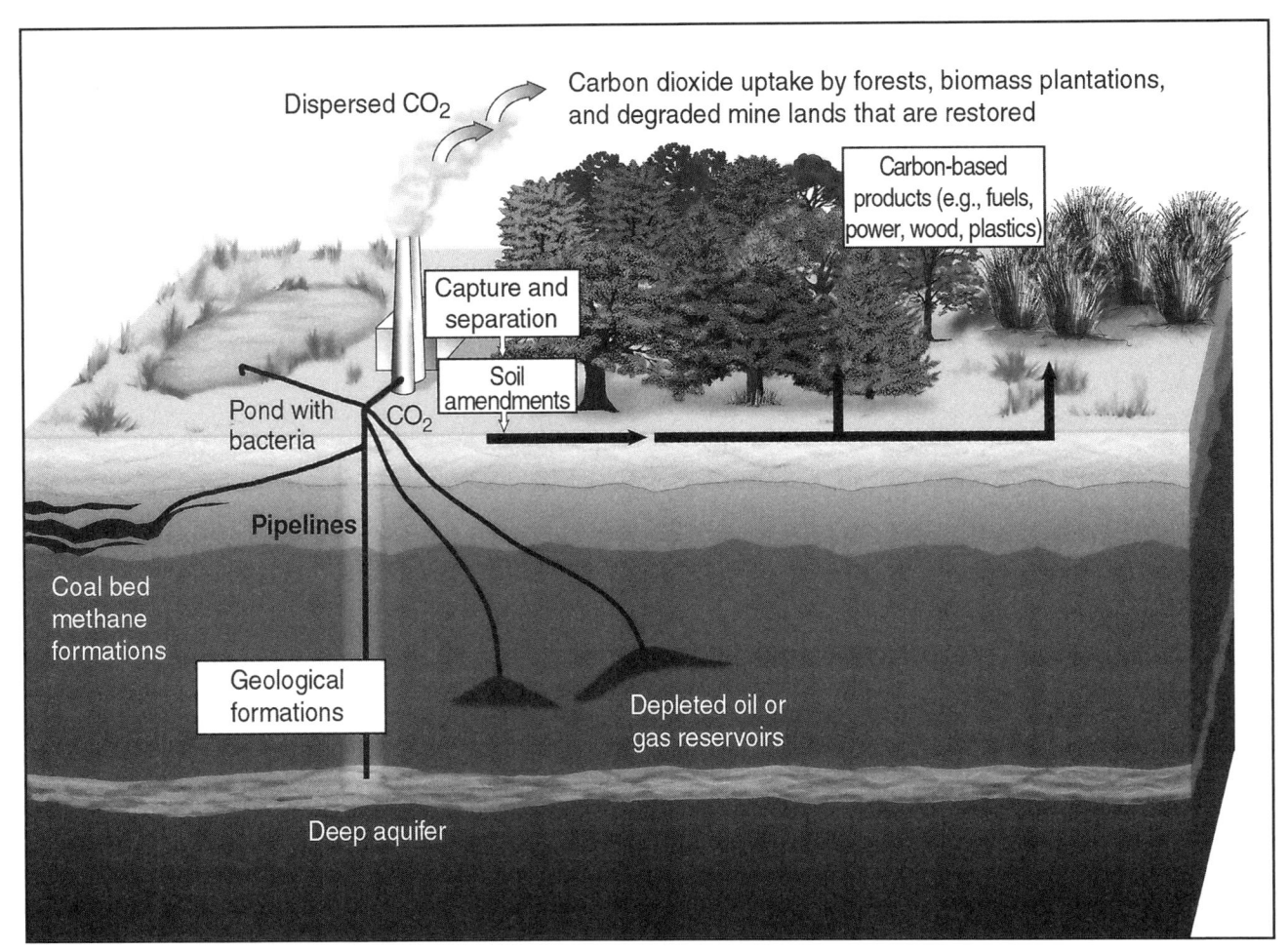

A drawing showing terresterial and geological sequestration of carbon dioxide emissions from a coal-fired plant. Oak Ridge National Laboratory. (Science Source)

loss. Large animals, human interactions, and forest fires are some of the most common disturbances to terrestrial carbon storage. For example, changes in agricultural practices can affect carbon storage. Limiting the destruction of old growth forest to make way for agriculture will maintain the large carbon reservoir, and evidence suggests that retaining crop residue and increasing soil nutrients can increase carbon levels in the soil by 30 to 50 percent in land that is already dedicated to farming. Studies also have shown that limiting oxidation of organic carbon in the soil will increase the overall carbon storage, as oxidized carbon is more likely to degrade back to carbon dioxide. Limiting oxidation can be accomplished by the burial of plant matter, such as trees or crop residues, so they are not in contact with the atmosphere. The burial of the plant matter also allows for the overall carbon content of the soil to increase. Further research is necessary to quantify the best practices for retaining or increasing carbon storage in plant material and soil.

WATER AS A SOURCE OF CARBON SEQUESTRATION

Carbon dioxide is also very soluble in water under certain conditions. Depending on temperature (when the temperature is increased, carbon is less soluble) and pressure (when pressure is increased, carbon is more soluble), large amounts of carbon can be and are stored in water. The stored carbon exists in water as carbon dioxide, carbonic acid, carbonates, and bicarbonate compounds. Two major sources of aquatic carbon storage are being studied—both oceans, which account for significant carbon storage by sheer volume, and underground aquifers, which can contain high concentrations of carbon.

The ocean already is a large dynamic reservoir of carbon dioxide from the atmosphere. Researchers estimate that the ocean absorbs about 2.3 gigatonnes of carbon yearly. The dissolved carbon plays an important role in aquatic life, as many marine species depend on soluble carbonate compounds for growth. However, the increase in these acidic carbon compounds has led to an acidification of the ocean. The acidification can led to poor solubility of compounds such as calcium carbonate, which is an important compound for marine organisms.

Some ideas have been put forth to use the ocean in a more active manner to store carbon. One method suggested is to inject the carbon dioxide deep into the ocean. After being injected, the carbon dioxide is theorized to follow a number of different paths. The carbon dioxide may dissolve and be absorbed by the sea water, or create large, deep-water carbon dioxide lakes, or become trapped in sediments or hydrates. Gas hydrates (clathrate hydrates) are crystallized structures of trapped gases and water that form under certain conditions of temperature and pressure—pressures and temperatures that can be found deep in the ocean, below about 500 meters. The effect of large carbon dioxide injections is still unknown.

Another carbon storing idea is to increase the fixation of carbon dioxide by marine plants and animals. Often, it is the lack of proper nutrients that retard marine growth. Fertilization of the ocean could lead to a larger fixation by the ocean of carbon-based materials. By creating more organic compounds, some of the fixed carbon is thought to settle to the sediment and be sequestered away in long-term storage. Researchers at the Department of Energy's Center for Research on Ocean Carbon Sequestration and associated organizations are conducting studies on the fate of both organic and inorganic carbon in the ocean. Preliminary studies show mixed results.

A second water source of carbon storage is in deep saline aquifers. Because of the great pressures found in the depths and the secluded nature of these aquifers, researchers postulate that a large amount of carbon dioxide can be stored for long periods. Researchers are looking into some of the possible pitfalls of deep aquifer storage. It is known that upon dissolving in the water, there is an acidification and sequential dissolving of carbonate rocks; it is not known if the dissolved compounds remineralize as other carbon-containing compounds and what impact that would have on the flow patterns found in the aquifer. Also, not much is known about where the carbon dioxide goes once it is injected. Many of these questions are being addressed by some of the first injection test sites in Norway and Canada that are capturing carbon dioxide and injecting it into deep saline aquifers and studying the resulting data.

RE-INJECTION INTO OIL, GAS, AND COAL FIELDS

Much research has been already done on using depleted oil and gas deposits as carbon sequestration sites. Injecting carbon dioxide into the natural gas and oil wells will sequester the carbon and sometimes makes the gas or oil field more productive by displacing the removed material with carbon dioxide. Using carbon dioxide in this manner has been around since the late 1970's and is known as the enhanced oil recovery method. The carbon dioxide helps to maintain the pressure of the reservoir, which reduces subsidence. It is hypothesized that between 120 and 150 gigatonnes of carbon could be stored in depleted oil and gas fields. The concern with this injection strategy is the feasibility of bringing concentrated carbon dioxide to the site of injection.

Coal mines are also good storage points for carbon sequestration. Carbon dioxide binds strongly to coal and will displace methane when injected into a defunct coal bed. This process could be advantageous, as the displaced methane can be collected and used for energy production and the potential reservoir of coal beds is thought to be between 50 and 200 gigatonnes of carbon.

ANTHROPOGENIC CARBON CAPTURE AND STORAGE

An important step in reducing carbon dioxide is reducing emissions from large producers, such as energy plants and industry. The best strategy for carbon capture and storage is to create techniques to be used with point source emitters, where the carbon dioxide is produced. A major source of carbon dioxide release comes from the process of energy production. During power production, combustion of carbon-rich fuels leads to release of carbon dioxide. Different types of power plants have different amounts of carbon released into the atmosphere based on the type of plant and the fuel source used. Coal-based plants are the main producers of carbon dioxide because their fuel source is rich in carbon and they are not very efficient in converting the fuel into energy. Other fuel sources, such as natural gas, produce less carbon but capturing that carbon is more complex. A large cost with carbon-capture technology is tied to retrofitting old plants or the building new plants that use these advanced carbon-capture techniques.

New power plants can be built that use techniques that are both more energy- and carbon-capture efficient. For example, new coal gasification plants, known as integrated gasification combined-cycle (IGCC) plants, turn coal into hydrogen gas and carbon monoxide. The produced gas can then be scrubbed of the carbon monoxide in a process that makes carbon dioxide. Next, the two gases can be separated to produce a pure stream of hydrogen gas to be used for energy production and a pure stream of carbon dioxide to be concentrated and stored.

Both old and new point source emitters can use various techniques to capture carbon dioxide. For carbon capture, both pre- and post-combustion capture are important tools in reducing emissions. During pre-combustion capture, the fossil fuels or natural gas are scrubbed of carbon dioxide; post-combustion processes remove carbon dioxide from the waste stream. The captured carbon dioxide is then concentrated and compressed for storage and transportation. A number of techniques are used to capture carbon dioxide at the source of emission. Amine solvents and cold methane have a high affinity for carbon dioxide and will take up carbon dioxide and remove it from the gas stream. Various membranes and activated carbon are also used to capture carbon dioxide gas.

The mineralization process is also harnessed, starting with a material such as calcium oxide that will react with carbon dioxide to make calcium carbonate (limestone). All of these techniques produce a concentrated mixture of carbon along with the carrier molecule. Many of these mixtures can then be recycled by heating the mixture to drive off a stream of carbon dioxide that is then concentrated and compressed for transport. For example, after producing calcium carbonate, the compound can be transported to a kiln. At the kiln, the reaction can be reversed, with the carbon dioxide which is released being concentrated for transport and the calcium oxide being reused to capture more carbon dioxide. All of these carbon-capture processes have an energy cost, which depends on which capture technique is used and how concentrated the carbon is in the waste stream. Different strategies are being explored to reduce these costs; however, the cost to capture and store carbon should go down as new techniques are discovered.

The same carbon-capturing techniques can be implemented with other point source emitters. Industry produces about 15 percent of all carbon dioxide during the production process. Because many of these plants produce the carbon dioxide directly from energy production, similar solutions would work on these processes as well.

Transportation and Storage

Carbon-capture technologies are constantly being explored, but also of significant importance is the transportation and storage/use of the carbon dioxide. Fifty percent of the cost of carbon capture and storage comes from the capture step, but the other 50 percent of the cost comes from transportation and storage costs. Finding a way to transport large volumes of liquefied carbon dioxide has sparked ideas such as using old gas pipelines for transport. Having carbon storage options close to the point of emission is also of concern to researchers and can help to control costs.

Outside of storage, alternative uses for carbon dioxide need to be found. If carbon-capture techniques are adopted on a large scale, a large amount of liquefied carbon dioxide will be available for use. By creating economical uses of the stored carbon dioxide, the cost to capture will be offset by the economic benefit of the final material. For example, creating plastic from carbon dioxide as a starting material will fix the carbon dioxide into a useful form. Also, using carbon dioxide to directly create biomass, such as algae farms, will create biomaterials that can be used as fuels or as starting materials for other products. The pitfalls and limitations of these ideas, though, still hamper their implementation.

Summary

For carbon sequestration to be viable, it will need to come in many forms, as both natural and industrial processes. Researchers are continuing to develop ideas for enhancing natural processes by exploring the possibilities of the ocean, plants, and soil to absorb more carbon. Most important in this quest, though, is to ensure that this increased uptake of carbon does not produce a more dangerous environmental situation. Industrial methods of reducing carbon emissions, as well as recycling and storing the carbon produced, are also integral to this discussion.

Elizabeth A. Shugart

Further Reading

Anderson, Soren T., and Richard G. Newell. *Prospects for Carbon Capture and Storage Technologies.* Washington, D.C.: Resources for the Future, 2003. A good overview of techniques for carbon dioxide capture, the cost of capture, and where capture can best be employed. Covers major point source emitters of carbon dioxide and how the techniques can be employed to capture carbon at these sources.

Khatiwala, S., F. Primeau, and T. Hall. "Reconstruction of the History of Anthropogenic CO_2 Concentrations in the Ocean." *Nature* 462, no. 7271 (2009): 346-349. A detailed overview of a method to quantify and measure anthropogenic carbon dioxide in the ocean. These methods can be used to predict better what happens to carbon dioxide when it enters the ocean.

Lorenz, Klaus, and R. Lal. *Carbon Sequestration in Forest Ecosystems.* Dordrecht: Springer, 2010. An important resource covering the most recent research on the ability of forests to sequester carbon and the processes involved. Also addresses future research needs.

Michael, K., A. Golab, V. Shulakova, J. Ennis-King, G. Allinson, S. Sharma, and T. Aiken. "Geological Storage of CO_2 in Saline Aquifers: A Review of the Experience from Existing Storage Operations." *International Journal of Greenhouse Gas Control* 4, no. 4 (2010): 659-667. Review paper on the pilot projects' results for deep saline aquifers. Covers the data from ten years of pilot projects, indicates how well the monitoring of the site is working, and discusses future research that needs to be done.

National Academy of Engineering, National Research Council. *The Carbon Dioxide Dilemma: Promising Technologies and Policies.* Washington, D.C.: National Academies Press, 2003. A good source of industrial and government research on carbon sequestration and carbon capture. The book highlights the research being done on sequestration in geological formation, the ocean, and in terrestrial ecosystems, along with research on carbon capture strategies.

Rackley, Steve. *Carbon Capture and Storage.* Oxford: Academic, 2009. Detailed book with associated references that covers carbon capture methods, transportation, and storage. Also has a number of useful links to further sources of information.

See also: Biogeochemistry: Climate Change: Causes; Environmental Chemistry; Freshwater Chemistry; Oxygen-Hydrogen, and Carbon Ratios.

CLIMATE CHANGE: CAUSES

Climate change is a general term to describe a set of long-term, significant alterations in the earth's climate; these include such phenomena as alterations in the amount of rainfall received by particular regions, rising sea levels and changes in ocean currents, and an average increase in global atmospheric temperatures. Climate change has taken place many times in the planet's history, but the current rate and the severity of its impact are unprecedented. Global temperatures are trending toward warmer averages; this trend is known as global warming, which can affect many other climate factors. Scientific evidence suggests that global warming is caused by the greenhouse effect—carbon emissions produced in the course of human activity, such as the agricultural release of greenhouse gases and the burning of fossil fuels.

PRINCIPAL TERMS

- **anthropogenic:** caused by or resulting from human activities; usually refers to man-made carbon emissions
- **carbon sequestration:** the process by which carbon is removed from the atmosphere and stored—for example, through the photosynthesis of trees and plants
- **desertification:** the degradation, or loss of biological productivity, of lands in arid (dry) or semi-arid areas; may include such effects as soil erosion, loss of natural vegetation, and deterioration of soil quality
- **El Niño Southern Oscillation (ENSO):** a fluctuation in the surface temperature and pressure of the Indian and Pacific oceans that leads to extreme weather events like floods and droughts; named after a warm water current that flows along the coast of Ecuador and Peru every few years
- **feedback:** processes that respond to climate forcings (see below) in such a way as to magnify the effects of climate change; for example, melting sea ice creates darker oceans that absorb more heat from the sun
- **forcing:** an event or phenomenon that drives an initial shift in climate, such as an increase in solar output
- **fossil fuels:** carbon-based fuels that are derived from the fossilized remains of living organisms, including coal, oil, and natural gas
- **global warming:** rising temperatures caused by the buildup of carbon dioxide in the earth's atmosphere
- **greenhouse gases:** any gas that contributes to the greenhouse effect (build-up of heat in the atmosphere) by absorbing infrared radiation; examples include water vapor, carbon dioxide, methane, and chlorofluorocarbons (CFCs)
- **heat island:** a region, usually within a city, marked by higher temperatures than its surroundings; this effect is created as a result of solar energy being absorbed by urban building materials like asphalt
- **thermal expansion:** an increase in the volume of water that results from its having become warmer; in the ocean, it leads to rising sea levels
- **thermohaline circulation:** the flow of water in the oceans that is driven by differences in density, temperature, and salinity; carries heated surface water from the tropics to the North Atlantic and cooler water at the ocean floor in the opposite direction
- **troposphere:** the part of the atmosphere that is closest to the surface of the earth; the region where weather phenomena occur

CONTEMPORARY CLIMATE CHANGE IN CONTEXT

The earth's climate has undergone many significant changes over the course of its 4.5 billion years of history. At some points in the planet's past, it has been almost completely covered in glacial ice; at others, the climate has been much warmer and more humid. These kinds of shifts have occurred at least seven times in the past 650,000 years alone. Historically, climate changes have most often taken place as a result of tiny variations in the orbit of the earth; these affect the amount of solar radiation, or energy, the planet receives on its surface. Other forcings, or events that drive climate shifts, have included volcanic activity and changes in the intensity of the sun's output.

A careful analysis of the data suggests that over the past two millennia, there have been three major shifts in the earth's climate. One began to occur about 1,100 years ago and lasted about 400 years; it is commonly known as the Medieval Climate Anomaly. During this time, there seems to have been a relative warming of Europe, Greenland, and Asia, and an unusually dry climate over much of the western part of North America. Another shift began to occur about 500 years ago and lasted about 350 years; it is known as the Little Ice Age. During this time, there seems to have been a small drop in average global temperatures. The most recent shift in the planet's climate began to occur about 250 years ago, at a time which coincided with the start of the industrial era, and is still occurring.

This current trend toward overall warming of average global temperatures is commonly referred to simply as "climate change." Two things differentiate contemporary climate change from historical shifts. One is the unprecedented rate at which it is proceeding. The other is the way in which it is occurring. Unlike past changes, this one is not primarily caused by variations in the planet's orbit or in solar intensity, or by volcanic activity. Instead, the scientific evidence suggests that the current period of climate change is closely linked with anthropogenic emissions of greenhouse gases; because these gases absorb solar radiation, they are causing a rapid build-up of heat within the earth's atmosphere (the "greenhouse effect"), that in turn is leading to a cascade of effects on the planet's climate.

A number of different types of data serve as key indicators of recent climate changes. Globally, records of air surface temperature taken from ships, satellites, and other vessels indicate that the average global surface temperature has risen about 0.8 degree Celsius (about 1.4 degrees Fahrenheit) since 1880. In addition, the twenty years with the warmest temperatures on record have all occurred in the past three decades. Sea levels are rising along with temperatures. On average, the mean global sea level has risen about 1.7 millimeters (about 0.07 inch) every year during the previous century, producing a total increase of about 17 centimeters (6.7 inches). But since 1993, this rate has sharply accelerated. Currently, the average rise in global mean sea levels is closer to 3.27 millimeters (about 0.13 inch) per year.

So far, most of this rise is attributed to thermal expansion, or the increase in the volume of ocean waters as a result of its becoming warmer. But as air temperatures continue to rise, scientists anticipate that melting ice from mountain glaciers around the world will also contribute to rising sea levels. It is estimated that between 2002 and 2006, about 150 to 250 cubic kilometers (36 to 60 cubic miles) of ice in Greenland melted each year. In Antarctica, about 152 cubic kilometers (36 cubic miles) of ice melted each year between 2002 and 2005. Mountain glaciers in Africa, Asia, and North America are also retreating, and all across the Northern Hemisphere, the average annual duration of snow cover is falling.

Global Sea Level Rise 1993-2011

Estimated change of 3.27 mm per year

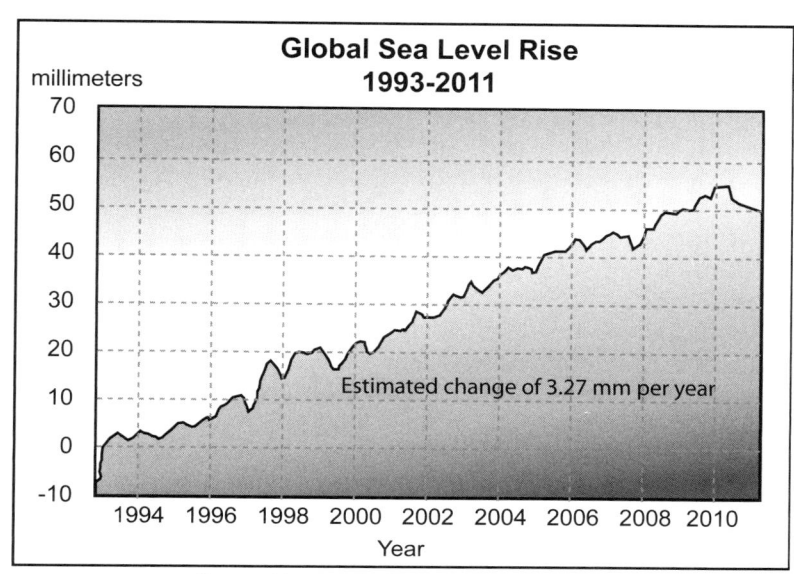

Source: Hansen, J., Goddard Institute for Space Studies, NASA, Table of Global-Mean Monthly, Annual, and Seasonal dTs Based on Met.station Data, 1866-present, http://climate.nasa.gov/keyIndicators/index.cfm#SeaLevel

NATURAL INFLUENCES ON CLIMATE CHANGE

Since solar radiation is the central source of heat energy within the earth's atmosphere, variations in the intensity of the sun's activity are a natural driver of climate change. There is evidence, for instance, that the Little Ice Age was caused by a temporary decrease in solar activity. However, changes in solar activity do not appear to be a major factor in the current trend of increasing global temperatures. One source of data about the sun's energy output is a collection of measurements that have been taken by instruments on board satellites since 1978. These figures do not reflect a rise in solar irradiance, or the amount of energy produced by the sun. In fact, they show a small decrease in the measurements over time. Another technique to measure solar irradiance indirectly involves looking at records of sunspots, or dark spots that appear temporarily on the surface of the sun. The number of sunspots that are visible at any given time is proportional to the amount of solar activity taking place at that time. Similarly, scientists can examine the amount of carbon contained in tree rings as a way of collecting a so-called proxy indicator of solar irradiance. This provides a fairly good reflection of the amount of solar energy a given tree received at various points in its lifetime.

None of these measurements suggests a strong role for solar irradiation in climate change. And statistical models of climate change that match the rises in temperature observed in the past one hundred years show that variations in solar irradiation can account for only a small fraction of the warming that has taken place. Finally, the pattern of warming does not correspond with what would be expected if the sun were simply putting out more energy into space. Instead of higher temperatures in every layer of the earth's atmosphere, only the troposphere, or the lowest layer of the atmosphere, is warming. The upper layers are actually cooling slightly. The best explanation for this phenomenon is that greenhouse gases are trapping heat energy from the sun close to the earth.

Another natural influence on climate change is volcanic activity. Volcanic eruptions can have an impact on the amount of solar radiation the earth's atmosphere retains. The carbon dioxide that is emitted during a volcanic event is a greenhouse gas, meaning that it absorbs heat from solar radiation and prevents it from escaping back into space. In the distant past, the earth was subject to periods of intense volcanic activity that are believed to have caused significant changes in the planet's climate during prehistoric times. However, the amount of carbon dioxide emitted as a result of volcanic eruptions today is 150 times less than the amount of carbon dioxide produced through human industrial activities. (Volcanic eruptions also emit aerosols, or suspensions of fine particles, like dust and ash, that become dispersed in the air. These aerosols tend to scatter solar radiation in such a way as to cause temporary drops in temperature.)

ANTHROPOGENIC CAUSES OF CLIMATE CHANGE

The earth's atmosphere is made up of a mixture of gases that act as a kind of "envelope" or "blanket" surrounding the planet. Some of these act as greenhouse gases. Gases that remain in the atmosphere for a long period of time act as forcings, or drivers of temperature rise. Gases that are short-lived, or that respond to rises in temperature, act as feedbacks. As long as a natural balance between "forcings" and "feedbacks" exists, the earth's temperature remains stable and life-supporting. For example, as the atmosphere becomes warmer, the amount of water vapor it contains increases—but at the same time, more clouds form and more precipitation falls, a feedback mechanism that helps to reduce water vapor back to a lower level.

What is driving global climate change is an intensification of this natural greenhouse effect, largely due to human emissions of greenhouse gases that stay in the atmosphere on a long-term basis and thus act as climate change forcings. Some of the most important of these are carbon dioxide, methane, nitrous oxide, and chlorofluorocarbons (CFCs).

Carbon dioxide is a gas that is released through many natural processes, including human respiration and the eruption of volcanoes. But since the beginning of the industrial era, a variety of human activities have dramatically increased the levels of carbon dioxide in the earth's atmosphere. For example, the burning of fossil fuels such as oil, coal, and natural gases causes a chemical reaction between the carbon that is stored in the fuels and the oxygen in the air, the product of which is carbon dioxide. The industrial activities that drive modern human society are estimated to have caused a sharp spike in atmospheric carbon dioxide levels within the past 150 years, from 280 parts per million to 379 parts per million.

Methane is another greenhouse gas that is produced as a result of human activities. Methane is emitted when waste materials in landfills decompose, and it is a major by-product of the digestion processes of livestock such as beef cattle and dairy cows. Both methane and nitrous oxide are also produced by various processes used to prepare soil for cultivation, including the use of heavy concentrations of nitrogen-based synthetic fertilizers.

Chlorofluorocarbons (CFCs) are artificial chemical compounds with a wide variety of industrial uses, including as refrigerants and as the propellants in aerosols (spray cans). CFCs are not only greenhouse gases, but they also destroy the ozone layer, a layer of the stratosphere that absorbs large amounts of solar radiation before it reaches the earth. Since these harmful effects were discovered, the production and use of CFCs has been phased out in many countries across the globe.

One other anthropogenic cause of climate change is deforestation, or the clearing of forested land to be used for agriculture, road-building, or the general urbanization of formerly rural areas of the world. Deforestation contributes to climate change in two ways. First, trees and other green plants remove carbon dioxide from the atmosphere through photosynthesis, a chemical reaction that takes place in green leaves. Photosynthesis transforms carbon dioxide into sugars and other organic compounds, and the carbon that is produced in this process is stored in the tissues of the plant or tree. This is known as carbon fixing, or carbon sequestration—and because of their ability to remove carbon dioxide from the air, forests are sometimes known as carbon sinks. The second way in which deforestation contributes to climate change is that most deforestation, especially in the tropics, is currently accomplished through burning. When trees are burned down to clear them, the carbon they have previously sequestered is released into the atmosphere.

In 1988, the United Nations established an international consortium of more than a thousand scientific experts from across the globe, known as the Intergovernmental Panel on Climate Change (IPCC). In 2007, the IPCC released its most recent report on climate change; the report concluded that based on all available scientific evidence, there is a greater than 90 percent probability that the rise in recorded global average temperatures over the past two and a half centuries has anthropogenic causes.

OCEAN CURRENTS AND CLIMATE CHANGE

A complex reciprocal relationship exists between global climate and ocean currents; both are affected by and affect each other in ways that scientists are still working to fully understand. One phenomenon that has received a great deal of recent research attention because of its potential to play a role in triggering "abrupt" climate change is thermohaline circulation. Thermohaline circulation describes a global flow of ocean currents that begins in the tropical waters of the South Atlantic Ocean. Intense solar heating warms the surface of the water in this region, causing evaporation that leaves the water saltier. The Gulf Stream carries a flow of warm, saline-heavy ocean water up to the North Atlantic. Here, the heat it has absorbed is released into the atmosphere, and the cooler, still salty surface water sinks to the ocean floor—where it then begins to travel south again to repeat the cycle.

Some scientists have suggested that as rising global temperatures cause melting of the ice sheet currently covering Greenland, as well as greater precipitation levels at higher latitudes, a large inflow of fresh water will be added to the North Atlantic Ocean. If this occurs, the surface waters in this region would become less dense—because the salinity of water is directly correlated to its density—and less likely to sink to the ocean floor. This could slow or even completely shut down the normal flow of water in the thermohaline circulation and potentially cause a series of sudden and catastrophic climate changes, including a significant cooling of temperatures in Europe, warming in the Southern Hemisphere, and an increase in the frequency and intensity of El Niño events (extreme weather events like floods and droughts, named after a warm water current that flows along the coast of Ecuador and Peru).

In the early years of the twenty-first century, a series of scientific studies seemed to suggest that a weakening of the thermohaline circulation was already happening, and that this slowdown was the result of anthropogenic causes. For example, a 2003 study found an increase in salinity in the tropical Atlantic and a decrease in salinity in the northern Atlantic, and a 2005 paper found a decrease in the strength of ocean currents in the Atlantic. More recent data, however, support the notion that these observations represent natural variations in the thermohaline circulation and are not directly related to anthropogenic climate change.

GLOBAL EFFECTS OF CLIMATE CHANGE

Climate change is already beginning to transform the earth's environment in a variety of ways, and scientists predict that these effects will accelerate as human activities continue to produce greenhouse gas emissions and average global temperatures continue to rise. The specific impact of climate change varies from region to region. In North America, for example, the snowpacks that cover the western mountains are expected to decrease in size. This is expected to result in higher precipitation and a greater likelihood of flooding during the wintertime, along with a reduction of water flow (and possible water shortages) during the summertime. The frequency and intensity of tropical storms hitting the eastern coastline of the continent is expected to increase, as well as the frequency and intensity of heat waves in urban areas that already experience them, such as Chicago and New York.

In Europe, scientists predict floods along the coast and flash floods in inland areas. Africa's current levels of water stress will increase as a result of more frequent droughts, and crop yields will decrease as both the length of the growing season and the amount of agriculturally viable land decline. In Latin America, significant decreases in precipitation levels are expected to cause water shortages, and areas of tropical forest are already being transformed to dry savannah as a result of changes in rainfall—which in turn leads to a high level of species extinctions. Asia's coasts are expected to experience greater numbers of floods, while higher water temperatures could lead to an increase in the number and severity of cholera infections. Finally, the current reductions in the size, longevity, and thickness of ice sheets, glaciers, permafrost, and sea ice in the polar regions will continue to worsen, causing ecosystem changes that will adversely affect animal habits and populations.

M. Lee

FURTHER READING

Boyce, Tammy, and Justin Lewis, eds. *Climate Change and the Media*. New York: Peter Lang, 2009. A collection of recent scholarly articles examining the relationship between the media, climate change research, and public understanding of climate change.

Broecker, Wallace. *The Great Ocean Conveyor: Discovering the Trigger for Abrupt Climate Change*. Princeton, N.J.: Princeton University Press, 2010. Broecker is a prominent voice in the field of abrupt global climate change. His book combines personal anecdotes drawn from decades of research and experience with sections containing rather dense technical information that assume a certain amount of prior knowledge on the reader's part.

Crowley, Thomas J. "Causes of Climate Change Over the Past 1,000 Years." *Science* 289, no. 5477 (July, 2000): 270-277. Crowley places recent changes in climate within a historical context by testing multiple mechanisms for global warming, and finds that the greenhouse effect has caused far higher than normal levels of variability in the earth's climate. Suitable for college students.

Cuff, David J., and Andrew S. Goudie. *The Oxford Companion to Global Change*. New York: Oxford University Press, 2009. This meticulously researched volume is organized like an encyclopedia; each substantial entry contains subheads, figures, tables, and a bibliography.

Faris, Stephen. *Forecast: The Surprising and Immediate Consequences of Climate Change*. New York: Henry Holt, 2009. This accessible book of popular science unpacks the effects of global warming across the world, from Darfur to the American coasts.

Henson, Robert. *The Rough Guide to Climate Change*. New York: Penguin, 2011. An up-to-the-minute overview of climate science for the beginner that is organized around a series of key questions and answers; highly readable, accurate, and well illustrated.

Kolbert, Elizabeth. *Field Notes from a Catastrophe: Man, Nature, and Climate Change*. New York: Bloomsbury, 2006. This sober but gripping book, written for a nonspecialist audience by former *New Yorker* reporter Kolbert, gains persuasive power from both the voices of experts and the personal stories of individuals, families, and villages affected by the consequences of climate change.

Rosenzweig, Cynthia, et al. "Attributing Physical and Biological Impacts to Anthropogenic Climate Change." *Nature* 453, no. 7193 (May, 2008): 353-357. The authors analyze a vast array of previously collected data on hundreds of physical systems and tens of thousands of plant and animal systems, showing that strong patterns of change are caused by human activity.

Stenchikov, Georgiy. "The Role of Volcanic Activity in Climate and Global Change." In *Climate Change: Observed Impacts on Planet Earth*, edited by Trevor M. Letcher. Boston: Elsevier, 2009. What does recent experimental and observational science have to say about the impact of volcanism on the earth's climate? This technical meta-analysis is best for students with an existing earth science background.

See also: Asteroid Impact Craters; Carbon Sequestration; Earth's Tides; Environmental Chemistry; Freshwater Chemistry; Geochemical Cycle; Glaciation and Azolla Event; Mass Extinction Theories; Oxygen, Hydrogen, and Carbon Ratios.

CONTINENTAL DRIFT

Continental drift, a theory first formally proposed in 1910, was proved only after the discovery of seafloor spreading more than fifty years later. Continental drift describes the dynamic movement of the continents as floating masses on the mantle material of the earth's interior. Plate tectonics, which encompasses the theory of continental drift, describes the mechanism whereby the convection of magmatic material drives continental movement and seafloor spreading.

PRINCIPAL TERMS

- **aluminosilicate:** rock and mineral molecular compositions of which aluminum and silicon are central atoms, primarily as various oxides
- **asthenosphere:** a zone of low seismic velocity between the lithosphere and the mantle
- **felsic:** descriptive of magma having both high silica content and high concentration of light-colored minerals like feldspar
- **isostasy:** the condition of equilibrium position of the continental plates atop the asthenosphere, equivalent to floating
- **lithosphere:** the solid material of the earth's crust and upper mantle, being 70-80 kilometers (40-50 miles) thick at the ocean floor and 100-150 kilometers (60-90 miles) thick elsewhere
- **magmatic:** composed of or originating in the molten rock, or magma, of the earth's mantle
- **paleomagnetism:** the record of the polarity of the planetary magnetic field embedded in the molecular structure of igneous rock material
- **subduction:** the process of old seafloor plate material being driven below the edge of a continental plate by the advancing seafloor as it spreads
- **terrane:** a coastal subsection of a continent, formed by the accretion of a small continental mass or plate to a larger one under the influence of seafloor subduction movement; generally, a fault-ridden rock mass relocated from its point of origin and unrelated to adjacent rock structures
- **zone refining:** a process of separation that may occur as mineral ores melt and freeze in a repetitive manner, thought to be one process responsible for the separation of specific elements and minerals in igneous rock formations

ORIGIN OF THE THEORY

In 1910, Alfred Wegener formally hypothesized that the continents had once been joined in a single supercontinent, but had since separated, producing the apparently matching coastal shapes of the continents as a result. Wegener proposed that the continental masses floated atop the denser molten material of the inner earth. Contemporary scientists believed that the surface of the planet was a static rather than a dynamic system, with earthquakes and volcanoes being artifacts of subterranean activity. Wegener's theory was summarily dismissed and ridiculed, and remained his only contribution outside of the field of meteorology. He died in 1930 while conducting field research on the glaciers of Greenland.

WEGENER'S THEORY FINDS SUPPORT

Several years after Wegener's death, geomagnetic phenomena were observed that directly supported the theory of continental drift. It was known that rock from magmatic sources retained an imprint of the planetary magnetic field as it cooled. Mapping this geomagnetic signature in the seafloor revealed bands in which the direction of the geomagnetic field had reversed. More importantly, the bands were not randomly oriented, but could be made to match up across fault lines. The only explanation for this observation was that the bands had moved relative to each other from the demarcation point of the fault line that crossed them. Such movement could occur only if the continental masses and the seafloor were able to move in accord with Wegener's theory. Similar mappings across the Mid-Atlantic Ridge revealed matching geomagnetic bands that progressed in opposite directions. The American geologists Harry Hess and Robert Dietz, working independently, concluded that the mid-ocean ridges were dynamic systems that produced new seafloor material through the mechanism of seafloor spreading. This added the final verification of Wegener's theory by demonstrating that eastern South America and western Africa had indeed been separating from each other for millions of years and must have once been a part of the same land mass.

Other evidence for the past connection of South America, Africa, and the other continents is found in the similarity of geological structures observed in those respective regions. Mapping of the continental shelves, the undersea edges of continental masses where they drop abruptly to the ocean floor, provides a closer matching of the continental shapes than do the land-sea coastal outlines. Paleontologists have found the remains of identical species in continents now widely separated by oceans. Taken altogether, the various data have led to the conclusion that some two hundred million years ago, the surface of the planet consisted of a single land mass (Pangaea) and a single ocean (Panthalassa), and under the influence of magmatic currents within the earth, the continental mass was split into sections that have been driven by those same forces into their present-day positions.

CONTINENTAL DRIFT AND PLATE TECTONICS

It is important to understand that "continental drift" and "plate tectonics" are two entirely different processes, though intimately related to each other. Continental drift refers specifically to the movement of the continental masses within the earth's crust and independently of it. Plate tectonics refers to the earth's crust as a whole and to the dynamic interactions of the various segments, or tectonic plates, of which it is composed. The difference can be visualized by covering the surface of a tub of water with a uniform layer of small Styrofoam pellets and a small number of large pieces of Styrofoam. If a heat source is applied to create convection currents in the water, it will be seen that the motion of the large pieces seems to be independent of the motions of the small pieces, while their motions nevertheless affect each other. The large pieces, representing the continental masses, move slowly, as they are most influenced by the underlying water convection currents. The small pieces move with much more vigor, moving away from upwelling convection currents and into broad circulatory motions. Plate tectonics and continental drift can be thought of as functioning in an analogous manner, though both processes are considerably more complex.

The floating Styrofoam analogy mimics the behavior of light crustal material on a much larger body of much denser material. To appreciate the nature of the processes that are taking place inside of the earth, however, it is necessary to have a workable description of the interior structure and dynamics of the planet. Seismic probing has indicated that the earth has at its center a large solid mass of nickel-iron, perhaps in the form of a compressed liquid. This inner core is surrounded by a more mobile, dynamic liquid layer. Recent models indicate that this liquid layer is not uniformly round, but is characterized by plumes that give it a shape more like a round ball with "spikes." Between this outer core and the crust is the mantle, composed of igneous rock material in a plastic state. The movement of this material by convection currents above the hot plumes of the outer core is believed to be the driving force for the eruption of magmatic flows from volcanic fissures, as well as for the displacement of crustal material in sea-floor spreading.

At the outer surface, the crust makes up only about 1 percent of the distance from the surface to the center of the planet. Crustal material is composed of lighter magmatic material, primarily of alumino-silicate rock, that has basically frozen. An apt visual

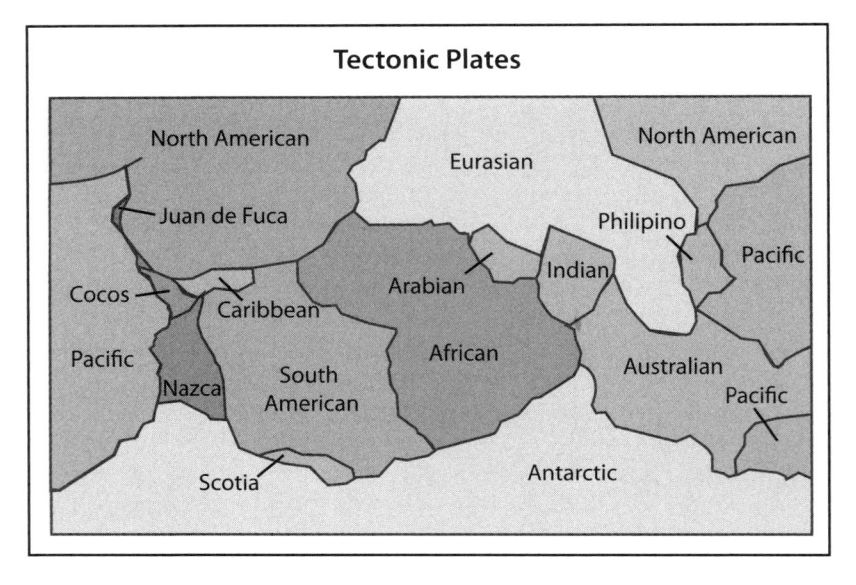

Primary and secondary tectonic plates; tertiary plates are incorporated into the major plates. Plate boundaries do not coincide with oceanic and continental boundaries, which have therefore been excluded.

representation of this system can be seen in video footage of certain lava lakes in which relatively slow movement of the molten rock allows the formation of a thin crust, still subject to fracturing and displacement, at the surface of the mass. The continental masses, or continental plates, float on the denser magmatic material of the mantle, moving slowly relative to the thin crustal material. This slow march of the continents is determined by a combination of factors, perhaps the most significant of these being that the crust entirely encloses the mantle like the shell of an egg. Resistance to the movement of a continental mass through this milieu requires that the crustal material be deflected. A closely related factor is that the crustal material moves toward the continental masses from opposing directions as it spreads from the mid-oceanic ridges of the Pacific and Atlantic Oceans. The new material of the seafloor wells up at the mid-ocean ridge and spreads outward, toward the continental masses, and works to hold the continental masses in place. Differential movement of the crustal material and the continental masses requires that the seafloor material must be deflected downward beneath the continental mass, or else it would continually accumulate. Subduction zones where this occurs are characterized by deep oceanic trenches. The continental side of such trenches is often characterized by mountainous ridges and seismic fault lines between the primary continental mass and various terranes, believed to have been smaller continental pieces driven into an original continent by seafloor spreading. The subduction of seafloor material places an upward force against the coastal regions of the continental masses that works to assist in maintaining the suspension of that mass on the mantle material. Torque and leverage effects from this upward force are felt far inland, as minor earthquakes and other seismic phenomena.

With this view of the dynamic nature of the planetary mass, geologists have come to accept Wegener's theory of "continental displacement" as a brilliant insight synthesized from observations in disparate fields.

Pangaea and Panthalassa

The ultimate conclusion of Wegener's theory of continental drift is that all of the current land masses were once conjoined in a single large body, which Wegner called "Pangaea," from the Greek words meaning "all land." Correspondingly, there was a single world-spanning ocean, which Wegener termed "Panthalassa," meaning "all sea." On the basis of the time scale of seafloor spreading, the age of various geological formations, and the paleomagnetic and paleontological records in detail, a time scale for continental drift places the existence of Pangaea to between two and three hundred million years ago. Pangaea is envisioned to have been driven slowly apart by seismic forces over millions of years into the two smaller supercontinents of Gondwana and Laurasia, and then into the smaller continental masses corresponding to the seven major continents known at the present time.

Using the present-day relative movement of the continents, and factoring in tectonic plate actions, the drift of continents has been modeled backward to provide an approximation of the appearance of Pangaea. It is not possible to know the coastal outline of that supercontinent with any degree of assurance, however, and any suggestion of how Pangaea was itself formed is sheer conjecture. At the fullest extent to which this may be known, Pangaea is supposed to have had rather an ovoid contour with the continent of Africa at its center. The northern region, Laurasia, contained the territories now known as North America, Europe and Asia. The southern region, Gondwana, conjoined South America, Africa, Australia, Antarctica and the Indian subcontinent in the same land mass. Support for this view is obtained from analysis of glacial scars in contemporary rock surfaces, and the identity of contemporary species found in the fossil record of those present-day regions.

The upwelling of magmatic material that occurs at the Mid-Atlantic Ridge is presumably the remnant of the process that broke apart Pangaea and continues to drive the separation of the continental masses. The effect pushes those regions to either side of Africa outward in a pivoting motion, for which the Eurasian subcontinent acts as the fulcrum or pivot point. Separation in this way resulted in the formation of the still-widening Atlantic Ocean, with northern Canada, Greenland, and Europe remaining in relatively close proximity as South America and Africa become more widely separated from each other and from Antarctica. At the other extreme, the movement threw the chain of Australian, Indian, and Antarctic subcontinental sections radially away from the African region to produce the Indian Ocean. Australia continues to drift north-eastward in isolation, while the Indian subcontinent was driven into the southern side of Asia, where it continues to push up the Himalayan mountain range.

One feature of Panthalassa that is recognizable to the present day from the geological record is the region known as the Tethys Sea. This body of water was in essence a gigantic bay that tapered inward between Laurasia and Gondwana to the eastern terminus of the North American subcontinent, separating regions that today comprise the Arctic shore of Asia, the north African shore of the Mediterranean Sea, and the seismically active fault zone between the Arabian Peninsula and the Middle East. Its present-day remnant is the Pacific Ocean.

The Study and Logic of Continental Drift

As an observable phenomenon, continental drift had to await technology that could detect it. Until the mid-1960's and the identification of the Mid-Atlantic Ridge, seafloor spreading was only a theory based on paleomagnetism data. Exploration of deep oceanic trenches provided the other half of the equation by identifying the process of oceanic crust subduction. Even taken together, however, these observations do not verify the movement of continents. The tool to measure such slow movement on a planetary scale did not exist until practical laser optics could be applied. The placement of laser monitoring devices that measure movement by the angular displacement of a laser beam have since allowed the direct measurement of the rates of seafloor spreading and the increase in separation between continents. The sensitivity of the electronics makes it possible to measure the time taken for an emitted light signal to be reflected back to a corresponding detector. In such an application, the greater the distance between light source and reflective surface, the more time is taken between emission of the signal and reception of the reflected signal, and the more accurately the time interval can be measured. Satellite relay of the signal provides an extraordinarily long distance for the signal to travel, and hence an extraordinary precision for the corresponding measurement. These are the techniques used to directly measure the movement of continents.

Coupled with paleomagnetism data, a complete picture of the dynamics of plate tectonics and continental drift is developing, in which Europe and North America grow farther apart by about 5 centimeters (2 inches) per year and other regions change relative positions at slightly different rates. Plate tectonics, though affecting the movement of continental masses, operate at rates seemingly independent of continental movement. Material upwelling in the Mid-Atlantic Ridge and the Mid-Pacific Rise pushes outward to build new seafloor at rates of between 1.8 and 4.1 centimeters per year for the former and between 3.7 and 18.3 centimeters per year for the latter. Subduction zones near the continental shelves direct old seafloor material downward below the continental masses, where it is reincorporated into the magmatic material of the mantle. At the same time, magma currents in the mantle, affected by the rotation of the planet, drive the large tectonic plates of the Pacific Ocean floor with a corresponding rotational force that engenders lateral slippage between plates to accompany the movement due to subduction. Given the age of the earth and the time over which Pangaea has split into the present distribution of continents, it is possible that this is a process that has repeated a number of times over the past two billion years. The geological record suggests that all ocean basins have closed and opened completely at least twice and as many as five times in that period of time.

Continental Drift and Human Civilization

The combination of forces embodied in continental drift and plate tectonics generates powerful earthquakes and volcanoes, sometimes with the horrendous aftereffects of tsunamis to devastate coastal regions. While the processes of plate tectonics and continental drift outwardly appear slow and methodical, their seismic effects are nonetheless sudden and unpredictable with any degree of accuracy. Accordingly, cities and other human constructions built on seashores and fault lines periodically suffer extensive damage from those effects, often with great loss of life. It is an unavoidable part of existence on a dynamic planet, and one to which proper attention should be paid to minimize the effects.

Richard M. Renneboog

Further Reading

Edwards, John. *Plate Tectonics and Continental Drift*. North Mankato, Minn.: Smart Apple Media, 2006.

This short book, intended for a general audience, provides an excellent overview of the basic principles and history of continental drift and plate tectonics, with excellent use of graphics to support the text.

Erickson, John. *Plate Tectonics: Unraveling the Mysteries of the Earth.* New York: Checkmark, 2001. This book provides a complete discussion of the evidence for continental drift and approaches the theory of plate tectonics as the means of understanding the geological dynamics of rocky planets.

Frankel, Henry. *The Continental Drift Controversy.* New York: Cambridge University Press, 2008. This three-volume set records the history and ultimate proof of Wegener's theory of continental displacement and the resolution of the debate about its validity.

Monroe, James S., Reed Wicander, and Richard Hazlett. *Physical Geology: Exploring the Earth.* 6th ed. Belmont, Calif.: Thomson, 2007. A textbook for college-level programs, this book provides a readable, in-depth discussion of plate tectonics, continental drift, and seafloor spreading set in a question-and-answer format.

Pinet, Paul R. *Invitation to Oceanography.* 5th ed. Sudbury, Mass.: Jones and Bartlett, 2009. This introductory college-level textbook provides a clear discussion of continental drift, seafloor spreading, and plate tectonics in the context of being the source of ocean basins.

Wegener, Alfred, and John Biram, trans. *The Origin of Continents and Oceans.* Mineola, N.Y.: Dover, 1966. This is a translation of Wegener's original publication of 1929, a succinct and erudite presentation of the theory of continental displacement.

Winchester, Simon. *Krakatoa: The Day the World Exploded, August 27, 1883.* New York: HarperCollins, 2003. Winchester does an excellent and entertaining job of describing and placing the violent eruption of Krakatoa within the context and theory of plate tectonics as a planetary process in this well-researched book.

Yount, Lisa. *Alfred Wegener, Creator of the Continental Drift Theory.* New York: Chelsea House, 2009. This biography of Alfred Wegener set against the context of contemporary geological science, describes the introduction of Wegener's revolutionary idea, its reception and the aftermath, and the end of Wegener's life.

See also: Creep; Deep-Earth Drilling Projects; Deep-Focus Earthquakes; Discontinuities; Earthquake Distribution; Earthquakes; Earth's Age; Earth's Core; Earth's Differentiation; Earth's Interior Structure; Earth's Magnetic Field; Earth's Mantle; Geodynamics; Geologic and Topographic Maps; Heat Sources and Heat Flow; Isostasy; Mantle Dynamics and Convection; Metamorphosis and Crustal Thickening; Mountain Building; Notable Earthquakes; Plate Motions; Plate Tectonics; Polar Wander; San Andreas Fault; Seismic Wave Studies; Seismometers; Slow Earthquakes; Stress and Stain; Subduction and Orogeny; Tectonic Plate Margins; Volcanism.

CREEP

Creep involves small deformations under small stresses acting over long periods of time. The effect of time on rock properties is important in understanding geologic processes as well as deformation and failure. In general, creep results in a decrease in strength and an increase in ductile or plastic flow.

PRINCIPAL TERMS

- **creep tests:** experiments that are conducted to assess the effects of time on rock properties, in which environmental conditions (surrounding pressure, temperature) and the deforming stress are held constant
- **dislocation:** a linear defect or imperfection in the atomic structure (arrangement) of rock-forming minerals; virtually all minerals and crystals contain dislocations
- **ductility:** the rock property that expresses total percent deformation prior to rupture; the maximum strain a rock can endure before it finally fails by fracturing or faulting
- **elastic deformation:** a nonpermanent deformation that disappears when the deforming stress is removed
- **plastic deformation:** a nonrecoverable deformation that does not disappear when the deforming stress is removed
- **strain:** the deformation resulting from the stress, calculated from displacements; it may involve change in volume, shape, or both
- **strain rate:** the rate at which deformation occurs, expressed as percent strain per unit time
- **stress:** the force per unit area acting at any point within a solid body such as rock, calculated from a knowledge of force and area
- **stress-strain test:** a common laboratory test utilized in the study of rock and soil deformation; stress is plotted versus strain throughout the test along the vertical and horizontal axes
- **ultimate strength:** the peak or maximum stress recorded in a stress-strain test

ENVIRONMENTAL FACTORS

Creep is an important geologic process related to rock deformation. It involves small displacements that occur under the influence of small but steady stresses that act over long periods of time. Scientists and engineers involved in experimental rock and soil deformation and the assessment of creep commonly perform stress-strain tests. These experiments are designed to deform earth materials in the laboratory under controlled conditions.

The effect of environmental factors such as surrounding (confining) pressure, temperature, pore-fluid pressure, and strain rate (or time) have been documented through the years based on countless tests. In essence, these factors dictate whether rocks will fracture as brittle substances or whether considerable ductile flow and creep strain will occur prior to rupture. The effect of increasing confining pressure on dry rocks (containing no appreciable amounts of liquid pore fluid) at room temperature is to increase both the ultimate strength and the ductility. Rocks tested under constant confining pressures tend to weaken and become more ductile as temperature increases. An increase in confining pressure on rocks saturated with pore fluids generally results in a decrease in both ultimate strength and ductility. This result is caused by the fact that part of the load (or stress) is carried by the pore fluid and less by grain-to-grain contacts. Decreasing the strain rate (or increasing the time during which the stress is applied) lowers ultimate strength and increases ductility—which basically defines the influence of creep strain on rock properties.

DEFORMATION STAGES

The mechanism of creep may be expressed as follows: Rocks subjected to the steady action of small stresses first undergo elastic deformation. After a given period of time, the elastic limit is exceeded. (The elastic limit is the point of no return beyond which deformation is permanent or nonrecoverable.) Following elastic deformation, rocks undergo strain hardening, a phenomenon characterized by a continuous rise in stress with increasing strain because of dislocations moving within individual mineral grains, interfering with one another and causing a literal "traffic jam" at the interatomic level. This initial stage of deformation comprising elastic behavior and strain hardening is termed transient creep; following the transient creep stage, steady-state creep is achieved. During this stage, rocks deform by plastic or ductile flow under a constant strain rate. Deformation mechanisms are characterized by gliding flow (intracrystalline movements) and by recrystallization. Gliding flow may take the form of translation or twin gliding. In translation gliding, layers of atoms slide one interatomic distance or a multiple thereof relative to adjacent layers. The overall mineral grain changes shape, but the interatomic lattice (arrangement) remains unchanged. In twin gliding layers, atoms slide a fraction of an interatomic distance relative to adjacent layers, distorting the interatomic lattice. Recrystallization involves rearrangements of the deforming minerals at the molecular scale through solution and redeposition by local melting or by solid diffusion. A common type of recrystallization occurs by local melting at those grain contacts experiencing the greatest stress and by precipitation (or redeposition) along grain contacts subjected to low stress. Recrystallization can also occur through mixing and rearrangement of the atoms and molecules in mineral grains by "spreading" into each other, analogous to the mixing of gases and liquids through the process of diffusion. Beyond steady-state creep, the final stage, known as accelerated creep, is reached. During accelerated creep, strain rate increases rapidly, ending in rock failure by fracturing or faulting. Deformation mechanisms during this final stage are characterized by cataclasis and formation of voids or pores. Cataclasis involves mechanical crushing, granulation, fracturing, and rotation of mineral grains. It results in intergranular movements.

Creep strain is equal to the sum of all of the stages of deformation, starting with elastic strain, followed by the transient stage, and culminating with steady-state and accelerated flows prior to failure by rupture. The rate at which creep strain occurs is very sensitive to temperature, with creep rates increasing rapidly as temperature rises. In fact, increasing temperature has been used as an alternative to experiments involving low strain rates or deformations over long periods of time. Increasing temperature or lowering strain rates affects rocks in a similar fashion by decreasing the ultimate strength and increasing the overall ductility.

LABORATORY AND FIELD STUDY

The study of creep is conducted in the laboratory in special experiments under controlled conditions. Environmental factors such as confining pressure, temperature, pore pressure, and strain rate are closely monitored and regulated. Among the methods that have been used to study creep are tension, bending, uniaxial compression, and triaxial compression. Pure tension has been utilized mainly to study creep in metals but has not been common in testing rocks. Bending is a simple method that has been used in creep studies of coal. By far, however, uniaxial and triaxial compression experiments have been utilized most often in testing creep behavior in rocks.

In uniaxial compression, rock samples are loaded with the stress directed vertically. The sample itself is generally unconfined laterally. The vertical or axial load is maintained at a constant level, and percent strain is plotted as a function of time.

In triaxial compression, a rock sample is loaded in a pressure chamber, and an all-around confining pressure is applied. The magnitude of the confining pressure can be significant, simulating pressure conditions expected several kilometers below the earth's surface. An axial or vertical load is then applied and maintained constant. The total stress along the vertical axis of the specimen is the sum of the axial load plus the confining pressure. The deforming stress therefore equates to the axial load. The latter is often referred to as differential stress or deviatoric stress because it is the stress that deviates from the all-around confining (or hydrostatic) pressure. The deviatoric stress is maintained at a constant level until failure occurs. Some pressure vessels are equipped with heating elements so as to increase the surrounding or ambient temperature; others have the additional capability of recording the increase in fluid pressure for samples saturated with pore fluids. Some of the recent designs have the capacity to subject samples to confining pressures of 20 kilobars (20,000 atmospheres), temperatures of 1,000 degrees Celsius, and strain rates as low as 10^{-10} per second.

Creep tests are not easy to run from a purely mechanical point of view. For example, the choice of magnitude of the deviatoric stress is a matter of difficulty and importance because each experiment may occupy an apparatus for a considerable time. (It is not unusual for creep tests to last for a period of one year.) In addition, if the stress is too low, little effect is produced; if it is too high, failure may occur too quickly. Temperature effects must be closely controlled because they can accelerate creep rates. Also, with many rocks, absorption of water produces effects similar to creep, so that humidity must be monitored and regulated.

Earth and soil creep that may eventually result in landsliding or damage to foundations and retaining structures can be studied in the field and laboratory. Evidence of creep strain along slopes may be detected by direct observation; bent or distorted tree trunks are common indicators. The rate of creep strain is recorded through installation and monitoring, or strain (displacement), gauges. The magnitude of pressures exerted on human-made structures resulting from creep flow can be predicted through laboratory experiments designed to record shear strength and shrink-swell (potential volume change) of soils and argillaceous (clay-rich) rocks. Specialized laboratory experiments simulating pressure-temperature conditions expected in the earth's mantle have been designed to study the effects of creep as a mechanism for releasing stored strain energy resulting in earthquakes.

Petrographic Study

Mechanisms common to creep (such as translation and twin gliding, recrystallization, and cataclastic flow) are routinely documented by studying thin sections of deformed rock specimens using the petrographic (polarizing) microscope and the universal stage. The petrographic microscope differs from a conventional model in that it is equipped with two polarizing elements and other accessories. When both polarizers (or nicols) are engaged, a ray of light transversing a mineral grain is generally refracted into two rays that vibrate in planes at right angles to each other. Analysis of the refraction of these rays makes it possible to identify the types of crystals and rocks that are involved and the nature of the creep process. In contrast, when the lower polarizer is the only one engaged, the light impinging on the mineral grain is plane-polarized. The universal stage allows the rock-forming minerals in the thin sections to be studied at different inclinations from vertical and horizontal axes.

Solifluction lobes on the side of a kame in the Nunatarssuaq region of Greenland. Solifluction is a special type of creep that occurs in cold climates where the soil is frozen most of the year. In summer, the ice in the upper layer of soil melts, and the soil becomes waterlogged and susceptible to downslope movement. (U.S. Geological Survey)

Applications for Structural and Engineering Geology

Understanding creep, or the effect of time on rock properties, is important to the structural geologist studying rock mechanics as well as to the engineering geologist concerned with landslide prediction and control or with the stability of earth-retaining structures. Deformation of earth materials may occur as brittle failures or after considerable ductile or plastic strain has resulted. Intuitively, it is easy to understand rock failure through fracturing or cracking, given that one tends to think of rocks as brittle substances. Under the influence of high confining pressures, elevated temperatures, and stresses acting over long periods of time, however, rocks can and do undergo considerable ductile or plastic deformation. Entire mountain chains of visibly folded rock are common throughout the planet.

Studying the process of folding, creep, and rock flowage has a number of practical applications. For example, it is common to find commercial quantities of oil and gas in folded structures known as anticlines. Therefore, understanding how and where rocks fold and which rock types are likely to develop the best porosity and permeability during the process is of critical significance in the search for new petroleum reserves. Similarly, quantifying creep strain and rates is very important in predicting, preventing, and correcting earth hazards such as landslides and in the proper design of foundations and retaining walls. On the subject of slope instability, creep can play a key role. Earth creep, or the slow, imperceptible downslope movement of soil and argillaceous rocks is the main cause of a specific type of landslide recognized worldwide. In this form of creep, considerable volumes of earth move as the sum of a very large number of minute displacements of individual particles and grains that do not necessarily strain at the same rate. This motion may be caused by expansion and contraction of clay-rich rocks in response to fluctuations in moisture content, which is especially critical in earth materials containing minerals from the smectite or montmorillonite family that expand considerably when wet and contract when dry. The end result of this creep strain is mass flow or landsliding. Similarly, soil creep can exert enormous stresses on retaining walls and foundations. Pressures exceeding 207,000 kilopascals or 2.1 kilobars (where one bar is basically equivalent to one atmosphere of pressure) have been recorded in north-central Texas.

Finally, creep is important in understanding earthquake mechanisms. Earthquakes are classified as shallow, intermediate, and deep based on their focal depth. Shallow earthquakes have focal depths not exceeding 70 kilometers. Intermediate earthquakes occur within a range of 70-300 kilometers. Deep earthquakes occur between 300 and 700 kilometers. The elastic rebound theory and the brittle failure of rock are accepted as the main mechanism giving rise to earthquakes—but only of the shallower types, because at depths where intermediate and deep earthquakes occur, the environmental conditions are conducive to ductile behavior. Convection currents in the earth's mantle and the thermal instability of creep have been proposed as the major mechanism responsible for the deeper earthquakes.

Robert G. Font

FURTHER READING

Cardno, Catherine A. "Reftrofit of Stadium Straddling Active Fault Moves Forward." *Civil Engineering.* 80 (2010): 12-14. This article presents the design development for a stadium on the Hayward Fault. Discussion of seismic creep and architecture meld in this text to offer a multidisciplinary view of creep.

Dennen, William H., and Bruce R. Moore. *Geology and Engineering.* Dubuque, Iowa: Wm. C. Brown, 1986. A complete reference dealing with the subject of engineering geology, this text is well illustrated and readable. Chapter 13 includes a discussion of slope stability and creep flow.

Font, Robert G. *Engineering Geology of the Slope Instability of Two Overconsolidated North-Central Texas Shales.* Vol. 3, Reviews in Engineering Geology. Washington, D.C.: Geological Society of America, 1977. A review of three distinct types of landslides, their causes, occurrence, and prevention along north-central Texas; one type is a classical example of creep strain and ductile flow. Although technical, the article is well illustrated and should be relatively easy for the nonscientist to follow.

Griggs, David T. "Creep of Rocks." *Journal of Geology* 47 (April/May 1939): 225-251. A classic reference on the subject of creep, this article is thorough and well illustrated. Covers the most important aspects of the subject as related to rock mechanics and structural geology. Although technical, it should not be too difficult for the nonscientist.

Heard, Hugh C. "Effect of Large Changes in Strain Rate in the Experimental Deformation of Yule Marble." *Journal of Geology* 71 (March 1963): 162-195. A very thorough and well-written article on the effects of time on rock properties. Technical, but definitely readable material for the nonscientist.

Hobbs, Bruce E., Winthrop D. Means, and Paul F. Williams. *An Outline of Structural Geology.* New York: John Wiley & Sons, 1976. Chapter 1 is a good review of mechanical properties of rocks, the concepts of stress and strain, and the response of rocks to stress. A fine discussion of ductile flow and creep strain is presented in the chapter, which is well illustrated.

Nabarro, Frank, and F. de Villiers. *The Physics of Creep: Creep and Creep-Resistant Alloys.* London: Taylor and Francis, 1995. This book introduces the physics of creep in language comprehensible to readers with little scientific background. The text begins with superalloys and ways to combat slope instability. Chapters 3 and 4 discuss the mechanics and types of creep.

Schulson, Erland M., and Paul Duval. *Creep and Fracture of Ice*. New York: Cambridge University Press, 2009. As the title suggests, this text focuses on creep models as they occur in ice. Presents the dynamics and characteristics of creep, comparing it with fractures. Chapter 13 provides a good description of the competition between creep and fracture under compression. Written for the advanced undergraduate or graduate student.

Spencer, Edgar W. *Introduction to the Structure of the Earth*. 3rd ed. New York: McGraw-Hill, 1988. This text has a good review of experimental study of rock deformation and is well illustrated. The effects of time on rock properties and the subject of creep are reviewed.

See also: Continental Drift; Cross-Borehole Seismology; Discontinuities; Earthquake Distribution; Earthquake Prediction; Earth's Age; Earth's Core; Earth's Interior Structure; Earth's Mantle; Elastic Waves; Experimental Rock Deformation; Faults: Normal; Faults: Strike-Slip; Faults: Thrust; Faults: Transform; Geodynamics; Petrographic Microscopes; Plate Motions; San Andreas Fault; Seismic Observatories; Seismic Tomography; Seismometers; Stress and Strain.

CROSS-BOREHOLE SEISMOLOGY

Cross-borehole seismology is a geophysical exploration technique involving the acquisition of data that can be used to image subsurface geology and determine the spatial distribution of the physical properties of geological materials. The data are acquired by placing seismic detectors in one borehole and a seismic source in an adjacent borehole. Seismic energy that propagates from the source travels through or reflects from subsurface geological materials. The total time of travel and amplitude of this energy is recorded by the detectors and is used to construct an image of the subsurface geology between the boreholes.

PRINCIPAL TERMS

- **amplitude:** the maximum departure (height) of a wave relative to its average value
- **imaging:** a computer method for constructing a picture of the subsurface geology from acquired seismic data
- **inversion (inverse problem):** using measured data to construct a geological model that describes the subsurface and is consistent with the measured data
- **lithology:** the description of rocks, such as rock type, mineral makeup, and fluid in the rock pores
- **reflectivity:** the ratio of the amplitude of the reflected wave to that of the incident wave
- **resolution:** the ability to separate two features that are very close together
- **seismic reflection profiling method:** measurements made of the travel times and amplitudes of events attributed to seismic waves that have been reflected from interfaces where seismic properties change
- **seismic tomography:** a processing technique for constructing a cross-sectional image of a slice of the subsurface from seismic data
- **seismology:** the study of seismic waves
- **travel time:** the amount of time it takes seismic energy to travel from the source into subsurface geology and arrive back at a seismic detector

CROSS-BOREHOLE VERSUS SURFACE EXPLORATION

Various methods exist that allow geoscientists to produce reasonable images of the earth's subsurface geology. Seismic data can be acquired using surface seismic exploration and cross-borehole seismology. When these data are properly processed and interpreted, an image of the subsurface is constructed.

In seismic reflection profiling, geophysicists typically arrange seismic detectors along a straight line at or near the surface of the earth and then generate sound waves by vibrating the ground. Seismic waves can be generated by detonating charges of dynamite, by dropping a weight on the ground, or by pounding the ground with a sledgehammer. To eliminate environmental risks associated with using explosives, a system called vibroseis is often used, in which a huge vibrator mounted on a special truck repeatedly strikes the earth to produce sound waves.

A seismograph records how long it takes sound waves to travel through or reflect from rock layers and arrive at seismic detectors. In seismic reflection profiling, the recorded data display the amplitudes of the reflected sound waves as a function of travel time. Such a graphic record is called a seismogram. The seismic source, detectors, and seismograph are then moved a short distance along the line, and the experiment is repeated. For seismic reflection profiling, the frequency range of investigation is limited to between 10 and 300 cycles per second, depending on the seismic velocity in the subsurface materials and the depth to the target of interest. This frequency range does not provide the necessary resolution of subsurface features that is typically required when making production decisions.

To help solve this problem, cross-borehole seismology was developed. In this method, a source of seismic energy is placed in one borehole, and appropriate seismic detectors are placed in an adjacent borehole. Boreholes can be vertical or horizontal. Between seven and twenty detectors are placed 2 to 5 meters apart and adjacent wells are at least 100 to 300 meters apart but always less than 1 kilometer apart. The source is fired, and the resulting seismic energy propagates through or reflects from the rock and is detected in the adjacent borehole. The travel times and amplitudes of seismic waves that have been transmitted or reflected through the rock mass between the drill holes are recorded. The source and detectors are then moved to another position in their respective boreholes, and the process is repeated. This procedure is continued until the region of interest is adequately covered by the propagating energy.

In contrast to seismic investigations conducted at the surface, cross-borehole surveys are performed at target depths; therefore, they do not suffer from the low-frequency resolution problems that are pervasive in surface seismic recordings. For cross-borehole seismology, the operating frequency range is from 400 to 30,000 cycles per second. Such data can provide high resolution of the subsurface geology between boreholes.

CROSS-BOREHOLE TECHNOLOGY

The idea of cross-borehole seismology has existed for many years. Field experiments were discussed as early as 1953 by Norman H. Ricker. Seismic surveys between boreholes were carried out in France by the Institute Français du Petrole in the early 1970's. Initially, the images produced from the data were fuzzy, filled with artifacts, and generally not worth the costs involved in acquiring them. Some geophysicists, however, saw in these images the proof of cross-borehole concepts and knew that with additional processing effort, coupled with advancements in seismic source and detector technology, the clarity and resolution would make the acquisition and processing of cross-borehole seismic data cost-effective.

Few disciplines affecting exploration geophysics developed more rapidly in the 1980's and 1990's than cross-borehole methods. Interaction among various scientific disciplines, along with advancements in seismic field data acquisition, imaging and inverse-problem theory, and computing speed, accelerated the development of cross-borehole seismology. Borehole source and detector technology, as well as computer-based data analysis algorithms, advanced to the point that routine application of the method became feasible by the late 1990's. Much of the progress in cross-borehole seismology has been driven by the development of powerful, nondestructive borehole seismic sources and by the increasing need for enhanced definition of oil-producing reservoirs.

Many borehole sources are made of piezoelectric ceramic materials that convert varying voltages in the material into mechanical vibrations, typically generating seismic signals in the range of 400 to 2,000 cycles per second. The same materials are also used in the seismic detectors because of their controllability, their high-frequency response, and their good impedance match to hard rock. The sources and detectors are packaged in metallic housings to form borehole probes that can operate in fluid-filled boreholes more than 1,000 meters deep. In the simplest borehole arrangement, the source and the detectors hang freely in the boreholes, but for better coupling or for operation in dry holes, more sophisticated sources and detectors with electrically powered clamping mechanisms lock the seismically active parts firmly against the rock in the borehole. For clamping detectors, the piezoelectric material is oriented to respond to seismic waves in three dimensions, allowing both compressional and shear waves to be recorded.

Another important borehole source was developed by the American oil company Texaco in the 1990's. A nondestructive, broadband air-gun array provided energy transmission over distances exceeding 600 meters. The broadband nature of the source provided the necessary spatial resolution of the subsurface.

Although the source waveform may be a pulse, transmitting a continuous, coded signal is a better way to achieve maximum transmission range. In the mid-1990's, a high-energy, broadband, clamped borehole vibrator was designed and successfully implemented to accomplish this task. Inside the borehole vibrator, a 114-kilogram mass is suspended below a hydraulic piston that is set into axial motion by a hydraulic valve. The vibrator clamp is coupled to the motion of the piston, which transmits stress to the borehole wall and sends seismic energy into the formation. This is analogous to surface vibroseis. The vibrator source transmits a continuous, controlled signal over a long period of time, keeping the stress low while transmitting considerable energy. Thus, a large amount of seismic energy is transmitted into the formation without harming the cement casing in a cased borehole.

SEISMIC TOMOGRAPHY

By applying the techniques of tomographic imaging, cross-borehole seismic data can be used to image the subsurface and to estimate subsurface physical properties. Tomographic imaging is a highly effective way of condensing and organizing the large amount of information contained in high-density cross-borehole seismograms. The medical community has used computed tomography (CT) scanning since the 1960's to generate imaged cross sections of different portions of the human body. In medical imaging techniques, such as X-ray tomography, the source and receiver rotate all the way around the object to be imaged. Cross-borehole seismology is similar to the medical case, except the angular coverage is not nearly as great.

In the early 1980's, geophysicists began applying techniques similar to those of medicine to earth science problems. These problems range from estimating the internal velocity structure of the subsurface to formulations that provide a complete three-dimensional image of the subsurface geology. In the early 1990's, tomographic reconstruction became a standard technique for analyzing cross-borehole and surface seismic data.

Obtaining information about subsurface geology from cross-borehole data constitutes a type of inverse problem. That is, measurements are first made of energy that has propagated through and reflected from within the subsurface. The received travel times and amplitudes of this energy are then used to estimate values of the physical parameters of the medium through which it has propagated. The parameters that are typically extracted are velocities and depths, from which a gross model of the subsurface structure can be derived. Initially, gross subsurface structure was considered the ultimate goal of seismic tomography, but it became obvious that an accurate set of velocities versus depth can effectively be used to constrain other types of seismic inversion, including the velocity control necessary for constructing an accurate depth image of the subsurface. During the 1990's, the goal for cross-borehole seismic data evolved into obtaining a reasonable estimate of the properties of the subsurface geology, particularly density, compressibility, shear rigidity, porosity (pore space in rocks), and permeability (ability for fluids to flow in rocks). To accomplish this goal, both compressional and shear wave data must be recorded.

CROSS-BOREHOLE DATA PROCESSING

Because seismic waves propagating in the earth's subsurface readily spread, refract, reflect, and diffract, algorithms for processing cross-borehole seismic data had to be developed to produce realistic subsurface images. Effective software and interactive graphics are required to pick and process the acquired cross-borehole data into a reasonable image. The received borehole signals are first filtered and digitized. By adding up many waveforms per detector, signal-to-noise ratios are enhanced. When using a vibrator source, impulse seismograms are obtained by cross-correlating the received data with the vibrator waveform, which further reduces random noise in the data. By implementing these signal enhancement techniques, the transmitted power in the borehole can be kept low enough to avoid damage to expensive boreholes and still record high-quality seismograms across distances of more than a few hundred meters in most rock types.

Using a variety of computer programs, cross-borehole seismograms are processed to yield seismic sections that represent the earth's reflectivity in time. However, since wells are drilled in depth, not in travel time, the seismic data need to be converted to depth in the imaging process. Furthermore, since reflectivity is a property associated with subsurface interfaces, a rock sample in the laboratory does not have any intrinsic reflectivity. Therefore, reflectivity is not an actual rock property, and this parameter must be converted to another parameter that really describes the rock. Typically, the chosen parameter is the internal velocity of seismic wave propagation through subsurface materials.

Internal velocity structure is estimated from cross-borehole and surface seismic data using seismic tomographic methods. Because subsurface velocities vary according to the physical properties of the rock through which the wave travels, geophysicists can use these velocities to determine the depth and structure of rock formations. In addition, since seismic waves change in amplitude when they are reflected from rocks that contain gas and other fluids, the fine details of amplitude changes in the seismograms can be used to infer the type of rocks in the subsurface. Thus, conventional seismic sections can yield far more information about the subsurface geology by using cross-borehole data and tomographic techniques to produce subsurface images as a function of depth and to estimate rock properties from the images.

The basic procedure for processing both cross-borehole and surface seismic data using tomography is iterative seismic tomography. The procedure involves picking transmitted or reflected events on raw seismograms and associating these events with the structure of a proposed subsurface geological model. Section of rock are then divided and imaged into a grid of pixels with local rock and fluid parameter values. The laws of physics are used to trace raypaths of seismic waves through the proposed geological model from the seismic source down through or from a subsurface boundary and back to the seismic detectors. Ray-traced travel times and amplitudes from the model are then compared with the travel times and amplitudes recorded on the seismogram. The medium parameters or geometry of the geological model are then updated to make the ray tracing consistent with the observed data. Corrections are made to the medium parameters or geometry systematically to reduce the differences between the observed and modeled travel times and amplitudes. After several iterations, the differences between the recorded and modeled data become acceptably small, and a "best" image of the subsurface geology that is self-consistent with the acquired data emerges.

Since the source-detector coverage of the object or area of interest is far from complete in cross-borehole seismology, non-uniqueness of solutions and lowered resolution of images result. Thus, tomographic images based on limited view angles must be interpreted with care. However, the tomographic parameter determination is still very useful, especially in areas of significant lateral velocity variations in the subsurface. By including cross-borehole, seismic reflection profiling, well-log, and any other available geophysical data, such as gravity, electrical, or radar, in the tomographic process, the resolution and certainty in describing the subsurface geology are greatly improved.

APPLICATIONS

Cross-borehole seismology has been used in a number of applications. The greater the degree of angular coverage around the rock mass of interest, the greater the reliability of the constructed subsurface image. By making numerous measurements for various source-detector positions in adjacent boreholes and analyzing the travel times and amplitudes for these source-detector locations, the velocity, elastic parameters, and attenuation of the intervening rock can be estimated from the transmitted and reflected energy. Cross-borehole seismology has been used for hydrocarbon exploration, mineral exploration, fault detection, stress monitoring in coal mines, delineation of the sides of a salt dome, investigation of dams, mapping dinosaur bone deposits, and nuclear waste site characterization.

In 1985, a research team from the Southwest Paleontology Foundation began excavating a giant sauropod dinosaur (45 to 60 meters long), later named *Seismosaurus*. The *Seismosaurus* skeleton was discovered after eight tailbones were exposed by weathering and erosion. By 1987, almost the entire tail had been excavated. To remove the rest of the skeleton, the location of the rest of the skeletal remains needed to be determined. In 1989, a number of boreholes were drilled in the area, and cross-borehole seismology provided data to construct vertical cross sections of where the *Seismosaurus* was located in the subsurface.

The Engineering Geoscience group at the University of California, Berkeley, used cross-borehole seismology in the 1990's to search for buried treasures, specifically the Victorio Peak treasure in southeastern New Mexico, said to consist of antique Spanish weapons, coins, and refined gold bars. The technique was also used to search for Yamashita's treasure of gold and jewels in the Philippines. The searchers' goal was to detect and delineate underground voids or cavities that could possibly contain the buried treasures. Although neither study resulted in the discovery of lost treasure, the ability of cross-borehole seismology to detect buried channels and voids was clearly demonstrated at both sites.

In 2009, the Federal Lands Highway Program of the U.S. Department of Transportation used cross-borehole seismic tomography to investigate an active sinkhole causing damage to property in a residential neighborhood of central Florida. Personnel were able to obtain images that showed the "throat" of the sinkhole at a depth of 24 meters.

One of the primary applications of cross-borehole seismology is the enhanced characterization of petroleum reservoirs in existing fields. Fewer frontier fields remain to be discovered, and existing fields may still hold up to two-thirds of their petroleum. The geologic detail needed to properly exploit most hydrocarbon reservoirs substantially exceeds the detail required to find them. For effective planning, drilling, and production, a complete understanding of the lateral extent, thickness, and depth of the reservoir is absolutely essential. This can be accomplished only from detailed seismic interpretation of three-dimensional cross-borehole data integrated with three-dimensional seismic reflection profiling.

Geologist at the U.S. Geological Survey, looking over seismograph readings from a California recording station. (Russell Curtis/Photo Researchers, Inc.)

A common practice in three-dimensional seismic reflection profiling is to place the seismic detectors at equal intervals and collect data from a grid of lines covering the area of interest. In addition, since many adjacent boreholes are typically available in existing petroleum fields, cross-borehole seismic data can be collected between numerous adjacent boreholes across the site. Integration of cross-borehole and surface seismic data using tomography-based imaging algorithms yields seismic depth sections and parameter characterization of the subsurface geology. The resulting depth sections assist in interpreting the structure (geometry), stratigraphy (depositional environment), and lithology (rock and fluid types) of established hydrocarbon reservoirs. A repeated sequence of cross-borehole surveys as a function of time can aid in monitoring the effectiveness of enhanced oil-recovery methods. Based on integrated tomographic models of the subsurface geology generated from the cross-borehole and surface seismic data, more wells can be drilled in the field at strategic locations, allowing a three-dimensional view of the subsurface to eventually emerge. These data provide petroleum companies with a continuously utilized and updated management tool that impacts reservoir planning and evaluation for years after the surface and cross-borehole seismic data were originally acquired and processed.

SIGNIFICANCE

The concept of estimating material properties from data collected around an object has a broad variety of applications. Cross-borehole seismology provides spatially continuous, high-resolution data that are necessary to image reservoir-scale features large distances from a well. These include faults, stratigraphic boundaries, unconformities, porosity, and fracturing. Coupled with seismic tomography, cross-borehole seismology provides an important methodology for investigating subsurface geology.

In addition to measuring one-way transmitted energy in the subsurface, seismic reflection waves can be recorded in cross-borehole surveys. These reflections are significantly better in resolving rock layers than are surface seismic reflections because of the fact that the cross-borehole frequencies are one to two orders of magnitude greater than those of surface surveys. Integration of cross-borehole and surface seismic surveys is especially valuable, particularly for analyzing subsurface velocity structure. Furthermore, the seismic problem of imaging subsurface geology in depth and estimating physical rock parameters can be solved by correlating and integrating cross-borehole and surface seismic data using tomographic algorithms. Numerous geophysical, engineering, and environmental applications have emerged.

Evaluation and exploitation of existing petroleum reservoirs is a major application of cross-borehole seismology. Reservoir complexity produced by spatial heterogeneities in porosity, permeability, clay content, fracture density, overburden pressure, pore pressure, fluid-phase behavior, and other related factors leads to large uncertainties in estimated total recovery. The most feasible approach for mapping these spatial variabilities comes from surface and cross-borehole geophysical measurements that are integrated through seismic tomographic processing. There is little doubt that cross-borehole seismology has played a major role in helping to solve not only exploration problems but also production and recovery problems in the petroleum industry.

In many areas of natural resource exploration and exploitation, numerous drill holes often exist, making cross-borehole seismology an ideal investigation tool. At these sites, a full suite of well-log data, rock-core analyses, and three-dimensional cross-borehole and surface seismic data can be acquired. Employing seismic tomography, these data can be processed, integrated, and interpreted to yield a geological model of subsurface lithology that closely approximates reality. The data acquired using cross-borehole seismology have been successfully processed and interpreted to yield three-dimensional images and lithological estimates of the subsurface geology.

Alvin K. Benson

FURTHER READING

Daily, William, and Abelardo Ramirez. "Electrical Resistance Tomography." *The Leading Edge* 23 (2004): 438-442. This article presents methodology for ERT and its use in the 1980's and 1990's. Provides a good background upon which recent research has been built.

Lines, L. R. "Cross-Borehole Seismology." *Geotimes* 40 (January 1995): 11. Applications of cross-borehole seismology and seismic tomography to the shallow subsurface.

_____, ed. *The Leading Edge* 17 (July 1998): 925-959. Contains five excellent papers that conceptually discuss advanced technology and a variety of applications of cross-borehole seismology.

Looms, M. C., K. H. Jensen, A. Binley, and L. Nielsen. "Monitoring Unsaturated Flow and Transport Using Cross-Borehole Geophysical Methods." *Vadose Zone Journal* 7 (2008), 227—237. Two methods of noninvasive borehole tomography are combined to estimate parameters in unsaturated zones. Each method provided data that were consistent with the other. This article is highly technical, but an interesting use of methods.

Nimmer, Robin E., et al. "Three-Dimensional Effects Causing Artifacts in Two-Dimensional, Cross-Borehole, Electrical Imaging." *Journal of Hydrology* 359 (2008): 59-70. Discusses techniques of cross-borehole electrical resistance tomography. Presents a significant issue of borehole inversion effects and provides possible solutions. A useful article for geologists using these techniques in their research.

Russell, B. H. *Introduction to Seismic Inversion Methods.* Tulsa, Okla.: Society of Exploration Geophysicists, 1988. Russell discusses techniques used for the inversion of cross-borehole and surface seismic data, including principles of seismic tomography. Contains some good data illustrations.

Sheriff, R. E., ed. *Reservoir Geophysics.* Tulsa, Okla.: Society of Exploration Geophysicists, 1992. Describes applications and shows examples of using cross-borehole seismology and seismic tomography to evaluate and exploit existing petroleum reservoirs. Reviews borehole source and detector technology.

Stewart, R. R. *Exploration Seismic Tomography.* Tulsa, Okla.: Society of Exploration Geophysicists, 1991. Includes some of the historical development of cross-borehole methods and tomography. Reviews the fundamentals of seismic tomographic techniques, and discusses applications of cross-borehole seismology, surface seismic profiling, and seismic tomography to exploration geophysics.

Tarantola, A. *Inverse Problem Theory: Methods for Data Fitting and Parameter Estimation.* Amsterdam: Elsevier, 1987. An in-depth treatise on determining subsurface information from surface and cross-borehole seismic data. Includes qualitative discussions, as well as technical details.

Telford, W. M., L. P. Geldart, and R. E. Sheriff. *Applied Geophysics.* 2d ed. Cambridge, England: Cambridge University Press, 1990. Contains a basic overview of cross-borehole seismology and how it relates to other seismic methods of exploration.

See also: Creep; Discontinuities; Earth's Age; Earth's Core; Earth's Differentiation; Earth's Interior Structure; Earth's Mantle; Elastic Waves; Experimental Rock Deformation; Faults: Normal; Faults: Strike-Slip; Faults: Thrust; Faults: Transform; San Andreas Fault; Seismic Observatories; Seismic Reflection Profiling; Seismic Tomography; Seismometers; Stress and Strain.

D

DEEP-EARTH DRILLING PROJECTS

Deep-earth drilling projects represent one of the most ambitious attempts by earth scientists to investigate the origins, structure, and nature of planet Earth. Scientists believe that these projects can reveal information concerning the planet unobtainable by other methods.

PRINCIPAL TERMS

- **core drilling:** a method of extracting samples of the materials being drilled through in a deep-drilling project
- **crust:** the outer layer of the earth, averaging 35 kilometers in thickness on land and 5 kilometers on the ocean bottoms
- **mantle:** the area of basaltic rocks separating the earth's crust from its core; it is estimated to be about 2,900 kilometers thick
- **Mohorovičić discontinuity (Moho) discontinuity:** an area of undetermined composition and depth between the earth's crust and mantle
- **rotary drilling:** a method of drilling holes to great depths using a rotating drill bit

ORIGINS OF SCIENTIFIC DRILLING PROJECTS

The idea of drilling deep holes in the earth's crust so as to determine its nature and history originated with nineteenth century geologists and naturalists. Charles Darwin was among the first scientists to call for a deep drilling project for purely scientific purposes. In 1881, Darwin proposed that a shaft be sunk to a depth of 150-180 meters on a Pacific atoll to test his theory concerning the origins and growth of coral islands.

Eighteen years later, the Royal Society of London financed the drilling of a 348-meter-deep hole on one of the Ellice Islands (Tuvalu) in the South Pacific to test Darwin's theories. That was perhaps the first deep-earth drilling project ever undertaken for purely scientific purposes.

THE MOHOLE PROJECT

Many earth scientists have proposed deep-earth drilling projects to advance scientific knowledge of the earth's interior. Many proposals have been met with indifference because of the great expense and the technological problems involved. In the 1950's, a number of well-known geologists, geophysicists, and oceanographers from several countries began corresponding with one another about deep-drilling projects. A major topic of their correspondence was whether they might be able to capture the imagination of the public with a deep-drilling project and consequently claim a larger share of research funds from government agencies. These scientists believed the public to be over-focused on space research. These scientists eventually formed an unofficial organization they called the American Miscellaneous Society (AMSOC), which held informal meetings during official scientific conferences at which they discussed the desirability of a deep-drilling project that would penetrate the Mohorovičić (Moho) discontinuity in the earth's mantle. Ideally, they agreed, there would be two deep-drilling projects, one on land and one on the ocean bottom.

In 1958, the members of AMSOC became the Deep Drilling Committee (DDC) of the National Academy of Sciences (NAS). The NAS is a private organization that was chartered by U.S. President Abraham Lincoln in 1865 to act as an adviser to the federal government on scientific matters.

The NAS, on the recommendation of the DDC, proposed in 1958 that the federal government fund a deep-drilling project to penetrate the Moho. Rumors that the Soviet Union was about to begin a similar project may have provided a catalyst for the recommendation. It was feared that the Soviets might learn the secrets of the earth's interior before the United States just as the Soviet Union had begun to learn the secrets of outer space before the United States with the launch of Sputnik in 1957.

The original proposal envisioned a hole to be drilled on land to a depth of perhaps 10,500 meters

35

as a training project to develop the technology necessary to penetrate the Moho at the ocean's bottom, where the crust is thinner. The DDC subsequently scrapped the ground-hole idea when it received a grant of $15,000 for a feasibility study of the deep-sea drilling project. A DDC member summarized the result of the feasibility study in an article in *Scientific American* (April 1959) entitled "The Mohole," which stirred immediate industrial and public interest.

The Mohole project became a source of considerable international embarrassment for the American scientific community. It failed to meet most of its objectives, cost considerably more than originally estimated, and discredited the idea of deep drilling in the minds of the public and many members of Congress. It was not until 1968 that oceanographers were able to convince Congress to finance another deep-sea drilling project. Another decade passed before federal funds were forthcoming for continental deep-drilling projects.

THE KOLA PROJECT

Although rumors concerning a Soviet deep-drilling project were at least partially responsible for the urgency with which the U.S. scientific community and government embraced the ill-fated Mohole project, historical evidence would later reveal that these rumors were unfounded. Scientists in the Soviet Union did not begin a deep-drilling project until twelve years after Mohole. In 1970, Soviet geophysicists launched a project on Kola Peninsula near Murmansk, 240 kilometers north of the Arctic Circle. The project reached a depth of more than 12,000 meters, almost twice the depth of any preceding hole. Under the direction of a Soviet government agency called the Interdepartmental Council for the Study of the Earth's Interior and Superdeep Drilling (formed in 1962), the Kola project became the first of several proposed deep holes meant to explore the structure of the earth's crust and mantle.

The Kola drillers penetrated through almost 3 billion years of earth's geologic history, into rock from the Archean eon. Along the way, they discovered large quantities of hot, highly mineralized water at greater depths than geophysicists and geologists had previously thought possible. Scientists at the Kola project concluded from this and other unexpected findings that enormous mineral deposits may be located at great depths, waiting for humankind to reach them. Soviet scientists expected that as they developed a better understanding of the deeper layers of the earth's crust from the Kola project, they would concurrently find ways to discover and exploit available supplies of petroleum, gas, and minerals located at great depths. However, temperature increases beyond 12,000 meters made the project unfeasible, and drilling was stopped in 1992, about a third of the way through the Baltic continental crust. Lack of funding stopped the project in 2005 and the site was abandoned in 2008.

THE APPALACHIAN PROJECT

Spurred in large part by Soviet Union's successes in deep continental drilling and the related propaganda created by the Soviet government, scientists and government agencies in several Western nations began developing similar programs in the 1970's. The U.S. Geodynamics Committee of NAS held a workshop on deep continental drilling near Los Alamos, New Mexico, in 1978. The members of the workshop convinced NAS to form its own permanent Continental Drilling Committee (CDC) that same year. The members of the CDC identified a number of geophysical objectives that could be addressed through deep-earth drilling projects, including those related to crust structure, geothermal systems, mineral resources, and earthquake research. The CDC also solicited proposals from U.S. geologists and geophysicists for specific projects that would address those objectives. After examining the many proposals received, the CDC assigned highest priority to two of them: a project to drill a hole 3.7 kilometers deep through the highly mineralized area near Creede, Colorado, and a core hole in the southern Appalachian region 8-10 kilometers deep.

At its 1983 meeting, the CDC unanimously endorsed the Appalachian project as the most promising for America's first deep continental drilling project. It also endorsed two other drilling projects in Creede and at Cajon Pass in California. The following year, the CDC convened a workshop in New York to consider exactly how the project should be approached. After the workshop, the committee organized the Continental Scientific Drilling Program, established with the aid of a grant from the National Science Foundation (NSF). NSF gave the grant to a management group called Direct Observation and Sampling of the Earth's Continental Crust

(DOSECC). DOSECC, coordinating its activities with the U.S. Geological Survey and the Department of Energy, prioritized deep-drilling projects and issued contracts for drilling deep continental holes for scientific purposes. In 1985, the White House's Office of Science and Technology Policy (OSTP) recommended to the NSF that it appropriate $2 million for preliminary studies of the Appalachian project. OSTP also recommended that by 1990 the various deep-earth drilling projects be funded at a level of $20 million per year. In 1988, the Cajon Pass project, overseen by DOSECC, got under way; within two years it reached its targeted depth of 4,875 meters. DOSECC began the Appalachian project in the northwest corner of South Carolina in 1989. Its goal was a continuously cored hole 15,250 meters deep.

The DOSECC undertook the Snake River Scientific Drilling Project in September 2010 to explore the interaction between the earth's mantle and crust. In addition, the Dead Sea drilling project using the Deep Lake Drilling System (DLDS) began in November 2010 to explore climate history from core samples at depth.

QUEST FOR PETROLEUM

The technology employed in deep-earth drilling projects is directly derived from the petroleum industry. The modern origins of that industry date back to the middle of the nineteenth century, when petroleum fields were discovered and exploited in regions all over the world, including Romania, Myanmar, Sumatra, Iran, the Caucasus, and the U.S.. The oil wells in those areas were initially shallow and drilled with tools designed to drill water wells.

As the petroleum industry grew in economic significance, the search for petroleum went deeper and deeper into the earth and required ever more sophisticated drilling machinery. In addition, oil companies began to employ geologists and geophysicists in their quest to meet the skyrocketing demand for petroleum of an increasingly industrialized society. These earth scientists, in part to become more efficient in their effort to locate significant quantities of petroleum and in part as a by-product of that endeavor, began to learn more and more about the nature and history of the earth's crust from the drilling process itself. This new knowledge led to the modern deep-earth drilling projects that seek to drill deep and superdeep holes to learn more about the earth.

Such project are not necessarily undertaken to locate mineral resources, although the discovery of such resources is often a by-product of the undertaking.

The quest for petroleum in the twentieth century caused earth scientists to penetrate ever deeper into the earth's crust. By the latter part of the century, wells in the U.S. were producing oil from depths of more than 6,000 meters (the record being a well in Louisiana, producing oil at a depth of 6,527 meters).

DRILLING METHODS

There are two basic methods of drilling deep wells: cable-tool and rotary. Cable-tool drilling utilizes a heavy drill bit and drill that are suspended by a cable and raised about a meter above the bottom of the hole, then dropped. Workers add mud and water to the hole to hold the rock chips produced by the concussion in suspension. Periodically, drilling crews extract the tools from the hole and pump out the mud, water, and rock chips. Cable-tool drilling is slow and has been largely superseded by the rotary drill, which has proven to be much more effective in deep-earth drilling.

Rotary drilling, though much faster than the cable-tool method, is also more expensive. The rotary drilling method requires hundreds and often thousands of meters of drill pipe and well casing, a derrick, drill bits of several kinds (depending on the type of rock being drilled through), drilling muds and chemicals, a power source (usually one or more diesel engines), and a sizable crew of workers. In rotary drilling, workers attach a drill bit to a string of drill pipe that has at its top a square cross section called the kelly. The kelly passes through a square hole in a powered turntable, which rotates the drill pipe and bit. Workers add new sections of drill pipe just below the kelly as the bit progresses downward. The rock cut by the rotating drill bit is removed by pumping chemically treated mud down the drill pipe through the drill bit, then back up through the space outside the pipe (the pipe being somewhat smaller in circumference than the hole in which it rests) to a settling pit on the surface.

One drill bit developed for the rotary process, the hollow or core bit, is of particular importance for deep-earth drilling projects. Because petroleum is often found in readily identified geological formations, companies exploring for oil found it expedient to bring to the surface intact samples of the rock being drilled through for examination by geologists. To extract the samples, technicians developed a hollow bit that would allow a cylindrical section of the rock, called a core, to be extracted from the hole without otherwise damaging it. These cores allow geologists to determine not only the petroleum potential of the rock, but also the geological history of the earth in the area of the hole. Core drilling has become an integral part of all deep-earth drilling projects.

MODIFICATION OF METHODS

In the years following their inception, deep-earth drilling projects have required considerable modifications as they have become more focused on uncovering sources of petroleum. Soviet scientists at the Kola project learned that at depths exceeding 9,000 meters, conventional rotary drilling methods encounter virtually insurmountable problems. The rotary method uses a power source to turn the drill bit by rotating the entire string of pipe connected to it and to the kelly. At 9,000 meters, the pipe weighs in excess of 800 metric tons, the rotation of which creates enormous stress at the kelly-power source interface and multiplies the friction resistance of the rock through which the scientists are drilling. The Soviets overcame this problem by developing a bottom-hole turbine to rotate the drilling bit, driven by the flow of the drilling mud being injected into the hole. The necessity of rotating the pipe was thus completely eliminated.

To reduce the weight of the huge lengths of drill pipe necessary for deep-earth drilling, Soviet scientists at Kola began to utilize a high-strength aluminum alloy pipe weighing only about half as much as conventional drill pipe. This innovation considerably reduced the burden on the derrick and the power required to lift the pipe periodically to replace the drill bits and remove the core samples.

The process of core sampling also underwent modification at Kola. In conventional wells, core samples several centimeters in diameter enter the hollow drill tube as the bit cuts a ring of rock at the bottom of the hole. The core remains in the tube until it is brought to the surface and removed. At depths of more than 2,100 meters, however, the rock is under such tremendous pressure that it bursts when the drill bit relieves the compression of overlying rock strata. Soviet drillers developed a new core-sampling device that diverted some of the mud into the core tube and caught the pieces of burst rock, then carried them to the surface in a special chamber. This technique also offered the advantage of clearing the tube for new core samples without the necessity of bringing the tube to the surface.

It is likely that other modifications in drilling technology will be necessary as continental deep-drilling projects proceed to great depths. Thus far, technological innovations have solved the problems encountered by adequately funded deep-earth drilling projects.

EXPLOITING EARTH'S ENERGY RESOURCES

The mineral and fossil fuel resources of the earth are finite. They are non-replaceable commodities. As industrialization continues to spread to developing countries and continues to intensify in countries with developed economies, the demand for those irreplaceable and limited commodities continues to increase. Sometime in the twenty-first century, the presently known reserves of fossil fuels may be exhausted. If modern lifestyles are to be maintained in industrialized regions of the world and living standards raised for burgeoning populations in India and China, new reserves of fossil fuels and sources of minerals must be discovered and exploited and new sources of energy must be developed. Deep-earth drilling projects represent one approach to both these necessities.

Strong evidence exists that significant quantities of fossil fuels, both petroleum and natural gas, might be found at depths that are currently unreachable, necessitating projects that are economically unfeasible. However, developing technology in the deep-earth drilling program has the potential to put those resources at the disposal of humankind.

Deep-earth drilling projects under way in the U.S. and elsewhere in the world are exploring the possibility of exploiting on a large scale a relatively new way of producing energy by utilizing the earth's internal heat. The temperature of the earth increases by about 1 degree Celsius per hundred meters to a depth of 3 kilometers and, as shown at Kola, by 2.5 degrees Celsius thereafter. At 15 kilometers beneath the surface, the earth's temperature is more than 300 degrees Celsius. The U.S. Department of Energy has concentrated efforts in deep-drilling projects aimed at investigating the possibility of commercially exploiting this source of virtually inexhaustible energy.

Disposing of Toxic Wastes

Some scientists suggest that deep-earth drilling might also offer a solution to the problem of toxic waste disposal. The safe disposal of the wastes produced by modern industry and by the production of atomic energy is a major concern of contemporary scientists. Before industries and atomic energy plants inject these materials into the deep crust of the planet, however, extensive studies must be conducted into the potential ecological consequences of such an action. Nevertheless, it is possible that deep-earth drilling can help to solve the problem of toxic waste disposal.

Paul Madden

Further Reading

Anderson, Ian. "Drilling Deep for Geothermal Power and Science." *New Scientist* 111 (July 24, 1986): 22-23. This brief article presents a nontechnical overview of current deep-drilling projects exploring the possibility of exploiting the earth's own heat to produce energy. The author is optimistic that many of the energy needs of the future may be met if governments are willing to make the necessary expenditures for research.

Bar-Cohen, Yoseph, and Kris Zacny, eds. *Drilling in Extreme Environments: Penetration and Sampling on Earth and Other Planets.* Weinheim: Wiley-VCH. 2009. A comprehensive presentation of current drilling techniques and technology. This book covers everything from ocean to extraterrestrial drilling and coring. Written by academics, researcher scientists, industrial engineers, geologists, and astronauts, in a manner accessible to the layperson, but also useful for scientists and engineers.

Bascom, Willard. "Deep Hole Story." *Modern Machine Shop* 54 (March 1982): 92-111. A thorough review of the technology of deep drilling, including Soviet innovations. Assumes considerable technical knowledge on the part of the reader but very informative for those wanting to know more about the technology involved in deep-drilling projects.

_____. "Drilling the World's Deepest Hole." *Engineering Digest* 34 (June 1988): 16-24. This article is primarily concerned with technical drilling problems. Readers should have a strong technical background. Will be of interest only to those readers wanting to know more about the actual mechanics of deep-earth drilling.

_____. "Geothermal Boreholes Make Drilling History." *Machine Design* 53 (October 22, 1981): 8. This article contains a rather technical discussion of the problems encountered in deep drilling for geothermal purposes and their often ingenious solutions. The reader should have a dictionary handy.

_____. *A Hole in the Bottom of the Sea: The Story of the Mohole Project.* Garden City, N.Y.: Doubleday, 1961. Although the main topic of Bascom's book is the ill-fated Mohole project, it contains a considerable amount of information about early proposals for scientific drilling projects and deep continental drilling projects. Also contains fascinating insights into the workings of the "scientific establishment" in the U.S. and the ways in which it interacts with the federal bureaucracy. Written for a nontechnical audience.

Dettmer, R. "Geo-energy Firms Work on Deep-Drilling Breakthrough." *Engineering & Technology* 5 (2010): 12. A concise article presenting new technology for deep drilling. Provides a brief overview of drilling technique, and issues.

Fridleifsson, Gudmundur O., and Wilfred A. Elders. "The Iceland Deep Drilling Project: A Search for Deep Unconventional Geothermal Resources." *Geothermics* 34 (2005): 269-285. This article is written for a reader with a background in engineering or geophysics. The authors provide a good overview of common issues in deep-drilling and then focus the article on the benefits of harnessing geothermal power.

Grotzinger, John, et al. *Understanding Earth.* 5th ed. New York: W. H. Freeman, 2006. This text includes one of the most complete descriptions of the causes of earthquakes, their measurement, where they occur, how they can be predicted, and how they affect humans. A map of the major plates is on the inside back cover. The glossary is huge and indispensable. Senior high school and college-level students should find this text suitable for general background information.

Gurnis, Michael, et al., eds. *The Core-Mantle Boundary Region.* Washington, D.C.: American Geophysical Union, 1998. This collection of articles is one volume of the American Geophysical Union's Geodynamics series. Although intended for the specialist, the essays contain plenty of information suitable for the careful college-level reader.

Bibliography.

Heath, Michael J. "Deep Digging for Nuclear Waste Disposal." *New Scientist* 108 (October 31, 1985): 30. This article briefly explores the possibilities of using deep-drilling technology to solve the perplexing problem of disposing of nuclear wastes. The ideas presented are interesting but not entirely convincing, as they do not address the environmental problems that might be created by such a program.

Kearey, Phillip, Keith A. Klepeis, and Frederick, J. Vine. *Global Tectonics*. 3rd ed. Cambridge: Wiley-Blackwell, 2009. This college text gives the reader a solid understanding of the history of global tectonics, along with current processes and activities. The book is filled with colorful illustrations and maps.

Kerr, Richard A. "Continental Drilling Heads Deeper." *Science* 224 (June 29, 1984): 1418-1420. An excellent account of the then-current status of U.S. deep-earth drilling projects and proposals. Offers considerable information about the organizations and individuals who were forming the policies for the U.S. continental drilling program.

Kious, Jacquelyne W. *This Dynamic Earth: The Story of Plate Tectonics*. Washington, D.C.: U.S. Department of the Interior, U.S. Geological Survey, 1996. Kious is able to explain plate tectonics in a way suitable for the layperson. The book deals with early development of the theory. Illustrations and maps are plentiful.

Kozlovsky, Yephrim A. "The World's Deepest Well." *Scientific American* 251 (December, 1984): 98-104. The best and most complete account in English of the Kola deep-drilling project in the Soviet Union. The reader should keep a geological dictionary nearby but will find a wealth of information about the origins, purpose, scientific findings, and future of the Soviet continental drilling program.

Sacher, Hubert, and Rene Schiemann. "When Do Deep Drilling Geothermal Projects Make Good Economic Sense?" *Renewable Energy Focus* 11 (2010): 30-31. A detailed article providing technical aspects of deep drilling projects and its economic feasibility.

Tarbuck, Edward J., Frederick K. Lutgens, and Dennis Tasa. *Earth: An Introduction to Physical Geology*. 10th ed. Upper Saddle River, N.J.: Prentice Hall, 2010. This college text provides a clear picture of the earth's systems and processes that is suitable for the high school or college reader. It has excellent illustrations and graphics. Bibliography and index.

See also: Earth's Age; Earth's Core; Earth's Internal Structure; Earth's Mantle; Engineering Geophysics; Ocean Drilling Program; Ocean-Floor Drilling Programs; Seismic Reflection Profiling.

DEEP-FOCUS EARTHQUAKES

Deep-focus earthquakes occur at depths ranging from 70 to 700 kilometers below the earth's surface. This range of depths represents the zone from the base of the earth's crust to approximately one-fourth of the distance into the earth's mantle. Deep-focus earthquakes provide scientists information about the planet's interior structure, its composition, and seismicity. Observation of deep-focus earthquakes has played a fundamental role in the discovery and understanding of plate tectonics.

PRINCIPAL TERMS

- **aftershock:** an earthquake that follows a larger earthquake and originates at or near the focus of the latter; many aftershocks may follow a major earthquake, decreasing in frequency and magnitude with time
- **brittle fracture:** rock that fractures at less than 3 to 5 percent compressional or tensional strain
- **ductile fracture:** rock that is able to sustain, under a given set of conditions, 5 to 10 percent deformation before fracturing or faulting
- **epicenter:** the point on the earth's surface that is directly above the focus of an earthquake
- **focus:** the place within the earth where an earthquake commences and from which the first P waves arrive; also called the hypocenter
- **lithosphere:** the solid portion of the earth used in plate tectonics as a layer of strength relative to the underlying plastic-like asthenosphere; encompassing the earth's crust and part of the upper mantle, it is about 100 kilometers in thickness
- **primary wave (P wave):** the primary or fastest wave traveling away from an earthquake through the solid rock; P waves also are capable of moving through liquids
- **secondary wave (S wave):** the secondary wave that travels more slowly through solid rock than the P wave; S waves cannot penetrate a liquid
- **shear stress:** stress that causes different parts of an object to slide past each other across a plane
- **strike-slip fault:** a fault along which movement is horizontal only; the movement is parallel to the trend of the fault
- **subduction:** the process in which a dense lithospheric plate descends into the mantle beneath another, less dense plate in a subduction zone
- **Wadati-Benioff zone:** a narrow zone of earthquake foci that seismically delineate an inclined subduction zone; they are generally tens of kilometers thick

EARTHQUAKES AND EARTH'S INTERIOR

Because direct physical access to the earth's interior is restricted, most knowledge about it is derived from the study of earthquake waves that travel through the planet and vibrate the earth's surface at some distant point. Earthquakes recorded on seismograms allow scientists to accurately measure the time required for earthquake waves (seismic waves) to travel from the focus of an earthquake to a seismographic station. Primary waves (P waves) travel the fastest, and this information is used to determine distances to earthquake epicenters. The time required for P waves and secondary waves (S waves) to travel through the earth depends on the physical properties encountered as the waves pass through the subsurface. Seismologists, therefore, search for variations in travel times that cannot be explained by differences in the distance traveled. These differences correspond to changes in the properties of the subsurface earth material encountered. Changes in rock properties indicate that the earth has four major layers: the crust, a thin outer layer; the mantle, a rocky layer beneath the crust with a depth of 2,885 kilometers; a 2,770-kilometer outer core exhibiting characteristics of a mobile liquid composition; and a solid inner core, a metallic sphere with a radius of 1,216 kilometers.

EARTHQUAKE ZONES

Most earthquakes occur in narrow zones that globally connect to form a continuous seismic network. Characteristic surface features of seismic zones are rift valleys, oceanic ridges, mountain belts, volcanic chains, and deep-ocean trenches. These global seismic zones represent the boundaries of the lithospheric plates. The interior regions of the lithospheric plates are largely free of earthquakes. The lithosphere is made up of twelve rigid plates that cover the entire globe. The depths of the lithospheric plates range between 60 and 100 kilometers. They are composed of either the entire continental crust and a portion of the upper mantle or the entire oceanic crust and a portion of the upper mantle. The lithospheric plates are in constant motion relative to each other and can diverge away from each other, forming ridge axes and new oceanic crust material; converge toward each other, where one plate subducts under the leading edge of its neighbor plate; or transform, where plates slide horizontally past each other.

Within the global seismic network, there are four types of seismic areas that are recognized by their form, structure, and geology. The first type is represented by narrow zones of high surface-heat flow and basaltic volcanic activity along the axes of mid-ocean ridges where the earthquake focus is shallow (70 kilometers deep or less). Here, molten rock material is welling up from the mantle area and is emplaced on either side of the ridge, adding to the ocean crust. Mid-ocean ridges are active areas of sea-floor spreading and are found in all ocean basins. A primary example is the Mid-Atlantic Ridge trending north to south in the Atlantic Ocean basin. This ridge rises above sea level in Iceland and delineates the boundary between the South American and African lithospheric plates.

The second type of seismic zone is identified by large surface displacements occurring parallel to the fault, shallow-focus earthquakes, and absence of volcanoes. Excellent examples of this seismic zone are the San Andreas fault in California and the Anatolian fault in Turkey, both of which demonstrate strike-slip movement between plate boundaries.

The third type of seismic zone is a widespread continental area ranging from the Mediterranean Sea to Myanmar; it is associated with high mountain ranges created by converging plate margins. Although this zone is usually characterized by shallow-focus earthquakes, earthquakes of intermediate focus (70 to 300 kilometers deep) have occurred in the Hindu Kush and Romania; deep-focus earthquakes (300 to 700 kilometers deep) have been recorded in a few places north of Sicily under the Tyrrhenian volcanoes.

The fourth type of seismic zone is physically connected to volcanic island-arc-trench systems, such as the Japan Trench and the Kermadec-Tonga Trench in the South Pacific Ocean. Earthquakes associated with trenches can be shallow, intermediate, or deep focus depending on where they are located in the steeply converging lithospheric plate adjacent to the trench. The earthquake foci define the plate being carried into the earth's interior and away from the trench. These inclined earthquake zones, called Wadati-Benioff zones, underlie active volcanic island arcs and have an assortment of complex shapes. The Wadati-Benioff zone also marks the downflowing portion of the convection cell as identified by geophysicists through calculations of moving oceanic plates.

SUBDUCTION OF THE LITHOSPHERE

New lithosphere is created at mid-ocean ridges by upwelling and cooling of magma from the earth's mantle. With new lithospheric material constantly created and no measurable expansion of the earth occurring, some of the lithosphere must be removed globally. Older oceanic lithosphere is removed by subduction of the oceanic plate into the earth's mantle at island-arc-trench systems. The originally rigid plate slowly descends and heats; over millions of years, it may be fully absorbed into the earth's mantle. The subduction of lithospheric plates accounts for many of the processes that shape the earth's surface, such as volcanoes and earthquakes, including nearly all of the earthquakes with deep and intermediate foci. Deep earthquakes cannot occur except in subducting plates, so the presence of a deep-focus earthquake implies the presence of subducting oceanic plate material.

The proposed driving force for movement of the lithospheric plates is a convection cell arrangement where molten mantle material rises into a crustal rift zone from the subsurface mantle zone. Portions of the molten mass that do not move into the central rift zone move to either side of the rift and travel away from it, below and parallel to the lithospheric plate. The convection current travels beneath the plate for some distance. It is in this zone that the convection current is thought to move and carry the lithospheric plate with it. As the convection cell cools because of its interaction with the cool oceanic plate, the outer area of the cell sinks into the mantle to be rewarmed by mantle convection currents. Again, the downflowing portion of the convection cell is represented by the subducting oceanic plate.

ISLAND-ARC-TRENCH SYSTEMS

Major island chains are known to geologists as island-arc-trench systems. The island chains are a surface expression of the oceanic subduction process, and the associated deep trenches are a surface expression of the seaward boundary of subduction zones. Examples of deep trenches appearing in connection with island arcs are the Java and Tonga Trenches.

When foci of earthquakes near island arcs and ocean trenches are compared, a notable pattern emerges that is well illustrated in the Tonga island arc in the South Pacific. To the east of the volcanic islands of Tonga lies the Tonga Trench, which is approximately 10 kilometers deep. Beginning at the Tonga Trench and moving from east to west, the earthquake foci lie in a narrow but well-defined zone, which slopes from east to west from the Tonga Trench to beneath the Tonga Island Arc at an angle of about 45 degrees. The earthquake record in the arc-trench zone reveals that the earthquake foci are extremely shallow at the trench; however, moving to the west away from the trench, the earthquake foci register at a depth of 400 kilometers. An additional move to the west reveals deep-focus earthquakes at below 600 kilometers. In other regions of deep earthquake activity, some variation in the angle of dip and distribution of foci is recorded, but the common feature, that of a sloping seismic zone, is characteristic of island arcs and deep-ocean trenches. Other regions associated with island arc and deep-ocean trenches are the Kurile Ridge and Trench and the Mariana Islands and Trench.

What the deepening earthquake foci are defining at island-arc-trench systems, and especially the Tonga Trench and Island Arc system, is the movement of the descending oceanic plate (the downgoing slab) into the inclined seismic zone, known to scientists as the Wadati-Benioff zone. In the process of the plate's gravity-driven downward movement, additional force is created, causing further deformation and fracturing and deep-focus earthquakes. At mantle depths of 650 to 680 kilometers, either the plate may have been absorbed into the mantle interior, or its properties may have been altered to the extent that earthquake energy cannot be released.

OCEANIC AND CONTINENTAL CRUST SYSTEMS

Major mountain belts, such as the Andes in South America, have been raised by the convergence and subduction of oceanic lithospheric plates beneath continental lithospheric plates. As the two plates converge, the denser oceanic plate is subducted under the less-dense continental plate (South America). The zone of subduction is identified by an oceanic trench, the Peru-Chile Trench. Initially, in this setting, the angle of descent of the plate into the subduction zone is low, but the angle gradually steepens into a downward curve as revealed by intermediate- and deep-focus earthquakes. On June 8, 1994, a magnitude 8.2 deep-focus earthquake occurred 600 kilometers below Bolivia in the Andes Mountains. This earthquake released a tremendous amount of energy, but because of its deep focus, the seismic waves were slow to reach the earth's surface, preventing damage to populations and structures.

Deep-focus earthquakes have proved especially useful to researchers because they do not produce many surface seismic waves, and they provide information about the density of the earth's mantle—a factor that ultimately controls how convection currents of rock inside the planet move the continents around the surface. Because the core and mantle remain hidden from view, researchers interested in these deep regions must wait for large deep earthquakes to provide indirect information about the planet's interior. The 1994 Bolivian earthquake was well recorded because of global seismic detectors placed by scientists some twenty years in advance. The seismic equipment was placed in an effort to fully capture deep-focus earthquake activity as a result of a large deep-focus earthquake under Colombia in 1970.

THERMAL PROPERTIES OF THE DESCENDING LITHOSPHERE

Temperatures near the surface of the earth increase rapidly with depth, reaching about 1,200 degrees Celsius at the depth of 100 kilometers. Here the minerals in peridotite, an olivine-rich major mineral constituent of the upper mantle, begin to melt. The temperatures then increase more gradually to 2,000 degrees Celsius at approximately 500 kilometers depth.

As the lithospheric plate descends into the mantle, it is heated primarily by heat flowing into the cooler lithospheric plate from the enclosing hotter mantle. Since the conductivity of the rock increases with temperature, conductive heating becomes more efficient with increasing depth, further warming the subducting plate. Heat within the earth is generated by the energy released when minerals in the mantle change to denser phases or more compact crystalline structures with the higher pressures present at depth.

Additional heat sources in the mantle are radioactivity and the heat of compression, which is activated by increasing pressure at depth.

Despite the elevated temperatures at depth, the interior of the descending plate remains cooler than the surrounding mantle until the plate reaches a depth of about 600 kilometers. As the plate continues to subside into the deep mantle interior, it may heat rapidly because of efficient heat transfer by conduction. At 700 kilometers, however, the lithosphere plate is difficult to decipher as a separate structural unit. Complicating the detection of the plate properties is the low number of earthquakes at or below the 700-kilometer depth, which could possibly reveal information about any remaining plate material. The subduction zone under the Japanese island of Honshu, under the Kuriles, and under the Tonga-Kermadec area (north of New Zealand) represent almost ideal subduction zones without major complications caused by the age or thermal properties of the plate.

However, not all subduction zones behave ideally. The descending oceanic lithospheric material can be assimilated before reaching deep mantle zones: A slow-moving plate may achieve thermal assimilation before reaching 700 kilometers, such as the Mediterranean plate under the Aegean Sea. In younger subduction zones, as found in the Aleutian Islands and Trench, the descending plate may have penetrated far less than 700 kilometers owing to the warmer, more buoyant oceanic plate.

STUDY OF DEEP-FOCUS EARTHQUAKES

The deep-focus earthquake problem has been one of the leading scientific problems of solid-earth geophysics since the 1920's, when Kiyoo Wadati, a Japanese seismologist, demonstrated that some earthquakes occur hundreds of kilometers beneath the earth's crust. Laboratory experiments attempting to replicate the temperatures and pressures of the earth's interior have confirmed that rock under stress at the higher temperatures and pressures of 70 kilometers fails suddenly by brittle fracture—that is, shallow-focus earthquakes fail by brittle fracture. At even higher temperatures and pressures, similar to what is present in the deep interior of the earth (300 to 700 kilometers), shear stress should deform rocks by ductile flow, even in the colder regions of the mantle beneath subduction zones. Yet seismic data indicate that rocks at these depths are apparently also failing by brittle fracture.

Research has been directed toward understanding the triggering mechanism for brittle fracture in deep-focus earthquakes. The first area of research concerns the mechanism for brittle failure on intermediate-depth earthquakes. Research seems to indicate that the brittle failure is driven by water subducted along with the oceanic lithospheric plate or by water released from the molecular structure of oceanic crustal minerals. Thus, down to a depth of 350 kilometers or so (many scientists consider any earthquake below 70 kilometers a deep-focus earthquake), the triggering mechanism is much like that of shallow earthquakes, with the available water reducing the normal stresses on faults at depth and allowing failure (an earthquake) to occur.

The second research position holds that olivine, a primary constituent mineral of the earth's upper mantle, transforms under stress to spinel, a denser mineral in which the atoms are more closely packed and display a rearranged crystal structure. The newly formed spinel is deposited in many elongated, bead-like structures that eventually become thin, shear faulting zones. Faulting occurs along planes of greatest stress in these thin, shear zones.

However, olivine could remain cool because of its presence within the cooler subducting plate and not transform until some greater depth. At this greater depth, the newly formed, fine-grained spinel could be the lubricating material that makes deep-focus earthquake faulting possible. As olivine undergoes the transformation to spinel, heat is released that may augment the catastrophic faulting.

Neither of these two positions has been proved conclusively, nor are they accepted by all scientists. Yet most research focuses on exploring the detailed geophysics and geochemistry of each position. For example, thermal calculations suggest that olivine may remain present even at great depth in subduction zones where the plate is old and cold (cooling over time) and, therefore, subducts rapidly. These findings might explain why there are deep-focus earthquakes in Tonga and the Kuriles where the plate material may be older, whereas warmer, younger subducting slabs may manifest only intermediate-depth events as in the Aleutians.

Another factor that may need consideration is that analysis of seismograms from intermediate or deep-focus earthquakes reveals little or no gross differences between the seismic properties for intermediate and deep earthquakes. In fact, the only gross mechanical difference between shallow earthquakes and deeper ones is that aftershocks are much rarer for deep and intermediate earthquakes. Therefore, if the seismograms are similar, then should not the source for intermediate- and deep-focus earthquakes be similar? Thus, one of the two premises may be the triggering mechanism for all earthquakes below 70 kilometers—or perhaps the triggering mechanism is a combination of the two positions.

SIGNIFICANCE

Deep earthquakes are significant for at least four reasons. First, they are exceedingly common, constituting almost 25 percent of all earthquakes occurring during the period of 1964 to 1986. Second, deep earthquakes most often occur in association with deep-ocean trenches and volcanic island arcs in subduction zones. One of the great achievements of twentieth-century geophysics was the recognition that the occurrence of deep earthquakes in Wadati-Benioff zones apparently delineates not only the subduction of the lithospheric plate but also the cold downflowing cores of convection cells in the uppermost mantle. Third, scientists use seismograms of deep earthquakes rather than those of shallow earthquakes to investigate core, mantle, and crustal structures. Deep-focus earthquakes are mechanically different from shallow-focus earthquakes because their body wave phases traverse the uppermost mantle only once from the focus to the seismic station, thus producing simpler seismographs. Finally, information on how deep-earth materials process and handle stress can be derived from the seismograms of deep earthquakes.

Mariana Rhoades

FURTHER READING

Bolt, Bruce A. *Earthquakes*. 5th ed. New York: W. H. Freeman, 2005. A concise account of earthquake knowledge for general readership. Readers will be able to answer questions about the cause, location, and occurrence of earthquakes. A glossary and bibliography are provided, along with information on the Northridge (1994) and Kobe (1995) earthquakes; an earthquake quiz with answers is included.

_____. *Earthquakes and Geological Discovery*. New York: Scientific American Library, 1993. This well-illustrated text may give the impression of a simpler format, but the text is technical and deserves thoughtful reading and study. The sections on predicting and forecasting earthquakes and reducing seismic risk are excellent.

_____. *Inside the Earth: Evidence from Earthquakes*. Rev. ed. San Francisco: W. H. Freeman, 1994. Includes a full explanation of seismology; the temperatures, densities, and elastic properties of Earth; and the interiors of Earth, the moon, and Mars. For college-level readers.

Chester, Roy. *Furnace of Creation, Cradle of Destruction*. New York: AMACOM Books, 2008. This text discusses the turbulent processes of the planet. It covers earthquakes, volcanoes, and tsunamis in reference to plate tectonics, natural disasters, and predicting and mitigating their effects. Multiple chapters explore seafloor spreading. The author also discusses hydrothermal activity. This text takes on an immense range of content, but still explains concepts clearly and with detail.

Gubbins, David. *Seismology and Plate Tectonics*. Cambridge: Cambridge University Press, 1990. Although highly technical, the text portions are easily read and understood, especially the section on ocean trenches and plate subduction. Describes seismograph interpretation techniques and includes practice seismograms for the reader to locate earthquakes and determine their source mechanisms. Suitable for college-level science courses.

Jiao, Wenjie, et al. "Do Intermediate- and Deep-Focus Earthquakes Occur on Preexisting Weak Zones? An Examination of the Tonga Subduction Zone." *Journal of Geophysical Research* 105 (2000): 125-128. This article examines the title question. Discusses the fault patterns leading to deep earthquakes. Provides a good overview of deep-focus earthquakes.

Wilson, J. Tuzo, ed. *Continents Adrift and Continents Aground*. San Francisco: W. H. Freeman, 1976. Wilson brings together seventeen classic articles from *Scientific American* that describe the plate tectonic revolution of the late 1960's. Articles are grouped under four topics: mobility in the earth; seafloor spreading, transform faults, and subduction; plate motion; and applications of plate tectonics. Illustrations and maps are clear and concise. Provides a historical perspective of plate tectonics.

Yeats, Robert, Kerry Sieh, and Clarence Allen. *The Geology of Earthquakes*. New York: Oxford University Press, 1997. The authors provide an interdisciplinary approach to the study of well-known earthquakes. Although it welcomes a wide range of readers, the text was written for engineers, geophysicists, and government planners. Includes a comprehensive glossary, a bibliography, and a global appendix that lists more than three hundred historical earthquakes with surface rupture. Well illustrated with photographs, maps, and cross sections.

See also: Continental Drift; Earthquake Distribution; Earthquake Engineering; Earthquake Hazards; Earthquake Locating; Earthquake Magnitudes and Intensities; Earthquake Prediction; Earthquakes; Earth's Interior Structure; Earth's Mantle; Geodynamics; Lithospheric Plates; Mantle Dynamics and Convection; Notable Earthquakes; Plate Motions; Plate Tectonics; Slow Earthquakes; Soil Liquefaction; Subduction and Orogeny; Tectonic Plate Margins; Tsunamis and Earthquakes; Volcanism.

DISCONTINUITIES

Discontinuities are boundaries within the earth that divide the crust from the mantle, the mantle from the core, and the outer core from the inner core. The term is also used to describe the less dramatic boundaries within layers.

PRINCIPAL TERMS

- **crust:** the top layer of the earth, composed largely of the igneous rock granite; it ranges from 3 to 42 miles in thickness
- **earthquake:** a tremor caused by the release of energy when one section of the earth rapidly slips past another; earthquakes occur along faults or cracks in the earth's crust
- **earthquake waves:** vibrations that emanate from an earthquake; earthquake waves can be measured with a seismograph
- **inner core:** the innermost layer of the earth; the inner core is a solid ball with a radius of about 900 miles
- **mantle:** the largest layer of the earth, about 1,800 miles in thickness; the mantle is within 3 miles of the earth's surface at some locations
- **outer core:** the outer portion of the core, about 1,300 miles in thickness; it is believed to be composed of molten iron
- **seismograph:** a device that measures earthquake waves

EARTH'S INTERIOR

Discontinuities are underground boundaries between layers of the earth. The closest discontinuity to the earth's surface is the Mohorovičić, which divides the earth's crust from the mantle underneath. Other discontinuities divide the mantle from the outer core and the outer core from the inner core. Minor discontinuities are found within these layers.

The interior of the earth has been the object of speculation and interest for thousands of years. Because direct observation of the earth's interior is usually impossible, however, inferences about its structure and characteristics must be made from phenomena seen and felt at or near the earth's surface. Several phenomena do give indications of the subsurface earth: caves that are often cool and damp, cool water emanating from springs and artesian wells, hot water spewing upward from geysers, and volcanoes from which extremely hot lava erupts. These phenomena give a mixed and incomplete picture of the earth beneath the surface.

SEISMIC WAVES

The structure and composition of the interior of the earth can, however, be inferred from the study of earthquake waves. Seismographs can detect three types of vibrations: surface waves (the ones that can cause damage when there is an earthquake), P (primary) waves, and S (secondary) waves, which are also generated by every quake. P waves are compressional (pushing) waves, in which earth or rock particles move forward in the direction of wave movement; S waves are shear waves, in which the particle motion is sideways or perpendicular to the direction of wave movement. The more efficient P waves travel twice as fast as S waves and thus are always detected first by a seismograph. Seismographs record these waves on charts, called "seismograms," attached to moving drums. By noting the arrival times of the various waves, seismologists can determine the distance to an earthquake and can see the effects on these waves caused by the type of rock through which the waves have moved.

Seismic waves travel through rock layers at specific speeds, which are different for each type of mineral or rock. For example, waves travel through basalt at 5 miles per second and through peridotite at 8 miles per second. Seismogram study has shown that the earth's interior is not homogeneous, but rather is composed of several major layers and many sublayers.

GUTENBERG DISCONTINUITY

In 1906, British geologist Richard Dixon Oldham discovered that S waves are never detected on the opposite side of the earth from any earthquake. As he already knew that S waves cannot travel through liquid substances, Oldham postulated that the center of the earth must be composed of a molten core and that the materials above this core are not molten. The depth of the boundary between this core and the material above it was discovered eight years later by Beno Gutenberg. Now called the Gutenberg discontinuity, it is located about 1,800 miles beneath the earth's surface.

MOHOROVIČIĆ DISCONTINUITY

When Oldham made his discovery of a central core, Andrija Mohorovičić was the director of the Royal Regional Center for Meteorology and subduction at Zagreb, one of the leading seismological observatories in Europe. In 1909, his meticulous study of a Croatian earthquake showed that some of the P waves from that quake had traveled faster than others. He already knew that other waves speed up or slow down when they move from one medium into another (as when light moves from air into water) and that this change in speed can result both in reflection, a bouncing back of waves, and in refraction, or a change in wave direction through the new medium. He deduced that the faster-moving P waves had traversed down through the earth, through a discontinuity to a material of a different density, and then had come back up to the surface. Deep in the earth was a material that allowed for faster transmission of P waves. Above this discontinuity, seismic waves travel at about 4.2 miles per second; below the boundary, they travel at about 4.9 miles per second.

When Mohorovičić's results were replicated by other seismologists, it was concluded that the discontinuity was a global phenomenon. Data from these studies showed that there were two very distinct layers of the earth: an upper, less-dense layer now called the "crust," and a denser layer below called the "mantle." Thus Andrija Mohorovičić had discovered what is now called the Mohorovičić discontinuity, the boundary between the earth's crust and mantle (it is often called the "Moho").

The crust of the earth is made up of continents and ocean basins that are very different from one another. Continental crust is made primarily of granite. Covering this granite over much of the earth's continents may be found layers of younger sedimentary rock such as sandstone, limestone, and shale. Ocean basins, in contrast, are composed of the dark, heavy rock basalt.

Mohorovičić believed the discontinuity between the crust and the mantle to be about 30 to 35 miles below the surface of the earth. Subsequent studies have shown that it is usually at a depth of about 21 miles. However, the Moho has an irregular shape that is roughly a mirror image of the surface of the earth.

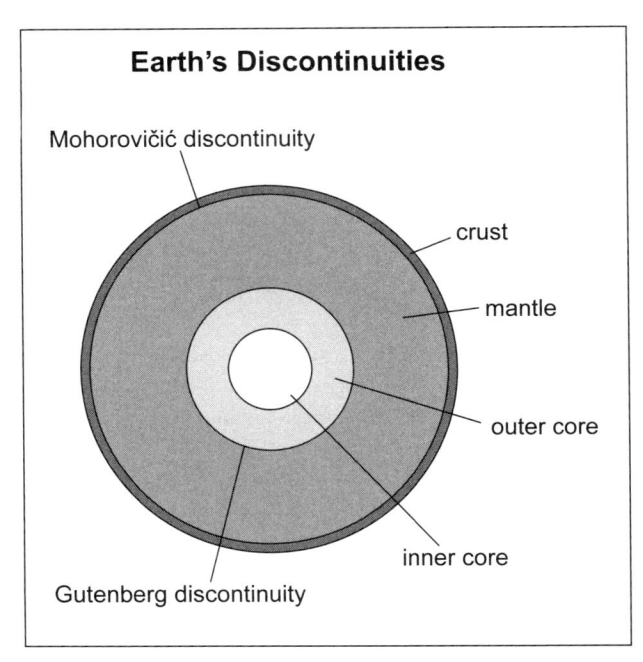

The Mohorovičić discontinuity is located at the boundary between the earth's crust and mantle. The Gutenberg discontinuity is located about 1,800 miles beneath the earth's surface.

Under the continents, the Moho is much deeper; under the oceans, the crust is very thin, and the Moho is as close as from 3 to 5 miles from the surface. The greatest depth of the Moho is probably beneath the Tibetan Plateau, where it reaches a depth of 42 miles.

EARTH'S MANTLE

The continents are higher because they are composed of granite, which is a lower-density rock than basalt or the materials of the mantle. Even though the mantle is composed of solid rock, under long-term stresses, the rock moves slowly like a liquid. Thus, just as ice floats in water, the continents actually are floating upon the heavier mantle rock. The Moho is the boundary between continental granitic rocks and the denser peridotite rock of the mantle.

The mantle extends from the bottom of the crust to a depth of 1,800 miles. It appears to be made of the rocks somewhat similar chemically to those in the earth's crust but more "basic"—that is, having more of the heavy iron and magnesium minerals such as olivine, and less lightweight aluminum. The mantle also appears to be composed of layers with discontinuities about 220 and 400 kilometers beneath the earth's surface. Although mantle rock is solid, it can, under certain conditions, behave somewhat like a liquid. Under long-term pressures, the molecules of this solid rock can move like liquids, but under sharp, short-term stresses, mantle rock fractures like a brittle solid.

In 1957, a project was conceived to drill a hole through the thin oceanic crust down past the Mohorovičić discontinuity to bring up rock from the mantle. Although the "Mohole" project was approved and funded by the National Science Foundation, funds for it were cut off by the U.S. Congress in 1966.

MANTLE CONVECTION CURRENTS

Heat within the earth is created through the decay of radioactive isotopes. Although this generated heat is very small compared to the heat received from the sun, it is well insulated and is enough to create volcanoes and the convection currents of the mantle. Mantle rock is extremely hot; because of the pressure on it from the crust above, however, it cannot melt, except where there is a decrease in this pressure.

Studies at the surface of the earth have revealed areas where great heat flow comes from the mantle. Near the center of the Atlantic Ocean, the basaltic ocean bottom has split; the two sides are being pulled away from each other as Europe and Africa move away from the American continents. At this split, a decrease in pressure allows the hot mantle rock to melt and well upward, filling the gap between the dividing ocean bottoms. Thus the new ocean basin is made of material directly from the mantle. Within the mantle are large, slow-moving convection currents where hot mantle rock moves upward, cools off, and slowly sinks. These currents are believed to be the driving forces of continental drift.

INNER EARTH DISCONTINUITIES

At the bottom of the mantle, beneath the Gutenberg discontinuity, is the earth's core. Seismic studies have shown that the outer core, which extends from roughly 1,800 to 3,100 miles beneath the earth's surface, is not a perfect sphere. The core rises in areas where hot mantle rock is moving upward and is depressed where cooler mantle rock is moving downward. The density of the core is much greater than that of the mantle. It is believed that this core is made of molten nickel and iron and that its motion generates the earth's magnetic field, which produces the aurora borealis.

In 1936, Danish seismologist Inge Lehmann discovered evidence for a solid core within the molten center of the earth by detecting seismic waves that had been deflected back to the surface from within the core. When she realized that these waves, though very weak, travel faster through this most-central part of the earth than through the rest of the molten core, she was able to infer that this inner core was

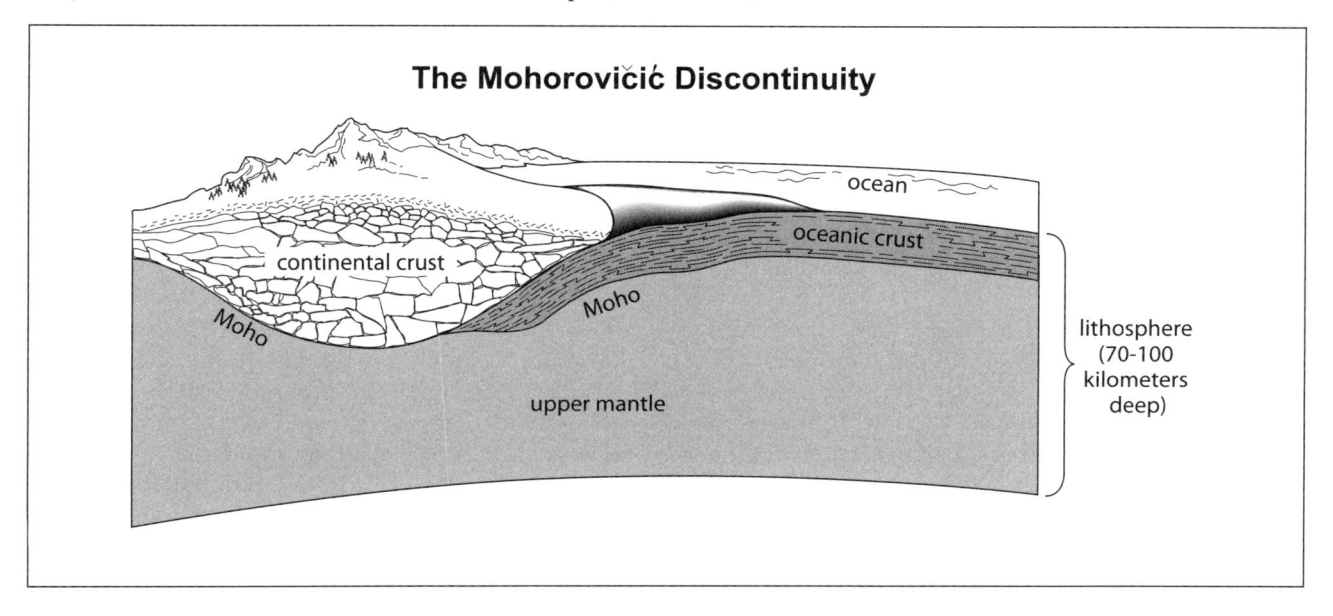

The Mohorovičić Discontinuity

composed of solid material completely surrounded by the molten outer core. This most central layer of the earth extends from 3,100 miles beneath the earth's surface to the center of the earth, 4,000 miles down.

SEISMOGRAPHS

The seismoscope is an ancient instrument that shows earth movements. A Chinese seismoscope of the second century C.E. had eight dragon figures each with a ball in its mouth. When the earth trembled, a ball would fall from the dragon's mouth into the mouth of a frog figure underneath it. European seismoscopes often used bowls of water that would spill when agitated. In 1853, Italian physicist Luigi Palmieri designed a seismometer that used mercury-filled tubes that would close an electric circuit and prompt a recording device to start moving when the earth vibrated.

In 1880, British seismologist John Milne invented the first modern seismograph, which employed a heavy mass suspended from a horizontal bar. When the earth would quake, the bar would move, and that movement would be recorded on light-sensitive paper beneath. Most seismographs employ a pendulum, which, because of inertia, remains still as the earth moves underneath it.

When seismographs measure shock waves from nearby earthquakes, they first receive the P waves, which vibrate in the direction in which the waves are moving. S waves, which vibrate perpendicular to the direction in which the waves are moving, then arrive, followed by surface waves. When seismographs record more distant earthquakes, the results are complicated by the reflection and refraction of seismic waves resulting from the various discontinuities underground. As the complications were deciphered, seismologists realized that the recordings described the rock layers below and between the quake and the seismograph.

Once geologists realized that they could learn about the earth by examining seismograms, some researchers became impatient when they wanted to study a particular area but had to wait for an earthquake to occur. This became particularly difficult in areas where earthquakes did not occur frequently. Milne solved the problem by dropping a one-ton weight from a height of about 25 feet. The impact of this weight on the ground generated seismic waves that were weaker than, but similar to, those generated by earthquakes. To create stronger waves, seismologists explode charges of dynamite. These artificially induced shock waves have enough energy to reach deep into the planet. Since the 1970's pistons on large trucks have been used to strike the earth and create artificial seismic waves.

When charges are exploded and the vibrations recorded by several nearby seismographs, a detailed description of rock layers can be detected. Since 1923, when a seismograph was first used to locate a large underground pool of petroleum, seismology has played a large part in the oil and gas industries. Earthquake waves artificially produced by explosions are also able to determine the location of underground geologic structures that may contain mineral deposits.

With the advent of the space age, seismographs connected to radio transmitters have been placed on the surfaces of the moon and Mars. There are more than a thousand seismographs in constant operation gathering seismic data around the world. Data from the National Earthquake Information Center are updated daily and are available on the Internet.

EARTHQUAKE STUDY AND PREDICTION

The same technology that has indicated the location of discontinuities deep within the earth has also provided a greater knowledge about the crust. Whereas ancient civilizations feared earthquakes as manifestations of angry gods, quakes are now seen as the results of energy being released when plates of the earth's crust move past one another.

Although earthquakes do occur in many places on the earth's crust, they are most common in certain areas such as the "Ring of Fire" around the Pacific Ocean. Most earthquakes are linked directly to the movement of the earth "plates," or sections of the crust. The Pacific plate and the North American plate meet along the San Andreas fault, which runs from western Mexico through California to the Pacific Ocean. The two plates are moving past each other along this fault. Each time there is movement along the fault, tremendous amounts of energy are released, and the earth quakes. Quakes along this fault and others have caused untold damage.

One of the primary goals of seismologists is to determine a way to predict exactly when earthquakes will occur. If this information were known in advance, people could prepare for quakes, and far fewer deaths would occur. Many phenomena have been observed before quakes, such as increased strains upon bedrock, changes in the earth's magnetic field, changes in seismic wave velocity, strange movements of animals, changes in groundwater levels, increased concentrations of rare gases in well water, geoelectric phenomena, and changes in ground elevation. Because none of these dependably occurs before every quake, these signs have not become reliable indicators.

Seismologists cannot prevent earthquakes from occurring, nor can they yet predict the exact time of a major quake, but they can predict where earthquakes are likely to occur. It is believed that certain active faults where there has been no earthquake activity for thirty years or so are about ready for an earthquake. With this information, urban and regional planners can provide for quake-resistant roads, bridges, and buildings.

Search for Oil and Gas Reservoirs

Seismology and the search for minor discontinuities play a great part in the search for oil, gas, and mineral resources. Since much petroleum is retrieved from off-shore locations where the crust of the earth is thinner, knowing the location of the Mohorovičić discontinuity sets the lower boundary for exploration.

Seismic studies are used regularly to assist in the search for oil and gas reservoirs. Natural gas and petroleum both can become trapped under some geologic formations. Seismologists routinely create a survey of an area before drilling to find minor discontinuities or boundaries between two different rock types, such as shale and sandstone. These surveys are made by measuring the reflection of seismic waves from the underlying rock layers. Geologic structures that can contain petroleum, natural gas, or mineral deposits can be identified from these surveys. Seismic surveys can show the distance and direction to these structures.

Kenneth J. Schoon

Further Reading

Calder, Nigel. *The Restless Earth: A Report on the New Geology*. New York: Viking Press, 1972. A companion book to the television program *The Restless Earth*, this text emphasizes how geologists came to the conclusion that the continents are moving. Illustrated with black-and-white and color photographs and diagrams. Indexed.

Cromie, William J. *Why the Mohole?* Boston: Little, Brown, 1964. A 1960's view of the never-finished American and Soviet plans to drill holes through the entire crust of the earth in order to reach the mantle. The author was public-information officer for the American project. Several diagrams and photographs, a bibliography, and an index.

Dahlen, F. A., and Jeroen Tromp. *Theoretical Global Seismology*. Princeton: Princeton University Press, 1998. Intended for the college-level reader, this book describes seismology processes and theories in great detail. The book contains many illustrations and maps. Bibliography and index.

Doyle, Hugh A. *Seismology*. New York: John Wiley, 1995. A good introduction to the study of earthquakes and the earth's lithosphere. Written for the layperson, the book contains many useful illustrations.

Emiliani, Cesare. *Planet Earth: Cosmology, Geology, and the Evolution of Life and Environment*. Cambridge, England: Cambridge University Press, 1992. A large, comprehensive book containing basic information about matter and energy, many aspects of the physical and historical earth, and a large section about the earth's relationship to the universe. The last section is a brief history of the earth sciences.

Erickson, Jon. *Rock Formations and Unusual Geologic Structures*. Rev. ed. New York: Facts on File, 2001. An easy-to-read description of the earth's crust, including the creation, deformation, and erosion of rock. Clear black-and-white photographs, diagrams, and maps along with a large glossary, bibliography, and index.

Jackson, Ian, ed. *The Earth's Mantle: Composition, Structure, and Evolution*. Cambridge: Cambridge University Press, 1998. Intended for the college student, *The Earth's Mantle* provides a clear and complete description of the elements that make up the earth's mantle and the process of change that it has undergone since its formation. Includes bibliography and index.

Lambert, D., and the Diagram Group. *Field Guide to Geology*. New York: Facts on File, 1988. A profusely illustrated book about the earth, its seasons, rocks, erosional forces, and geological history. Contains a list of "great" geologists and a list of geologic museums, mines, and spectacular geologic features. Indexed.

Miller, Russell. *Continents in Collision.* Alexandria, Va. Time-Life, 1983. A thorough text describing how earth motions have created geologic features. Profusely illustrated with color and black-and-white illustrations. Bibliography and index.

Plummer, Charles C., and Diane Carlson. *Physical Geology.* 12th ed. Boston: McGraw-Hill, 2007. A college-level introductory geology textbook that is clearly written and wonderfully illustrated. An excellent sourcebook of basic information on geologic terminology and fundamentals of geologic processes. An excellent glossary.

Reynolds, John M. *An Introduction to Applied and Environmental Geophysics.* 2d ed. New York: John Wiley, 2011. An excellent introduction to seismology, geophysics, tectonics, and the lithosphere. Appropriate for those with minimal scientific background. Includes maps, illustrations, and bibliography.

Shearer, Peter M. "Upper Mantle Seismic Discontinuities." In *Earth's Deep Interior: Mineral Physics and Tomography from the Atomic to the Global Scale.* Edited by Shun-Ichiro Karato, et al. American Geophysical Union, 2000. Provides a detailed analysis of the seismic discontinuities of the upper mantle. Discusses discontinuity properties and seismic observations. The author suggests constraints applied to properties to produce best results.

Tarling, D., and M. Tarling. *Continental Drift: A Study of the Earth's Moving Surface.* Garden City, N.J.: Anchor Press, 1971. A small paperback book with black-and-white photographs and diagrams that help the reader to understand the principles of earth structure and plate tectonics.

Vogt, Gregory. *Predicting Earthquakes.* New York: Franklin Watts, 1989. A good text on the earth's interior and on how earthquakes are generated, detected, and measured. The last chapter discusses the prediction of earthquakes and efforts to control their effects. Black-and-white photographs and diagrams, glossary, and index.

Weiner, Jonathan. *Planet Earth.* New York: Bantam Books, 1986. A companion volume to the television series *Planet Earth,* this book is well illustrated with both black-and-white and color pictures and diagrams. No glossary, but a comprehensive bibliography and index.

See also: Creep; Cross-Borehole Seismology; Earthquake Distribution; Earthquake Prediction; Earth's Core; Earth's Differentiation; Earth's Interior Structure; Earth's Mantle; Elastic Waves; Experimental Rock Deformation; Faults: Normal; Faults: Strike-Slip; Faults: Thrust; Faults: Transform; Heat Sources and Heat Flow; Plate Tectonics; San Andreas Fault; Seismic Observatories; Seismic Reflection Profiling; Seismic Tomography; Seismometers; Stress and Strain.

E

EARTH-MOON INTERACTIONS

Interactions between Earth and its moon have led to many effects on both bodies. Some of these effects are ocean tides, rates of rotation and revolution, and other basic forces, especially gravity.

PRINCIPAL TERMS

- **axis:** the point around which a planet rotates
- **dynamo:** the actions of a moving, liquid core of a planetary body
- **neap tides:** smaller than normal tides that occur when the sun and moon are 90 degrees apart
- **nutation:** a periodic wobble or oscillation in an axis of rotation
- **precession:** the axial movement of a spin axis of a planet, moon, or other astronomical body in a circular motion around a cone shape
- **precession of the equinoxes:** the slow precession of the earth over 2,600 years
- **spring tides:** larger than normal tides that occur when the sun and moon are approximately aligned
- **synchronous rotation:** when a planetary body revolves at the same rate as it rotates; synchronous rotation of the moon around Earth is why only one side of the moon is visible from Earth
- **tidal bulge:** the areas of a planet's surface and bodies of water that become "stretched out" when acted upon by gravitational pull
- **tidal force:** the gravitational pull of an astronomical body that causes a squashed, stretched-out effect of a planet's surface (and, in Earth's case, bodies of water)
- **tidally locked:** the phenomenon in which the gravitational pull between Earth and the moon results in the moon revolving around Earth at the same rate as it revolves on its axis so that the same side of the moon always faces Earth
- **tides:** the rising and falling of water levels, usually twice a day, caused by the moon's gravitational pull

GRAVITATIONAL INTERPLAY BETWEEN THE MOON AND EARTH

The moon, Earth's closest celestial neighbor, exerts a variety of forces on the earth. These forces affect planetary motions and activities on the bodies' surfaces. Even though the moon is far smaller than the sun, it applies regular and measurable forces that affect the movement of the earth and activities on the earth's surface because it is so close to Earth.

The moon orbits Earth approximately once every 27.3 Earth days. At an average distance from Earth of about 384,400 kilometers (km), or 238,900 miles (mi), the moon's mass has a gravitational force that pulls Earth toward it as Earth, likewise, tugs at the moon. The physics of rotation and orbital motion prevents the two objects from crashing into each other; rather, they are "falling together" toward and around the sun. These interacting forces have a number of effects.

One of the most noticeable effects of the moon on Earth is tides, or the daily rising and subsiding of sea levels worldwide. The side of the earth that is closest to the moon at a given time will feel a stronger pull from the moon's gravity than the sides not facing the moon. This uneven pull, called a tidal force, causes Earth to stretch very slightly, like a squashed globe. The parts of the globe that stretch out on the side facing and opposite the moon are called tidal bulges, and are where ocean tides rise. As the earth rotates on its axis once every twenty-four hours, the tidal bulges try to follow the pull of the moon, causing ocean levels to rise and fall twice a day in most places on Earth. Ocean levels can change as much as 40 feet in some places because of high tides.

The sun also affects Earth's tides. Even though it is far larger than the moon, its greater distance makes the tidal force from the sun weaker than that of the moon. Generally, how close an object is rather than how large an object is will increase its tidal force. However, when the moon and sun are roughly aligned on the same side of the planet (during the full or new moon), their forces combine to cause slightly bigger tides than normal; these tides are dubbed spring tides. When the sun and moon are 90 degrees apart and when their forces are counteracting each other, Earth experiences smaller tides than normal, which are called neap tides.

The gravitational pull between Earth and the moon also cause the moon to become tidally locked, synching the moon's rotation around its axis and its revolution around the earth. This synchronous rotation causes the same side of the moon to always face Earth, and the other half (dubbed the "dark" side of the moon because it is never seen from Earth) to always face away.

Motions of the Earth and Moon

The tidal effect has additional consequences besides rising and lowering water levels on Earth. For one, the moon is accelerating away from Earth and appears to be farther away. This occurs because Earth's tidal bulges have a gravitational force that tries to pull the moon along with them. The gravitational pull from the tidal bulges attempts to speed up the moon in its orbit around Earth, adding more torque to the moon's orbit and increasing its angular momentum. The overall effect results in the moon accelerating forward in a larger orbit, moving it away from Earth at about 3 to 4 centimeters per year.

This tidal tug-of-war also has an effect on Earth's motion, gradually slowing its rotation and lengthening its days. According to Newton's third law, objects will exert equal and opposite forces on each other, so energy is conserved within the Earth-moon system. The angular momentum transferred to the moon's orbital movement, and energy added to its orbit from Earth's rotation, causes Earth to slow down. Just as Earth tries to speed up the moon, the moon's own gravitational pull is trying to slow down the rotation of the tidal bulges.

Because the moon is a relatively substantial mass (larger than Pluto and about 1/81 of the mass of Earth), its pull causes Earth's rotation to slow very slightly. This effect is great enough that Earth's days are slowly getting longer at about 1 second every 50,000 years. So, in one hundred years, Earth days will be 1.6 milliseconds longer. This slowing-down effect was first indirectly suggested in 1695 by British astronomer Edmond Halley, who said the moon was moving faster than before.

The Wobble Effect

In addition to affecting Earth's rotation, the moon helps the earth maintain a fairly stable rotation on its axis. The earth does not rotate in a straight-up or straight-down fashion, like a spinning top. Rather, it tilts on its axis at 23.5 degrees. This tilt means that over time, its axis, pointing to the sky, will swing in a small circle. Like a spinning top that is tapped from the side, Earth wobbles a bit around its tilted axis as it rotates. It stays at its 23.5-degree tilt, but it points in different directions.

Earth's axis moves in a circular motion called precession, a phenomenon first noted by ancient Greek astronomers. Earth completes this axis wobble, or a precession, once every 2,600 years. This means that Earth's axis will point in a different direction over time, so the star that appears over the North Pole will change. Earth's north now points to Polaris in the constellation Ursa Minor, but two precessions ago, Thuban in the constellation Draco was the North

Alma Harbor located in the town of Alma, New Brunswick, Canada, at low tide. It is located in the Bay of Fundy. The Bay of Fundy has the greatest tidal range in the world (up to 55 ft/16.75 m). (Ted Kinsman/Photo Researchers, Inc.)

Star, and in 13,000 years, Vega in the constellation Lyra will be the new North Star. This motion is also called the precession of the equinoxes.

The wobble occurs because Earth is not a perfect sphere, but rather is flattened slightly with a bulge around its equator that acts like a thick belt. The moon and the sun pull on the equatorial bulge, which causes the wobble. These gravitational forces cause nutations, or slight periodic vibrations, within the larger precession movement. The most notable nutation that Earth experiences is an 18.6-year cycle (identical to the precession of the moon's orbit, which lies at an angle of 5 degrees to that of Earth).

In addition to contributing to Earth's precession, the moon helps to stabilize Earth's tilt by acting as a constant force. The planet Mars, for example, wobbles dramatically on its axis because it is pulled in different directions by large celestial bodies. If Earth's wobble was more extreme, climate cycles on Earth would be much more drastic and life as it is presently known might not exist in its current form. By stabilizing Earth's wobble, the moon has helped contribute to Earth's climate and, consequently, its biodiversity.

BROADER LUNAR EFFECTS

The moon also influences Earth's magnetism, though how it does so is not yet completely understood. Earth's magnetic field, or magnetosphere, surrounds Earth, arising as a result of the motion of its molten core. The magnetosphere is not a circular bubble, and it is pushed away from Earth's sun-facing side because of the solar wind, which is a constant flow of high-speed, charged electrons, protons, and other particles emitted from the sun. The solar wind pushes back on Earth's magnetic field, creating a long, magnetic "tail" facing away from the sun, which the moon crosses into during its full-moon period.

Earth's magnetic tail likely charges the moon's surface and can cause a host of activity on its surface, such as dust storms and potentially floating lunar rocks. Before the solar wind reaches Earth, however, it will interact with the moon's weak, varying magnetic field, perhaps affecting how the solar wind reaches Earth's magnetosphere. It is not entirely known how the solar wind affects Earth or how it electrifies Earth's atmosphere. Space missions such as the National Aeronautics and Space Administration's Artemis (Acceleration, Reconnection, Turbulence, and Electrodynamics of the Moon's Interaction with the Sun) and LRO (Lunar Reconnaissance Orbiter) are working to provide more insight. Artemis, for example, aims to glean understandings about the effects of the solar winds on the two bodies by remaining outside Earth's magnetosphere and monitoring solar wind streams.

Tidal interactions between Earth and the moon may also be the cause of the moon's mysterious magnetic source. Normally, a planetary body needs the churning of a dynamically moving liquid core, called a dynamo, to generate a magnetic field. The moon's core dynamo likely ceased around 4.2 billion years ago, but magnetic lunar rocks much younger than that have been found, suggesting another, external source of local lunar magnetism. Some researchers theorize that Earth's pushing of the moon into a farther orbit created the churning needed to generate magnetism in these rocks. It may also be that a large impact on the moon created its faint, uneven magnetism.

The moon also has played a large role in shaping human life on Earth culturally, mathematically, and scientifically. Since the beginning of human history, humans have been inspired by the moon and have sometimes even feared it. Countless mythological beings, such as the goddess figures Hecate and Artemis (ancient Greece), Isis (ancient Egypt), and Shing-Moo (ancient China), are associated with the moon. Stories of werewolves and wild behavior have long been associated with the moon. (The word "lunatic" is derived from the moon's Latin name, *luna*.) Lunar (and solar) eclipses have frightened and excited people for millennia and have inspired a great deal of thought and discussion.

The moon also has inspired scientific insights. By observing its motions and phases relative to other celestial bodies, humans could begin to calculate astronomical distances, such as how far Earth is from the moon and sun, and to calculate the approximate size of Earth. The moon provided a way to keep time on a scale larger than that of days by providing a longer rhythmic cycle, that of months.

The moon also prompted technological advancements as nations raced to be the first to reach the moon. The Soviet Union sent the first satellites to the moon in 1959, culminating in the first human landing on the moon by the United States in 1969. Rocks retrieved from the moon and the many orbiting satellites and landing spacecraft that have studied its surface have helped humans to understand astrogeology, the history of the solar system, and Earth itself.

Most astronomers believe the moon formed when a large celestial body crashed into Earth about 4.5 billion years ago, spitting up debris that circled Earth and eventually became the moon. This theory may also explain how other planets came to have moons. In the unprotected surface of the moon's asteroid-pocketed landscape scientists, can see a history of the earlier universe and of cosmic events such as asteroid showers that might have enveloped both Earth and its moon.

Kristina Grifantini

FURTHER READING

Bertotti, Bruno, Paolo Farinella, and David Vokrouhlicky. *Physics of the Solar System: Dynamics and Evolution, Space Physics, and Spacetime Structure.* New York: Springer, 2003. This text focuses on gravitational dynamics, particularly the larger picture. Suitable for college level and beyond.

Lowrie, William. *Fundamentals of Geophysics.* 2d ed. New York: Cambridge University Press, 2010. A basic primer for geology students. Covers both applied and theoretical aspects of geophysics. A good basic text to prepare students for further studies in geophysics.

Pater, Imke de, and Jonathan Lissauer. *Planetary Sciences.* 2d ed. New York: Cambridge University Press, 2011. A comprehensive overview that serves as a primer for students serious about geology study.

Stacey, Frank D., and Paul M. Davis. *Physics of the Earth.* 4th ed. New York: Cambridge University Press, 2009. A textbook on the basics of earth geophysics, with illustrations. Suitable for students up to the graduate level.

Telford, W. M., L. P. Geldart, and R. E. Sheriff. *Applied Geophysics.* 2d ed. New York: Cambridge University Press, 2010. A classic textbook for geophysical studies. Good for undergraduates and graduates alike, and a solid reference for geologists.

Turcotte, Donald L., and Gerald Schubert. *Geodynamics.* 2d ed. New York: Cambridge University Press, 2010. First published in 1982, this standard textbook discusses physical processes on Earth and phenomena such as gravity field fluctuations. College-level and beyond.

See also: Earth's Differentiation; Earth's Magnetic Field; Earth Tides; Gravity Anomalies; Importance of the Moon for Earth Life; Jupiter's Effect on the Earth; Lunar Origin Theories; Magnetic Reversals; Polar Wander; Solar Wind Interactions.

EARTHQUAKE DISTRIBUTION

Seismologists have been monitoring global earthquake activity for approximately one century. These studies have led to an understanding of earthquake frequency and distribution that have contributed dramatically to the confirmation of plate tectonics theory.

PRINCIPAL TERMS

- **epicenter:** the point on the earth's surface directly above an earthquake's focus
- **focus:** also known as the hypocenter, the actual place of rupture inside the earth's crust
- **P wave:** the primary or fastest wave traveling away from a seismic event through the rock, consisting of a series of compressions and expansions of the earth material
- **plate boundary:** a region where the earth's crustal plates meet, as a converging (subduction zone), diverging (mid-ocean ridge), transform fault, or collisional interaction
- **S wave:** the secondary seismic wave, traveling more slowly than the P wave and consisting of elastic vibrations transverse to the direction of travel; S waves cannot propagate in a liquid medium
- **seismic belt:** a region of relatively high seismicity, globally distributed; seismic belts mark regions of plate interactions
- **seismic wave:** an elastic wave in the earth usually generated by an earthquake source or explosion
- **seismicity:** the occurrence of earthquakes as a function of location and time
- **seismograph:** an instrument used for recording the motions of the earth's surface caused by seismic waves, as a function of time
- **subduction zone:** a dipping ocean plate descending into the earth away from an ocean trench

PLATE TECTONICS

Although seismic instruments can record them from virtually anywhere on the globe, earthquakes occur primarily along active tectonic regions of the earth's crust where mountain building, folding, and faulting are occurring. More temporal and often less severe earthquakes also accompany volcanic activity.

The key to understanding earthquakes lies in the powerful theory of plate tectonics. The earth is far from a geologically "dead" world like its moon. The earth's crust is broken into several large slabs of crust, or lithospheric plates, and convection currents caused by the planet's internal heat drive the plates into motion, like a bunch of small rafts crowded onto the surface of a boiling pot of viscous jelly. At the mid-ocean ridges, new sea floor is created by magmatic eruptions—pushing two plates away from each other. These divergent boundaries are characterized by the Mid-Atlantic Ridge, where the North American and European continents (riding on the lithospheric plates) are moving away from each other. Along the Alpine belt, two continents are literally crashing into each other as Africa is pushing into and subducting under the Eurasian plate. The subduction zone is marked by a complex series of transform and thrust faults, which give the region its high seismicity.

Earthquakes are predominantly distributed along plate boundaries. Another converging boundary is found along the Hindu Kush and Himalayan mountain ranges, where the subcontinent of India is slowly crashing into and thrusting under the huge Eurasian plate at the rate of some 5 centimeters per year. This ongoing collision has created the world's highest mountains and makes this area of the world an earthquake-prone region.

SEISMIC BELTS

By mapping earthquake epicenters, scientists are able to map the seismicity (earthquake activity as a function of time) of the planet. Most earthquakes occur along three main belts: the Mid-Atlantic Ridge system; the Alpine Tethys belt, which extends from the Mediterranean Sea through Turkey and Armenia all the way into Asia, where it merges with the third main belt; and the infamous circum-Pacific "Ring of Fire." The least threatening of these is the mid-ocean ridge system such as that found in the Atlantic Ocean, along which new ocean crust is being created. As the sea floor spreads from the volcanic activity occurring at the spreading ridges, earthquakes occur along transform faults that bound the offset ridges. Although population centers are sparse along the mid-ocean ridges, Iceland, the Azores, and other small mid-Atlantic islands are regions of potential quake hazard. Owing to the steady rate of spreading at a few centimeters annually, earthquakes occurring along the ridge offsets tend to be frequent and of relatively low magnitude.

A far more dangerous region of earthquake activity is the Alpine Tethys belt, which extends across southern Europe and Asia. A listing of only a few of the major earthquakes along this belt reads like a litany of destruction and human suffering: Persia in 1505; Calabria in southern Italy in 1509, 1783, and 1832; Lisbon in 1755; the Neapolitan in Italy in 1857; numerous quakes in recent decades in Yugoslavia, Romania, Greece, and Turkey; and the 1988 disaster in Soviet Armenia. Volcanic activity also occurs in the region, with notable examples including Mounts Etna and Vesuvius and the island of Thera.

Perhaps the most seismically active region of the world lies on the eastern end of the Mediterranean-Himalayan belt. Stretching across Tibet and into China, this colossal zone of high seismicity threatens all who live along its 4,000-kilometer length. More than a dozen earthquakes of Richter magnitude 8.0 or greater have caused well in excess of a million human lives to be lost in this notoriously seismic region. It is believed that the Shaanxi region earthquake of 1556, which resulted in an estimated 830,000 casualties, was one of the worst earthquakes in history. The Gansu earthquake of 1920 caused approximately 200,000 casualties.

"RING OF FIRE"

The trans-Asian earthquake belt passes through Burma and Indonesia, ending in the southern Philippines. This transitional region marks the border of the earth's greatest seismic belt—the circum-Pacific, or "Ring of Fire." A region of complex plate interactions, the Pacific Rim is no stranger to earthquakes and volcanoes. Perhaps no region characterizes the circum-Pacific belt better than the islands of Japan. In geologic terms, the Japanese islands are an island arc, formed by a subduction zone off the coast of the country's landmass. As the sea floor spreads from the ridge systems, it collides with the Asian continental plate. The dense, water-soaked sea floor is subducted at a deep ocean trench. As the oceanic plate descends, the slab grinds and shudders in resistance before finally being swallowed by the mantle. Accounting for 90 percent of the world's earthquakes, trench subduction zones have a seismic fingerprint of ever-deepening quakes that can cause very severe shocks.

Sea floor earthquakes can cause tsunamis (sometimes incorrectly called "tidal waves") along the coastline. A 7.7-magnitude shock hit the Oga Peninsula in 1983 and brought on a tsunami that caused extensive damage. The Great Tokyo earthquake (magnitude 8.2) and fire of 1923 caused 143,000 casualties. The Fukui earthquake of 1948 killed more than 3,700 people. The 9.2-magnitude Indian Ocean earthquake on December 26, 2004, off the coast of Sumatra, produced a tsunami that killed more than 230,000 people in 14 countries. The Great East Japan Earthquake of 2011 and the tsunami it caused resulted in widespread destruction throughout northeastern Japan and killed an estimated 20,400 people. All of these events serve as stark testimonies to the danger of living near active plate subduction zones. In addition to the trench quakes, the volcanic islands are crisscrossed by numerous faults. Thousands of quakes have been recorded by Japanese historians, dating to well before the birth of Christ.

To the north and east along the circum-Pacific belt, the Aleutian Island arc reaches into the North American continent in Alaska. A complex system of faults and a subduction zone off the coast make Alaska a region of dangerous seismicity. In 1964, one of the most severe earthquakes ever recorded (magnitude 8.6) struck near the port of Valdez and generated a series of tsunamis that wracked the coast. On Alaska's southern coast is the Fairweather fault, a transform fracture on which a 1958 temblor shook loose 90 million tons of rock, which cascaded into Lituya Bay, raising a wave exceeding 500 meters in height. The Fairweather fault is a northern extension of a system of transform faults (so named because the fault is transformed into a ridge or trench at its ends) that includes the most famous of all—the San Andreas fault.

SAN ANDREAS AND ALPINE FAULTS

Extending from the Mendocino fracture zone 700 miles south to Mexico, the San Andreas and its attendant system of faults make the state of California an earthquake-prone territory. A unique type of plate boundary, the San Andreas represents a fracture line along which the oceanic Pacific plate is slowly but inexorably sliding north with respect to the North American continent at a rate of roughly 5 centimeters per year. In places where the fault is displacing smoothly, small tremors regularly shake the

View southwest along Hanning Bay fault scarp, reactivated during the 1964 Alaska earthquake. The 10- to 15-foot-high bedrock scarp trends from the right foreground to the left background. The uplifted wave-cut surface to the right is coated with desiccated marine calcareous organisms. (U.S. Geological Survey)

landscape. Yet, in regions where the fault is believed locked, major quakes are impending, placing the large population centers of San Francisco and Los Angeles at risk. On October 17, 1989, a strong earthquake shook the San Francisco Bay area, causing extensive damage, 63 deaths, and more than 3,750 injuries. In addition to the San Andreas, the Hayward fault in the northern California and a multitude of faults in southern California further increase the state's seismicity. The Whittier, San Fernando, and Inglewood-Newport faults are among those that threaten the city of Los Angeles.

As seismically active as California is, its seismicity pales by comparison with Latin America. The 1985 Mexico City earthquake (magnitude 8.1) was a grim reminder of the subduction zone off the coast of western Mexico. Central America is in a particularly precarious position. It is sandwiched between the Cocos, North American, South American, and Caribbean plates, and, therefore, is a zone of active volcanoes and numerous severe, deep-trench earthquakes. El Salvador, Guatemala, and Nicaragua are among the nations in greatest earthquake danger. As the Nazca plate bumps into South America's plate, the oceanic plate is subducted and the melting slab has caused the volcanism and massive uplifting of the towering Andes mountain range. While the eastern part of South America is seismically quiet, the west coastal regions of Peru and Chile are known to experience severe quakes.

The circum-Pacific belt continues along the South Pacific through New Zealand. Analogous to the San Andreas region, New Zealand is regularly shaken by tremors occurring along the Alpine fault. Continuing up through Indonesia, the ring of earthquake and volcanic activity completes its loop.

VOLCANIC ACTIVITY

Other earthquake regions of note include the American Northwest's Cascade volcanic chain, where a series of tremors often indicate pending eruptions. The 1980 explosive eruption of Mount St. Helens in Washington was caused by an earthquake that triggered a landslide, initiating the lateral blast, or nuée ardente.

In California, the San Andreas is not the only region of earthquake activity. An area with an explosive volcanic past in recent geologic times, the Mammoth-Mono Lake region was hit by four magnitude 6.0 temblors in 1980, occurring along the northern perimeter of the Long Valley caldera. South of this site, the Sierra Nevada mountain range continues to undergo periodic spasms of uplift, like the one that caused the Owens Valley quake of 1852.

In the Caribbean, earthquakes and volcanic activity are an ever-present threat along the borders of the Caribbean plate. This seismic belt is actually an extension of the Pacific belt, although it lies on the Atlantic side of the Americas. Examples of sites experiencing major shocks and activity include Port Royale, which plunged 50 feet underwater following a major earthquake in 1692, and Mount Pelée, which destroyed the town of Martinique with a nuée ardente in 1902. Regions of hot spot volcanism are also zones of especially high seismicity. The Hawaiian Islands lie on top of a mantle plume of magma, and the same forces that built the island chain are working on the main island of Hawaii today. A similar region lies beneath the North American continent, the site of Yellowstone National Park in Wyoming, Montana, and Idaho. Both Hawaii and Yellowstone are earthquake-prone regions.

CONTINENTAL INTERIORS

For the most part, continental interiors, especially the Precambrian metamorphic basement and its thin veneer of sedimentary rocks that make up the craton, are regions of low seismicity. Such regions include parts of the United States and Canada in the Great Lakes region, virtually all of South America except for the Pacific coast and Andes belt, and most of Africa.

While continental interiors are seismically quiet compared with the active plate margins, there are exceptional regions. Although most earthquakes in the United States take place west of the Rockies, the Mississippi Valley has been the site of some of the most severe earthquakes ever, the New Madrid quakes of 1811-1812. New England and South Carolina have also experienced powerful shocks in the past. The western two-thirds of Africa is seismically inactive, but the East African rift valley is a zone of earthquake and volcanic activity where a geologically new plate boundary is rifting the eastern edge of the continent apart. Curiously, the only continent on earth that is seismically quiet is Antarctica.

EARLY RECORDS

In a sense, the study of where earthquakes occur traces back to the roots of western and eastern culture roughly four thousand years ago in Mesopotamia and in Asia. The Bible's Old Testament and other ancient Middle Eastern documents are filled with accounts of earthquakes toppling cities. The most complete historical records of seismic activity are, appropriately enough, those of Japan and China. Chinese earthquake records date back thirty centuries, with exhaustive accounts of tragic earthquakes striking the Asian mainland. Japan, which experiences up to one thousand noticeable shocks per year, has been keeping detailed earthquake records on Tokyo's tremors since 818 C.E.

Modern scientific views on earthquake distribution perhaps began in response to the tragic All Saints Day earthquake and tsunamis that wrecked Lisbon in 1755. While a horrified western civilization reeled at the scope of the disaster, which killed many at Lisbon's numerous downtown churches, one of the more insightful minds of the scientific revolution looked at the event more objectively. Immanuel Kant advised that learning about where and why earthquakes occur was a more reasoned approach than blaming the disaster on divine causes.

GLOBAL SEISMIC STUDY

Some one hundred years after Lisbon's fateful quake, Irishman Robert Mallet published a study of the Neapolitan earthquake of 1857 in which he produced a seismographic map of the world that, with the exception of the mid-ocean ridge systems, remains an accurate seismic diagram. Teaching in seismically active Japan, British geology professor John Milne invented the modern seismograph. By the time of the 1906 quake in San Francisco, global seismic observatories were in place; they recorded the jolts in California. As the distance to an epicenter (the surface site above the actual fracture or focus of the quake) could now be determined, a set of three properly placed seismographs, Milne reasoned, could pinpoint an epicenter anywhere on the globe.

Seismic waves generated by an earthquake produce different types of waves. The primary wave, or P wave, is compressional, while the secondary or S wave is sheering as it travels through the earth. Quakes also produce surface waves, which cause the shaking motions that occur during the most destructive part of a seismic event. By measuring the ratio of the arrival time of the primary and secondary waves and the size or amplitude of the waves on the seismograph recording, Charles Richter, in 1935, was able to establish a scale for measuring the energy released in a quake, its magnitude. Richter and his colleague at the California Institute of Technology, Beno Gutenberg, published high-quality maps of worldwide earthquake distribution in 1954.

In the 1950's and 1960's, the United States helped to organize enough of the world's seismic observatories to establish a global monitoring network called the World Wide Standardized Seismograph Network. Data from decades of shocks recorded by the network and oceanographic research vessels led to the revolutionary theory of plate tectonics and its acceptance by the vast majority of earth scientists.

Identifying Regions of Seismic Hazard

Although plate tectonics theory and plate boundaries are invoked to explain the vast majority of earthquakes, scientists are still puzzled by earthquakes that occur far from active margins. Examples are the Mississippi Valley region and Charleston, South Carolina, which shook violently in 1886. In these regions, seismologists are alarmed by public perception that the land east of the Rockies is "solid bedrock." The most plausible cause of the Mississippi Valley quake activity is the enormous weight of sediments the great river system has deposited on a weak part of the continental crust. South Carolina is a region riddled with faults, yet it is far from an active plate margin. Seismologists use historical accounts and recent seismic records to predict regions of earthquake hazard. One theory of seismic hazard involves identifying regions where earthquakes have not occurred along active fault regions. Such seismically quiet regions are called "seismic gaps" and represent regions of accumulated strain along which a major rupture may be anticipated.

Using seismicity data, seismologists produce maps that indicate seismic hazard. Not only useful for scientific purposes, such maps help public agencies to create building codes and other earthquake prevention methods appropriate to the earthquake hazards of the region. Studies conducted by the United States Geologic Survey and the National Oceanic and Atmospheric Administration have concluded that the areas of San Francisco and Los Angeles, California; Salt Lake City and Ogden, Utah; Puget Sound, Washington; Hawaii; St. Louis-Memphis, Tennessee; Anchorage and Fairbanks, Alaska; Boston, Massachusetts; Buffalo, New York; and Charleston, South Carolina, are at greatest seismic risk in the United States. The worldwide network of seismic stations make digital records which enable computers to analyze the seismic waves more swiftly, thereby improving observation of the planet's moving plates.

Dangers to Densely Populated Regions

Approximately once every 30 seconds, a million times each year, the earth's crust shivers. Most of these tremors are perceptible only by sensitive instruments, but more than three thousand are strong enough to be felt by those nearby. Roughly twenty quakes a year are strong enough to do catastrophic damage to populated areas. By coincidence, some of the earth's most active seismic regions are also among its most densely populated. The lands bordering the Mediterranean Sea and Pacific Rim, the mountainous Middle East, India, China, and Japan are all familiar with the havoc of a major shock. In China alone, the death toll from earthquakes in recorded history exceeds 13 million. Earthquakes and related phenomena claim up to 15,000 lives annually in these regions of dangerous seismicity.

All told, millions of people have been killed by seismic activity, with untold loss of property through the half-tick of geologic time comprising human history. Seismologists still lack the capability of precise prediction of earthquakes, but areas of high seismicity and the ominous seismic gaps warn of quake hazard. In cities such as Tokyo, Los Angeles, San Francisco, and Anchorage, citizens must be prepared for the next big quake, which is as sure to come as the slow but steady movement of the earth's crustal plates continues.

David M. Schlom

Further Reading

Bolt, Bruce A. *Earthquakes: 2006 Centennial Update.* 5th ed. New York: W. H. Freeman, 2005. A revision of the University of California, Berkeley, seismologist's classic, *Earthquakes: A Primer.* Chapter 1 deals with earthquake distribution, and also of special interest are the appendices on world earthquakes and seismicity rates and lists of important earthquakes in the Western Hemisphere. Suitable for the lay reader.

Collier, Michael. *A Land in Motion: California's San Andreas Fault.* San Francisco: Golden Gates National Parks Association, 1999. Filled with beautiful color photographs that accompany text intended for the nonscientist, *A Land in Motion* gives the reader excellent insight into earthquakes and their aftermaths. There are also many diagrams and graphs that explain subduction, faults, and orogeny.

Condie, Kent C. *Plate Tectonics and Crustal Evolution.* 4th ed. Oxford: Butterworth Heinemann, 1997. An excellent overview of modern plate tectonics theory that synthesizes data from geology, geochemistry, geophysics, and oceanography. Extensive coverage of plate boundary interactions and earthquake distribution. An excellent "Tectonic Map of the World" is enclosed. Nontechnical and suitable for a college-level reader. A useful

"Suggestions for Further Reading" is provided at the end of the chapters.

Doyle, Hugh A. *Seismology*. New York: John Wiley, 1995. A good introduction to the study of earthquakes and the earth's lithosphere. Written for the layperson, the book contains many useful illustrations.

Eiby, G. A. *Earthquakes*. New York: Van Nostrand Reinhold, 1980. Written by an experienced seismologist, this text is filled with charts, maps, and photographs that help demystify the science of seismology. Two chapters address seismic geography of the world. Especially useful for readers interested in the seismicity of New Zealand. A lucid account, suitable for a college-level reader.

Field, Ned, et al. "Earthquake Shaking: Finding the 'Hotspots'." *U.S. Geological Survey: FS 001-01*. 2001. This fact sheet provides information on the geological characteristics that are used to identify hotspots.

Halacy, D. S. *Earthquakes: A Natural History*. Indianapolis: Bobbs-Merrill, 1974. Excellent treatments of historical earthquakes and an extensive discussion of world seismicity patterns. Written for a lay audience, the book is a lively discussion of all aspects of seismology and earthquakes, along with volcanoes and tsunamis.

Jacobs, J. A. *Deep Interior of the Earth*. 1st ed. London: Chapman and Hall, 1992. Deals in detail with all aspects of the earth's inner and outer core. The origin of the core, its constitution, and its thermal and magnetic properties are discussed in detail. Well suited to the serious science student.

Lambert, David. *The Field Guide to Geology*, 2d ed. New York: Facts on File, 2007. An excellent reference for the beginning student of geology, it is filled with marvelous diagrams that make the concepts easy to understand. Several sections address earthquake distribution and related topics. Suitable for any level of reader from high school to adult.

Plummer, Charles C., and Diane Carlson. *Physical Geology*, 12th ed. Boston: McGraw-Hill, 2007. A college-level introductory geology textbook that is clearly written and wonderfully illustrated. An excellent sourcebook of basic information on geologic terminology and fundamentals of geologic processes. The chapters on structural geology and global plate tectonics are especially relevant to understanding the San Andreas fault in the context of large-scale geologic processes. Three full pages are devoted to the San Andreas fault. An excellent glossary.

Sullivan, Walter. *Continents in Motion*, 2d ed. New York: American Institute of Physics, 1993. Dedicated to Harry Hess and Maurice Ewing, two late pioneers of plate tectonics theory, this book is the classic popular work on moving crustal plates. Lucid explanations of seismic evidence for plate motions and historical vignettes on seismology and earthquakes. A highly readable book.

Verney, Peter. *The Earthquake Handbook*. New York: Paddington Press, 1979. Superb historical accounts of humanity's struggle to understand earthquakes with easy-to-follow discussions of seismology and important sections on earthquake safety and preparedness. Extensive discussion on the causes and distribution of seismic events.

Walker, Bryce. *Earthquake*. Alexandria, Va.: Time-Life Books, 1982. A volume in the Planet Earth series, this book is filled with color photographs and diagrams, with an essay entitled "Grand Design of a Planet in Flux" addressing plate tectonics' role in earthquake distribution. Contains a map, index, and bibliography and is suitable for all readership levels.

Wilson, J. Tuzo, ed. *Continents Adrift and Continents Aground*. San Francisco: W. H. Freeman, 1976. Selected, classic readings from *Scientific American* magazine that are introduced with commentary by Wilson, a leading figure in the history of plate tectonics. Includes discussion of trench earthquakes, transform faults, and collisional boundaries. A classic post-San Fernando earthquake (1971) article by Don L. Anderson on the San Andreas fault will interest students of California's seismicity. Provides a historical perspective of plate tectonics. Suitable for a general audience.

See also: Deep-Focus Earthquakes; Earthquake Engineering; Earthquake Hazards; Earthquake Locating; Earthquake Magnitudes and Intensities; Earthquake Prediction; Earthquakes; Elastic Waves; Faults: Normal; Faults: Strike-Slip; Faults: Thrust; Faults: Transform; Notable Earthquakes; Plate Motions; Plate Tectonics; San Andreas Fault; Seismometers; Slow Earthquakes; Soil Liquefaction; Subduction and Orogeny; Tsunamis and Earthquakes.

EARTHQUAKE ENGINEERING

Earthquake damage and injury are aggravated by the fact that neither the time nor the location of major tremors can be precisely predicted by earth scientists. Damage to human-made structures may be lessened, however, through the use of proper construction techniques. Earthquake engineering studies the effects of ground movement on buildings, bridges, underground pipes, and dams to determine ways to build future structures or reinforce existing ones so that they can withstand tremors.

PRINCIPAL TERMS

- **epicenter:** the central aboveground location of an earth tremor; that is, the point of the surface directly above the hypocenter
- **failure:** in engineering terms, the fracturing or giving way of an object under stress
- **fault:** a fracture or fracture zone in rock, along which the two sides have been displaced vertically or horizontally relative to each other
- **hypocenter:** the central underground location of an earth tremor; also called the focus
- **natural frequency:** the frequency at which an object or substance will vibrate when struck or shaken
- **natural period:** the length of time of a single vibration of an object or substance when vibrating at its natural frequency
- **shear:** a stress that forces two contiguous parts of an object in a direction parallel to their plane of contact, as opposed to a stretching, compressing, or twisting force; also called shear stress
- **unreinforced masonry (URM):** materials not constructed with reinforced steel (for example, bricks, hollow clay tile, adobe, concrete blocks, and stone)

SOIL CONDITIONS

Earthquake engineering attempts to minimize the effects of earthquakes on large structures. Engineers study earthquake motion and its effects on structures, concentrating on the materials and construction techniques used, and recommend design concepts and methods that best permit the structures to withstand the forces.

One might logically expect that the structures nearest an earthquake fault would suffer the most damage from the earthquake. Actually, structural damage seems to bear little direct relation to the faults or to their distance from the structure. It is true that buildings near the fault are subject to rapid horizontal or vertical motion and that if the fault runs immediately beneath a structure (which is more likely in the case of a road or pipe than a building) and displaces more than a few inches, the structure could easily fail. The degree of damage, however, has more to do with the nature of the local soil between the bedrock and the surface. If the soil is non-cohesive and sand-like, vibrations may cause it to compact and settle over a wide area. Compaction of the soil raises the pressure of underground water, which then flows upward and saturates the ground. This "liquefaction" of the soil causes it to flow like a fluid so that sand may become quicksand. Surface structures, and even upper layers of soil, may settle unevenly or drop suddenly. Sinkholes and landslides are possible effects.

NATURAL FREQUENCY

Ground vibration and most ground motion are caused by seismic waves. These waves are created at the earthquake's focus, where tectonic plates suddenly move along an underground fault. The waves radiate upward to the surface, causing the ground to vibrate. Wave vibrations are measured in terms of frequency—the number of waves that pass a given point per second.

Much earthquake damage depends on what is known as natural frequency. When any object is struck or vibrated by waves, it vibrates at its own frequency, regardless of the frequency of the incoming waves. All solid objects, including buildings, dams, and even the soil and bedrock of an area, have different natural frequencies. If the waves affecting the object happen to be vibrating at the object's frequency, the object's vibrations intensify dramatically—sometimes enough to shake the object apart. For this reason, an earthquake does the most damage when the predominant frequency of the ground corresponds to the natural frequency of the structures.

At one time it was thought that earthquake motion would be greater in soft ground and less in hard ground, but the truth is not that simple. Nineteenth-century seismographers discovered that the natural frequency of local ground depends on the ground's particular characteristics and may vary widely from one location to another. The predominant frequency of softer ground is comparatively low, and the maximum velocity and displacement of the ground are greater. In harder ground, the predominant frequency is higher, but the acceleration of the ground is greater. When the ground is of multiple layers of different compositions, the predominant frequency is quite complex.

Free- and Forced-Vibration Tests

In order to determine the effect of vibrations on a building, an engineer must do the obvious: shake it. Whereas the effects on a very simple structure such as a pipe or a four-walled shack may be computed theoretically, real-life structures are composed of widely diverse materials. By inducing vibrations in a structure and measuring them with a seismograph, one can easily determine properties such as the structure's natural frequency and its damping (the rate at which vibrations cease when the external force is removed).

The simplest type of test is the free vibration, and the oldest of these is the pull-back test. A cable is attached to the top of the test structure at one end and to the ground or the bottom of an adjacent structure at the other. The cable is pulled taut and suddenly released, causing the structure to vibrate freely. Other tests cause vibrations by striking the structure with falling weights or large pendulums or even by launching small rockets from the structure's top.

Forced-vibration tests subject test structures to an ongoing vibration, thereby giving more complete and accurate measurements of natural frequencies. In the steady-state sinusoidal excitation test, a motor-driven rotating weight is attached to the structure, subjecting it to a constant, unidirectional force of a fixed frequency. The building's movements are recorded, and the motor's speed is then changed to a new frequency. Measurements are taken for a wide range of different frequencies and forces. Surprisingly, the natural frequencies for large multistory buildings are so low that a 150-pound person rocking back and forth will generate measurable inertia in the structure, thereby providing an adequate substitute for relatively complex equipment.

Another useful device is the vibration table: a spring-mounted platform several meters long on each side. Although designed to hold and test model structures, some tables are large enough to hold full-scale structural components—or even small structures themselves. Useful forced vibrations are also provided by underground explosions, high winds, the microtremors that are always present in the ground, and even large earthquakes themselves.

Earthquake-Resistant Design

Structures can be designed to withstand some of the stresses put upon them by large ground vibrations. They must be able to resist bending, twisting, compression, tension, and shock. Two approaches are used in earthquake-resistant design. The first is to run dynamic tests to analyze the effects of given ground motions on test structures, determine the stresses on structural elements, and proportion the members and their connections to restrain these loads. This approach may be difficult if no record exists of a strong earthquake on the desired type of ground or if the research is done on simplified, idealized structures.

The other approach is to base the designs on the performance of past structures. Unfortunately, new buildings are often built with modern materials and techniques for which no corollary exists in older ones. It follows that earthquake-resistant design is easier to do for simple structures such as roads, shell structures, and one-story buildings than for complex skyscrapers and suspension bridges.

Basic Configuration of a Structure

The first concern in examining a structure is its basic configuration. Buildings with an irregular floor plan, such as an "L" or "I" shape, are more likely to twist and warp than are simple rectangles and squares. Warping also tends to occur when doors and windows are non-uniform in size and arrangement. Walls can fail as a result of shear stress, out-of-plane bending, or both. They may also collapse because of the failure of the connections between the walls and the ceiling or floor. In the case of bearing walls, which support the structure, failure may in turn allow the collapse of the roof and upper floors. Nonstructural walls and partitions can be damaged by drift, which occurs when a building's roof or the floor of a given story slides farther in one direction than the floor below it does. This relative displacement between consecutive stories can also damage plastering, veneer, and windowpanes.

Lateral (sideways) cross-bracing reduces drift, as do the walls that run parallel to the drift. Another way to avoid drift damage is to let the nonstructural walls "float." In this method, walls are attached only to the floor so that when the building moves laterally, the wall moves with the floor and slides freely against the ceiling. (Alternatively, floating walls may be affixed only to the ceiling.) Windows may be held in frames by non-rigid materials that allow the frames to move and twist without breaking the panes. The stiffness and durability of a wall can be improved by reinforcement. For reinforcement, steel or wooden beams are usually embedded in the wall, but other materials are used as well. If the exterior walls form a rectangular enclosure, they may be prevented from separating at the top corners by a continuous collar, or ring beam.

STRUCTURAL ELEMENTS

Frame buildings are those in which the structure is supported by internal beams and columns. These elements provide resistance against lateral forces. Frames can still fail if the columns are forced to bend too far or if the rigid joints fail. Unlike bearing walls, frame-building walls are generally non-structural; the strength of the frame, however, can be greatly enhanced if the walls are attached to, or built integrally into, the frame. This method is called "in-filled frame" construction. Roofs and upper floors can fail when their supports fail, as mentioned earlier, or when they are subjected to lateral stress. An effective way to avoid such failure is to reduce the weight of the roof, building it with light materials.

Another danger to walls is an earthquake-induced motion known as pounding, or hammering, which can occur when two adjacent walls vibrate against each other, damaging their common corners. The collision of two walls because of lateral movement or the toppling of either is also called pounding. Columns and other structural elements may pound each other if they are close enough; in fact, the elements pounding each other may even be adjacent buildings. If the natural vibrations of the two structures are similar enough, the structures may be tied together and thus forced to vibrate identically so that pounding is prevented. Because such closeness in vibration is rare, the best way to avoid pounding is simply to build the structures too far apart for it to occur.

Shell structures are those with only one or two exterior surfaces, such as hemispheres, flat-roofed cylinders, and dome-topped cylinders. Such shapes are very efficient, for curved walls and roofs possess inherent strength. For this reason, they are sometimes used in low-cost buildings, without reinforced walls. When failure does occur, it is at doors and other openings or near the wall's attachment to the ground or roof, where stress is the greatest.

Much earthquake damage could be prevented if the stresses on a structure as a whole could be reduced. One of the more practical methods of stress reduction uses very rigid, hollow columns in the basement to support the ground floor. Inside these columns are flexible columns that hold the rest of the building. This engineering technique succeeds in reducing stress, but the flexible columns increase the motion of the upper stories. More exotic methods to reduce stress involve separating the foundation columns from the ground by placing them on rollers or rubber pads. Structures with several of these lines of defense are much less likely to collapse; should a vital section of cross-bracing, bearing wall, or partition fail, the building can still withstand an aftershock. Overall, the earthquake resistance of a structure depends on the type of construction, geometry, mass distribution, and stiffness properties. Furthermore, any building can be weakened by improper maintenance or modification.

ALTERNATIVES TO UNREINFORCED MASONRY

Buildings using unreinforced masonry (URM) or having URM veneers have a poor history in past earthquakes. Because URM walls are neither reinforced nor structurally tied to the roof and floors, they move excessively during an earthquake and often collapse. Similarly, ground floors with open fronts and little crosswise bracing move and twist excessively, damaging the building. URM chimneys may fall to the ground or through the roof.

Buildings with URM bearing walls are now forbidden by California building codes, but URM is still common in many less developed areas of the world. There are several low-cost earthquake-resistant alternatives to such construction. Adobe walls may be reinforced with locally available bamboo, asphalt, wire mesh, or split cane. Low-cost buildings should be only one or two stories tall and should have a uniform arrangement of walls, partitions, and openings to obtain a uniform stress distribution. The floor plan should be square or rectangular or, alternatively,

The ATLSS Engineering Research Center's structural testing lab at Lehigh University in Bethlehem, Pennsylvania, is home to one of the largest structural testing facilities in the United States. Scientists at the lab test new materials which will help buildings survive earthquakes with little or no structural damage. (AP Photo)

have a shell shape such as a dome or cylinder. Roofs and upper floors should be made of lightweight materials—wood, cane, or even plastic, rather than mud or tile—whenever possible, and heavy structural elements should never be attached to nonstructural walls.

The Center for Planning and Development Research at the University of California at Berkeley noted certain features of modern wood-frame houses that make them especially susceptible to damage from strong ground motion. In addition to URM walls or foundations, such houses may have insufficient bracing of crawl spaces, unanchored water or gas heaters, and a lack of positive connections between the wooden frame and the underlying foundations. Porches, decks, and other protruding features may be poorly braced. Most of these deficiencies can be corrected.

Protection Against Injury and Property Damage

Earthquakes are arguably the most destructive natural disaster on the planet. No other force has the potential to devastate so large an area in a very short time. Not only can it not be predicted, but there is also even less advance warning for the earthquake than for other types of disaster. A hurricane can be seen coming by radar and a volcano may belch smoke before it erupts. An earthquake simply happens.

Yet the magnitude of the earthquake is not solely responsible for the destruction. Property damage and injury to humans also depend on the type and quality of construction, soil conditions, the nature of the ground motion, and distance from the epicenter. The tremor which struck Anchorage, Alaska, in 1964 measured 8.3 on the Richter scale and killed eleven people; by comparison, the earthquake that hit San Fernando, California, in 1971 measured only 6.6—less than a tenth of the force of the Anchorage quake—and fifty-nine people died. Most of the San Fernando deaths occurred in one building: a hospital that collapsed. It seems likely that the hospital had not been adequately constructed to withstand the stresses to which it was suddenly subjected. The higher damage toll resulted from the soil characteristics in San Fernando and an underground fault that had previously been unmapped.

The only protection earthquake-zone residents have against property damage is that given by the engineers who design and build their homes, workplaces, railway structures, dams, harbor facilities, and nuclear power plants and by the public officials who regulate them. Now that engineers can learn how ground movement affects engineering structures and can design new structures accordingly, many of the earthquake-prone regions have building codes mandating earthquake-resistant construction. In some communities, programs exist to determine which buildings are unsafe and how they may be made resistant. Unfortunately, not all quake regions have such rules and programs in place, because of apathy, high cost, or other reasons. The high costs of recovery after major quakes, however, provide a compelling rationale for better preparation.

Shawn V. Wilson

FURTHER READING

Case, James C., and Green, J. Annette. "How to Make Your Wyoming Home More Earthquake Resistant." *Wyoming State Geological Survey: Information Pamphlet 5* (rev.): 2001. A step-by-step guide to earthquake-resistant structural enhancements. Written for the general public, the introductions to each part provide an overview of common earthquake damage.

Center for Planning and Development Research. *An Earthquake Advisor's Handbook for Wood Frame Houses.* Berkeley: University of California, 1982. This slim book was a result of the Earthquake Advisory Service Project at the Center for Planning and Development Research at the University of California, Berkeley. The book is addressed both to the homeowner and to public policy personnel who want to plan an earthquake advisory project. Includes a long, diagrammed how-to section on repairs. Suitable for all readers.

Coburn, Andrew, and Robin Spence. *Earthquake Protection*, 2d ed. New York: Wiley, 2002. Coburn examines earthquake engineering and prediction, along with safety procedures. He also offers a historical look at seismic activity and earthquake hazard analysis.

Collier, Michael. *A Land in Motion: California's San Andreas Fault.* San Francisco: Golden Gates National Parks Association, 1999. Filled with beautiful color photographs that accompany text intended for the nonscientist, *A Land in Motion* gives the reader excellent insight into earthquakes and their aftermaths. There are also many diagrams and graphs that explain subduction, faults, and orogeny.

Consortium of Univeristies for Research in Earthquake Engineering. *Earthquake Engineering.* Department of Building Inspection, City & County of San Francisco. 2006. Covers a variety of topics related to earthquake engineering including effects, new technology, and research methods. Specific case studies and general overviews, glossary and detailed diagrams enhance this text.

Federal Emergency Management Agency. *Designing for Earthquakes: A Manual for Architects.* FEMA 454. 2006. A publication for the non-specialist reader, explaining the general principles of seismology and building design. Also includes chapters on the effects of earthquakes, site selection, and design regulations.

Kanai, Kiyoshi. *Engineering Seismology.* Tokyo: University of Tokyo Press, 1983. This book begins with a discussion of the seismograph and proceeds logically through seismic waves, ground vibrations, and their effects on structures. Concise, but the calculus is difficult in spots. Suitable for college-level students.

Kramer, Steven Lawrence. *Geotechnical Earthquake Engineering.* Upper Saddle River, N.J.: Prentice Hall, 1996. Intended for a person with some civil engineering background, this book can be quite technical at times. Kramer provides clear but advanced descriptions of the many aspects of seismic engineering.

Newmark, Nathan M., and Emilio Rosenblueth. *Fundamentals of Earthquake Engineering.* Englewood Cliffs, N.J.: Prentice-Hall, 1971. A highly technical, mostly theoretical graduate textbook and reference manual. Addresses basic dynamics, earthquake behavior, and recommended design concepts. Text is occasionally unclear, even where the language is not especially technical. Some diagrams are given inadequate explanation.

Okamoto, Shunzo. *Introduction to Earthquake Engineering*, 2d ed. New York: University of Tokyo Press, 1984. This textbook has chapters on earthquakes, earthquake-resistant design procedures, and earthquake resistance of roads, tunnels, railways, bridges, and various types of dams. The chapters on seismicity and on historical earthquakes are based on data from Japan, one of the world's most active seismic zones. Written for engineers and college-level students, but the many nontechnical passages are not too difficult to follow.

Powell, Robert, et al., eds. *The San Andreas Fault System: Displacement, Palinspastic Reconstruction, and Geological Evolution.* Boulder, Colo.: Geological Society of America, 1993. This book provides a clear description of the power and size of the San Andreas fault. It details the history of the fault, as well as its constant evolution. Illustrations and folded leaves; includes bibliography and index.

Reps, William F., and Emil Simiu. *Design, Siting, and Construction of Low-Cost Housing and Community Buildings to Better Withstand Earthquakes and Windstorms.* Washington, D.C.: U.S. Department of Commerce/National Bureau of Standards, 1974. A report prepared for the U.S. Agency for International Development on the construction of small buildings in earthquake- and windstorm-prone areas of the world. Explains concisely and accessibly the forces put on small buildings, the effects they have, and construction techniques to prevent them. It would be a good primer on earthquake engineering in general were it not for its necessary focus on small, low-cost buildings and ways to build them using inexpensive and locally available materials. Written for a general audience.

Tierney, Kathleen J. *Report of the Coalinga Earthquake of May 2, 1983*. Sacramento, Calif.: Seismic Safety Commission, 1985. This manual focuses on the Coalinga earthquake to show how an earthquake can affect a community physically, financially, and socially and how local and state governments can deal with the aftereffects. For all readers.

Villaverde, Roberto. *Fundamental Concepts of Earthquake Engineering*. CRC Press, 2009. Covers the newest topics in earthquake-resistant engineering. Includes details in earthquake mechanics, soil dynamics, and a great description of Fourier spectrum.

Wiegel, Robert L., ed. *Earthquake Engineering*. Englewood Cliffs, N.J.: Prentice-Hall, 1969. A large and comprehensive volume on the subject. Based originally on lectures given in a University of California course on engineering. Although dated, this book includes data on earthquake causes, ground motion and effects, mathematical modeling and theory, tests and effects on structures, and design of earthquake-resistant structures. Aimed at students and professionals.

See also: Deep-Focus Earthquakes; Earthquake Distribution; Earthquake Hazards; Earthquake Locating; Earthquake Magnitudes and Intensities; Earthquake Prediction; Earthquakes; Elastic Waves; Faults: Normal; Faults: Strike-Slip; Faults: Thrust; Faults: Transform; Notable Earthquakes; Plate Motions; Plate Tectonics; San Andreas Fault; Seismometers; Slow Earthquakes; Soil Liquefaction; Tsunamis and Earthquakes.

EARTHQUAKE HAZARDS

Over the past four thousand years, about 13 million persons have died as a result of earthquake activity, and an unknown amount of property destruction has occurred as well.

PRINCIPAL TERMS

- **creep:** the very slow downhill movement of soil and rock
- **epicenter:** the point on the surface of the earth directly above the focus
- **fault:** a fracture or zone of breakage in a rock mass which shows movement or displacement
- **focus:** the point below the surface of the ground where the earthquake originates and its energy is released
- **intensity:** an arbitrary measure of an earthquake's effect on people and buildings, based on the modified Mercalli scale
- **landslide:** the rapid downhill movement of soil and rock
- **liquefaction:** the loss in cohesiveness of water-saturated soil as a result of ground shaking caused by an earthquake
- **magnitude:** a measure of the amount of energy released by an earthquake, based on the Richter scale
- **subsidence:** the sinking of the surface of the land
- **tsunami:** a seismic sea wave created by an undersea earthquake, a violent volcanic eruption, or a landslide at sea

EARTHQUAKE OCCURRENCE

Earthquakes are the rapid motion and vibrations caused by movement of the ground along a fracture in a rock or along a fault. Movement occurs when rocks are unable to store any more stress, at which time they reach their breaking point, release energy, and create an earthquake. The point of origin of an earthquake below the surface where its energy is released is known as the focus. The focus can be located at either a shallow or a deep depth. The point on the surface of the earth directly above the focus is called the epicenter; it is the spot frequently cited by the news media as the location of the earthquake.

Throughout the earth's surface, numerous faults and fault systems exist. Larger faults are usually confined to specific areas. Earthquakes and faults are not hazardous in themselves, but they can become hazardous when they directly endanger humans and their immediate environment. Each year, the earth is subjected to at least 1 million earthquakes. Only a few, however, are strong enough to cause major structural damage or result in casualties. The major hazards directly created by an earthquake are ground shaking, ground rupture, and tsunamis. The major indirect hazards are fires, floods, building collapse, disruption of public services, and psychological effects.

GROUND SHAKING

Ground shaking occurs as energy released by the earthquake reaches the surface and causes the materials through which it passes to vibrate. The intensity of these vibrations and of the shaking at the surface depends on several factors: the amount of energy released, the depth of the focus, and the type of material through which the energy is moving. The closer the focus is to the surface, the more powerful the earthquake. Also, the denser the material, the more the vibrations will be felt. More vibrations result in stronger ground motion. There are a few documented cases in which very strong ground motion caused parked cars to bounce along the road, trees to become uprooted and snap, and the surface of the land to move in rippling waves. Yet, damage to open, uninhabited land is usually minimal.

The amount of damage to buildings subjected to strong ground motion is controlled by many complex and interacting factors: the buildings' method of construction, the types of building material used, the depth of the bedrock, the distance from the epicenter, and the duration and intensity of the shaking. During an earthquake, buildings constructed on thin, firm soil and solid bedrock fare much better than those constructed on thick, soft soil and deep bedrock. However, if the shaking's duration is great, even the most well-constructed building is likely to be destroyed. Such a situation was observed in the earthquake that struck Mexico City in 1985, in which more than 1,000 buildings were destroyed and 10,000 persons were killed. This city was built on ancient lakebed deposits of sand and silt that rapidly lost rigidity as a result of intense shaking. This caused tall buildings in the city to collapse vertically, one floor on top of the other.

There are four other important elements that determine the amount of destruction: the degree of compaction of the soil or bedrock on which the buildings' foundations are resting, the amount of water saturation of this material, its overall chemical composition, and the buildings' physical structure. If construction took place on or within solid bedrock, then the structures would move as a unit and would suffer much less damage. Some buildings may be able to withstand severe shaking for a few seconds, although prolonged shaking will completely destroy them. Ground shaking in the 1964 Alaskan earthquake lasted for about four minutes, causing major damage to the sturdiest of buildings. In contrast, a particular building may easily withstand the effects of shaking but be destroyed by other factors. In Soviet Armenia during December 1988, about 25,000 people were killed because of the effects of multiple aftershocks that shook apart poorly designed structures and buildings that were designed to withstand a lesser degree of ground motion. Moreover, very strong ground motion can knock a building completely off its bedrock foundation, rendering it unusable, and buildings may also fall prey to other types of ground failure triggered by an earthquake.

GROUND FAILURE

Ground failure includes landslides, avalanches, fault scarps, fissuring, subsidence, uplift, creep, sand boils, and liquefaction. Areas such as mountain valleys and regions surrounding an ocean bay can be subjected to these kinds of failure, since they usually consist of recently deposited, fine-grained materials that have not yet been completely compacted or have variable degrees of groundwater saturation.

Landslides occur when unstable soil and rock move rapidly downslope under the influence of gravity. Landslides regularly occur as a result of an earthquake. They commonly occur on steep slopes but can also move down gentle inclines. An avalanche is similar to a landslide but consists of snow and ice mixed with rock and soil. In either case, masses of material move with great rapidity and force, sometimes filling, burying, or excavating the land along its path. Some earthquake-induced avalanches have been clocked at velocities of more than 320 kilometers per hour (or nearly 200 miles per hour). The Tadzhik Soviet Socialist Republic, in late January 1988, was hit with an earthquake of a 5.4 magnitude on the Richter scale that shook the ground for almost 40 seconds, unleashing a massive landslide that was 8 kilometers wide. It buried the nearby village of Okuli-Bolo with mud to a depth of 15 meters, killing between 600 and 1,000 people.

Fault scarps are created when a fault intersects the surface of the earth and large chunks of the ground are uplifted or dropped. Within these chunks, deep ground cracks known as fissures appear. Despite what has been portrayed in movies, there is no chance that the earth will open, close, and "swallow" anything during an earthquake. Deep ground cracks commonly remain open, since the forces that created them operate only in one direction. Sometimes, however, animals fall into these cracks and appear to have been swallowed. In many places, buildings, roads, and other structures are constructed across an old fault scarp, and they undergo extensive structural damage from vertical or horizontal ground displacement. The largest measured vertical displacement along a fault scarp is 15 meters; the largest horizontal displacement that occurred at one time is 6 meters. The rate of displacement is variable, however, and some faults can show a slow but accumulated horizontal displacement of several kilometers.

The subsidence or depression of the surface of the land may occur when underground fluids such as oil or water are removed or drained by a nearby fault; the land sinks, creating water-filled sag ponds. This process occurs over a long period and does not result in casualties or injuries. Another slow form of ground failure is earthquake creep, which occurs more or less continuously along a fault. Creep is really an earthquake in slow motion; stored energy is gradually released in the form of very small earthquakes, or microearthquakes, causing the land to move in opposite horizontal directions. It results in the slow bending and breaking of underground pipes and railroad tracks and causes concrete building foundations to crack. The Hayward fault near San Francisco Bay, California, is a prime example.

Under the worst conditions—a low degree of compaction, thick and fine-grained sandy or silty soil, a high degree of water saturation, and intense ground shaking—solid land loses its cohesiveness and strength. It then begins to liquefy and flow by a

The collapsed Cypress viaduct on Interstate 880 after the 1989 Loma Prieta earthquake, Oakland, California. (U.S. Geological Survey)

process known as liquefaction. Tall buildings slowly move as if they were built on shifting sand. An unusual, localized ground failure event related to liquefaction is a sand boil. Rapid ground motion can cause a pressurized mixture of sand and concentrated groundwater to make its way toward the surface and create a small volcano-like mound of sand that spouts sandy water.

Tsunamis

A spectacular and extremely hazardous coastal event is the seismic sea wave, or tsunami (sometimes incorrectly called a "tidal wave"). Tsunamis are usually produced by undersea earthquakes; the sea floor undergoes rapid vertical motion along one or more active faults and energy is transmitted directly from solid rock into the seawater. Tsunamis can also originate with massive undersea landslides or the eruption of an oceanic volcano. The eruption of the volcanic island of Krakatau in 1883 created a series of waves more than 40 meters high that drowned an estimated 36,000 persons who lived along the low-lying coastal areas of Java and Sumatra. The great Indian Ocean earthquake on December 26, 2004, off the coast of Sumatra, produced a tsunami with waves up to 30 meters that killed more than 230,000 people in 14 countries, inundating coastal lands around most of the Indian Ocean. The earthquake had a magnitude of 9.2 and was caused by undersea subduction. The exact cause of all these large waves is not completely understood by geologists and oceanographers.

Regardless of their cause, tsunami waves begin to radiate away from their point of origin in a manner similar to when a stone is tossed into a quiet body of water. Generally, tsunamis have a wavelength, or distance between successive wave crests, of greater than 161 kilometers and move at a velocity of more than 966 kilometers per hour. However, these large and fast-moving walls of water are not observable as such in the deep ocean, becoming visible only as they enter shallow water. Here the trough of the wave encounters the sea floor and begins to slow down, allowing the crest to build in size. At this point, water is rapidly sucked out of inland bays and harbors to feed the increasing mass as the wave comes roaring into the mainland, drowning people and smashing buildings. In some cases, tsunamis have traveled as far inland as 3 kilometers.

For example, after tossing trains and fishing boats almost 1 kilometer inland, the seismic sea waves produced from the 1964 Alaskan earthquake traveled south many thousands of kilometers as far as Crescent City, California, where local surfers decided to challenge the 6-meter-high waves that resulted from the event. The east coast of Japan, the Hawaiian Islands, the west coast of the United States, Alaska, Chile, Peru, and most other Pacific coastal regions have suffered damage from these powerful waves. They are rare in the Atlantic Ocean. The last major Atlantic tsunami occurred in 1755, when an offshore earthquake (the All Saints Day earthquake) generated a series of waves that hit the coast of Portugal, killing an estimated 14,000 persons.

The 2011 Tōhoku earthquake (also known as the Great East Japan Earthquake) off the northeast coast of Japan resulted in a tsunami that killed more than 15,000 people and destroyed the city of Sendai and its surroundings. The disaster also caused a nuclear accident at the Fukashima Daiichi power facility.

Not as spectacular, but similar to tsunamis, are seiches. Although they also have several origins, they can be produced by an earthquake. Seiches are small, oscillating waves that may travel for several hours, back and forth within the enclosed basin of a lake or reservoir, sometimes causing flooding and minor structural damage to nearby buildings.

INDIRECT HAZARDS

One of the most deadly indirect hazards from an earthquake is fire. Fire has claimed the most number of victims and caused more property damage than all the direct hazards combined. In 1906, the San Francisco earthquake, better known as the San Francisco fire, killed 500 people, destroyed 25,000 buildings, and burned 12.2 square kilometers of the city. Fires are often started by the sparking of downed electrical wires, which can result in the ignition of ruptured gas lines. Such fires are difficult, if not impossible, to control, since water pressure in hydrants may be low or nonexistent because of the breakage of underground water pipes. Flooding is another indirect hazard of earthquakes. Although the risk of flooding is usually minimal in a seismically active area, the potential failure of large concrete or earthen dams and reservoirs poses a great threat to nearby life and property. During the 1971 San Fernando earthquake in Southern California, for example, the lower part of the Van Norman earthen dam partially gave way, threatening the 80,000 people who lived in the surrounding area. If the shaking had continued for another minute, it would have been disastrous for the local community.

The danger of being trapped in a collapsing building during an earthquake is real. No building is "earthquake proof," but construction innovations aimed at making existing buildings more secure continue to be developed. Most concrete and steel-reinforced buildings built on solid bedrock and most one- or two-story wooden frame houses suffer little or no damage in an earthquake, provided that the ground motion is not too protracted or severe. Generally, older buildings suffer much more damage than newer ones. If you are in a building while an earthquake occurs, one of the safest places to be is in a doorway.

STRUCTURAL DESIGN

Although earthquakes cannot be prevented, engineers and city planners want the dangers they pose to be either eliminated or reduced. Modern buildings are designed to withstand a certain amount of ground shaking. As structural design in construction has improved, the number of people killed in earthquakes has decreased.

However, it is very difficult to predict the effect of ground motion on a building design. To design a more flexible building, engineers perform tests on scale models using simulated or actual samples of the local bedrock and the construction materials. In the state of California, for example, legislation requires the removal of overhanging ledges and the reinforcement of key structural supports in older buildings, bridges, and highway overpasses. The installation of numerous cutoff valves on gas, water, and sewage lines can help localize and minimize utility disruptions, while aboveground pipes, roads, and power lines built across active faults are designed to anticipate fault motion.

LAND-USE POLICY

Geologists have the primary responsibility for the gathering of geologic information needed for the accurate assessment of the seismic risk for an area. This information can be obtained by the drafting of a geologic map of a region that includes an accurate tracing of all known faults and fault-derived topographic features, such as scarps and sag ponds, and the identity of the rock types involved. These maps help geologists to predict where an earthquake is most likely to occur. However, they cannot predict the earthquake's intensity, frequency, or time of occurrence without further data. Geologic information may also be gained through studies of the movement history of existing faults, the determining of their relative ages, the monitoring of current fault motion, and the detection of previously unknown faults. Once the information is assimilated, an estimate of a possible earthquake's magnitude on the Richter scale, its epicenter, its intensity on the modified Mercalli scale, and the amount of horizontal and vertical ground movement can be made.

If the geologic data indicate that an area may suffer an unacceptable amount of destruction, alternative land-use policies for that area are often adopted. Such land-use policies involve the establishment of a fault hazard easement, whereby construction is restricted to a certain minimum distance from the nearest fault trace or fault zone. Geologic hazard zoning also identifies areas affected by past landslides, floods, and seismic sea waves. In many regions, much urban or industrial development already exists in hazardous areas, simply because the danger was not recognized at the time of its construction.

As the recurrence interval between large earthquakes is very long or poorly known, the largest potential hazards exist in areas that have suffered little or no seismic activity. Local governments in areas such as New York City, South Carolina, and Missouri have given little thought to earthquake disaster planning. The New Madrid, Missouri, earthquake of 1811-1812 was felt over a sparsely populated area of 2.6 million square kilometers; large sections of the ground were uplifted, others sank, deep cracks appeared in the ground and bells were caused to ring in church towers as far away as Boston. It was the most powerful earthquake ever to strike the eastern half of the United States, and it occurred in an area that was thought to be earthquake-free. If a major earthquake were to strike an East Coast city today, the property damage and loss of life would be tremendous.

Humankind's Role in Earthquakes

As the population of the world continues to grow and people compete for living space, more and more areas once considered hazard-free because of their lack of development are becoming inhabited, increasing the potential that damage will be suffered from movements of the earth. Therefore, research in the area of earthquake control and prediction is growing more important.

Humans' ability to trigger an earthquake was discovered during the early 1970's at the Rocky Mountain Arsenal in Denver, Colorado. There, liquid wastes were disposed of by high-pressure injection into wells that were drilled to 3,600 meters below the surface. These liquids reduced the pressure along deep faults, allowing them to slip and causing an increase in the number of minor earthquakes in the area. When the pumping stopped, the number of earthquakes decreased, and when pumping was continued, the tremors began again. A careful study proved that the number of recorded earthquakes was statistically higher than that which would normally be expected. Human-made earthquakes were also created through underground nuclear explosions and the filling of water reservoirs behind major dams. Some geologists believe that these processes may relieve the pressure along faults and help to prevent a major earthquake from occurring.

Earthquake Prediction

The ability to predict an earthquake's time and place is based on the quality of geologic data available for a given area. Geologists look at historic evidence of a fault's movement and at its current activity. The problem lies in the brevity of the record-keeping period. The science of seismology is relatively new. The first seismometer was built around 1889 and data have been collected on some faults for less than one hundred years. Many more faults have been discovered since that time, and not enough information is available to give a reliable estimate of their seismic activity.

Most earthquakes are preceded by warning signs. Some of these indicators are local ground swelling, an increase in the number and frequency of minor tremors, an increase in the amount of radioactive radon gas in water wells, and unusual animal activity. The problem is that not all earthquakes have these precursors and that precursors are not always reliable. Moreover, new hazards are created by the prediction of an earthquake. A short-term prediction may cause panic, which could result in major traffic jams, riots, and looting. Long-term predictions may cause property values to drop, disrupt tourism, and cause the gradual abandonment or economic depression of cities thought to be at risk of being impacted by a seismic of event. In addition, incorrect earthquake predictions would impact public trust and may result in major lawsuits arising from injuries or damages from the evacuation of an area forecasted to be at risk.

Steven C. Okulewicz

Further Reading

Cargo, David N., and Bob F. Mallory. *Man and His Geologic Environment.* 2d ed. Reading, Mass.: Addison-Wesley, 1977. Chapter 10 of this college textbook covers earthquakes and volcanoes.

Chester, Roy. *Furnace of Creation, Cradle of Destruction.* New York: AMACOM Books, 2008. This text discusses the turbulent processes of the planet. It covers earthquakes, volcanoes, and tsunamis in reference to plate tectonics, natural disasters, predicting and mitigating. Multiple chapters explore sea floor spreading. The author also discusses hydrothermal activity. This text takes on an immense range of content, but still explains concepts clearly and with detail.

Clarke, Thurston. *California Fault: Searching for the Spirit of State Along the San Andreas.* New York: Ballantine Books, 1996. Clarke traveled the length of the San Andreas fault collecting first-hand accounts from earthquake survivors and predictors. Along with the entertaining stories, Clarke provides historical and scientific information about the fault.

Collier, Michael. *A Land in Motion: California's San Andreas Fault.* San Francisco: Golden Gates National Parks Association, 1999. Filled with beautiful color photographs that accompany text intended for the nonscientist, *A Land in Motion* gives the reader excellent insight into earthquakes and their aftermaths. There are also many diagrams and graphs that explain subduction, faults, and orogeny.

Federal Emergency Management Agency. *Designing for Earthquakes: A Manual for Architects.* FEMA 454. 2006. A publication for the non-specialist reader, explaining the general principles of seismology and building design. Also includes chapters on the effects of earthquakes, site selection, and design regulations.

Griggs, Gary B., and John A. Gilchrist. *Geologic Hazards, Resources, and Environmental Planning.* 2d ed. Belmont, Calif.: Wadsworth, 1983. Chapter 2 thoroughly covers geologic hazards related to earthquakes and faulting. Supplemented by an extensive bibliography and many photographs, charts, and maps. Highly recommended.

Howard, Arthur D., and Irwin Remson. *Geology in Environmental Planning.* New York: McGraw-Hill, 1978. Chapter 8, "Earthquakes and the Environment," contains a wide-ranging discussion of earthquakes and their distributions, hazards, prediction, and warning signs. Supplemented with many photographs and line drawings relating to famous earthquakes.

Keller, Edward A. *Environmental Geology.* 9th ed. Upper Saddle River, N.J.: Prentice Hall, 2010. A chapter covering earthquakes and related phenomena is included. Contains many well-illustrated and well-written discussions.

Sharpton, Virgil L., and Peter D. Ward, eds. *Global Catastrophes in Earth History.* Boulder, Colo.: Geological Society of America, 1990. A compilation of fifty-eight papers on various natural disasters. Accessible to readers with some scientific background. Bibliography and index.

Utgard, R. O., G. D. McKenzie, and D. Foley. *Geology in the Urban Environment.* Minneapolis: Burgess, 1978. Chapter 12, "Seismic Hazards and Land-Use Planning," is a brief outline of earthquake hazards. Chapter 13, "The Status of Earthquake Prediction," presents a quick overview of research in the field. Both chapters are written for the general reader.

See also: Deep-Focus Earthquakes; Earthquake Distribution; Earthquake Engineering; Earthquake Locating; Earthquake Magnitudes and Intensities; Earthquake Prediction; Earthquakes; Earth's Lithosphere; Elastic Waves; Faults: Normal; Faults: Strike-Slip; Faults: Thrust; Faults: Transform; Notable Earthquakes; Plate Motions; Plate Tectonics; San Andreas Fault; Seismometers; Slow Earthquakes; Soil Liquefaction; Tsunamis and Earthquakes.

EARTHQUAKE LOCATING

Earthquake locating requires a network of seismographs. The tremors felt at the earth's surface originate at depth by sudden jerky motions of faults under great pressures. Finding earthquakes requires measuring depths and determining geographic locations of the source zones within the earth where the rocks ruptured.

PRINCIPAL TERMS

- **core:** the innermost portion of the earth's interior; it measures 2,900 kilometers in diameter
- **crust:** the upper 30-60 kilometers of the earth's rocks in which shallow earthquakes occur; it is thickest under continents, thinnest beneath oceans
- **epicenter:** the point on the earth's surface directly above the focus of an earthquake
- **fault:** a crack in the earth's crust where one side has moved relative to the other
- **focus:** the point beneath the earth's surface where rocks have suddenly fractured along a fault zone, generating a train of seismic waves that travel through the earth and that are experienced at the surface as an earthquake
- **mantle:** the portion of the earth's interior between the Moho and the core, from 30-60 kilometers to 2,900 kilometers deep
- **Mohorovičić discontinuity (Moho):** the lower boundary of the crust, 30-60 kilometers deep on the average; the velocities of P waves and S waves both increase sharply just below the Moho
- **P wave:** a type of seismic wave generated at the focus of an earthquake traveling 6-8 kilometers per second, with a push-pull vibratory motion parallel to the direction of propagation; "P" stands for "primary," as P waves are the fastest and first to arrive at a seismic station
- **S wave:** a type of seismic wave generated at the focus of an earthquake traveling 3.5-4.8 kilometers per second, with a shear or transverse vibratory motion perpendicular to the direction of propagation; "S" stands for "secondary" because S waves are usually the second to arrive at a seismic station
- **seismograph:** a sensitive instrument that detects vibrations at the earth's surface and records their arrival times, amplitudes, and directions of motion

DETERMINING THE EARTHQUAKE SOURCE REGION

Even though tremors from a single earthquake may be felt for hundreds of kilometers and recorded throughout the world, they always begin at a point or very small region within the earth. Earthquakes are caused when pressures build up to the point that rocks within cannot withstand any more stress, and they snap and move to adjust to the stress. As they rupture, they either form a crack or move along an existing fault, with one side moving with crushing force against the other. This violent internal rubbing creates vibrations that propagate away from the disturbance and ripple across the surface of the earth.

When a violin bow is rubbed against a violin string at a point, the whole string vibrates; as it vibrates, energy is transmitted to the surrounding air, sending out sonic waves, or sound waves. Similarly, when rocks of the earth's interior rapidly rub against one another at a point, all the neighboring rocks vibrate. As they vibrate, they transmit energy upward and through the earth. This energy sends out seismic waves felt and measured as tremors of an earthquake throughout a region. Sometimes, if an earthquake is strong enough, seismographs in every area of the world will measure it.

In order to locate an earthquake, earth scientists need to find the region within the earth where the rocks ruptured and the vibrations started. For small quakes, the source region is no more than a few meters in size. For very large earthquakes, the source region may be hundreds of meters and even a kilometer or more in dimension. In the case of large quakes, however, the entire area of disturbed fault motion does not move at once. The earthquake still starts at some point in the fault region, then the disturbance moves away from point to point in a chain reaction until the entire stressed region has adjusted. The actual fault motion may occur in one or two seconds for small quakes or extend over a minute or more for the largest quakes.

FOCUS AND EPICENTER

The point within the earth where the fault began its motion is called the focus, or hypocenter, of the earthquake. The focus is where the rocks break. The geographic location of the point vertically above the focus on the earth's surface is called the epicenter. The epicenter is thus the point on the earth's surface nearest to the focus. It is also the place where the most intense ground motions are usually experienced—but not always. Sometimes, because of the peculiarities of underlying geology, the most violent vibrations experienced above ground are displaced a few kilometers from the epicenter. For this reason, an array of sensitive instruments called seismographs is needed to locate the true epicenter and focus. Surface expressions of an earthquake can be very misleading. The earthquake of September 19, 1985, that devastated Mexico City actually had its epicenter some 400 kilometers away beneath the Pacific Ocean. Because of the geologic peculiarities in that part of the world, Mexico City, at a considerable distance from the source, was more severely shaken than was Acapulco, which was much closer to the epicenter.

Earthquakes are generated from a point or small restricted region within the earth, but they send out waves that can be felt for hundreds of kilometers and recorded by seismographs throughout the entire world. For example, the great New Madrid, Missouri, earthquakes of 1811-1812 were felt from the Rocky Mountains to the Atlantic seaboard, but their epicenters were in the Mississippi River valley of the midwestern United States.

Seismic Stations

Among seismologists, "earthquake locating" means finding the focus and epicenter, which requires seismic instrumentation. Finding the regions of strongest shaking and heaviest damage is another kind of investigation in which seismologists also engage, but it does not require seismographs. Earthquakes can be located by well-calibrated seismographs by noting the times that various seismic waves arrive at different stations. By knowing the speeds at which P waves, S waves, and other seismic waves travel through the earth and the precise times they arrive at several stations, distances and directions can be calculated and the earthquake epicenter and depth determined. A minimum of three seismic stations scattered around an epicenter is needed to determine a location. At least one station near the epicenter is needed to estimate the depth accurately. For truly reliable and precise measurements within less than 0.1 kilometer, a dozen or more stations are needed at varying distances surrounding the focal region.

All seismic stations are timed to universal time referenced to the zeroth meridian passing through Greenwich, England. This time is broadcast by shortwave radio stations, such as WWV radio in Boulder, Colorado, which broadcasts every second of time, twenty-four hours a day. The time is accurate to within billionths of a second. Seismic stations are equipped with special radios to receive this information and to set their seismographs on a daily basis.

Networks of seismographs can be found throughout the world, including a worldwide net received by radio, satellite, and telephone lines in Golden, Colorado, by the U.S. Geological Survey's National Earthquake Information Service. Smaller, more local nets exist around the major faults and seismic zones in the country, including the San Andreas fault system of California; the Puget Sound area of Washington State; the Wasatch fault of Utah; the New Madrid fault of Arkansas, Missouri, and Tennessee; the Charleston, South Carolina, seismic zone; and several seismic areas of New England. Locating earthquakes is a cooperative effort between many seismic stations and scientists throughout the country and the world, including those associated with universities, government facilities, and private corporations. In earthquake seismology, the whole world is the laboratory. Sharing of information—irrespective of state, provincial, or national boundaries—is required to understand, study, and locate earthquakes.

P - S Delay Time Technique

The easiest way to locate an earthquake is when three stations are located in a triangle around its source regions. For example, imagine that an earthquake occurs at exactly 06:00 hours universal time. At the instant the fault rips, two kinds of waves are generated: P waves and S waves. Velocities of P waves in the upper crust of the earth are about 7 kilometers per second, while those of S waves are typically 40 percent less, or 4.2 kilometers per second. When the P wave has traveled 70 kilometers in 10 seconds, the S wave has traveled 42 kilometers. To reach the point the P wave reached in 10 seconds, the S wave will take 16.7 seconds, or 6.7 seconds longer. If a seismic station were at this site, it would record both waves and note that the S wave lost the "race" by 6.7 seconds. Another station 100 kilometers away in another direction would also note the arrival of the P wave and

S wave. To travel 100 kilometers, the P wave would take 14.3 seconds, while the S wave would take 23.8 seconds, or 9.5 seconds longer than the P wave. If, in a third direction from the earthquake, another seismic station were situated 160 kilometers away, it would receive both the P and S waves at later times: 06 hours 22.9 seconds and 06 hours 38.1 seconds, respectively. The third seismic station would note a "P minus S" (P - S) arrival time difference of 15.2 seconds. Although the three seismic stations recording the arrivals of P and S waves would not identify where the earthquake had occurred or the time of its origin, seismologists could calculate the exact time of arrival in universal units to the nearest tenth of a second or better. Even more useful, they would know precisely the difference in arrival times of the P and S waves, which could be read from each station's seismograms. The above examples also reveal a simple way to estimate the distance from a seismograph to an earthquake: multiply the P - S delay time by 10.5 in kilometers.

Each seismographic observatory has sets of empirically determined travel times for P and S waves for various distances from its station. The first station in the example, at 70 kilometers' distance, would refer to the travel time tables and see that for a difference of 6.7 seconds between P and S waves, the event had to be about 70 kilometers away. Seismologists would not know, however, in which direction the waves had come. The staff at the station could draw a circle on a map 70 kilometers in radius, with the location of the station at the center, knowing that somewhere on that circle the earthquake had occurred. By contacting the station located 100 kilometers from the epicenter, the staff at the first station would learn that the seismograph at the second station noted a 9.5-second P - S difference. The staff could then refer to the travel time charts to discover that a 9.5-delay time corresponds to 100 kilometers, thus defining a second circle, of that radius, centered on the second station. Similarly, the staff at the first station could contact the third station and obtain another P - S delay time and find that the 15.2-second difference observed there implied a 160-kilometer distance of that station from the earthquake focus.

Determining Depth

Plotting the three circles centered on the three station locations with the appropriate radii, the seismologists would find that all three intersect at a single point. That point would be the epicenter. To obtain the depth requires some analysis of the P - S delay times and how closely the three circles intersect precisely at a point. By assuming a depth, calculating the arrival times to the stations, and comparing them to the real arrival times, the seismologist can judge how realistic the assumed depth was. By assuming various depths and repeating the calculations until a close correspondence with real arrival times results, the depth can be said to have been determined.

The best way to determine a depth is by having a station very near the epicenter. In this way, the distance from the focus at depth to the station recording at the surface is an approximation of the depth below the epicenter. The most common way to determine depth, however, is by inputting measurements from many stations into a computer and repeating trial calculations until they best resemble the data for a depth determination. Epicenters are also located by the mathematical convergence of data from many stations, even though theoretically only three stations are required.

Limitations of P - S Delay Time Method

The P - S delay time technique is the original and most simple way to locate earthquakes. This method does not work in all cases, however; even when it does work, beyond 500 kilometers, the curvature of the earth becomes a contending factor. The simple triangulation method works best for shallow quakes less than one-fourth of the circumference of the world away, or less than 10,000 kilometers away.

A shallow earthquake is one that occurs in the crust less than 60 kilometers deep. Most active faults do not extend visibly up through the surface (as does the San Andreas fault in California) but lie buried beneath layers of rock and soil. The deepest earthquakes are 300-700 kilometers deep; those between 300 and 60 kilometers deep are termed "intermediate" by seismologists. Deep earthquakes occur only in certain places, mostly around the perimeter of the Pacific Ocean. For a deep earthquake, even a station at the epicenter would be hundreds of kilometers away, and the seismic energy might take an entire minute to propagate from the focus to the surface.

There are several reasons the P - S delay time triangulation method cannot always work. One reason is that the core of the earth acts like a liquid, not a solid.

Liquids do not allow S waves to propagate. Hence, beyond a certain distance around the earth, direct S waves cannot be received. The core also bends and diffracts P waves, even though P waves do propagate readily through the core. Because of the inner structure of the earth, receiving P and S waves directly from the focal source zone at a seismic recorder is not possible. Instead of P waves and S waves being the first and second waves to arrive at a station, other kinds of seismic waves—including P and S waves that have been reflected, refracted, and diffracted along complex pathways—will arrive first. When the direct P and S waves are not received by a seismic station, other waves can be used to determine the epicentral distance in an analogous fashion. Hence, seismologists have tables and charts not only for direct P and S wave travel times to their station but also for more than a dozen other ray paths and wave types. By reading all such data from their seismograms and applying multiple travel time calculations, seismologists can determine more precisely depths and epicentral distances.

In the early days of seismology, near the beginning of the twentieth century, earthquake epicenters were found by using large spherical globes, thumb tacks, and pieces of string to strike off intersecting radii. Presently, epicenters and depths are found by sophisticated computer programs that consider the data of numerous stations and numerous phases of the various kinds of seismic waves and their possible wave paths through the earth.

APPLICATION OF SEISMOGRAPHIC DATA

The science of locating earthquakes, as it has developed over the past one hundred years, has been responsible for providing most of what is known about the earth's interior. The Moho, the thickness of the crust, the earth's mantle, the liquid core, and even the inner solid core floating within the fluid outer core—all of these have been deduced from seismograms taken from all over the world. The motions of the earth's crustal plates are also observed by the analysis of seismograms. Locating an earthquake by a seismogram determines much more than just when and where the rocks ruptured. The amount of energy released and the relative directions of motion that occurred on both sides of the fault can also be determined.

The installation by the United States of a worldwide network of seismographs has made it possible over the decades to monitor the underground nuclear experiments of the Soviet Union and other countries. An underground nuclear blast has many of the earmarks of an earthquake and can be located by the same methodologies, but it also has distinctively different seismic characteristics. Studying seismograms of nuclear blasts has helped refine understanding of the earth's inner makeup. Another practical application of earthquake locating is measurement of the tremors associated with volcanic magma moving toward the surface prior to an eruption. Volcanic eruption predictions are based on such data.

David Stewart

FURTHER READING

Bolt, Bruce A. *Earthquakes*, 5th ed. New York: W. H. Freeman, 2005. An elementary treatment of earthquakes in general. Could be used by lower-level college students who are nonscience majors.

_____. *Inside the Earth*. Fairfax, Va. Techbooks, 1991. An excellent introduction to seismic waves and how they behave within and define the interior of the earth. College level.

Clarke, Thurston. *California Fault: Searching for the Spirit of State Along the San Andreas*. New York: Ballantine Books, 1996. Clarke traveled the length of the San Andreas fault collecting first-hand accounts from earthquake survivors and predictors. Along with the entertaining stories, Clarke provides historical and scientific information about the fault.

Collier, Michael. *A Land in Motion: California's San Andreas Fault*. San Francisco: Golden Gates National Parks Association, 1999. Filled with beautiful color photographs that accompany text intended for the nonscientist, *A Land in Motion* gives the reader excellent insight into earthquakes and their aftermaths. There are also many diagrams and graphs that explain subduction, faults, and orogeny.

Field, Ned, et al. "Earthquake Shaking: Finding the 'Hotspots'." U.S. Geological Survey: FS 001-01. 2001. This fact sheet provides information on the geological characteristics that are used to identify hotspots.

Fradkin, Philip L. *Magnitude 8: Earthquakes and Life Along the San Andreas Fault*. Berkeley: University of California Press, 1999. Written for the layperson, this book can sometimes read overdramatic or unscientific. However, *Magnitude 8* traces the seismic history, mythology, and literature associated with the San Andreas fault.

Garland, G. D. *Introduction to Geophysics: Mantle, Core, and Crust*. Philadelphia: W. B. Saunders, 1971. Excellent and thorough treatment of the earth's interior and how its character is deduced from seismic wave behavior as measured by seismographs. Understandable to college science majors.

Richter, Charles F. *Elementary Seismology*. San Francisco: W. H. Freeman, 1958. The author of this classic 768-page text, who was a seismologist for many years at the California Institute of Technology, developed the Richter scale for measuring the magnitude of earthquakes. Judging from his book, Dr. Richter must have been an excellent teacher. Even though this source is outdated, its lucid explanations of basic principles make it a worthwhile reference. Contains excellent and detailed chapters on the complexities of earthquake locating, along with examples, charts, diagrams, and travel-time curves. Some sections using differential equations would be for upper-level college students, but most of the book, including the parts on earthquake locating, would be quite readable to any advanced high school student.

Simon, Ruth B. *Earthquake Interpretation*. Golden: Colorado School of Mines, 1968. Basic primer on seismogram interpretation. Aimed at lower-level college science students.

_____. *Earthquake Interpretations: A Manual for Reading Seismographs*. Golden: Colorado School of Mines, 1981. A continuation of Simon's previous work from 1968.

Tarbuck, Edward J., Frederick K. Lutgens, and Dennis Tasa. *Earth: An Introduction to Physical Geology*. 10th ed. Upper Saddle River, N.J.: Prentice Hall, 2010. This college text provides a clear picture of the earth's systems and processes that is suitable for the high school or college reader. It has excellent illustrations and graphics. Bibliography and index.

_____. *Earth Science*. 12th ed. Westerville, Ohio: Charles E. Merrill, 2008. Freshman college text. Covers spectrum of earth sciences with many full-color, excellent illustrations.

Verhoogen, John, et al. *The Earth*. New York: Holt, Rinehart and Winston, 1970. Provides introductory earth science material. Freshman college level.

See also: Creep; Deep-Focus Earthquakes; Earthquake Distribution; Earthquake Engineering; Earthquake Hazards; Earthquake Magnitudes and Intensities; Earthquake Prediction; Earthquakes; Earth's Lithosphere; Elastic Waves; Faults: Normal; Faults: Strike-Slip; Faults: Thrust; Faults: Transform; Notable Earthquakes; Plate Motions; Plate Tectonics; San Andreas Fault; Seismic Reflection Profiling; Seismometers; Slow Earthquakes; Soil Liquefaction; Subduction and Orogeny; Tsunamis and Earthquakes.

EARTHQUAKE MAGNITUDES AND INTENSITIES

The measurement of earthquake intensity (based on observed effects of earthquakes) and magnitude (based on instrument readings) is useful not only for scientists who study and predict earthquakes but also for land-use planning and other aspects of public policy.

PRINCIPAL TERMS

- **amplitude:** the displacement of the tracings of the recording pen (or light beam) on a seismogram from its normal position
- **deep-focus earthquakes:** earthquakes whose focus is greater than 300 kilometers below the surface
- **epicenter:** the point on the earth's surface directly above the focus of an earthquake
- **focus:** the point within the earth that is the source of the seismic waves generated by an earthquake
- **high-frequency seismic waves:** those earthquake waves that shake the rock through which they travel most rapidly
- **low-frequency seismic waves:** those earthquake waves that shake the rock through which they travel most slowly (also called long-period waves)
- **seismogram:** an image of earthquake wave vibrations recorded on paper, photographic film, or a video screen
- **seismograph:** the mechanical or mechanical-electrical instrument that detects and records passing earthquake waves
- **shallow-focus earthquakes:** earthquakes having a focus less than 60 kilometers below the surface

INTENSITY SCALES

Magnitude is a numerical rating of the size or strength of an earthquake, based on instrument readings. Intensity is a different kind of numerical rating having to do with the actual effects of an earthquake on people, buildings, and the landscape. Magnitude rating values allow comparison between earthquakes on a worldwide basis, whereas intensity ratings are more useful for comparing relative effects in areas surrounding the epicenter.

Because only human judgment is required for an intensity rating, intensity scales were developed first. Many different scales of earthquake effects have been devised in different countries since Renaissance times. The earliest known scale was developed in Italy and had only four rating values. The earliest widely accepted intensity scale, in use after 1883, was the Rossi-Forel scale, which had ten value ratings. In the United States, a revision of a later European scale, the 1931 modified Mercalli scale, has become the standard. This scale was modified by Giuseppe Mercalli in 1884, and after several further revisions was expanded to a twelve-value scale. The values of intensity range from I, which barely would be felt, to XII, which would be the most violent. Each earthquake has a zone of maximum intensity, surrounded by zones of successively lesser intensity.

INTENSITY RATINGS

Intensity I on the modified Mercalli scale would not be felt except by very few. It might trigger nausea or dizziness if it occurs in the marginal zone of a large earthquake. In an area experiencing intensity II effects, ground vibration may be felt by some persons at rest, especially on upper floors, where building motion may exaggerate ground motion. Regions experiencing level III intensity are characterized by a brief period of vibration like that of a passing loaded truck. Many do not recognize it as an earthquake. Zones of intensity IV are felt indoors by many, outdoors by few. Buildings may shudder slightly; windows and doors of older homes may rattle, and glassware in cupboards may start clinking. In an intensity V earthquake, which is widely felt, some people may be awakened. A few may be frightened; some run outside. Windows and glassware may break, and cracks may appear in plaster.

Intensity VI earthquakes are felt by all, frightening the general public and resulting in panic. Some plaster may fall, some brick chimneys may be damaged, and some furniture moves. Objects are often thrown from shelves, and trees shake noticeably. An intensity VII earthquake brings strong shaking that may last for many seconds, causing considerable damage to older brick buildings and slight to moderate damage to well-built wood- or steel-frame structures. These quakes are noticed by persons driving vehicles. In an intensity VIII earthquake, damage to buildings is considerable. Specially designed earthquake-resistant structures may hold up, but many older brick buildings collapse totally. Branches and trunks of trees may break. Damage to most buildings—ranging from collapse to being thrown out of plumb or off the foundation—is caused by intensity IX earthquakes. Conspicuous ground cracks appear, and buried water and gas pipes break. Intensity x earthquakes cause most buildings to collapse partially, some totally. Railroad tracks are bent, and buried pipes buckle or break. Landsliding along riverbanks and steep slopes is triggered and obvious ground cracks are widespread. Strong shaking may last for many tens of seconds. After intensity XI quakes, few structures remain standing. Broad fissures appear in the ground. The earthquake may cause a large tsunami if it occurs near a coastal area. The strong shaking may last a minute or more. Finally, in an intensity XII earthquake, objects are thrown in the air. Waves are "frozen" in the ground surface. Fewer than a half-dozen earthquakes have been rated at this level of intensity.

ASSIGNING INTENSITY RATINGS

Assignment of the lower values on the modified Mercalli scale is possible only if people are present to experience the effects. In the middle and upper values, effects on structures are a primary basis for the assignment of ratings, although earthquakes of greatest intensity may produce long-term geologic effects on the landscape, including ground fissures, fault scarps (cliff-like features visible at the earth's surface), landslides, and sandblows (small volcano-like mounds of sediment that erupt from water-saturated ground as a result of severe shaking). Thus, earthquakes that occur in uninhabited areas of the earth cannot be rated unless the shaking was sufficiently strong to produce geologic effects; similarly, rating is impossible for quakes occurring below large areas of the ocean with few or no populated islands, although in some cases, intensity can be estimated if a tsunami is generated.

Because effects on buildings are an important means of differentiating between ratings in the middle and upper part of the scale, the nature of construction becomes an important differentiating criterion. In earthquake-prone California, for example, earthquake-resistant design practices mandated by law have made the average new building less susceptible to damage or destruction than are older buildings. Unless this factor is taken into account, equal-sized earthquakes would be rated lower over time because of less damage as older buildings are replaced.

ISOSEISMAL LINES

Despite such problems, the modified Mercalli scale is still quite useful. There are many more people (each of whom is a potential observer of earthquake effects) distributed around the world than there are earthquake-recording instruments. Earthquake intensity ratings begin to be gathered soon after earthquakes large enough to be felt by more than a few people. A government agency (for example, the U.S. Geological Survey) sends questionnaires to people in the area where the earthquake was felt. The forms contain questions related to the specific location and activity of the observer at the time of the earthquake as well as details of what happened before, during, and after the quake. The responses are then rated, using the modified Mercalli scale, and the ratings are plotted on a map. Lines separating the various values, or "isoseismal lines," can then be drawn to show areas having equal intensity. Frequently an irregularly shaped bull's-eye pattern emerges, with the highest rating zone in the middle. This zone of maximum intensity usually contains the earthquake's epicenter.

The size and shape of the pattern of isoseismal lines can be invaluable in land-use planning or zoning of areas that experience frequent earthquake activity. The pattern may give clues about the distribution of land that would make a poor foundation for buildings because of greater susceptibility to seismic shaking. Certain types of sediment overlying bedrock can actually amplify seismic vibrations.

The highest intensity value determined for a specific earthquake is only a rough indicator of the quake's real strength. Quite large shocks can occur at many hundreds of kilometers of depth (deep-focus earthquakes), but because their most damaging vibrations are largely absorbed by rock before arriving at the surface, their maximum intensity ratings are low.

MAGNITUDE: THE RICHTER SCALE

Unlike earthquake intensity determinations, which vary with distance from the epicenter and depend on factors such as the depth of focus and soil depth, the magnitude rating of an earthquake is reported as a single number that is *usually* calculated from an instrumental recording of ground vibrations. ("Usually" is emphasized because an alternative method for magnitude determination has been developed, known as moment magnitude.) Every earthquake large enough to be detected and recorded by seismographs can be assigned a magnitude value. The Richter scale of earthquake magnitude was the first to be used widely. It is named for its originator, Charles F. Richter, professor of seismology at the California Institute of Technology.

When an earthquake occurs, it generates ground vibrations that radiate out to surrounding areas, much as ripples do when a small pebble is dropped into a quiet pond. A seismograph is an instrument designed to detect and record, over time, even very small ground vibrations. It does this by amplifying the motion of the ground on which the seismograph sits when seismic waves pass through it. For some seismographs, the amplified recording (seismogram) is simply a piece of paper wrapped around a cylindrical, clock-driven drum, on which the ground vibrations are recorded as a zigzag line drawn by a recording pen. Larger ground vibrations result in larger zigzags or displacement on the paper. The amount of displacement from the normal or average position is called the amplitude, and it can be measured accurately with a simple fine-scale ruler.

In 1935, Richter first published his method for determining the relative size of the many earthquakes for which he had recordings in southern California. Most of these were small and moderate-sized shallow-focus earthquakes. Eventually, after various modifications, the Richter scale came to be used around the world to measure and compare earthquakes.

MAGNITUDE RATINGS

Richter's original definition of the magnitude of an earthquake is as follows: "The magnitude of any shock is taken as the logarithm of the maximum trace amplitude, expressed in microns, with which the standard short-period [Wood-Anderson] seismometer . . . would register that shock at an epicentral distance of 100 kilometers." Simply stated, the Richter scale magnitude of an earthquake is determined by measuring the greatest horizontal displacement from the average of the recording-pen tracing on the seismogram of a standard instrument at a standard distance from the earthquake epicenter (earthquake ground motions weaken as they travel out from the source). The width of the largest amplitude is measured in microns (thousandths of a millimeter). The logarithm of this number (to the base 10) is calculated and becomes the rating number "on the Richter scale." If an earthquake produced a maximum recording-pen displacement of 1 centimeter, that would be equivalent to 10,000 microns (which is equal to 10^4 microns). The logarithm of 10^4 is 4 (the exponent); thus, the earthquake would be rated at 4 on the Richter scale. If the maximum displacement had been 10^5 microns, it would be rated 5.

Because of the logarithmic nature of the Richter scale, each higher value in whole numbers actually represents a tenfold increase in the amount of ground motion recorded on the seismogram. Compared with an earthquake rated at 4, one rated at 5 involves 10 times more ground-shaking displacement; one rated at 6 would involve 100 times more displacement than 4, and 7 would be 1,000 times more. In other words, the largest earthquakes produce more than 1,000 times more ground motion than those that just begin to damage average American homes.

Another way to compare earthquakes is to determine the amount of energy they release. The energy released by earthquakes is about 30 times greater for each higher Richter scale value. For example, an earthquake rated at 6 would be about 900 times (30 H 30) more energetic than one rated at 4.

Looking west along the Motagua fault after the 1976 Guatemala earthquake. At intensity IV on the Mercalli scale, conspicuous ground cracks appear, and buried water and gas pipes break. (U.S. Geological Survey)

OTHER MAGNITUDE SCALES

Methods for magnitude determination have advanced substantially since the original magnitude scale was developed by Richter. His original scale was valid only for shallow-focus earthquakes occurring in the local region of southern California, because it depended on a Wood-Anderson type of seismograph. That instrument is "tuned" to pick up the higher-frequency vibrations typical of the small and medium-sized earthquakes of southern California. Later work by Richter and others extended his original scale (which could not be applied much above magnitude 6) for use with other instruments tuned to pick up earthquakes that occurred at greater distance (more than 1,000 kilometers from a seismograph station) and had a deeper focus. These magnitude extensions were designed to coincide, as much as possible, with the values of the original scale. The extensional scales, however, use bases that are different from those of the original. One extensional scale, for example, requires a seismograph tuned to pick up only low-frequency vibrations. As a result, magnitude values differ somewhat and are not strictly "on the Richter scale." Technical literature actually limits the use of the term "Richter scale" to magnitude values determined essentially according to the original specifications.

For the very largest earthquakes, even the extensional scales become inadequate for ranking accurately the relative strength of earthquakes, because the sensitive instruments are said to become "saturated"—essentially thrown off scale. To solve this problem, a "moment magnitude" scale was developed; it is based not on instrument readings but on data obtained in the field, along the earthquake-generating fault. The average amount of fault offset, the length and width of slippage along the fault, and rock rigidity data are used to calculate the moment magnitude. Because of its greater accuracy, the moment magnitude scale is becoming more commonly used, particularly for medium-sized and larger earthquakes. Although the values of the moment magnitude scale essentially merge with those of the Richter scale for medium-sized earthquakes, at higher values there can be substantial differences. Thus, the largest earthquake rated by moment magnitude is 9.5.

PREDICTION OF EARTHQUAKE RECURRENCE

A key element to prediction of earthquake recurrence, particularly for the less frequent but larger and more destructive earthquakes, is a detailed record of the relative sizes of earthquakes occurring along a particular fault (or fault segment) through time. The most accurate and consistent indicator of size is the magnitude value. Because instruments needed for magnitude determination have been in existence less than one hundred years, magnitudes of earlier events must be estimated. This can be done using several approaches.

Maximum intensity values correlate with different but known magnitude values in areas where the depth of focus is thought to be consistent through time and where the nature of bedrock absorption of seismic waves is known. For determining maximum intensity values for unrecorded earthquakes of the past, archives and historical documents sometimes yield useful data. For prehistoric earthquakes, newly developed techniques are proving successful in defining the occurrence of large earthquakes and, under the right circumstances, even of their relative sizes. Such old events may be judged by the nature and extent of geologic traces preserved in radiocarbon-datable buried sediments.

Another avenue of study of earthquake recurrence is based on the amount of stored-up energy that is released by earthquakes worldwide in a year. A curve of the energy release can be compared to the occurrences of every possible recurrent earthquake-triggering mechanism—for example, tidal forces. To determine energy released, the moment magnitude must be determined for as many earthquakes as possible, but especially for the largest ones, because they release much more energy overall than do the more numerous smaller ones. One study, for example, revealed a strong correlation between times of higher-than-average earthquake activity (around 1910 and again around 1960) and the extent of "wobbling" of the earth's axis of rotation.

DETERMINATION OF SEISMIC RISK

In earthquake-prone areas, seismic risk must be considered in urban and regional planning. The effects of ground shaking at any location are dependent primarily on magnitude, distance from the source of the earthquake waves, and the nature of the bedrock and the type and thickness of materials

Comparison of Magnitude and Intensity

Magnitude	Intensity (Mercalli)	Description
1.0-3.0	I	Not felt except by a very few under especially favorable conditions.
3.0-3.9	II-III	II felt only by a few persons at rest, especially on upper floors of buildings. III felt quite noticeably by persons indoors, especially on upper floors of buildings. Many people do not recognize it as an earthquake. Standing motor cars may rock slightly. Vibrations similar to the passing of a truck. Duration estimated.
4.0-4.9	IV-V	IV felt indoors by many, outdoors by few during the day. At night, some awakened. Dishes, windows, doors disturbed; walls make cracking sound. Sensation like heavy truck striking building. Standing motor cars rocked noticeably. V felt by nearly everyone; many awakened. Some dishes, windows broken. Unstable objects overturned. Pendulum clocks may stop.
5.0-5.9	VI-VII	VI felt by all, many frightened. Some heavy furniture moved; a few instances of fallen plaster. Damage slight. VII. Damage negligible in buildings of good design and construction; slight to moderate in well-built ordinary structures; considerable damage in poorly built or badly designed structures; some chimneys broken.
6.0-6.9	VII-IX	VIII. Damage slight in specially designed structures; considerable damage in ordinary substantial buildings with partial collapse. Damage great in poorly built structures. Fall of chimneys, factory stacks, columns, monuments, walls. Heavy furniture overturned. IX. Damage considerable in specially designed structures; well-designed frame structures thrown out of plumb. Damage great in substantial buildings, with partial collapse. Buildings shifted off foundations.
7.0+	X-XI	X. Some well-built wooden structures destroyed; most masonry and frame structures destroyed with foundations. Rails bent. XI. Few, if any (masonry) structures remain standing. Bridges destroyed. Rails bent greatly. XII. Damage total. Lines of sight and level are distorted. Objects thrown into the air.

Source: U.S. Geological Survey, National Earthquake Information Center; URL: http://neic.usgs.gov/neis.

above it. After historical patterns of earthquake recurrence are determined, particularly their characteristic (or most typical) size, it is possible to make estimates of probable intensity patterns in the vicinity of faults. Detailed maps have been prepared of the areas along the San Andreas fault and for many miles on either side of it, outlining the zones of greatest potential seismic risk. Such maps can be of great value when decisions are being made regarding sites for potential secondary earthquake hazards such as nuclear power plants, fuel storage depots, and dams.

DETECTION OF UNDERGROUND NUCLEAR EXPLOSIONS

Additionally, seismic detection and characterization of distant underground nuclear explosions is of considerable political importance. One method of attempting to discriminate between a nuclear explosion and a natural seismic event is by analysis of its magnitude as recorded by several types of seismographs, each "tuned" to pick up different frequencies of ground vibrations. Ratios between such magnitude values appear to be quite useful for this purpose.

SIGNIFICANCE OF NUMBERS ON THE RICHTER SCALE

Nearly every time a news broadcast makes reference to a damaging earthquake somewhere in the world, a number on the Richter scale is mentioned to give the listener some idea of the relative size of the event. If the earthquake has happened in a remote part of the globe and has not caused significant damage, it becomes merely another of the many facts that are soon forgotten. If the earthquake has happened where one's relative or a friend lives, however, that number on the Richter scale becomes extremely important because it is one of the first available indicators of possible severity. It can be determined within minutes after earthquake waves have been detected at seismological observatories, whereas direct communication from and damage assessment at the site of the earthquake may be very slow in coming. Exactly what does a value of 7.3, for example, mean to those who were near the epicenter? How does that figure compare to the value associated with damage to homes? (That number is about 5 on the Richter scale.)

Many of the world's large cities are located close to active earthquake-generating faults. What effect will a major earthquake in such a city have on the economy of that area, and how in turn will its misfortune affect the rest of the world? What would happen if the flow of goods between the United States and Japan were suddenly disrupted for a long period because of an 8.5-magnitude earthquake near Tokyo or Los Angeles? The initial answers to these and many other questions may someday hinge on that critical number on the Richter scale.

VALUE OF SEISMIC MAPS

Aspects of earthquake intensity are less likely to be mentioned in the media except, perhaps, when covering local aspects of seismic risk along a certain fault or the risk of earthquakes in various areas of the United States. Maps identifying zones of potential seismic hazard are likely to become more common as their need becomes more apparent. The lack of such knowledge and of the will to act on it could be costly in terms of lives and property.

One of the most instructive illustrations of seismic intensity patterns is a map comparison of the 1906 San Francisco earthquake and a similar-sized earthquake in southern Missouri. The seismic wave absorptive properties of the bedrock along the western margin of North America are much greater than that of the central and eastern states. As a result, except in California, little damage is expected to occur from earthquakes in the United States—even from major ones. A great earthquake in the New Madrid fault zone of southern Missouri, however, is likely to have a wide zone of maximum intensity, resulting in severe damage in cities hundreds of miles from the epicenter.

Valentine J. Ansfield

FURTHER READING

Bolt, Bruce A. *Earthquakes*, 5th ed. New York: W. H. Freeman, 2005. An authoritative introduction to most aspects of earthquakes, including magnitude and intensity. Includes lists of important earthquakes, a glossary, a bibliography of titles suitable for the general reader, and an interesting "earthquake quiz" with answers. Suitable for high school or college-level readers.

Chester, Roy. *Furnace of Creation, Cradle of Destruction*. New York: AMACOM Books, 2008. This text discusses the turbulent processes of the planet. It covers earthquakes, volcanoes, and tsunamis in reference to plate tectonics, natural disasters, predicting and mitigating. Multiple chapters explore sea floor spreading. The author also discusses hydrothermal activity. This text takes on an immense range of content, but still explains concepts clearly and with detail.

Doyle, Hugh A. *Seismology*. New York: John Wiley, 1995. A good introduction to the study of earthquakes and the earth's lithosphere. Written for the layperson, the book contains many useful illustrations.

Eiby, G. A. *Earthquakes*. Auckland, New Zealand: Heineman, 1980. A relatively technical discussion of earthquakes and seismology. Suitable for college-level readers.

Emergency Management BC. *A Simple Explanation of Earthquake Magnitude and Intensity*. Ministry of Public Safety and Solicitor General: Provincial Emergency Program. 2007. Provides fundamental material for the layperson.

Fradkin, Philip L. *Magnitude 8: Earthquakes and Life Along the San Andreas Fault.* Berkeley: University of California Press, 1999. Written for the layperson, this book can sometimes read overdramatic or unscientific. However, *Magnitude 8* traces the seismic history, mythology, and literature associated with the San Andreas fault.

Gere, James M., and Haresh C. Shah. *Terra Non Firma: Understanding and Preparing for Earthquakes.* New York: W. H. Freeman, 1984. Similar in many respects to Bolt's book, described earlier. Includes a table relating maximum intensity values of the modified Mercalli scale to Richter scale magnitude values (page 87) and a table relating the duration of strong motion to Richter scale values (page 173). Suitable for high school or college-level readers.

Monastersky, Richard. "Abandoning Richter." *Science News* 146 (October 1994): 250-252. The author describes why the Richter scale of measuring earthquake size is outdated. Current reports of earthquake magnitude are based on a number of factors quite different from that of the simpler Richter scale as originally proposed.

Nance, John, and Howard Cady. *On Shaky Ground.* New York: William Morrow, 1988. One of the best-written and easiest-to-understand accounts of the actual effects of some of the most significant earthquakes of various magnitudes. Based on interviews with survivors as well as with top researchers in seismology. Strongly recommended.

Richter, Charles F. *Elementary Seismology.* San Francisco: W. H. Freeman, 1958. The author of this classic 768-page text, who was a seismologist for many years at the California Institute of Technology, developed the Richter scale for measuring the intensity of earthquakes. Judging from his book, Dr. Richter must have been an excellent teacher. Even though this source is outdated, its lucid explanations of basic principles make it a worthwhile reference. Contains excellent and detailed chapters on the complexities of earthquake locating, along with examples, charts, diagrams, and travel-time curves. Some sections using differential equations would be for upper-level college students, but most of the book, including the parts on earthquake locating, would be quite readable to any advanced high school student.

Walker, Bryce. *Earthquake.* Alexandria, Va.: Time-Life Books, 1982. A well-written popular account of earthquakes; includes many good illustrations and a map. Suitable for high school-level readers.

See also: Deep-Focus Earthquakes; Earthquake Distribution; Earthquake Engineering; Earthquake Hazards; Earthquake Locating; Earthquake Prediction; Earthquakes; Elastic Waves; Faults: Normal; Faults: Strike-Slip; Faults: Thrust; Faults: Transform; Notable Earthquakes; Plate Motions; Plate Tectonics; San Andreas Fault; Seismic Observatories; Seismic Wave Studies; Seismometers; Slow Earthquakes; Soil Liquefaction; Subduction and Orogeny; Tsunamis and Earthquakes.

EARTHQUAKE PREDICTION

Predicting the location and timing of earthquakes is an active area of research in many countries throughout the world. Although significant progress has been made in understanding the causes and consequences of earthquakes, scientists are still unable to predict with sufficient accuracy the occurrence of major temblors.

PRINCIPAL TERMS

- **crust:** the outermost layer of the earth, which consists of materials that are relatively light
- **elastic rebound theory:** the theory that states that rocks across a fault remain attached while accumulating energy and deforming; the energy is released in a sudden slip, which produces an earthquake
- **faulting:** the process of fracturing the earth such that rocks on opposite sides of the fracture move relative to each other; faults are the structures produced during the process
- **lithosphere:** the earth's rigid outer layer, which is composed of the crust and uppermost mantle
- **seismicity:** the temporal and spatial distribution of earthquakes
- **seismology:** the study of earthquakes and their causes
- **stress:** a force acting in a specified direction over a given area

Earthquake Occurrence

Chinese scientists pioneered the study of earthquakes hundreds of years ago. Since that time, predicting the location and time of major earthquakes has been an important part of seismology. Earthquakes occur, with varying frequency, in diffuse belts in nearly every region of the globe. The distribution of earthquakes is explained easily by the modern theory of plate tectonics, which holds that the surface of the earth is composed of a mosaic of interlocking rigid plates that move relative to one another at speeds up to 12 centimeters per year. Motions along the boundaries of the plates produce earthquakes; if the plates are not able to accommodate the motions easily, then large earthquakes may accompany the relative motion between the plates. The widespread occurrence of earthquakes makes them important to everyone, and for people living near plate boundaries, earthquakes play an even more potentially destructive role in shaping the environment.

Earthquakes are generated when some portion of the earth's rigid outer layer, called the lithosphere, ruptures catastrophically along a sharp discontinuity or fault. This creates significant ground motion near the source of the rupture. Earthquakes occur most commonly at the three types of boundaries of lithospheric plates, which are known as convergent or destructive, divergent or constructive, and transcurrent or transform. The largest number of earthquakes are at divergent plate boundaries located along mountain ridges in the ocean basins. Because these earthquakes are small and far from population centers on the continents, little effort is expended to predict ruptures along divergent plate boundaries. In contrast, earthquakes at convergent or transcurrent plate boundaries, although fewer in number, are larger. Most convergent and transcurrent boundaries coincide with continental margins along which the majority of the global population lives. For this reason, earthquake prediction research focuses on convergent and transcurrent plate boundaries such as those in Japan and California.

Slip on fault planes of large earthquakes is on the order of 10-20 meters, and the forces responsible for faulting are simply the result of the relative motions of the plates at the plate boundaries. Rocks near a region of an impending earthquake may accumulate motion and change volume and shape for hundreds of years prior to causing a rupture. When the lithosphere does finally break, energy stored by the rocks is released suddenly as seismic waves that travel through and around the surface of the earth. These waves generate the intense vibrations associated with an earthquake. For great earthquakes, the rupture may extend for as much as 1,000 kilometers, and it may propagate at speeds in excess of 10,000 kilometers per hour.

Earthquake Categorization

Seismologists categorize earthquakes by several different features, but the two most important for earthquake prediction are an earthquake's location and its magnitude or size. The location is described by an epicenter, which is the projection of the earthquake's focus within the lithosphere onto the earth's surface.

The magnitude is a number from 1 to 10 on a scale devised by Charles Richter that describes the relative changes in ground motion recorded on a seismometer. The so-called Richter scale is logarithmic; an increase from a value of 1 to 2 corresponds to a tenfold increase in ground motion and to an approximately thirtyfold increase in the amount of energy released during the rupture. Contrary to popular belief, there is no upper limit to the Richter scale. The Richter scale is based on a "standard seismometer" placed at a "standard distance" from the epicenter of the earthquake. The traditional Richter scale magnitudes are denoted by "M" to distinguish them from other more recent magnitude scales. Richter originally devised his magnitude scale to be most appropriate for describing moderate local earthquakes in California. Unfortunately, despite its widespread use, the scale is not a good measure of the energy released from very small or very large earthquakes.

POTENTIAL TRIGGERS

Seismologists have learned much about the rupture process that causes earthquakes by studying the ground motion close to and far away from the source. The development of modern seismological instruments and procedures in the early twentieth century led seismologists to the discovery that different rupture mechanisms are at work in different plate tectonic settings. Early attempts at earthquake prediction used analysis of the frequency of major earthquakes in specific regions of the globe to determine whether any significant pattern was apparent. This approach proved to be fruitless. With the acceptance of elastic rebound theory to describe the rupture mechanism for earthquakes, seismologists shifted their attention away from statistical analysis of earthquake occurrences toward developing methods to understand the "trigger" of major earthquakes. Most seismologists agree that the energy necessary to produce a major earthquake is accumulated slowly relative to the time it takes for a rupture to occur. If no trigger were involved in the rupture process, prediction of earthquakes would be extremely difficult, if not impossible. Modern seismologists interested in earthquake prediction primarily rely on developing methods to understand any precursory phenomena that would enable them to predict at least several days or weeks in advance the location of large (M > 6.0) to great (M > 8.0) earthquakes. Of course, if an impending earthquake is far removed—for example, more than 1,000 kilometers—from any population center, the need to alert the public is minimal.

Over the years, investigators have suggested a variety of potential triggers to earthquakes. These include rapidly changing or severe weather conditions; variations in the gravitational forces among the moon, sun, Earth, and other planets in the solar system; and volcanic activity. Scientists have searched historical seismicity records, including extensive catalogs for California, for relationships between the suggested triggers and the occurrence of earthquakes, without much success. For example, every 179 years, the planets of the solar system align. Some researchers suggested that this alignment would increase the gravitational forces acting upon earth and thereby trigger an increase in seismicity. The last such planetary alignment was in 1982, and no significant increase in earth seismicity occurred.

PRIOR CHANGES IN PHYSICAL PROPERTIES

Because the research on earthquake triggers has been largely unsuccessful, seismologists have turned their attention away from potential triggers of major earthquakes toward the role of changing physical properties prior to an earthquake. Some promising properties include shifts in ground elevation near the site of an impending earthquake; variations in the velocity of certain types of seismic waves as they traverse regions that may produce a major quake; increased escape of radon, helium, and other gases from vents and cracks in the earth's surface prior to the earthquake; changes in the electrical conductivity of rocks near the region of impending rupture; and fluctuations in pore fluid pressure in the rocks near major fault zones. In addition, seismologists have focused on recognizing certain premonitory swarms of smaller quakes, called foreshocks, that may foreshadow a major earthquake.

Another technique is assessing the time between major earthquakes in a specific region. If an area that is expected to produce earthquakes is seismically quiet—that is, a gap exists in its seismic activity—then the area may be more likely to experience an earthquake in the near future. This is referred to as "seismic gap" theory.

Finally, strange animal behavior has been linked to periods of several days to several hours prior to an earthquake. Some researchers have claimed that cats and dogs tend to run away from home or exhibit unusual behavior, such as seeking out special hiding places, before the onset of an earthquake. Individual reports of odd animal behavior are substantiated by the increase in advertisements for lost pets in local newspapers during the days before a major earthquake. Scientists have suggested that some animals are sensitive to minute changes in their environment, which allows them to "sense" an earthquake prior to onset of severe ground shaking. Research in this area is actively pursued in China and Italy. Most workers, however, are interested in developing instruments that would be able to measure the same effects that disturb animals. Although many of these effects are known to occur prior to a major earthquake, scientists still must develop highly sensitive devices that will alert them in enough time to evacuate or prepare the region near an impending earthquake.

SEISMOGRAM ANALYSIS

Several methods are used by seismologists to study earthquake prediction. The techniques include analysis of seismograms to identify either foreshock or aftershock patterns that signal an impending large earthquake, examination of active fault zones in the field to determine the frequency of great earthquakes over the last tens of thousands of years, and investigation of deep boreholes to characterize the orientations and magnitudes of stresses associated with active faults. In addition, elaborate arrays of sophisticated instruments are frequently deployed near active faults to collect geophysical data that may shed light on earthquakes.

The energy carried by seismic waves is recorded on seismometers or seismographs, which are instruments that monitor ground motion. Seismometers are composed of a mass attached to a pendulum. During an earthquake, the mass remains still, and the amount that the earth moves around it is measured. Ground motion is recorded on a chart as a series of sharp peaks and valleys that deviate from the background value, measured during times of no earthquake activity. The arrival of the waves at different times at different places allows seismologists to calculate the epicenter of an earthquake. The height or amplitude relative to the background noise of the first peak in a long series of peaks associated with a particular earthquake is an estimate of the magnitude of that earthquake.

Since the early twentieth century, seismometers have recorded hundreds of thousands of earthquakes per year worldwide. Seismologists have carefully cataloged many seismograms, the actual paper records of ground motion from a particular location, so that they may be easily compared. Examining these records in detail has allowed seismologists to deduce certain characteristics of major earthquakes. They have noted, for example, that most large earthquakes are followed by a series of smaller earthquakes in the same region. These smaller earthquakes are called aftershocks, and they allow seismologists to constrain the orientation and dimensions of the rupture or fault plane that produced the main earthquake. With the development of modern seismometers and digital recording networks in the 1970's and 1980's, seismologists began to recognize certain precursory seismicity patterns in addition to aftershock sequences. These precursory phenomena are referred to as foreshock sequences and, as yet, are poorly understood. Seismologists hope that with enough data on the overall seismicity of an area, they will be able to note deviations from normal patterns that would signal the onset of a major earthquake.

FIELD ANALYSIS

Information about prehistoric seismic activity must be obtained by examining ancient fault zones exposed at the surface of the earth. Large motions between two rock masses produce characteristic features that may be identified in the field. Geologists are now examining the recent rock record near the San Andreas fault in California. Careful mapping of areas that have been excavated across the fault zone has yielded evidence for large earthquakes prior to historical and seismological records. The now well-established technique of carbon-14 dating was applied to organic material trapped in the fault zone to determine the approximate age, location, and intensity of several ancient earthquakes. The data, although sparse, seem to suggest that great earthquakes occur every fifty to three hundred years. In addition, there is some indication that great earthquakes may occur closely spaced in time with significant periods of quiescence between them. This type of analysis is similar to the seismic gap theory, where catalogs of seismograms are examined to determine which known faults or regions that have been active previously are currently inactive and perhaps are ready for renewed activity.

Many countries are carrying out elaborate experiments in areas of repeated seismic activity. One example is in central California on the San Andreas fault near the town of Parkfield. There, geophysicists arrayed a variety of instruments, including seismometers, tiltmeters, gravimeters, and laser surveying equipment, to measure ground motion, elevation changes, gravity variations, and minute amounts of slip on the fault. Based on seismic records, scientists discovered that earthquakes with magnitudes of approximately 5.0 M occur with predictable frequency. Since 2004 Parkfield has been the site of the San Andreas Fault Observatory at Depth (SAFOD), with extensive instrumentation in a 3-kilometer borehole angled into the fault. The experiments are designed to learn as much as possible about the changes that occur in the region prior to, during, and after an earthquake of moderate size. Scientists hope that these data will allow them to know what features to monitor to predict much larger earthquakes in other areas.

INVESTIGATION OF DEEP BOREHOLES

Another method used by scientists to understand precursory phenomena associated with earthquakes is drilling deep boreholes into the earth's crust near major fault zones. One such hole is in Fort Cajon, California. At this site, researchers examined changes in pore fluid pressure and electrical conductivity in the borehole. In addition, instruments measured the orientation of fractures in the borehole's walls. These fractures are related to the forces acting on the rocks deep in the crust, and some researchers attempted to relate these manifestations of stress to earthquake fault orientation. Geophysicists hope that the newest techniques will measure these stresses in real time, so they will be able to compare these data to those obtained from studies of seismicity. Understanding the behavior of a major fault zone at depth may prove useful in predicting earthquakes in the future.

Earthquake damage to the cathedral in downtown Port-au-Prince, Haiti, from the 7.0 earthquake that hit on January 12, 2010. Improper construction practices in densely populated areas resulted in catostrophic building collapses and loss of lives. (Julie Dermansky/Photo Researchers, Inc.)

PROGRESS IN PREDICTION

Many countries are actively involved in earthquake prediction research. Since the early 1960's, these efforts have been particularly active in Japan, the People's Republic of China, and the United States. The overall goal of these research efforts is to attain the same level of reliability in earthquake prediction as in weather prediction. Although the majority of effort has been focused on predicting the exact time and place of a major earthquake, an equally important, though often overlooked, aspect of earthquake prediction is an assessment of the severity of ground shaking for a specific site. This information is crucial for public policy discussions on the location of dams, hospitals, schools, and nuclear reactors, all of which may be at significant risk during a major earthquake.

Despite the numerous uncertainties that enter into forecasting an earthquake, some have been successfully predicted. The most spectacular was the Haicheng earthquake of northeast China in February 1975. Five hours before the earthquake, warnings were issued and several million people from towns in the vicinity of the predicted epicenter were evacuated. Devastation was widespread, but loss of human life was minimal. Scientists who later visited the area estimated that hundreds of thousands of lives were saved. Unfortunately, the Chinese were only able to predict that a great earthquake was to strike the Tangshan region within five years. In August 1976, a very strong earthquake struck this area without warning, leaving 700,000 people dead.

MINIMIZING EARTHQUAKE HAZARDS

Earthquake prediction remains critical to modern society because most of the world's population lives along convergent plate boundaries and, therefore, within the destructive reach of a great earthquake. The purpose of earthquake prediction, then, is to prepare a society for any earthquakes with magnitudes capable of disturbing normal life. This may include warning and evacuation of the local population or assessing the risks of severe ground motion on current or future structures. In addition to strong ground motion, earthquakes may be responsible for hazards such as tsunamis, avalanches, and fires. In both the great San Francisco earthquake of 1906 and the Tokyo earthquake of 1923, many of the fatalities attributed to the earthquakes were actually the result of the subsequent fires that consumed the cities. Another danger of earthquakes is soil liquefaction, which occurs when the seismic waves cause the soil to lose rigidity and slide away. When this happens, the soil can no longer support structures. Although the structures may be strong enough to withstand the shaking associated with an earthquake, their foundations may be undermined, causing the buildings to topple.

Perhaps the most promising aspect of earthquake prediction is the development of stringent building codes. After each major earthquake in Southern California, for example, municipal, county, and state statutes are changed to reflect new data concerning the behavior of building materials during strong ground motion. Engineers now know that unreinforced masonry buildings are likely to be destroyed in even a moderate earthquake. Because the seismic risk is high in Japan and California, these areas now have the most stringent building codes in the world. That these codes can prevent much unnecessary loss of property and human life unfortunately can be demonstrated by a comparison of the 1971 San Fernando and 1988 Armenian earthquakes. The earthquakes were of similar magnitude—slightly greater than 6.0 M—yet the San Fernando earthquake resulted in about fifty deaths, most of which were in one wing of an old hospital building, while tens of thousands perished in the Armenian earthquake because of the collapse of poorly constructed masonry buildings.

Humans will never be able to prevent earthquakes. Understanding how, why, when, and where earthquakes occur, therefore, is extremely important to society. Earthquake prediction, like weather prediction, is one way that society seeks to minimize harmful effects of these complex natural phenomena.

Glen S. Mattioli and Pamela Jansma

FURTHER READING

Berlin, G. Lennis. *Earthquakes and the Urban Environment.* Vols. 2 and 3. Boca Raton, Fla.: CRC Press, 1980. These books are two volumes of a three-part series written by a geographer who is primarily concerned with effective land-use planning in seismically active areas. Volume 3 concentrates on strategies to minimize the effects of great earthquakes, such as disaster planning and improved building codes. Many of the social aspects of earthquakes in an urban environment are presented, including human response and insurance. Volume 2 addresses earthquake prediction and building codes. Both volumes contain an extensive reference list of more than 1,400 articles and books. These volumes can be quite technical and are recommended for college-level readers.

Bolt, Bruce A. *Earthquakes and Geological Discovery.* New York: Scientific American Library, 1993. As the title suggests, an excellent introductory text on earthquakes. Earthquake prediction is discussed extensively in one chapter. The illustrations and photographs of the effects of earthquakes add considerably to the text. Anyone interested in earthquakes will find this an invaluable source.

Doyle, Hugh A. *Seismology.* New York: John Wiley, 1995. A good introduction to the study of earthquakes and the earth's lithosphere. Written for the layperson, the book contains many useful illustrations.

Eiby, G. A. *Earthquakes.* New York: Van Nostrand Reinhold, 1980. A reference aimed at beginning college-level students. Well illustrated and addresses all topics relevant to earthquakes and seismology. Earthquake prediction and the effect of large earthquakes on human-made structures are discussed in two separate chapters.

Farley, John E. *Earthquake Fears, Predictions, and Preparations in Mid-America.* Carbondale: Southern

Illinois University Press, 1998. This book examines seismic activity, the practice of predicting earthquakes, and the hazards associated with them, focusing on the American Midwest. Bibliography, charts, and index.

Grotzinger, John, et al. *Understanding Earth.* 5th ed. New York: W. H. Freeman, 2006. One of the finest illustrated introductory texts on geology. The book has chapters focusing on plate tectonics, seismology, and earthquakes. A map of the major plates is on the inside back cover. The glossary is huge and indispensable. Senior high school and college-level students should find this text suitable for general background information.

Hough, Susan. *Predicting the Unpredictable: The Tumultuous Science of Earthquake Prediction.* Princeton, N.J.: Princeton University Press. 2010. The author provides a detailed, but non-technical description of the history of earthquake predictions. She identifies unresolved issues scientists have in making predictions. This text has a wide range of information.

Iacopi, Robert. *Earthquake Country.* 4th ed. Menlo Park, Calif.: Lane Books, 1996. Part of the Sunset Book series, this source is directed toward the lay reader. Discusses California geology in relation to seismic risk in a straightforward and nontechnical way. Contains many photographs of the effects of earthquakes on both human-made structures and the natural landscape. Also has a foreword by Charles F. Richter, the inventor of the famous Richter scale.

Lomnitz, Cinna. *Fundamentals of Earthquake Prediction.* New York: John Wiley & Sons, 1994. Lomnitz examines the principles and mythologies of earthquake prediction. Illustrations, maps, bibliography, and index.

Mogi, Kiyoo. *Earthquake Prediction.* San Diego, Calif.: Academic Press, 1985. A comprehensive and highly technical text that discusses most aspects of earthquake prediction. The majority of prediction experiments that are described are from Japan. The reader is expected to have a considerable background in earth science and mathematics. The text is suitable for senior college-level students.

National Research Council (U.S.), Panel on the Public Policy Implications of Earthquake Prediction. *Earthquake Prediction and Public Policy.* Washington, D.C.: Government Printing Office, 1975. This book was prepared by a National Research Council panel on the public policy implications of earthquake prediction. The panel was composed of outstanding scientists and engineers involved in all fields of earthquake research, in addition to sociologists and other public figures. They specifically evaluate seismic risk in certain regions of the United States and propose action to prepare these regions for significant earthquakes. Guidelines for earthquake prediction research are discussed. This text is suitable for anyone.

Rikitake, Tsuneji. *Earthquake Forecasting and Warning.* Norwell, Mass.: Kluwer Academic Publishers, 1982. Focuses on advances in earthquake prediction in Japan, California, the Soviet Union, and the People's Republic of China. The major programs for earthquake study in each of these regions are discussed. Case studies in Japan are presented in detail. The reader is required to have an excellent understanding of geophysics. Suitable for senior college-level students.

Tarbuck, Edward J., Frederick K. Lutgens, and Dennis Tasa. *Earth: An Introduction to Physical Geology.* 10th ed. Upper Saddle River, N.J.: Prentice Hall, 2010. This college text provides a clear picture of the earth's systems and processes that is suitable for the high school or college reader. It has excellent illustrations and graphics. Bibliography and index.

Uyeda, Seiya. *The New View of the Earth.* Translated by Masako Ohnuki. San Francisco: W. H. Freeman, 1978. The historical development of the theory of plate tectonics is presented, in addition to an excellent explanation of the theory itself. The text is well illustrated and is non-technical. Designed primarily for the nonscientist.

See also: Deep-Focus Earthquakes; Earthquake Distribution; Earthquake Engineering; Earthquake Hazards; Earthquake Locating; Earthquake Magnitudes and Intensities; Earthquakes; Earth's Lithosphere; Elastic Waves; Faults: Normal; Faults: Strike-Slip; Faults: Thrust; Faults: Transform; Notable Earthquakes; Plate Motions; Plate Tectonics; San Andreas Fault; Seismic Reflection Profiling; Seismic Wave Studies; Seismometers; Slow Earthquakes; Soil Liquefaction; Subduction and Orogeny; Tsunamis and Earthquakes.

EARTHQUAKES

An earthquake is the sudden movement of the ground caused by the rapid release of energy that has accumulated along fault zones in the earth's crust. The earth's fundamental structure and composition are revealed by earthquakes through the study of waves that are both reflected and refracted from the interior of the earth.

PRINCIPAL TERMS

- **crust:** the uppermost 5-40 kilometers of the earth
- **deformation:** a change in the shape of a rock
- **elastic rebound:** the process whereby rocks snap back to their original shape after they have been broken along a fault as a result of an applied stress
- **lithosphere:** the solid part of the upper mantle and the crust where earthquakes occur
- **mantle:** the thick layer under the crust that contains convection currents that move the crustal plates
- **strain:** the percentage of deformation resulting from a given stress
- **stress:** a force per unit area

STRESS

Earthquakes are sudden vibrational movements of the earth's crust and are caused by a rapid release of energy within the earth. They are of critical importance to humans because they reveal much about the interior of the earth and because they are one of the most destructive naturally occurring forces found on earth.

The outermost skin of the earth, called the crust, is in constant motion as a result of large convection cells within the upper mantle that circulate heat from the interior of the earth toward the surface. The crust of the earth is about 5 kilometers thick in the oceanic basins and about 40 kilometers thick in the continental masses, while the upper mantle is about 700 kilometers thick. Because the crust is relatively thin compared to the upper mantle, the crust is broken up into several plates that float along the top of each convection cell in the upper mantle. Most earthquakes occur along the boundaries separating the individual plates and are represented by faults that may be thousands of kilometers long and tens of kilometers deep. Although the vast majority of earthquakes occur along these plate boundaries, some also occur within the plate interior. The rocks on either side of the fault fit tightly together and produce great resistance to movement. As the blocks of rock attempt to move against one another, the resistance of movement causes stress, which is a force per unit area, to build up along the fault. As the stress continues to build, the rocks in the immediate vicinity slowly deform, or bend, until the strength of the rock is exceeded at some point along the fault. Suddenly, the rocks break violently and return to their underformed state, much as a rubber band snaps to its original shape when it breaks. This rapid release of stress is called elastic rebound. The point at which the stress is released is called the focus of an earthquake, and that point at the earth's surface directly above the focus is called the epicenter.

SEISMIC WAVE MOTION

The release of energy associated with elastic rebound manifests itself as waves propagating away from the focus. When these waves of energy reach the surface of the earth, the land will oscillate, causing an earthquake. These waves move through the earth in two ways.

P (primary) waves move in a back-and-forth motion in which the motion of the rock is in the same direction as the direction of energy propagation. This type of wave motion is analogous to placing a spring in a tube and pushing on one end of the spring. The motion of the spring in the tube is in the same direction as is the motion of the energy. These waves are called primary because they move through the earth faster than do other waves—up to about 25 kilometers per second. Thus, P waves are the first waves to be received at a seismic recording station. Because the individual atoms in a rock move back and forth along the direction of energy movement, P waves can move through solids and liquids and, for this reason, do not tell geologists much about the state (solid or liquid) of a given rock at depth.

In contrast to P waves, for S waves, the rock motion is perpendicular to the direction of energy propagation. Guitar strings vibrate in a similar manner. Each part of the guitar string moves back and forth while the energy moves along the string to the ends. S waves are the second waves to be received at a seismic recording station and derive their name from this fact. Unlike P waves, S waves cannot move through liquids but can move through solids. Thus, when a P wave is received by a seismic station but is not followed by an S wave, seismologists know that a liquid layer is between the focus of the earthquake and the receiving seismic station.

Both S and P waves are bent, or refracted, as they move in the earth's interior. This refraction occurs as the result of the increase in density of rocks at greater depths. Furthermore, both types of waves are reflected off sharp boundaries, representing a change in rock type located within the earth. Thus, by using these properties of S and P waves, geologists have mapped the interior of the earth and know whether a given region is solid or liquid.

Although S and P waves represent the way seismic energy moves through the earth, once this energy reaches the earth's surface, much of it is converted to another type of wave. L (Love) waves move in the same manner as do S waves, but they are restricted to surface propagation of energy. L waves have a longer wavelength and are usually restricted to within a few kilometers of the epicenter of an earthquake. These waves cause more damage to structures than do P and S waves because the longer wavelength causes larger vibrations of the earth's surface.

EARTHQUAKE INTENSITY

The amount of energy released by an earthquake is of vital importance to humans. Many active fault zones, such as the famous San Andreas Fault in California, produce earthquakes on an almost daily basis, although most of these earthquakes are not felt and cause no damage to human-made structures. These minor earthquakes indicate that the stress that is accumulating along some portion of a fault is continuously being released. It is only when the stresses accumulate without continual release that large, devastating earthquakes occur. The intensity of an earthquake is dependent not only on the energy released by the earthquake but also on the nature of rocks or sediments at the earth's surface. Softer sediments such as the thick muds that underlie Mexico City will vibrate with a greater magnitude than will the very rigid rocks, such as granites, found in other parts of the world. Thus, the great earthquake that devastated Mexico City in 1985 was in part the result of the nature of the sediments upon which the city is built.

For a given locality, earthquakes occur in cycles. Stress accumulates over a period of time until the forces exceed the strength of the rocks, causing an increase in minor earthquake activity. Shortly thereafter, several foreshocks, or small earthquakes, occur immediately before a large earthquake. When a large earthquake occurs, it is usually followed by many aftershocks, which may also be rather intense. These aftershocks occur as the surrounding rocks along the fault plane readjust to the release of stress by the major earthquake. The cycle then repeats itself with a renewed increase in stress along the fault. Although seismologists can usually tell which part of the seismic cycle a region is experiencing, it is difficult to predict the duration of each of these cycles; thus, it is impossible to predict precisely when an earthquake will occur.

SEISMOGRAPHS

Seismographs are the primary instruments used to study earthquakes. All seismographs consist of five fundamental elements: a support structure, a pivot, an inertial mass, a recording device, and a clock. The support structure for a seismograph is always solidly attached to the ground in such a fashion that it will oscillate with the earth during an earthquake. A pivot, consisting of a bar attached to the support structure via a low-friction hinge, separates a large mass from the rest of the seismograph. This pivot allows the inertial mass to remain stationary during an earthquake while the rest of the instrument moves with the ground. The recording device consists of a pen attached to the inertial mass and a roll of paper that is attached to the support structure. Finally, the clock records the exact time on the paper so that the time of arrival of each wave type is noted. When an earthquake wave arrives at a seismic station, the support structure moves with the ground. The inertial mass and the pen, however, remain stationary. As the paper is unrolled, usually by a very accurate motor, the wave is recorded on the paper by the stationary pen. Modern seismographs, however complex in design, always contain these basic elements. The clock, which each minute places a small tick mark on the recording, is calibrated on a daily basis by a technician using international time signals from atomic clocks. The recording pen often consists of an electromagnet that converts movement of the inertial mass relative to the support structure to an electrical current that drives a light pen. The light pen emits a narrow beam of light onto long strips of photographic film that are developed at a later date. Digital seismographs record measurements electronically using computers. A global network of these machines is overseen by the FDSN, the International Federation of Digital Seismograph Networks.

RICHTER AND MERCALLI SCALES

Seismologists have adopted two widely used scales, which are called the Richter and Mercalli scales, to measure the energy released by an earthquake. The Richter magnitude scale is based on the amplitude of seismic waves that are recorded at seismic stations. Because seismic stations are rarely located at the epicenter of earthquakes, the amplitude of the seismic wave must be corrected for the amount of energy lost over the distance that the wave traveled. Thus, the Richter magnitude reported by any seismic station for a given earthquake will be approximately the same. Richter magnitudes are open-ended, meaning that any amount of seismic energy can be calculated. The weakest earthquakes have Richter magnitudes less than 3.0 and release energy less than 10^{14} ergs. These earthquakes are not usually felt but are recorded by seismic stations. Earthquakes between magnitudes 4.0 and 5.5 are felt but usually cause no damage to structures; they release energy between 10^{15} and 10^{16} ergs. Earthquakes that have magnitudes between 5.5 and 7.0 cause slight to considerable damage to buildings and release energy between 10^{18} and 10^{24} ergs. Earthquakes that are greater than 7.5 on the Richter scale generate energy up to 10^{25} ergs—as much as a small nuclear bomb.

The Mercalli intensity scale is based not on the energy released by an earthquake but rather on the amount of shaking that is felt on the ground; it rates earthquakes from Roman numerals I to XII. Unlike the Richter scale, the Mercalli scale provides descriptions of sensations felt by observers and of the amount of damage that results from an earthquake. Thus, an earthquake of Mercalli intensity I is felt by only a very few persons, while an earthquake of intensity XII causes total destruction of virtually all buildings.

Both the Mercalli and Richter scales have advantages and disadvantages. The Mercalli scale provides the public with a more descriptive understanding of the intensity of an earthquake than does the Richter scale. The damage caused by an earthquake is a function not only of the energy released by such an event but also of the nature of the sediments or rocks upon which the buildings in the vicinity are constructed. The Richter scale is best used to study specifically the amount of energy released by an earthquake. Finally, the Richter scale, which is purely quantitative, does not rely on subjective observations such as those required by the Mercalli scale.

TRIANGULATION TECHNIQUES

The exact location of an earthquake epicenter can be deduced from three seismographic stations using triangulation techniques. Because the P and S waves travel at different velocities in the earth, seismologists can determine the distance from the station to the epicenter. They calculate the difference in time between the first arrival of the P and S waves, respectively, at the station. They then multiply this time difference by the product of the P and S velocities and divide by the difference in wave velocities to obtain the distance to the epicenter. The earthquake must have occurred along a circle whose radius is the distance so calculated and whose center is the seismographic station; any three stations that record the event can be used to draw three such circles, which will intersect at a single point. This point is the epicenter.

EARTHQUAKE PREDICTION

Earthquakes are one of the most important processes that occur within the earth because they have such a profound effect on how and where people should develop cities. Geologists understand how and where earthquakes occur, yet despite their best efforts, they still cannot accurately determine when an earthquake will happen. They are merely able to predict that a large earthquake will occur in a particular region "in the near future." Very great earthquakes of magnitude 8 or greater, such as the San Francisco earthquake of 1906, occur about every five to ten years throughout the world. Industrialized societies, such as Japan, the United States, and many European countries, have developed buildings that are capable of withstanding devastating seismic catastrophes, but other countries are not as fortunate. Furthermore, some great earthquakes occur in regions that are not considered seismically active. The great Charleston, South Carolina, earthquake of 1886 and the Tangshan, China, earthquake of 1976 are examples of seismic events that could not have been easily predicted using modern technology. In such regions, buildings are not designed to withstand devastating earthquakes. Finally, many regions of the world do not experience earthquakes on a daily basis and, therefore, their governments lack the motivation to plan adequately for such potentially catastrophic events.

A. Kem Fronabarger

FURTHER READING

Abaimov, S. G., et al. "Earthquakes: Recurrence and Interoccurrence Times." *Pure and Applied Geophysics* 165 (2008): 777-795. Provides an analysis of the statistical probability of earthquake recurrence at the San Andreas Fault. Background in statistics needed to fully comprehend equations.

Bolt, Bruce A. *Earthquakes and Geological Discovery.* New York: Scientific American Library, 1993. As the title suggests, an excellent introductory text on earthquakes. Earthquake prediction is discussed extensively in one chapter. The illustrations and photographs of the effects of earthquakes add considerably to the text. Anyone interested in earthquakes will find this an invaluable source.

Doyle, Hugh A. *Seismology.* New York: John Wiley, 1995. A good introduction to the study of earthquakes and the earth's lithosphere. Written for the layperson, the book contains many useful illustrations.

Emergency Management BC. *A Simple Explanation of Earthquake Magnitude and Intensity.* Ministry of Public Safety and Solicitor General, Provincial Emergency Program. 2007. Provides fundamental content on earthquakes, written for the layperson.

Farley, John E. *Earthquake Fears, Predictions, and Preparations in Mid-America.* Carbondale: Southern Illinois University Press, 1998. This book examines seismic activity, the practice of predicting earthquakes, and the hazards associated with them, focusing on the American Midwest. Bibliography, charts, and index.

Grotzinger, John, et al. *Understanding Earth.* 5th ed. New York: W. H. Freeman, 2006. This text includes one of the most complete descriptions of the causes of earthquakes, their measurement, where they occur, how they can be predicted, and how they affect humans. A map of the major plates is on the inside back cover. The glossary is huge and indispensable. Senior high school and college-level students should find this text suitable for general background information.

Hodgson, John H. *Earthquakes and Earth Structure.* Englewood Cliffs, N.J.: Prentice-Hall, 1964. This source provides the reader with an understanding of how earthquakes have been used to determine the structure and composition of the interior of the earth.

Hough, Susan. *Predicting the Unpredictable: The Tumultuous Science of Earthquake Prediction.* Princeton, NJ; Princeton University Press. 2010. The author provides a detailed, but non-technical description of the history of earthquake predictions. She identifies unresolved issues scientists have in making predictions. This text has a wide range of information.

McKenzie, D. P. "The Earth's Mantle." *Scientific American* 249 (September 1983): 66-78. This article, written at the college undergraduate level, is a very complete description of current scientific understanding of the interior of the earth.

Nichols, D. R., and J. M. Buchanan-Banks. *Seismic Hazards and Land-Use Planning.* U.S. Geological Survey Circular 690. Washington, D.C.: Government Printing Office, 1974. The effect of earthquakes on human-made structures is discussed in this short bulletin. Written explicitly for the layperson by the United States government, it provides additional sources of information for land-use planning.

Press, Frank. "Earthquake Prediction." *Scientific American* 232 (May 1975): 14-23. Press's article details geologists' current understanding of earthquake prediction. Also provides a discussion of the methods by which earthquakes can be predicted. Written at the college undergraduate level.

Prothero, Donald R. *Catastrophes!: Earthquakes, Tsunamis, Tornadoes, and Other Earth-Shattering Disasters.* Baltimore: Johns Hopkins University Press, 2011. This text provides a detailed and clear explanation of the many natural and anthropogenic disasters facing our planet. Each chapter is devoted to a different catastrophe, including earthquakes, volcanoes, hurricanes, ice ages, and current climate changes.

Tarbuck, Edward J., Frederick K. Lutgens, and Dennis Tasa. *Earth: An Introduction to Physical Geology.* 10th ed. Upper Saddle River, N.J.: Prentice

Hall, 2010. This college text provides a clear picture of the earth's systems and processes that is suitable for the high school or college reader. It has excellent illustrations and graphics. Bibliography and index.

United States Department of the Interior. *Earthquake Information Bulletin*. Washington, D.C.: Government Printing Office. This bimonthly bulletin provides the reader with a concise understanding of where earthquakes occur in the United States and which regions are likely to be affected in the future. Also lists other sources of information on earthquakes. For general and specialized readers.

See also: Continental Drift; Creep; Deep-Focus Earthquakes; Earthquake Distribution; Earthquake Engineering; Earthquake Hazards; Earthquake Locating; Earthquake Magnitudes and Intensities; Earthquake Prediction; Earth's Interior Structure; Earth's Lithosphere; Earth's Mantle; Elastic Waves; Experimental Rock Deformation; Faults: Normal; Faults: Strike-Slip; Faults: Thrust; Faults: Transform; Heat Sources and Heat Flow; Lithospheric Plates; Mantle Dynamics and Convection; Notable Earthquakes; Plate Motions; Plate Tectonics; San Andreas Fault; Seismic Observatories; Seismic Wave Studies; Seismometers; Slow Earthquakes; Soil Liquefaction; Stress and Strain; Subduction and Orogeny; Tectonic Plate Margins; Tsunamis and Earthquakes; Volcanism.

EARTH'S AGE

Humans have tried to determine the age of Earth for thousands of years by various means. Only when a clock-like mechanism, such as tree rings, relating the present day to some past time was discovered, could a reasonably accurate determination of Earth's age be made.

PRINCIPAL TERMS

- **activity:** the number of transmutations that occur in a specific process in a specific period of time, such as counts per minute
- **alpha (α) particle:** the equivalent of the nucleus of a helium atom consisting of two protons and two neutrons ejected from the nucleus of some radionuclides as the mechanism of radioactive decay
- **beta (β) particle:** a particle produced by the decomposition of a neutron to a proton, emitted from a nucleus as an electron during radioactive decay processes
- **electron capture:** retention of the electron emitted from the nucleus as a β particle to balance the positive charge of the newly formed proton and maintain electrical neutrality of the nuclide atom
- **half-life:** the length of time required for one-half of a given quantity of a radionuclide to decay to another nuclide in an exponential decay process
- **isotope:** atoms of an element having the same number of protons but different numbers of neutrons in the nucleus
- **radionuclide:** a radioactive isotope of an element
- **rate-determining step:** the step in a multistep process that determines the maximum rate at which the overall change is observed to take place
- **transmutation:** the conversion of an atom of one element into a corresponding atom of another element by altering the number of protons and neutrons within the nucleus

RADIOMETRIC DATING AND ABSOLUTE AGE

Determining the age of a material requires that there be some feature usable as a clock to count time backward from the present. The best and most familiar example of such a natural clock is the annual growth rings of trees. Counting the rings provides the absolute age of the tree because definite starting points and ending points exist for counting the number of years that have passed since the tree began to grow. The natural transmutation of certain radionuclides during millions of years permits the determination of the absolute age of rocks by providing the clock mechanism that ties the present day to a past starting point in time.

The structure of atoms consists primarily of three basic particles in a specific structural arrangement. These particles are the negatively charged electron, the positively charged proton, and the electrically neutral neutron. The protons and neutrons together form a small dense nucleus that is surrounded by a diffuse cloud of electrons. An equal number of electrons and protons also must exist to maintain electrical neutrality. Each element may contain atoms that have the same number of protons but different numbers of neutrons. Such atoms are known as isotopes, or nuclides.

Some nuclides are unstable and undergo fission processes that eject parts of the nuclear structure. Such processes alter the elemental identity of the particular atom by changing the number of protons in the nucleus. In each case, the decay of a naturally occurring nuclide produces atoms of a stable element.

NATURAL TRANSMUTATION

The symbols for the elements of the periodic table follow a strict international convention that unequivocally identifies the particular atom and specifies the isotopic form, or nuclide. According to this convention, the atomic number of the element (the number of protons in the nucleus) is shown as a preceding subscript and the atomic mass of the specific nuclide is shown as a preceding superscript. This can be confusing because the manner in which the element and nuclide weight are identified in normal speech is to state the name of the element and then the nuclide mass. Thus, to identify the particular nuclide of uranium having a mass of 238 units, one would write ^{238}U but would pronounce the same as U-238 or uranium-238. When writing the equations of nuclear transmutation processes, including the nuclide mass is essential.

There are two main mechanisms by which an atom of one radionuclide transforms into an atom of a different nuclide. In one mechanism, a neutron in an unstable nucleus may decompose into a proton with the elimination of a β (beta) particle. The β particle is ejected as an electron. This is the mechanism whereby the isotope of carbon known as ^{14}C transmutes to the corresponding nuclide of nitrogen, ^{14}N. In the process, the nucleus of the ^{14}C atom changes from the combination of six protons and eight neutrons to the combination of seven protons and seven neutrons. The ejected electron does not just disappear; it is quickly captured as energy by surrounding atoms. Although the elemental identity of the atom changes, its mass is not affected.

The other major mechanism of nuclear transmutation is through the ejection of an α (alpha) particle. The α particle is identical to the nucleus, of a helium atom, consisting of two protons and two neutrons. Ejection of an α particle from a nucleus, therefore, changes the elemental identity of the atom by two places in the periodic table and reduces the atomic mass by four units. Transmutation by the ejection of α particles is a process reserved for large atoms such as uranium, and the process can take place as a cascade in which the product nucleus that is immediately formed may itself eject an α particle and continue to do so until a stable atomic nucleus is formed. For example, the transmutation of uranium-238 (^{238}U) to lead-206 (^{206}Pb) occurs by this sequential process. The intermediate fission steps are unimportant, however, because they occur at a faster rate than the initial fission of ^{238}U and so do not affect the half-life of the process.

HALF-LIFE

Radionuclides undergo fission at an exponential rate described by the mathematical equation $A = A_o e^{kt}$, where A is the amount of material at time t, A_o is the amount at time $t = 0$, and k is the rate constant for the process. A special relationship exists for any time at which the quantity A is exactly one-half of the quantity A_o. The value of t is constant for all values of A and A_o for which this condition is true, and is referred to as the half-life of the process. Accordingly, it takes exactly the same amount of time for one million kilograms (kg) of a specific radionuclide to decay to one half-million kg as it does for one gram (g) to decay to one-half g.

The value of this relationship in regard to radiometric dating and the age of the earth is that it ties a past starting point to the present time and so acts as the clock necessary for counting time backward. Suppose that a mass of rock containing one kg of ^{238}U was produced in the crust of the earth at some point in time. After one half-life has passed from that point in time, only one-half kg of ^{238}U remains. After a second half-life has passed, one-half of that one-half kg, or one-quarter of kg, remains. After a third half-life has passed, another half of the existing ^{238}U, or one-eighth of the original amount, remains. The process continues with the amount remaining each time being given by the equation $A = A_o(1/2^n)$, where n is the number of half-lives that have elapsed. In this way, knowing the amount of material present, and relating it to the amount of product nuclide, permits the calculation with reasonable certainty of the number of half-lives that have passed, and hence the age, of the material.

The length of a half-life for different processes covers an exceedingly broad range. Tables found in the *CRC Handbook of Chemistry and Physics* and other reference books list known half-lives for different radionuclides ranging from 10^{-16} seconds to more than 10^9 years. In the majority of cases, the radionuclide sequence is not amenable to use as a timekeeping mechanism in radiometric dating. Measurement of the process may be problematic because of the actual duration of the half-life or because of the presence of materials that interfere with the analysis; these factors make uncertain the starting conditions of the source material.

MASS SPECTROMETRY

The principal technique of determining nuclide ratios is through accurate mass analysis using mass spectrometry. A mass spectrometer functions using precise mathematical relationships that describe the circular trajectory of an ion having a certain mass and electrical charge as it travels through a magnetic field.

A typical mass spectrometric analysis begins by extracting the desired nuclide species from the matrix of interest. A sample of the extract is introduced into the injection port of the spectrometer, where it is given an electrical charge and directed into the magnetic field sector of the machine. Precise manipulation of the field strength directs ions to a detector in order according to mass. The detection circuitry counts the number of ions detected per unit time, and the results are displayed graphically and numerically. From the measured data, the relative proportions of nuclides and molecular fragments are determined, and the resulting values are used with the exponential rate equations to determine the number of half-lives that have passed for the process; from this one can determine the age of the material from which the nuclides were extracted.

RADIOMETRIC AGE OF EARTH

Radiometric dating is a precise calculation because of the unequivocal structure of the atoms involved. Careful analysis of uranium ore such as pitchblende (U_3O_8) reveals the presence of a small amount of lead as the ^{206}Pb nuclide, but no other lead isotopes. In lead ores that do not contain uranium, this isotope accounts for only 26 percent of the lead present. The stable ^{206}Pb nuclide results only from the breakdown of ^{238}U through a complex series of intermediate steps having much shorter half-lives. The initial fission step that is actually measured is the decay of ^{238}U to the thorium radionuclide ^{234}Th by ejection of an α particle, with a measured half-life of 4.6 billion years.

The subsequent steps of the process resulting in ^{206}Pb are well known. The ^{234}Th ejects a β particle to produce ^{234}Pa (half-life = 24.1 days), which subsequently ejects another β particle to produce the uranium isotope ^{234}U (half-life = 1.14 minutes). In β particle emission, the atomic number of the element increases by one unit while the atomic mass remains the same, but in α particle emission the atomic number decreases by two units and the atomic mass decreases by four units. Once formed, the ^{234}U undergoes five successive α particle emissions to produce ^{230}Th (half-life = 2.7×10^5 years), ^{226}Ra (half-life = 8.3×10^4 years), ^{222}Rn (half-life = 1.6×10^3 years), ^{218}Po (half-life = 3.8 days), and ^{214}Pb (half-life = 3.1 minutes). This radionuclide of lead emits two successive β particles to produce first ^{214}Bi (half-life = 27 minutes) and then ^{214}Po (half-life = 20 minutes). The ^{214}Po emits an α particle to produce the ^{210}Pb radionuclide (half-life = 1.5×10^{-4} seconds), then ^{210}Bi (half-life = 22 years) and ^{210}Po (half-life = 5 days) by two successive β particle emissions. A final α particle emission transmutes the ^{210}Po into the stable nuclide ^{206}Pb (half-life = 140 days). The specificity of the process is apparent, and no other known natural radionuclide decay process results in the formation of ^{206}Pb.

Thus, assuming that no ^{206}Pb existed in the pitchblende ore when it was initially formed, all of the ^{206}Pb present in the ore originated through the series of transmutations just described. The assumption is verified by the observation that lead ores that do not contain uranium also do not contain ^{206}Pb. Of the four stable isotopes of lead, only ^{204}Pb is not produced through the transmutations of radionuclides. The ratio of ^{204}Pb to ^{206}Pb determines the amount of the latter nuclide present in excess of its natural abundance. The combination of these determinations from the examination of samples of ancient rock formations has indicated their age to be about 5 billion years. Analysis of other uranium-containing minerals provides a similar result.

Other radionuclide decay processes are known and have been used as a cross-check of Earth's age as determined by the ^{238}U-^{206}Pb process. The rubidium radionuclide ^{87}Rb transmutes to the stable strontium isotope ^{87}Sr with a half-life of 46 billion years, providing a much longer time frame in which to measure the age of the planet. Determinations based on this transformation have produced almost identical results for the age of Earth and for several meteorites that have been studied. Similarly, the decay of the potassium radionuclide ^{40}K to the inert and stable ^{40}Ar nuclide of argon (half-life = 1.3×10^9 years) has also yielded similar results.

ERRORS

As with any empirical method of analysis, precision and accuracy of radiometric dating techniques are restricted by practical limitations. In any quantitative scientific determination, the value of the variable being measured can only be known within the limits of the least accurate measurement. The error limits of a technique determine the range of values within which the actual value may be found. For example, when volumes are measured using a standard titration burette, each reading has an error limit of, generally, 0.01 milliliter (mL). Errors from multiple readings accumulate in such a way that a volume determined by difference can be known only within 0.02 mL. This translates throughout subsequent calculations to the final range of values that necessarily includes the actual value of the quantity being determined.

The same logic applies in radiometric determinations such that only an approximate range of ages can be determined. Given that a determination may have a total error of ± 1 percent based on the limitations of the actual material manipulations involved in the laboratory, subsequent determination of the age can also be calculated only to the same error limit of ± 1 percent. Over the span of 4.6×10^9 years, an age could thus be determined only as being within a span of 4.6×10^7 years. Ages determined radiometrically are, therefore, generally stated for the midpoint of the range. Accordingly, many estimates of age determined by radiometric means are often called into question for verification by more precise measurements and techniques.

A source of error that becomes increasingly important as the length of time increases is the alteration of composition by the impingement of cosmic rays on Earth. Cosmic rays consist of a variety of high-energy particles, including protons, electrons, neutrons, x rays, and α particles. The interaction of these, and especially of neutrons, with terrestrial nuclides brings about transmutations at a rate that increases the uncertainty of authentic nuclide counts as the time span increases. While this effect is most pronounced for ^{14}C or radiocarbon dating, it also plays a role in increasing the uncertainty or error associated with other radionuclide transmutations.

HISTORICAL ESTIMATES

Before the discovery of radioactivity in 1895 and before the more recent realization of precise measurement techniques by mass spectrometry, people attempted to determine the age of Earth by different means. Biblical scholars and Creation fundamentalists, even in the present day, relied on the tabulation of generations in the Bible to arrive at an absolute date for the formation of Earth. By their calculation, the planet could not be any more than about six thousand years old (having been formed, they figure, at 9:00 a.m. on October 23 in the year 4004 b.c.e.).

More scientific minds, however, have looked at natural phenomena, including erosion rates of soil and rock formations, the rate of deposit of sediments, and calculation of the time that would be required to achieve the level of salinity observed in the world's oceans. Such methods lack the clockwork mechanism that was provided by the exponential decay rate of radionuclide transmutations, and universally proved unsatisfactory as methods to determine the age of Earth.

Richard M. Renneboog

FURTHER READING

Cotner, Sehoya, and Randy Moore. *Arguing for Evolution: An Encyclopedia for Understanding Science.* Santa Barbara, Calif.: Greenwood, 2011. The second chapter of this well-researched book presents a thorough and detailed discussion of the many historical estimates of Earth's age that have been made during the past two thousand years.

Dalrymple, G. Brent. *Ancient Earth, Ancient Skies: The Age of Earth and Its Cosmic Surroundings.* Stanford, Calif.: Stanford University Press, 2004. The fourth chapter, "How Radiometric Dating Works," is devoted to the principles of radiometric dating and also provides a list of isotopic materials used for dating rocks and minerals.

De Pater, Imke, and Jack J. Lissauer. *Planetary Sciences.* 2d ed. New York: Cambridge University Press, 2010. This book discusses radiometric dating in the context of Earth and planetary sciences. It also provides detailed breakdown sequences for some important nuclide pairs.

Leddra, Michael. *Time Matters: Geology's Legacy to Scientific Thought.* Hoboken, N.J.: Wiley-Blackwell, 2010. This book contains a thorough examination of the methods and meaning of geological time measurement.

Van Kranendonk, Martin J., R. Hugh Smithies, and Vickie C. Bennett, eds. *Earth's Oldest Rocks.* Boston: Elsevier, 2007. This book is a specialist treatise on the age determination and properties of the oldest known rock structures of Earth.

See also: Earth's Core; Earth's Differentiation; Earth's Interior Structure; Earth's Mantle; Earth's Oldest Rocks; Lithospheric Plates; Radioactive Decay; Radiocarbon Dating; Relative Dating of Strata; Uranium-Thorium-Lead Dating.

EARTH'S CORE

Earth's core cannot be studied directly because of its high heat and pressure and because of its unreachable location. Research into Earth's formation and structure requires seismology and other fields of study to make appropriate and logical inferences. As seismological technology improves, scientists' understanding of the core will increase. Earth's core is a critical element in understanding of Earth's past, present, and future, especially because of the core's effect on Earth's magnetic field.

PRINCIPAL TERMS

- **accretion:** the growth of a planet through gravitational attraction of matter in space; the process by which Earth and other large planetary bodies formed
- **chondrite meteorite:** the most common type of meteorite to fall to Earth; much core research relies on the assumption that Earth is like a chondrite meteorite
- **convection:** transfer of heat via the movement of molecules in liquids and gases
- **Coriolis effect:** an inertial force that makes objects in motion deflect to a side; caused by Earth's rotation
- **dynamo theory:** explains the creation of long-lived magnetic fields by a fluid that convects, conducts electricity, and rotates
- **iron catastrophe:** an early Earth event in which iron and other dense elements sank to the center of the planet, leading to the formation of Earth's core
- **magnetic field:** an invisible area produced by an electric field that can exert a magnetic force on certain things in or around it
- **pain-out model:** the stage of planetary differentiation when the core is formed during the "raining out" of iron from the molten NiFe (nickel and iron) from silicate emulsion toward the center of the planet
- **planetary differentiation:** a process in which a planetary body separates into distinct layers because of the different chemical and physical characteristics of the substances that make up its composition
- **radioactive decay:** the spontaneous emission of ionizing particles from an unstable atom; the rate of radioactive decay is described as the "half-life" of a given substance
- **seismic waves:** energy waves that travel through a planetary body because of processes, such as earthquakes, which produce low-frequency acoustic energy
- **seismology:** the study of the propagation of waves through a planet; mostly focused on earthquake research

CHARACTERISTICS OF EARTH'S CORE

Whether the structure of Earth is divided in rheological or chemical terms, the center is defined as a solid inner core surrounded by a liquid outer core. The existence of the inner core as a separate entity from the outer core was discovered in 1936 by Danish seismologist Inge Lehmann.

Deeper than the crust and the mantle, the outer core begins at a depth of 2,890 kilometers (km), or 1,790 miles (mi), below Earth's surface and has a thickness of roughly 2,266 km (1,408 mi). Earth's inner core begins at a depth of 5,150 km (3,160 mi) below the surface, or about the length of 50,000 football fields. With a radius of about 1,226 km (760 mi), the inner core is roughly 70 percent the size of Earth's moon. Together, the inner and outer core have a diameter about twice that of the moon.

Most substances can exist in different physical states of matter depending on the pressure and temperature to which they are subjected. Earth's core is a perfect example. Whereas both the inner and the outer core are composed mainly of iron and nickel, pressure and temperature differences contribute to their different states: the inner core is solid; the outer core, however, is molten liquid.

Iron is thought to account for about 80 percent of the inner core's composition, with nickel making up most of the remainder. Research suggests the presence of several siderophiles ("iron-loving" transition elements such as gold and platinum) in the innermost part of the core. These dense metals tend to bond readily with solid or molten iron. Trace amounts of lighter elements also are assumed to be present in both the inner and outer cores.

Temperature increases with distance from the surface; the core's temperature ranges from about 4,673 kelvins (K), or 4,400 degrees Celsius (C), at the outer edge of the outer core to about 6,373 K (6,100

Pohutu (big splash in Maori) Geyser is the largest of seven active geysers on Geyser Flat, the main part of the thermal area of Whakarewarewa Thermal Village & Reserve, located in New Zealand's North Island. It erupts 10 to 25 times a day up to 30 m high. This awesome display of energy demonstrates the powerful forces within earth's core as superheated steam escapes from a vast chamber of boiling water through narrow vents in roaring towers of spray and steam. (Time Life Pictures/Getty Images)

degrees C) at the center of Earth, approximately the same as the surface of the sun. The extreme heat at the core results from three main sources: first, the residual heat from the processes that formed the earth 4.5 billion years ago; second, frictional heat created as denser elements such as iron sank toward the center of the earth during a major Earth-formation event known as the iron catastrophe; and third, heat released as a by-product of the decay of radioactive materials. The third source continues perpetually and likely accounts for the vast majority of the core's heat. The first two sources occurred billions of years ago but still contribute because Earth loses heat extremely slowly.

Since 1996, scientists have speculated that the inner core rotates faster than the rest of the earth. Subsequent research confirms this, but adds that the rotation is slower than initially thought. It is believed now that the inner core spins at a rate between 0.1 and 1 degree more than the rest of the earth every one million years, much slower than earlier estimates of 1 degree each year.

FORMATION OF EARTH'S CORE

The solar system was born from the big bang 13.7 billion years ago. At first, the solar system was a solar nebula, a rotating cloud of helium, hydrogen, and other materials. An event 4.6 billion years ago, possibly a supernova, triggered the solar nebula to contract and flatten into a disc-like shape as it started rotating faster. Eventually the sun formed in the center of this system, partly as a result of nuclear fusion of hydrogen and helium. The remaining debris collided to form protoplanets, including one that would become Earth. These protoplanets continued to grow by the process of accretion. When the center of the earth reached a high enough temperature, it was time for the interior of the planet to become more organized.

Earth's inner layers were created through a process called planetary differentiation, whereby the different substances that make up a planetary body separate into distinct layers because of different chemical and physical characteristics, particularly density. The core's formation took tens of thousands of years, which is considered fairly quick in planetary science. Modern isotopic studies suggest that this completion of the core layer occurred within 30 million years of Earth's beginning.

Before the core's formation, Earth comprises a mostly uniform blend of silicates and NiFe and other materials. Residual heat from Earth's formation and heat from the decay of radioactive materials in the earth increased Earth's temperature to a point where the silicates began to melt. NiFe, which has a higher melting point and density than silicates, started sinking to Earth's center because of gravity. Even as the temperature became high enough to melt the NiFe, the material remained distinct from the silicates and "rained out" of the emulsion to continue sinking, ultimately forming Earth's core. This is referred to as the "rain-out model" when applied to Earth and other planetary bodies. In the earth's planetary history, the formation of the core through the rain-out model is part of an event known as the "iron catastrophe," which ultimately explains the core's effect on Earth's magnetic field.

SEISMOLOGICAL RESEARCH OF EARTH'S CORE

Earth's core is too hot and too deep to be studied by humans directly, and even sending down equipment to make remote observations has proven to be impossible: Scientists have been able to drill 12 km (7.5 mi), only 0.2 percent toward the center. A Russian team of researchers started a drilling project in 1962, planning and building for eight years before breaking ground, and had to halt the project twenty-four years later as the drill reached the 12-km mark. The temperature was hotter than anticipated (180 degrees C; 356 degrees Fahrenheit [F]) instead of the originally estimated 100 degrees C (212 degrees F). The rocks in the crust began to behave like heated plastic, melting and blocking the progress of the drill.

Much of what is known about Earth's core comes from seismological research, although other approaches have included pressure and temperature research with crystalline solids, measurements of Earth's gravitational field, and the assumption that Earth's core is like a chondrite meteorite. Seismology, the study of the propagation of waves through a planetary body (particularly waves caused by earthquakes), allows scientists to peer into the virtual black box of the earth to gather data, and gain insight into its interior structure.

The fluidity of the outer core and the boundary between the mantle and core (2,890 km or 1,790 mi below the surface) were both discovered by observation of how seismic waves, particularly P ("pressure" and "primary") waves and S ("shear" and "secondary") waves, travel during earthquakes. P waves and S waves are types of body waves, which are a type of seismic wave that travels through substances. P waves involve direct longitudinal motion in the direction of propagation, whereas S waves behave transversely, moving perpendicular to the direction of propagation. P waves can travel through solids and fluids, but S waves can travel only through solids; the viscosity of fluid is too low to support the shear stresses of S waves. Scientists have observed that the waves behave differently in the core and mantle: P waves travel more slowly in the outer core than in the mantle, and S waves do not exist in the core at all. These observations helped clarify the location of the core-mantle boundary and helped to show that the outer core is liquid.

Seismology, the study of seismic waves, also accounts for the discovery of the two distinct parts of the core: the solid inner portion and the liquid outer portion. This discovery was made by Inge Lehmann in 1936. Lehmann had analyzed the data from a large earthquake, a magnitude 7.8 quake that hit New Zealand in 1929 and killed fifteen people. Lehmann noticed that the data showed the quake had strange P-wave behavior: Some waves were detected at unexpected observation points on Earth's surface, as if they had been deflected within the earth's interior. This would not have happened had the core been homogeneously liquid throughout, so Lehmann hypothesized about the existence of a solid inner core. (Lehmann later found another seismic discontinuity closer to Earth's crust within the mantle; this discontinuity is generally called the Lehmann discontinuity.)

Lehmann's discovery of the inner/outer core boundary is similar to earlier work with P waves and S waves done by Croatian seismologist Andrija Mohorovičić, who discovered the boundary between Earth's crust and Earth's mantle. The boundary is now called the Mohorovičić discontinuity.

Seismology research continues to be one of the most useful studies of Earth's interior. Seismological instruments are becoming more and more sensitive, leading to a greater body of data.

OTHER RESEARCH METHODS USED TO STUDY EARTH'S CORE

Some properties of the earth's core are determined mainly by inference. It is known that the core is highly dense, for example, because of the discrepancy between the earth's average density (5,515 kg per cubic meter, or 344 pounds per cubic foot) and the average density of materials at the earth's surface (3,000 kg per cubic meter or 187 pounds per cubic foot), which can be observed directly. (The earth's average density is known by calculating the earth's mass based on the force of its gravitational pull and then estimating its volume.)

Geophysicists also infer much about the core based on the assumption that Earth is like an ordinary chondrite meteorite, an assumption initially put forth by American geophysicist Francis Birch in

1940. Chondrites are the most common type of meteorite to land on Earth, and about 90 percent of those are classified as "ordinary." Ordinary chondrites are composed mainly of NiFe alloys and iron sulfide (FeS), also called troilite. Chondrite meteorites are born from asteroids that were too small to melt or undergo planetary differentiation during the formation of the solar system.

Modern laboratory experimentation provides further insight into the core's structure and properties. In August 2011, for example, Kei Hirose, a scientist at the Tokyo Institute of Technology, became the first to examine what conditions in the center of the core must be like by re-creating these conditions in his laboratory. After using a diamond-tipped vise to subject a NiFe alloy to three million times atmospheric pressure and 4,500 degrees C (8,132 degrees F), Hirose found that the alloy structure realigned to form a "forest" of large crystals. Computer models in the mid-1990's had suggested the presence of one giant iron crystal in the earth's core, but subsequent research leaned toward an alignment of smaller crystals, even going so far as to pinpoint the alignment in a way that allows north-south seismic waves to travel faster through the earth than do east-west waves. Hirose's research adds support to the hypothesis of many smaller crystals.

Effect of the Core on Earth's Magnetic Field

Earth's magnetic field protects the planet from the effects of solar wind, a stream of charged particles that escapes from the sun's surface. While most solar wind particles are deflected outward by the magnetic field, some are trapped by the Van Allen radiation belt, a region held in place by the magnetic field. Some particles even make it through to the earth's upper atmosphere, causing geomagnetic storms and aurorae (such as the northern lights).

Earth's magnetic field moves over time, even completely reversing direction every few hundred thousand years. The geographic North Pole and geographic South Pole are, therefore, not the same as the magnetic North Pole and magnetic South Pole, respectively.

The outer core is largely responsible for maintaining Earth's magnetic field through a phenomenon described by the dynamo theory, which explains the existence of long-lived magnetic fields that do not collapse from ohmic decay over time. According to the dynamo theory, these fields can be created and maintained by a fluid that meets three requirements: It rotates (providing kinetic energy), convection is present within it (caused by an internal source of heat), and it can conduct electricity. The liquid outer core of the earth fulfills all of these requirements.

The Coriolis effect (an inertial force powered by Earth's rotation) produces the necessary rotation within the core fluid, the liquid composition conducts electricity, and a variety of sources provide the energy for convection, including gravitational energy still released in the aftermath of core formation and radioactive decay of trace elements found in the core. The particulars of convection within the outer core (imagine a turbulently churning molten sea of metal) is a widely researched topic. The strength of the magnetic field within the outer core is estimated to be about fifty times that of the earth's magnetic field at the surface.

The solid inner core is too hot to maintain a magnetic field on its own. Under high heat, the orientation of iron's molecules becomes randomized, which prevents it from exerting a magnetic force. Additionally, the inner core cannot meet the fluidity requirements set forth by the dynamo theory (because it is solid), but the inner core is thought to somehow help the outer core stabilize Earth's magnetic field.

Continuing Research About Earth's Core

Modern research continues to examine Earth's core to try to confirm past observations and to get a more detailed picture of its formation, composition, and effects on Earth's magnetic field. In 2010, for example, a team at Université Joseph Fourier in Grenoble, France, used models of fluid flow in the core to solve the discrepancies among various past inferences of the strength of the magnetic field within the outer core. Meanwhile, collaboration between Université Paris Diderot (in France) and the Lawrence Livermore National Laboratory (in Livermore, California) confirmed the percentages of nickel and silicon assumed to be present in the inner core (about 5 percent nickel and 2 percent silicon). Earlier estimates were inferred by seismological research; the contemporary team took high-pressure sound velocity measurements from a crystalline iron-nickel-silicon alloy to arrive at the same results.

Outer core convection is a popular topic of core-related research. In February 2010, for example, a team of researchers from Japan examined the process of convection in the outer core, part of what maintains the earth's magnetic field, and found evidence of zonal flows: The convection system in the core actually contains two types of motion (sheet-like radial plumes and the newly discovered zonal flows). Also in 2010, a Chinese team of computer scientists developed an improved "solver" for modeling the core's convection. In September 2011, a research team in California examined the thermal and chemical buoyancy forces driving convection in the outer core; the team created a buoyancy model to be used with geodynamo models.

An actual "journey to the center of the earth" is still impossible by all modern standards. However, scientists can continue to improve upon understandings of Earth's core with the development of better computer models and other research strategies.

Rachel Leah Blumenthal

FURTHER READING

Dickey, John S. *On the Rocks: Earth Science for Everyone.* New York: Wiley, 1996. From stardust to Earth's formation to planetary neighbors, this book is ideal for Earth science novices. Easy to understand but comprehensive, *On the Rocks* is an enjoyable overview of Earth's history.

McBride, Neil, and Iain Gilmour. *An Introduction to the Solar System.* New York: Cambridge University Press, 2004. A textbook for introductory college courses, this work is an excellent resource for exploring the formation of the earth and the other planets of the solar system. Includes summaries and a glossary.

Sorokhtin, O. G., et al. *Evolution of Earth and Its Climate: Birth, Life, and Death of Earth.* Boston: Elsevier, 2011. This text provides ample information on the earth's formation, composition, and magnetic activity. It also explores the formation of the moon and its influence on Earth.

Tarbuck, Edward J., and Frederick K. Lutgens. *Earth Science.* 13th ed. Upper Saddle River, N.J.: Pearson Education, 2012. Originally published in 1976, *Earth Science* is a time-proven and updated introductory textbook geared to undergraduates, including those without a science background. Covers geology and astronomy and other Earth science topics.

Williams, Linda D. *Earth Science Demystified.* New York: McGraw-Hill, 2004. A quick and easy self-teaching guide aimed at readers without formal science training, *Earth Science Demystified* covers the basic range of Earth science topics and includes summaries.

Wu, Chun-Chieh. *Solid Earth.* Advances in Geosciences, vol. 26. London: World Scientific, 2011. This volume highlights the results of research papers in seismology, planetary exploration, the solar system, and other topics relevant to Earth science.

See also: Deep-Focus Earthquakes; Discontinuities; Earthquakes; Earth's Age; Earth's Differentiation; Earth's Interior Structure; Earth's Lithosphere; Earth's Magnetic Field; Earth's Mantle; Earth's Oldest Rocks; Geodynamics; Heat Sources and Heat Flow; Lithospheric Plates; Magnetic Reversals; Mantle Dynamics and Convection; Metamorphism and Crustal Thickening; Plate Tectonics; Rock Magnetism; Seismic Wave Studies; Stress and Strain; Volcanism.

EARTH'S DIFFERENTIATION

Earth formed from the debris of the solar nebula and the differentiation of layers in the earth's interior was initially driven by heat. This continuing process has had many parallels in other planetary and celestial bodies throughout the universe, particularly through the activities of tectonic plates.

PRINCIPAL TERMS

- **accretion:** a process by which planetary bodies grow by gravitationally pulling in matter
- **chemical stratification:** differentiation of layers of a planetary body based on chemical affinities and characteristics rather than physical properties
- **fractional crystallization:** the process of minerals precipitating from a molten substance
- **giant impact hypothesis:** the widely accepted hypothesis about the moon's formation, which is believed to have occurred when a protoplanet collided with the earth and ejected parts of the earth's mantle and crust, which entered orbit and formed the moon by accretion
- **gravitational compression:** the process by which an object becomes smaller and denser because of the force of gravity acting on it; it produces heat in the center of a planetary body
- **iron catastrophe:** the event leading up to the formation of the earth's iron core, when iron precipitated out of the early earth's molten silicate mixture and sank to the center of the planet
- **planetary differentiation:** the separation of interior layers of a planetary body because of chemical and physical differences
- **protoplanet:** the early stage of a planet, often referred to as a planetary embryo; it often collides with other protoplanets to form planets
- **radioactive decay:** spontaneous emission of ionizing radiation from the nucleus of an unstable atom
- **tectonic plate:** a slowly moving chunk of the earth's uppermost layer, the lithosphere

EARTH'S FORMATION

Before the earth could differentiate, it had to form. The most widely accepted timeline involves the formation of the universe by the big bang, which occurred roughly 13.7 billion years ago. More than 9 billion years later, the solar system began to form from the solar nebula, a giant rotating cloud of big bang debris. Protoplanets and asteroids began to form from that debris by a process called accretion: the adding on of mass by pulling that mass in gravitationally. One of these protoplanets became Earth, whose temperature eventually increased to a critical point so that differentiation of interior layers could occur.

The big bang theory was first proposed in 1927 by Georges Lemaître, a Belgian astronomer, physicist, and priest, who called his theory a "hypothesis of the primeval atom." (The term "big bang" was coined during a 1949 radio broadcast by Sir Fred Hoyle, an English mathematician and astronomer who supported an opposing theory of the universe's formation.)

According to the big bang theory, the universe was in an extremely hot, dense state 13.7 billion years ago and began rapidly expanding, resulting in cooling and in continuous expansion, even into the present day. Protons, neutrons, and electrons formed early on in this expansion, followed by nuclei and atoms. With these pieces of matter in existence, galaxies and stars began to form, some beginning only 480 million years after the big bang.

According to the nebular hypothesis, first described in 1734 by Swedish scientist Emanuel Swedenborg, the solar system and its sun formed from the gravitational collapse of a "tiny" portion of a giant molecular cloud (GMC) 8 to 9 billion years after the big bang. ("Tiny," in this sense, means having a diameter of about 3.25 light-years.) GMCs are huge, dense, and gravitationally unstable collections of molecular hydrogen molecules; these clouds collapse to form stars. The mass of this particular GMC collided to form the presolar nebula, a rotating disc-shaped cloud of debris and gas, at the center of which was the sun in its earliest stages of formation. The bulk of the nebula's composition was hydrogen and helium, along with small amounts of lithium and other heavier elements.

As the nebula spun, it began compressing and spinning faster. The matter in the center, even more compressed than it had been earlier, collided more frequently, raising the temperature and bringing the proto-sun to the state of a T Tauri star, a young star not yet in the main-sequence, hydrogen-burning stage of life. During the next 50 million years, the sun's core temperature and pressure became high enough to launch it into its main sequence, propelled by hydrogen-to-helium nuclear fusion. The leftover debris from the sun's formation continued rotating as a solar nebula, and asteroids and protoplanets began to form within it, growing by accretion. One of these—a rocky, terrestrial protoplanet within the inner part of the solar system—was Earth.

Initially, the young earth's interior was a mostly heterogeneous mixture of silicates, nickel, and iron. When the center of the earth reached a critical temperature, one at which the silicates could melt, differentiation of internal layers, especially the core, began. Earth was under a period of bombardment by meteorites and other protoplanets at this time, and according to the giant impact hypothesis, one particularly large impact blew the barely formed mantle and crust from Earth, ultimately forming the moon. The forming core escaped unscathed. Most of the earth's existing continental crust is probably only 2 billion years old.

Sources of Heat in Earth's Interior

Heat provided the impetus for the differentiation of the earth's layers: As temperatures within the earth rose high enough to melt the substances within the early earth's interior, the molten substances were able to separate out because of density and other differences. The necessary heat came from multiple sources, including the decay of radioactive elements contained within the planet, gravitational compression as the earth continued to become more compact, and the ongoing impacts of meteorites.

Radioactive decay refers to the spontaneous emission of ionizing particles from unstable atoms, atoms whose nuclei have excess energy to release. The process likely contributed the most heat, leading up to the differentiation of the earth's layers. (Radioactive decay continues to occur within the earth.) At the time of the earth's differentiation, radioactive decay of potassium-40, uranium, and thorium drove this heating, and the decay of these elements continues to help drive convection in the molten outer core. Despite the high density of uranium and thorium, both elements are found abundantly in the crust rather than in the core.

Although most substances migrated based on density during the differentiation of the earth's layers, some migrated based on chemical affinity; uranium and thorium interact more readily with the silicates in the earth's upper layers than with the densely packed iron in the core. This process is called chemical stratification.

Another contributor to the earth's internal heat was gravitational compression, a process by which gravity forced the earth to become smaller and denser as it cooled, releasing pressure. Mathematically described by the Kelvin-Helmholtz mechanism, this process, which also occurs in other planets and stars, results in an increased temperature in the core. The gravitational compression of a planet ends when an opposing pressure gradient balances that compression. This same equilibrium is evident in the earth's atmospheric layers: Pressure keeps them from collapsing onto the earth and gravity holds them down enough so that they do not disappear into space.

The third factor was external: the bombardment of Earth by meteors, which added heat locally through shock waves and impact melts. Upon impact, meteorites send physical energy through the earth in the form of propagating shock waves, causing a rapid increase in temperature and pressure at and around the point of impact. This resulted in the temporary melting of rock (impact melts).

With all three processes in place, especially radioactive decay, the earth's temperature increased past the melting point of the silicates and eventually past the melting point of iron (1,538 degrees Celsius [C] or 2,800 degrees Fahrenheit), allowing the iron catastrophe to begin; this triggered the formation of the earth's core and ultimately the differentiation of all of earth's interior layers.

Differentiation of Earth's Layers

Core formation began roughly 10 million years after Earth started to form; this time period is within the Hadean eon, which spans from Earth's formation approximately 4.6 billion years ago until an arbitrary point generally pinned at 3.8 billion years ago. While a primitive mantle and core likely differentiated around the same time, the giant impact hypothesis suggests that 4.52 billion years ago, a nearly Mars-sized protoplanet smashed into Earth, ejecting much of the crust and mantle (but not the core) from the planet and into orbit, ultimately forming the moon.

Over the next 150 million years, a new mantle and crust formed on Earth.

Brought about by heat, particularly radioactive decay, the core's formation is often referred to as the iron catastrophe because iron was involved significantly in the process. As the temperature rose, the fairly well-mixed substances within the earth's interior—the silicates, the nickel-iron mixture (NiFe), and other elements—began to melt and separate from one another, mostly on the basis of density differences. In a process referred to as the rain-out model, the heavy NiFe rained out from the molten emulsion of silicates and fell toward the center of the earth, where it accumulated into what became known as the core.

Because temperature and pressure increase toward the center of the earth, further differentiation occurs: a solid inner core and a molten liquid outer core. The core spans from the earth's center to about 2,890 kilometers (1,790 miles) below the earth's surface, giving it a diameter about twice as large as the diameter of the moon. The temperature at the center of the earth is an estimated 6,373 kelvins (6,100 degrees C), about the same as at the surface of the sun.

However, not all elements differentiated by density. As in the case of radioactive decay, chemical stratification was most significant in some cases, such as with heavy elements like uranium. Because of its high affinity for silicates and its difficulty fitting within iron's dense structure, uranium rose toward the surface rather than sinking to the core.

With the differentiation of the earth's core also came the generation of the earth's magnetic field, which is sustained by the outer core. This is explained by the dynamo theory, which describes a magnetic field generated by a rotating, convecting, electricity-conducting liquid. The molten outer core's rotation is provided by the earth's Coriolis effect. Convection is a result of internal heat sources and it conducts electricity. The magnetic field is critical for many reasons, particularly protecting the earth's atmosphere from solar wind.

The mantle and the crust continue to differentiate, with old material being destroyed and new material being created at tectonic plate boundaries and faults. The mantle, mainly composed of silicates, magnesium, and some iron that did not end up in the core, is considered a solid layer, although its upper portion does move very slowly over time. Making up 84 percent of the volume of the earth, the upper mantle provides a moving platform for the crust's tectonic plates to ride upon.

The crust, which accounts for about 1 percent of the earth's volume, is a rocky, solid layer. There are two types of crust: continental crust and oceanic crust. Continental crust is thick, with a relatively low density, and consists of rocks such as granite. Oceanic crust is thinner, is higher in density, and consists mainly of rocks such as basalt. Continental crust has an average age of about 2 billion years, with some samples dating back as far as 4.3 billion years; most existing oceanic crust is 200 million years old at most.

PARALLELS IN OTHER PLANETARY BODIES

In basic terms, planetary differentiation merely refers to the separation of layers in a planetary body caused by physical and chemical properties; density is typically a main factor in the separation. Planetary differentiation generally results in a distinct core and mantle, and sometimes in a crust as well. It is not necessarily a process with a definite end; the earth, for example, continues to differentiate as crust is destroyed and created through tectonic activity.

The formation and differentiation processes of Earth are not unique; indeed, Earth is just one planetary body among many that exists under similar external factors, so there are many examples of parallel formation and differentiation processes that have occurred or are still occurring throughout the universe. Earth's own moon, dwarf and regular planets, and asteroids are just a few examples of planetary and celestial bodies that share early similarities with Earth.

The moon, which is Earth's only natural satellite, is one of several hundred celestial bodies in the solar system that orbit a larger body. The giant impact hypothesis suggests that the formation of the moon occurred around 4.53 billion years ago, in a process that parallels Earth's formation by accretion from the debris of the solar nebula.

The moon's differentiated layers are also similar to those of Earth: a solid core rich in iron, a molten outer core also mainly composed of iron, a mantle, and a crust. Scientists believe that these layers were formed by a process called fractional crystallization, which also occurs now in the earth's upper layers as magma cools. Fractional crystallization refers to minerals precipitating from a melted substance, much like the formation of the earth's core during the iron catastrophe. When the moon accreted, the energy released by the impact probably caused the early moon to be covered by partially melted rock (a type of magma ocean), from which minerals precipitated, leading to the differentiation of internal layers.

The differentiation of one of the solar system's largest asteroids, 4 Vesta, also bears some similarities to Earth's differentiation. Asteroid 4 Vesta accounts for about 9 percent of the total mass of the asteroid belt, an asteroid-heavy region between Mars and Jupiter. Much can be inferred about 4 Vesta's differentiated interior because some of its material has fallen to Earth in the form of meteorites from an impact that hit 4 Vesta probably within the last 1 billion years. These meteorites (called Howardite-Eucrite-Diogenite [HED] meteorites) provide ample evidence of igneous processes, which involve the cooling of melted substances like magma.

From the HED meteorites, scientists have determined a rough timeline of 4 Vesta's differentiation. About 2 million years after forming by accretion, the asteroid likely experienced a process like Earth's iron catastrophe: Radioactive decay of an aluminum isotope provided enough heat to partially or fully melt the forming body, leading to the differentiation of a heavy metal core and a molten convecting mantle, most of which slowly crystallized as the asteroid cooled. The remaining molten material extruded to the surface, either by flowing or erupting, ultimately cooling to form a crust.

EARTH'S DIFFERENTIATION CONTINUES

Earth's differentiation did not end billions of years ago. It continues in the present, as parts of the lithosphere (crust and upper mantle) continually break down and rebuild. The lithosphere contains about seven major "chunks" called tectonic plates, and many minor chunks. These plates move very slowly, about one centimeter each year, and interact at transform faults, convergent boundaries, and divergent boundaries. This movement causes the formation of mountains and ocean trenches and also causes volcanic and seismic activity. These processes make the earth a living, ever-differentiating planetary body.

To understand plate tectonics, one must understand the three main types of boundaries that occur between plates. Convergent boundaries (also known as collision boundaries or destructive plate boundaries) are areas in which tectonic plates move toward each other, resulting in either a direct collision or subduction (when one plate slides under another). Because of density differences, plates made of oceanic crust tend to slide under continental crust plates, whereas a convergence of two continental crust plates often, but not always, results in a collision. Any of these actions is accompanied by a great deal of pressure, friction, and melting, resulting in earthquakes, volcanic activity, and the formation of mountain ranges.

Tectonic plates move from each other at divergent boundaries (also known as extensional or constructive boundaries). Molten magma flows up from the convecting mantle to fill the space. When boundaries of this type exist between two plates of continental crust, rift valleys form. Between oceanic plates, one can find the mid-ocean ridges (essentially underwater mountains) and volcanic islands.

Transform faults, the third boundary type, are also known as conservative plate boundaries. Unlike the other boundary types, transform faults do not involve the creation or destruction of crust. Instead, transform faults relieve stress caused by other boundaries. Typically present in a zigzag shape, they are often found within mid-ocean ridges and between continents.

To interact at these boundaries, tectonic plates ride on the slow-moving, denser asthenosphere, driven by a variety of forces, particularly convection within the mantle. These convection currents slowly bring heat from the interior of the earth, a process that exerts frictional and gravitational forces on tectonic plates. Other gravitational forces also are involved. At spreading ridges, for example, oceanic lithosphere is created in a way that the new material adds on to the ridge side of the plate. As that side becomes thicker than the other side of the plate, it sinks to create a slight lateral incline, resulting in gravity-driven sliding. The moon and sun also create a tidal drag that influences plate movement, and the Coriolis effect exerts some influence, too. Finally, the small wobbles that occur in Earth's imperfect rotation affect the movement of tectonic plates.

Some other terrestrial planets show evidence of plate tectonics as well. In general, it is expected that dry planets and celestial bodies larger than Earth have tectonic plate activity, while bodies around Earth's size might have activity if water is present.

Tectonic activity is closely tied to other processes on Earth, particularly sea level rise.

Rachel Leah Blumenthal

FURTHER READING

Blakey, Ronald C., Wolfgang Frisch, and Martin Meschede. *Plate Tectonics: Continental Drift and Mountain Building.* New York: Springer, 2011. This book provides an introduction to plate tectonics, covering the earth's early history to the present day. Topics include subduction zones, mid-ocean ridges, and the formation of mountains.

Chilingar, George V., et al. *Evolution of Earth and Its Climate: Birth, Life, and Death of Earth.* Boston: Elsevier, 2011. This text provides ample information on the earth's formation, composition, and the differentiation of its layers. It also explores the formation of the moon and its influence on the earth.

Dickey, John S. *On the Rocks: Earth Science for Everyone.* New York: Wiley, 1996. From stardust to Earth's formation to its planetary neighbors, this book is ideal for earth science novices. Easy to understand but comprehensive, *On the Rocks* is an enjoyable overview of the history of the earth.

Lutgens, Frederick K., and Edward J. Tarbuck. *Earth Science.* 13th ed. Upper Saddle River, N.J.: Prentice Hall/Pearson, 2012. Originally published in 1976, *Earth Science* is an updated introductory textbook geared at undergraduates, including those without a science background. It covers geology, astronomy, and other earth science topics.

Monroe, James S., and Reed Wicander. *The Changing Earth: Exploring Geology and Evolution.* 5th ed. Belmont, Calif.: Brooks/Cole, Cengage Learning, 2009. This textbook offers a solid introduction to geology in an easily readable format, supplemented with many relevant photographs, diagrams, and real-world examples. This edition has been condensed from previous editions.

Wu, Chun-Chieh. *Solid Earth.* Vol. 26 in *Advances in Geosciences.* London: World Scientific, 2011. This book features research papers on the fields of seismology, planetary exploration, the solar system, and other topics relevant to earth science.

See also: Asteroid Impact Craters; Earthquakes; Earth's Age; Earth's Core; Earth's Interior Structure; Earth's Lithosphere; Earth's Magnetic Field; Earth's Mantle; Geodynamics; Heat Sources and Heat Flow; Lithospheric Plates; Plate Motions; Plate Tectonics; Rare Earth Hypothesis; Seismic Wave Studies; Stress and Strain; Subduction and Orogeny; Tectonic Plate Margins; Volcanism.

EARTH'S INTERIOR STRUCTURE

To begin to understand planet Earth, one must understand what is below the earth's surface. Although much knowledge of the earth's interior structure is based on inference, seismological research has provided a detailed picture of the earth's layers and their physical and chemical characteristics.

PRINCIPAL TERMS

- **accretion:** the growth of a planet through gravitational attraction of matter in space; the process by which Earth and other large planetary bodies formed
- **asthenosphere:** the second layer (from the top) of the earth's interior; corresponds with part of the upper mantle (100 to 200 kilometers [62-125 miles] below the earth's surface)
- **differentiation:** the process of a planetary body's substances separating into distinct layers caused by different chemical and physical characteristics
- **discontinuity:** an area within the earth's surface in which seismic wave activity abruptly changes; in some cases, discontinuity corresponds with a border between two of the earth's layers
- **lithosphere:** the uppermost layer of the earth's interior; a hard, rigid, rocky layer that includes the crust and the very top of the mantle
- **mesosphere:** the third layer (from the top) of the earth's interior; a dense, rigid layer that corresponds with most of the mantle
- **rheology:** the study of the flow of matter; focuses on liquids and soft solids that behave like plastic
- **seismic wave:** a moving, energy-transferring disturbance that occurs because of an event, such as an earthquake, that releases low-frequency acoustic energy
- **silicate:** a compound with a negatively charged silicon ion; a main component of the earth's crust
- **wave velocity:** the rate of propagation of a wave; one factor in seismological research that can provide useful information about the structure of the earth's interior, which includes regions in which wave velocity markedly increases or decreases

LAYERS OF THE EARTH'S INTERIOR: CHEMICAL DIVISIONS

The earth's interior is made up of three chemically distinct layers: the crust, the mantle, and the core. The mantle is further divided into the upper mantle and the lower mantle, and the core is divided into the outer core and the inner core.

The earth's solid crust spans from the surface to a depth of about 35 kilometers (22 miles), although the depth varies with location, sometimes reaching as far as 70 kilometers (43 miles). In general, the crust is thicker at continents and thinner at the ocean floor. Also, the continental crust is different from the oceanic crust in terms of composition: Continental crust is made of low-density rocks such as granite, whereas oceanic crust contains mainly high-density rocks such as basalt, a volcanic rock.

The crust is also described as having an upper layer called sial, which corresponds with the continental crust, and a lower level called sima, which corresponds with the oceanic crust. Sial is a reference to two of the main types of minerals found in the rocks of the continental crust: silicates and aluminum. Sima refers to the magnesium silicates in the oceanic crust.

The mantle stretches from a depth of about 35 to 2,890 kilometers (22 to 1,790 miles) below the earth's surface and is composed of silicates containing large amounts of magnesium and iron. Although the mantle is considered to be a solid layer, the layer's high temperature allows it to flow slowly over time (less so in the lower mantle, which is under more pressure). The mantle makes up about 84 percent of the earth's volume, and its plastic upper layer provides a slow-moving "sea" of rocks on which the earth's tectonic plates ride.

The lower mantle is more rigid and filled with denser materials, because of increased pressure and temperature. The lowest 200 kilometers (124 miles) of the mantle is referred to as the D" zone, an area in which the velocity of seismic waves decreases abruptly. This zone is also called the Gutenberg discontinuity, and it leads into the boundary between the mantle and the core.

The center of the earth is known as the core, and it is made up of two sections: a molten outer core and a solid inner core. The outer core starts 2,890 kilometers (1,790 miles) below the surface of the earth and is roughly 2,266 kilometers (1,408 miles) deep. Then, at about 5,150 kilometers (3,160 miles) below the surface, the inner core begins. The core's diameter is approximately twice the diameter of the moon.

Both sections of the core are composed mainly of iron and nickel, along with trace amounts of lighter elements. Substances behave differently under different amounts of pressure, which accounts for the inner core's solidity and the outer core's liquidity. The center of the core reaches a temperature of about 6,373 kelvins (6,100 degrees Celsius), which is roughly the same as the temperature at the sun's surface.

LAYERS OF THE EARTH'S INTERIOR: RHEOLOGICAL DIVISIONS

Whereas the crust, mantle, and core are considered chemical divisions, the earth's interior is also described rheologically, with layers divided by their physical characteristics and further defined by the flow and elasticity of matter, which behaves differently depending on temperature, pressure, and other variables. Rheology focuses on fluids as well as soft solids, in which solids behave like plastics under certain conditions. From the surface to the center, the earth's rheological layers are the lithosphere, asthenosphere, mesosphere, and the outer and inner cores (which correspond to the chemical core divisions).

The lithosphere—a hard, rigid, rocky layer—encompasses the crust and the very top of the mantle. Like the crust, the lithosphere has two distinct types—oceanic and continental—broken up into tectonic plates. These plates ride on top of the next layer, the asthenosphere, and play a major role in earthquakes and volcanic activity. The lithosphere is slow-moving and considered to behave elastically on a scale of thousands of years. The lithosphere plays a heat-conductive role atop the convecting mantle.

The boundary between the lithosphere and asthenosphere is evident based on response to pressure: The former is brittle while the latter is more viscous. The asthenosphere corresponds with part of the upper mantle, spanning a depth of about 100 to 200 kilometers (62-125 miles), although parts of the asthenosphere are thought to extend as deep as 700 kilometers (435 miles). This layer is characterized by low density, high viscosity, and mechanical weakness. (In fact, its name comes from the Greek word for "weak.") The asthenosphere flows on the order of millimeters per year (up to about one centimeter at most) and convects heat up from the inner layers of the earth.

The mesosphere (not to be confused with the atmospheric layer of the same name) encompasses the remainder of the mantle, spanning from about 600 kilometers (373 miles) deep to the core-mantle boundary, and it is denser and more rigid than the asthenosphere. Finally, the core divisions are the same chemically and rheologically.

DIFFERENTIATION OF EARTH'S INTERIOR LAYERS

Before the earth's interior layers could differentiate, the earth itself had to form. The planet began as a solar nebula, a giant spinning cloud of helium, hydrogen, and other elements and debris from the big bang, which likely occurred 13.7 billion years ago. More than 9 billion years later, the nebula began to contract into a disc, causing it to spin faster. Nuclear fusion of the hydrogen and helium led to the formation of the sun in the center of the nebula, and other debris collided, forming asteroids and larger protoplanets; this growth by addition of matter is a process called accretion. One of these protoplanets became Earth.

As accretion continued, the earth's temperature increased until it reached a point in which the interior began to differentiate. Within 10 million years, the substances that made up the primitive earth's mostly homogeneous interior started separating into layers based on their physical and chemical properties.

An event called the iron catastrophe describes the formation of the earth's core, in a process that took tens of thousands of years, a relatively short time on the geologic time scale. During the iron catastrophe, NiFe (a dense alloy of nickel and iron) separated from its molten emulsion with other interior materials, particularly lighter silicates, and traveled toward the center of the earth. This process was driven by density differences between NiFe and the other materials, and it was facilitated by the earth's rising temperature, which allowed the once-homogeneous mix of materials to begin to melt. A primitive mantle, a layer that included what would eventually become the crust, also settled during the iron catastrophe.

Around 4.3 billion years ago, a crust formed with a basaltic composition similar to the oceanic crust. During this time, high temperatures in the mantle drove convection faster than it occurs in the present, and this initial crust likely did not last long. Four billion years ago, the continental crust as we know it began to form.

The upper mantle continues to differentiate through a process called plate tectonics. As mentioned, the lithosphere features tectonic plates, about seven or eight major ones and many more minor ones, that move about one centimeter per year and interact with each other at three types of boundaries: convergent, divergent, and transform faults. These interactions can result in earthquakes, volcanic activity, the formation of mountains, and the formation of trenches in oceans.

The movement of tectonic plates is influenced by a variety of factors, but the main driver is thought to be convection within the mantle, which exerts both gravitational and frictional forces on the plates. Mantle convection is the process of heat rising slowly from the interior of the earth through currents. To a lesser extent, gravitational forces not related to convection also help drive the plate movements. One such force relates to the creation of an oceanic lithosphere at spreading ridges: The new material adds on in a way that plates become thicker near the ridge, creating a lateral incline that results in a gravitational sliding effect. Other influences on plate movement include tidal drag from the moon's and the sun's gravitational pulls, the Coriolis effect, and minor wobbles created by the earth's rotation.

BOUNDARIES BETWEEN EARTH'S LAYERS

Seismology, the study of seismic waves, provides a clear picture of the location and characteristics of the boundaries between the earth's interior layers. A seismic wave is a disturbance that carries energy through the earth from an event that releases low-frequency acoustic energy; an earthquake is a good example of one such event. While there are multiple types of seismic waves, one subtype, body waves, vastly contributed to seismological research of the earth's interior. Body waves travel through the interior (as opposed to the surface) of the earth and are divided into two classes: P waves ("primary" or "pressure" waves) and S waves ("shear" or "secondary" waves). P waves move faster, as their motion is direct. S waves, in contrast, involve motion that moves perpendicularly to the overall propagation of the wave. The viscosity of fluids is too low to support the perpendicular movement of S waves, but P waves can travel through both fluids and solids.

With an understanding of S waves and P waves, scientists have been able to use seismic data from earthquakes to map out discontinuities in the earth's interior; these discontinuities are areas of abrupt change of seismic activity, such as an increased or decreased P-wave velocity. Discontinuities are indicative of a compositional change and, therefore, help pinpoint the boundary between different layers of the earth.

Closest to the surface, the Mohorovičić discontinuity (often referred to as the Moho) indicates the border between the crust and the mantle. Because the crust has variable thickness, the Moho's relative depth varies: Although it can be found 5-10 kilometers (3-6 miles) under the ocean floor, it is much farther below continents, about 20-90 kilometers (10-60 miles) deep, at an average depth of 35 kilometers (22 miles). In the rheological division of layers, the Moho lies within the lithosphere except at mid-ocean ridges, where it indicates the boundary between the lithosphere and the asthenosphere.

The Moho was discovered by its namesake, Croatian seismologist Andrija Mohorovičić, in 1909. Mohorovičić had noticed that in seismograms of shallow earthquakes, there were actually two sets of P waves and S waves being recorded, not one as expected. One set moved directly from one point to another, fairly close to the surface, but the other set was refracted, much like a beam of light hitting a prism.

At a depth of about 220 kilometers (137 miles), within the upper mantle, lies another discontinuity. Called the Lehmann discontinuity, it was described in 1958 by Danish seismologist Inge Lehmann, who found that P-wave and S-wave velocities sharply increased upon reaching this depth. Later research showed conflicting results as to whether this discontinuity is present only under continents or only under the ocean floor. (Lehmann also is known for her discovery of the boundary between the earth's inner and outer cores, a boundary often called the Lehmann discontinuity.)

Farther into the mantle, in the deepest 200 kilometers (124 miles), which is referred to as the D" zone, there exists another discontinuity (the Gutenberg discontinuity), at the bottom of which is the core-mantle boundary. Named for German-born American seismologist Beno Gutenberg, this discontinuity is marked by decreased P-wave velocity and the disappearance of S waves, which cannot travel through fluid. This is the boundary between the solid mantle and the molten outer core. It is an uneven boundary, caused in part by convected heat from the core to the mantle and by turbulent eddies produced by the Coriolis effect within the liquid outer core.

Drilling to Earth's Center

Much of what is known about the interior structure of the earth is conjecture, based on the behavior of seismic waves. Because of the interior's extreme heat and pressure, sending humans or equipment far below the surface of the earth is impossible. Attempts have been made, however, to drill partway through the crust.

One of the largest-scale drilling attempts occurred in 1962. A Soviet research team began a project called the Kola Superdeep Borehole. The team's goal was to drill to the Mohorovičić discontinuity (at a depth of about 15 kilometers, or 9.3 miles) to learn more about the earth's interior. After three years of searching for a suitable location and after five years of constructing the drill and planning the operation, the drilling finally began in 1970 at the Kola Peninsula in the northwest region of the Soviet Union. Because the project aimed to go much deeper than ever before, the standard rotating drill shaft used in most deep-drilling projects would not work. Instead, the researchers built a drill in which the drill bit rotated independent of the drill pipe, greatly reducing the amount of friction resistance. The drill was powered by a pressurized lubricant that was pumped down the shaft.

The 9 inch borehole grew slowly, hitting the 12-kilometer (7.5-mile) mark twenty-four years later. It was here that the team had to give up. The temperature in the borehole was a sweltering 180 degrees Celsius (356 degrees Fahrenheit), much hotter than the team's initial estimate of 100 degrees Celsius (212 degrees Fahrenheit). This temperature was more than the drill could handle, because the heat at this level of the crust causes rocky material to behave like plastic, melting and blocking the drill's path. Had the drill made it to the 15-kilometer (9.3-mile) point, the temperature likely would have reached 300 degrees Celsius (572 degrees Fahrenheit).

Though the Kola project did not reach its depth goal, it was able to collect many core samples along the way, providing valuable insight into the earth's interior structure and geological history. The oldest samples were estimated to date from 2.7 billion years ago. Even fossils (twenty-four species of plankton) were found in the upper half of the borehole.

Researchers continue to attempt to learn more about the earth's interior. One group, the Integrated Ocean Drilling Program (IODP), is a collaboration of researchers from twenty-four countries seeking to advance this field of study. The program involves deep drilling through the oceanic crust, which is thinner than the continental crust. This drilling allows for the collection of ancient rock samples that tell a detailed story of the earth's geological history.

Deep-sea drilling has far-reaching research implications that go beyond understanding the earth's interior structure. One of the major objectives of the IODP, for example, is to monitor climate change, both on a small time scale and in larger cycles over time. The IODP builds on decades of work done by two older projects, which are now defunct: the Deep Sea Drilling Project and the Ocean Drilling Program.

Through seismology, drilling, and other research, scientists have moved well beyond the "hollow earth" hypotheses of the seventeenth and eighteenth centuries. Some of these hypotheses even included the idea of more suns and atmospheres inside the earth.

Rachel Leah Blumenthal

Further Reading

Blakey, Ronald C., Wolfgang Frisch, and Martin Meschede. *Plate Tectonics: Continental Drift and Mountain Building.* New York: Springer, 2011. With great relevance to the characteristics of the earth's crust, this book provides an introduction to plate tectonics, covering the earth's early history to the present day. Topics include subduction zones, mid-ocean ridges, and the formation of mountains.

Chilingar, George V., et al. *Evolution of Earth and Its Climate: Birth, Life, and Death of Earth.* Boston: Elsevier, 2011. This text provides ample information on the earth's formation, composition, and magnetic activity. It also explores the formation of the moon and its influence on the earth.

Dickey, John S. *On the Rocks: Earth Science for Everyone.* New York: Wiley, 1996. From stardust to Earth's and other planets' formation, this book is ideal for earth science novices. Easy to understand but comprehensive, *On the Rocks* is an enjoyable overview.

Lutgens, Frederick K., and Edward J. Tarbuck. *Earth Science*. 13th ed. Upper Saddle River, N.J.: Prentice Hall/Pearson, 2012. Originally published in 1976, *Earth Science* is an updated introductory textbook geared to undergraduates, including those without a science background. It covers geology and astronomy and other earth science topics.

Monroe, James S., and Reed Wicander. *The Changing Earth: Exploring Geology and Evolution*. 5th ed. Belmont, Calif.: Brooks/Cole, Cengage Learning, 2009. This textbook offers a solid introduction to geology in an easily readable format, supplemented with many relevant photographs, diagrams, and real-world examples.

Wu, Chun-Chieh. *Solid Earth*. Vol. 26 in *Advances in Geosciences*. London: World Scientific, 2011. This book discusses research papers in the fields of seismology, planetary exploration, the solar system, and other topics relevant to earth science.

See also: Continental Drift; Creep; Cross-Borehole Seismology; Deep-Earth Drilling; Discontinuities; Earthquake Magnitudes and Intensities; Earthquake Prediction; Earthquakes; Earth's Age; Earth's Core; Earth's Differentiation; Earth's Lithosphere; Earth's Magnetic Field; Earth's Mantle; Earth's Oldest Rocks; Geodynamics; Faults: Normal; Faults: Strike-Slip; Faults: Thrust; Faults: Transform; Heat Sources and Heat Flow; Lithospheric Plates; Magnetic Stratigraphy; Mantle Dynamics and Convection; Metamorphism and Crustal Thickening; Notable Earthquakes; Plate Motions; Plate Tectonics; Rock Magnetism; Seismic Wave Studies; Seismometers; Slow Earthquakes; Soil Liquefaction; Stress and Strain; Subduction and Orogeny; Tsunamis and Earthquakes; Volcanism.

EARTH'S LITHOSPHERE

Within the lithosphere, earthquakes occur, volcanoes erupt, mountains are built, and new oceans are formed. An understanding of the lithosphere's structure is needed in the search for oil and gas, for the prediction of earthquakes, and for the verification of nuclear test ban treaties.

PRINCIPAL TERMS

- **asthenosphere:** the partially molten weak zone in the mantle directly below the lithosphere
- **basalt:** a dark-colored igneous rock containing minerals, such as feldspar and pyroxene, high in iron and magnesium
- **crust:** the rocky, outer "skin" of the earth, made up of the continents and ocean floor
- **granite:** a light-colored igneous rock containing feldspar, quartz, and small amounts of darker minerals
- **mantle:** the thick, middle layer of the earth between the crust and the core
- **Mohorovičić discontinuity (Moho):** the boundary between the crust and the mantle, named after the Croatian seismologist Andrija Mohorovičić, who discovered it in 1909
- **peridotite:** an igneous rock made up of iron- and magnesium-rich olivine, with some pyroxene but lacking feldspar
- **reflected wave:** a wave that is bounced off the interface between two materials of differing wave speeds
- **refracted wave:** a wave that is transmitted through the interface between two materials of differing wave speeds, causing a change in the direction of travel

DEFINING TERMS

The lithosphere is the rigid outer shell of the earth. It extends to a depth of 100 kilometers and is broken into about ten major lithospheric plates. These plates "float" upon an underlying zone of weakness called the asthenosphere. The phenomenon is somewhat like blocks of ice floating in a lake. As lake currents push the ice blocks around the lake, so do currents in the asthenosphere push the lithospheric plates. The plates carry continents and oceans with them as they form a continually changing jigsaw puzzle on the face of the earth.

The word "lithosphere" is derived from the Greek *lithos*, meaning "stone" Historically, the lithosphere was considered to be the solid crust of the earth, as distinguished from the atmosphere and the hydrosphere. The words "crust" and "lithosphere" were used interchangeably to mean the unmoving, rocky portions of the earth's surface. Advances in the understanding of the structure of the earth's interior, resulting mostly from seismology, have forced the redefinition of old terms. "Crust" presently refers to the rocky, outer "skin" of the earth, containing the continents and ocean floor. "Lithosphere" is a more comprehensive term that includes the crust within a thicker, rigid unit of the earth's outer shell. To appreciate the reason for this redefinition, it is necessary to learn about the nature of the earth's interior. American geologist Joseph Barrell investigated the lithosphere and asthenosphere in a series of papers in 1914.

EARTH'S INTERIOR

Except for the upper 3 or 4 kilometers, the earth's interior is inaccessible to humans. Therefore, indirect methods, such as studying earthquakes and explosions, are used to learn about the inside of the earth. Earthquakes and explosions, both conventional and nuclear, generate two types of energy waves: compressional (P) waves and shear (S) waves. P waves travel faster than do S waves and are generally the first waves to arrive at an observation station. The speed of a wave, however, depends on the rock through which it travels. When seismic waves encounter a boundary between two different rocks, some energy is reflected back, and some is transmitted across the boundary. If the rock properties are very different, the transmitted waves travel at a different speed and their travel path is bent, or refracted. This phenomenon can be illustrated by placing a pencil in a glass of water. Light in water travels at a speed different from that of light in air, so light is refracted, or bent, as it travels from water to air. Thus, the pencil appears to be bent. P and S waves are reflected and refracted as they travel through the earth. Waves following different paths travel at different speeds.

Since 1900, seismologists have studied P and S waves arriving at different locations from the same earthquake. They discovered three distinct layers in the earth: the crust, the mantle, and the core. The boundaries separating these layers show abrupt changes in both P- and S-wave speeds. These changes in wave speeds provide information about the earth's interior. Scientists studying the theory of traveling elastic waves, such as earthquake waves, related the speed of waves to the physical properties of the material through which they travel. It was found that S waves do not travel through liquids. From this finding, scientists concluded that the earth's core had a liquid outer region and a solid inner region. Other scientists measured the P- and S-wave speeds of many different rocks and provided clues to the kind of rocks found inside the earth.

The continental crust averages 30-40 kilometers thick and is divided into two main seismic layers. One layer, the upper two-thirds of the crust, has P- and S-wave speeds corresponding to those of granitic rocks. The speeds increase slightly in the bottom third of the continent, corresponding to rocks of basaltic composition. The average oceanic crust is 11 kilometers thick and is of basaltic composition. Beneath both continental and oceanic crust, the P- and S-wave speeds increase sharply. This boundary between the crust and mantle is called the Mohorovičić discontinuity, or Moho. The Moho marks a compositional change to a dense, ultramafic rock called peridotite.

The Asthenosphere

At an average depth of 100 kilometers, the S-wave speed decreases abruptly. It remains low for about 100-150 kilometers. This region is called the low velocity zone (LVZ). Laboratory experiments have shown that seismic-wave speeds, particularly those of S waves, decrease in rocks containing some liquid. The LVZ in the mantle indicates a zone of partial melting, perhaps 1-10 percent melt. The presence of the melt reduces the overall strength of the rock, giving the region its name, "asthenosphere," from the Greek *asthenes*, meaning "without strength."

The partially molten asthenosphere is very mobile, allowing the more rigid lithosphere above it to move about the earth's surface. The boundary between the lithosphere and the asthenosphere does not mark a change in composition; it marks a change in the physical properties of the rocks. The lithosphere defines this region of crust and mantle from the mantle region below by its seismic-wave speeds and its physical properties.

Lithospheric Plates

Seismic-wave speeds and earthquake distribution provide information about the lithospheric plates and the boundaries between them. Like the earth's crust, lithospheric plates are not the same everywhere. For example, the Pacific plate contains primarily oceanic crust, the Eurasian plate is mostly continental, and the North American plate contains both continental and oceanic crust. The lithosphere is thinnest at spreading centers, or regions where two plates are moving away from each other, such as the Mid-Atlantic Ridge and the East Pacific Rise. Here, the asthenosphere is close to the surface and the melt portion pushes upward, separating the plates and creating new lithosphere. Shallow earthquakes occur as the new crust is cracked apart. In areas such as western South America or southern Alaska, two plates are coming together, with the oceanic lithosphere being thrust under the continental plate. Earthquakes occur as deep as 700 kilometers as one plate slides under the other. Along the California coast, two plates slide past each other along faults that cut through the lithosphere. Earthquakes are common, and the faults can move several meters at a time. Where two continental plates, India and Eurasia, have collided, the crust is highly faulted and 65 kilometers thick. Earthquakes in and near the Himalaya are numerous, often occurring along deep fault zones.

Although earthquakes are most common along plate boundaries, they can also occur within lithospheric plates. Some earthquakes are related to newly forming boundaries. The Red Sea is believed to be a recently formed spreading center pushing the Arabian Peninsula and Africa apart. Some earthquakes result from the movement along ancient geologic faults buried within the crust. The causes of some earthquakes, however, such as the one in 1886 in Charleston, South Carolina, remain unknown.

Upper and Lower Lithosphere

Structural details within plate regions cannot be determined by earthquake studies alone. P and S waves generated by explosions are reflected and refracted by layers within the lithospheric plates. Regional studies show the upper lithosphere to be highly variable. In mountainous regions, such as the Appalachians or the Rocky Mountains, the continental crust is thicker than average and shows much layering. In the midcontinent and the Gulf of Mexico regions, the crust consists of thick layers of sediments and sedimentary rocks. Oil companies, combining the data from many controlled explosions, discovered petroleum and natural gas within these layers from the changes in P- and S-wave speeds. Other regional seismic studies have found ancient geological features deep within the crust. Similarities in the seismic structure between these and other known features can uncover potential sites of much-needed natural resources. The discovery of the oil fields of northern Alaska was prompted by the area's structural similarity to the Gulf of Mexico, a known source of oil and gas.

The seismic structure of the lower lithosphere is less well known. Early studies show that it is also highly variable and that crustal structures are often related to features deep in the lithosphere. Much work, however, remains in unraveling the details of the lithosphere.

STUDY OF SEISMIC WAVES

Scientists use a number of seismic techniques to study the lithosphere. They use P and S waves generated by earthquakes that travel through the earth (body waves) and along the earth's surface (surface waves). Reflection and refraction seismology use seismic waves generated by explosions to study the continental and oceanic lithosphere. Data from experimental studies of rocks are used to relate seismic speeds to specific kinds of rocks. Computers help analyze the vast amounts of seismic data and are used to develop models to aid in the understanding of the earth.

The use of P and S waves from earthquakes is the oldest method of studying earth's structure. The times at which P and S body waves, reflected and refracted by the layers in the earth, arrive at different distances from the same earthquake are related to the average speed at which the waves travel. The arrival times of surface waves also depend on the layer speeds. Using seismic waves from many earthquakes, seismologists can determine the seismic structure of the lithosphere.

In regions with numerous earthquakes, seismologists record P and S waves using many portable seismographs, instruments that record seismic-wave arrivals. The scientists can then determine a more detailed regional structure. Earthquakes, however, do not occur regularly everywhere on the earth. Until an average regional structure is known, it will be difficult to determine the precise location and time of an earthquake.

Explosions as a source of seismic waves to study crustal structure have been developed and used extensively by the oil industry. With an explosive source, the location and time of detonation can be precisely controlled. Two basic techniques using artificial sources are reflection seismology and refraction seismology. Refraction seismology studies the arrivals of waves that are refracted, or bent, by the layers in the crust. The scientist determines an average velocity structure for an area by recording the time the first waves arrive at receivers located varying distances from the explosion. To determine deep structure, the distance between the explosion and the receivers must be very large. Reflection seismology allows a deeper look into the crust by studying reflections from many different layers. The seismic-wave receivers do not need to be placed as far from the source as they must in refraction studies. The reflection technique combines the results from many explosions, producing a picture of the earth's layers. This method is used extensively in the search for oil and gas. The techniques of reflection and refraction seismology have been applied to the lithosphere. Long reflection and refraction profiles have been acquired over geologically interesting but little-understood regions.

STUDY OF ROCK PROPERTIES

Seismic waves are vibrations traveling around and through the earth. Because of friction, these vibrations eventually stop, and seismic waves no longer travel. Earthquakes and explosions generate waves that vibrate at many frequencies. The earth slows each frequency differently. As a seismic wave travels through different rocks, the shape of its vibrations recorded on a seismograph is related to the properties of the rocks through which it travels. The analysis of seismic waveforms has shown differences between waves generated by earthquakes and by explosions.

To understand the lithosphere, it is necessary to know about rocks. Using a hydraulic press, scientists squeeze rocks in the laboratory to pressures and heat them to temperatures present deep within the earth. They then measure the rocks' physical properties at these conditions. Experimentally measured P and S speeds are compared to wave speeds determined from earthquakes and explosions to infer the kind of rocks and the conditions that exist within the earth. The complexity of the lithosphere, however, does not allow simple answers.

Computer Modeling

To aid the scientists in their studies, computers are used to develop models—simplified representations—of the earth. By making changes in the model, the scientist can study changes in computed seismic properties and compare them to the observed earth properties. Changes in the model are made to resemble the earth more closely. In modeling the lithosphere, scientists incorporate data from a wide range of sources, such as earthquake studies, experimental rock studies, and geologic maps. The computer allows the earth scientist to test more complex models in an effort to provide a better understanding of the lithosphere.

Significance

For the earth scientist, increased knowledge of the seismic structure of the lithosphere helps in unraveling the processes by which geologic features are formed. The movement of the lithospheric plates about the earth creates mountain ranges, causes earthquakes, and devours or creates ocean basins. Because much of the earth is inaccessible, seismic waves generated by earthquakes and explosions are used to look deep within the earth to provide a picture of the earth's structure.

Increased knowledge of the lithosphere is important to the average person for three reasons: First, earthquakes are caused by movements between and within the lithospheric plates. Every year, lives are lost and millions of dollars in damage occur because of earthquakes and earthquake-related phenomena. Detailed knowledge of the lithosphere helps scientists understand where and how earthquakes occur. This information can lead to regional assessment of the potential for earthquakes and earthquake-related damage. Knowledge of the earthquake potential of a region can result in the improvement of local building codes and the evaluation of existing emergency preparedness plans. Earthquake-hazard assessment can also aid in prediction by determining the probability of future earthquake occurrence. Some success in long-term predictions has been seen in Japan and China. Eventually, the increased understanding of the lithosphere may lead to the short-term prediction of earthquakes.

Second, detailed knowledge of lithospheric structure will lead to the discovery of potential sites of needed natural resources, such as oil, gas, and coal; metals, such as iron, aluminum, copper, and zinc; and nonmetal resources, such as stone, gravel, clay, and salt. Scientists are beginning to unravel the relationship of tectonic features to the formation of many mineral deposits. Detailed knowledge of the structure of the lithosphere from seismic studies can uncover deeply buried features that may provide new sources for critically needed resources.

Finally, scientists require detailed information on the seismic structure of the lithosphere to locate and identify earthquakes and nuclear explosions. More structural information will also lead to better identification of the differences between these two types of seismic wave sources. An accurate and reliable means of distinguishing between earthquakes and nuclear explosions is critical for the verification of any nuclear test ban treaty.

Pamela R. Justice

Further Reading

Bakun, William A., et al. "Seismology." *Reviews of Geophysics* 25 (July 1987): 1131-1214. A series of articles summarizing research in seismology in the United States from 1983 to 1986. Reviews findings and unresolved problems in all areas of seismology from the early 1980's. Articles are somewhat technical but suitable for the informed reader. Extensive bibliographies.

Bolt, Bruce A. *Earthquakes*, 5th ed. New York: W. H. Freeman, 2005. A popular, illustrated book on the many features of earthquakes. Chapter topics include the use of earthquake waves to study the earth's interior and earthquake prediction. A bibliography and an index are included. Suitable for the layperson.

Bullen, K. E., and B. A. Bolt. *An Introduction to the Theory of Seismology.* 4th ed. New York: Cambridge

University Press, 1985. Introductory sections of most chapters provide historical and nonmathematical insight into the subject, suitable for the general reader. Contains a selected bibliography, references, and an index. (Designed as a text for the advanced student with a mathematics background.)

Burger, H. Robert, Anne F. Sheehan, and Craig H. Jones. *Introduction to Applied Geophysics: Exploring the Shallow Subsurface.* W. W. Norton & Company, 2006. This text was written for use in an advanced undergraduate or graduate course. Contains detailed explanations of methodologies; each chapter has problem sets, examples, and applications. A CD-ROM accompanies the text with software covering seismology, gravity, and magnetism content.

Dahlen, F. A., and Jeroen Tromp. *Theoretical Global Seismology.* Princeton: Princeton University Press, 1998. Intended for the college-level reader, this book describes seismology processes and theories in great detail. The book contains many illustrations and maps. Bibliography and index.

Doyle, Hugh A. *Seismology.* New York: John Wiley, 1995. A good introduction to the study of earthquakes and the earth's lithosphere. Written for the layperson, the book contains many useful illustrations.

Grotzinger, John, et al. *Understanding Earth.* 5th ed. New York: W. H. Freeman, 2006. This comprehensive physical geology text covers the formation and development of the earth. Readable by high school students, as well as by general readers. Includes an index and a glossary of terms.

Langel, R. A. *The Magnetic Field of the Earth's Lithosphere: The Satellite Perspective.* Cambridge, England: Cambridge University Press, 1998. Focusing on remote sensing, Langel's book describes what has been learned about geomagnetism through the use of artificial satellite technology. Includes illustrations.

Lutgens, Frederick K., Edward J. Tarbuck, and Dennis Tasa. *Essentials of Geology*, 11th ed. Prentice Hall, 2011. Lutgens, Tarbuck, and Tasa focus the text on the physical geology of the lithosphere. The book examines current issues in geology with a thorough look at environmental issues. This text contains many exceptional images, diagrams, and color photos.

Mutter, John C. "Seismic Images of Plate Boundaries." *Scientific American* 254 (February 1986): 66-75. An article on the application of explosion seismology to the study of plate boundaries. Summarizes the method of seismic reflection profiling. Shows results of studies across different plate boundaries.

Ogawa, Yujiro, Ryo Anma, and Yildirim Dilek. *Accretionary Prisms and Convergent Margin Tectonics in the Northwest Pacific Basin.* New York: Springer Science+Business Media, 2011. Discusses new techniques in plate tectonic studies. One volume of the series Modern Approaches in Solid Earth Sciences: Accretionary Prisms, Tectonics, and Pacific Ocean Events.

Pitman, Walter C. "Plate Tectonics." In *McGraw-Hill Encyclopedia of the Geological Sciences*, 2d ed. New York: McGraw-Hill, 1988. A brief summary of plate tectonics, discussing evidence for the theory and an explanation of causes of present-day features. Cross-referenced, illustrated, with bibliography. Suitable for the general reader.

Press, Frank, and Raymond Siever. *Earth.* 4th ed. San Francisco: W. H. Freeman, 1986. A book for the beginning reader in geology. Of interest are Chapter 17, "Seismology and the Earth's Interior," and Chapter 19, "Global Plate Tectonics: The Unifying Model," for an overall understanding of the importance of the lithosphere. Illustrated and supplemented with numerous marginal notes.

Reynolds, John M. *An Introduction to Applied and Environmental Geophysics*, 2d ed. New York: John Wiley, 2011. An excellent introduction to seismology, geophysics, tectonics, and the lithosphere. Includes maps, illustrations, and bibliography.

Smith, Peter J., ed. *The Earth.* New York: Macmillan, 1986. A well-illustrated, comprehensive guide to the earth sciences. Chapter 3, "Internal Structure," describes historical development of the current view of the earth's lithosphere. Chapters 1, 2, and 5 provide related material. Includes glossary of terms.

See also: Earthquake Hazards; Earthquake Locating; Earthquake Prediction; Earthquakes; Earth's Core; Earth's Differentiation; Earth's Interior Structure; Earth's Mantle; Elastic Waves; Heat Sources and Heat Flow; Lithospheric Plates; Plate Tectonics; Plumes and Megaplumes; Seismic Reflection Profiling; Seismometers.

EARTH'S MAGNETIC FIELD

Earth's magnetic field is generated by rotation, convection, and electrical currents in the earth's outer core, and it acts as a protective shield against solar wind, which stripped away the atmospheres of other planetary bodies, like Venus and Mars. As electromagnetism is one of nature's four fundamental interactions, the magnetic field has wide-ranging applications to our lives and to the future of the planet.

PRINCIPAL TERMS

- **coronal mass ejection:** a sudden, large burst of solar wind that can cause geomagnetic storms and aurorae in Earth's upper atmosphere
- **dynamo theory:** a set of three conditions that allow for a body of fluid, such as Earth's outer core, to generate a magnetic field that does not collapse from ohmic decay over time
- **electromagnetism:** the relationship between electric fields and magnetic fields; one of four fundamental interactions in nature
- **geographic poles:** the north and south "ends" of Earth; the spot on either end of Earth's rotational axis where the longitude lines meet
- **geomagnetic poles:** hypothetical magnetic poles located at the points where the axis in the simplified dipole-like model of Earth's magnetic field intersects with Earth's surface; not to be confused with magnetic poles
- **geomagnetic storm:** a type of space weather that occurs when solar wind particles penetrate Earth's magnetosphere
- **interplanetary magnetic field:** embedded bits of magnetic field that are carried with charged particles in the solar wind
- **magnetic dipole:** a pair of magnetic poles with equal magnitude and opposite signs (generally referred to as "north" and "south")
- **magnetic field:** an invisible area produced by an electric field that can exert a magnetic force on certain things in or around it
- **magnetic poles:** the two spots where Earth's magnetic field becomes vertical; not to be confused with geomagnetic poles
- **magnetosphere:** the area around a planetary body where the influence of a planet's magnetic field is felt
- **solar wind:** a stream of charged particles (mostly protons and electrons) that are pushed out of the sun's upper atmosphere; can affect Earth's atmosphere when not fully blocked by the magnetosphere

PROPERTIES OF EARTH'S MAGNETIC FIELD

With any electrical current comes a magnetic field. Electromagnetism, defined as the relationship between electrical and magnetic fields, is a fundamental interaction in nature, one of four such phenomena. (The other fundamental interactions are strong interactions, weak interactions, and gravitation.) French physicist and mathematician André-Marie Ampère derived an equation to describe this relationship; the equation became known as Ampère's law and, in recognition of Ampère's work in electromagnetism, the standard unit for measuring electrical current (the ampere) was named for him.

Like Mercury, Saturn, and several other planets in the solar system, Earth has its own magnetic field. Generated by Earth's outer core, the magnetic field is an estimated 3.5 billion years old, according to a 1980 paleomagnetic study of basalt found in Australia. The shape of Earth's magnetic field can be conceptualized by imagining the field produced by a standard magnetic dipole, such as a basic bar magnet with a "north" end and a "south" end. If one were to draw a diagram with field lines, the lines would curve from the South Pole to the North Pole, becoming virtually vertical at the poles.

Earth's magnetic field is generally tilted approximately 11 degrees from the planet's rotational axis, although it should be noted that the field is not stationary: It moves slowly and even reverses direction completely every few hundred thousand years. At the surface of the earth, the magnetic field strength averages 5.0×10^{-5} tesla (0.5 gauss), varying locally between 3.0×10^{-5} and 6.0×10^{-5} tesla (0.3 and 0.6 gauss). Measurements suggest that the strength has decreased about 10 percent in the past 150 years.

Like other planets that generate their own magnetic fields, Earth is surrounded by a magnetosphere, a region that encompasses the area of influence of the magnetic field. The magnetosphere is present above Earth's ionosphere. The shape of the magnetosphere is formed by the interactions between

Earth's magnetic field and the magnetic field embedded in the solar winds that bombard it. The solar wind is a plasma stream pushed from the sun's upper atmosphere, and it is filled with charged particles, mostly protons. The magnetic field that travels with it is called the interplanetary magnetic field, or IMF, and it has a strength of approximately 2.0×10^{-9} to 5.0×10^{-9} tesla. As solar wind approaches the magnetosphere, it abruptly loses velocity when it hits a region called the bow shock. One can imagine the bow shock as a sort of invisible armor that cushions the blow of the solar wind hitting the magnetosphere.

Because of the influence of the sun, the outer edge of the "sunny" side of the magnetosphere is much closer to Earth's surface than is the magnetosphere's opposite side (about six to ten times Earth's radii compared with an estimated two hundred or more times Earth's radii on the opposite side. The longer side is referred to as the magnetotail because it extends from the planet in a tail-like manner. The magnetosphere's border, which takes on a bullet-like shape, is called the magnetopause.

EARTH'S MAGNETIC FIELD: A PROTECTIVE BARRIER

Earth's magnetic field serves an important protective purpose: It partially blocks the solar wind, a stream of charged particles from the sun, from stripping away Earth's upper atmosphere. The magnetic field cannot deflect everything, though; some particles do make their way in. Of these, some become trapped within the Van Allen radiation belt, others cause geomagnetic storms within the magnetosphere, and others reach the earth's thermosphere, causing beautiful aurorae, such as the aurora borealis, or northern lights.

Mars and Venus are good models for the harmful effects of solar wind on a planet's atmosphere. Mars shows signs of having had water billions of years ago, but it is now an empty "desert" with a low-density atmosphere. Evidence from the National Aeronautics and Space Administration's Mars Global Surveyor and older probes suggests that because Mars lacks a full protective magnetosphere, solar wind has eroded the planet's atmosphere over time. It seems that Mars had a dynamo-powered magnetic field 4 billion years ago, but for reasons unknown, that field collapsed.

Venus, a younger planet, faces a similar problem. The Venus Express, an orbiter sent by the European Space Agency in 2005, has been analyzing Venus's

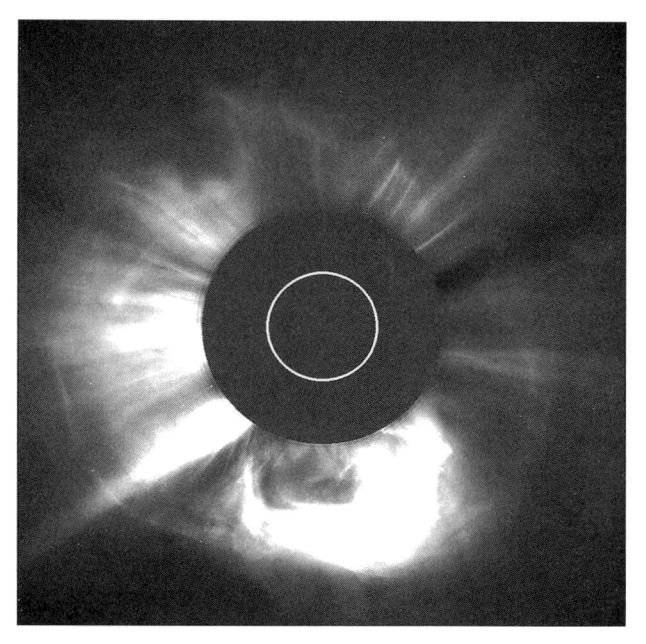

Solar eruption. SOHO (Solar and Heliospheric Observatory) image of a huge coronal mass ejection (CME, lower right) erupting from the sun. If the stream of charged particles is directed toward Earth, it may cause electrical blackouts and widespread aurora displays. This flare was the third most powerful ever detected. It erupted from a sunspot group called 486. The frequency of sunspots and eruptions varies on an 11-year cycle. The maximum of this cycle was in 2000. (Science Source)

atmospheric ions, which are being swept away by the solar wind; many of these ions are hydrogen and oxygen. In effect, Venus is losing water to the solar wind. Earth, however, is mostly protected by its magnetosphere, which is able to deflect many of the solar wind's incoming charged particles.

The Van Allen radiation belt (actually two belts, an inner and an outer) catches charged particles from solar wind and from cosmic rays. The belts can be a nuisance, as satellites must be shielded appropriately from radiation if they will be orbiting within a belt for too long. The inner belt is located within 1.5 Earth radii of Earth's surface, whereas the outer belt spans from approximately 3 to 10 Earth radii.

Geomagnetic storms can result from coronal mass ejections (a sudden flare-up of solar wind) or other disturbances beyond and within the magnetosphere. These storms occur regularly, and depending on their severity, they can affect Earth in a variety of ways. For example, the radiation produced during a geomagnetic storm is hypothetically a lethal hazard to humans, but realistically, the radiation could affect only astronauts and, to a much lesser extent, the crews of high-altitude airplane flights. The storms also appear to affect animals, particularly pigeons, dolphins, and whales, which use magnetoception for navigation.

Some communication and navigation systems, particularly those that send signals through the ionosphere, also can be disrupted by geomagnetic storms, a particular hazard for airliners. In the days of telegraph communication, these signals could be disrupted. During some extreme geomagnetic storms, telegraph operators were shocked and receivers caught fire. When the sun is at a peak in its solar cycle between the years 2012 and 2013, phone, television, radio, and Internet signals that depend on satellites could undergo significant service disruptions. The most extreme geomagnetic storm in recorded history was in September 1859, resulting in aurorae visible through much of the world; the storm disrupted telegraph lines and had other effects.

The solar wind does provide one arguably positive phenomenon: the notoriously beautiful northern lights and other aurorae, displays of colored lights in the sky made of charged solar wind particles interacting with the earth's magnetic field. Most aurorae span from about 100 kilometers (60 miles) to 200 or 300 kilometers (120-200 miles) above Earth's surface, although some are higher or lower. The best-known example is perhaps the aurora borealis, which appears in the skies of the Northern Hemisphere.

Dynamo Theory: Earth's Core and Magnetic Field Formation

Earth's molten outer core forms and maintains Earth's magnetic field. This process is explained by the dynamo theory, which was set forth in 1946 by German-born American physicist Walter M. Elsasser. In the dynamo theory, a rotating, convecting, electricity-conducting fluid generates a long-lasting magnetic field. Convection is key; without it, a magnetic field would collapse from ohmic decay in just tens of thousands of years. Good evidence also exists for the necessity of speedy rotation. Venus's core is thought to have an iron content similar to that of Earth's core, but Venus's core does not produce a magnetic field, likely because it just does not rotate fast enough. (One Venus day equals 243 Earth days.)

The dynamo theory sets forth three requirements for a fluid to generate a magnetic field: Planetary rotation must occur to create kinetic energy; an internal energy source must cause convection; and the fluid must conduct electricity. The outer core of Earth fulfills all of the necessary requirements. First, Earth's rotation powers the core's rotation through the Coriolis effect. (The outer core is a turbulent, molten sea.) Second, convection occurs within the outer core due to several heat sources, including the radioactive decay of trace elements and the presence of residual heat that is still being slowly released after the formation of the core billions of years earlier. (Compositional and thermal convection are both thought to play a role.) Third, the fluid conducts electricity.

The inner core is unable to create a magnetic field on its own. The temperature is too high, prohibiting magnetization by causing the molecules in the core's iron to adopt a randomized, rather than orderly, orientation. Some scientists believe that the inner core does play a stabilizing role to support the magnetic field.

Earth's Magnetic, Geomagnetic, and Geographic Poles: Location, Movement, and Reversal

The terms "North Pole" and "South Pole" are vague, as one must clarify whether one is referring to Earth's north and south geographic poles, the magnetic poles, or the geomagnetic poles. The geographic poles are (mostly) fixed, representing the spot on either end of Earth's rotational axis where the longitude lines meet. Because of some "wobbling" in the earth's axis, the geographic poles occasionally shift slightly on the order of a few meters. In simple terms, the geographic North Pole and the geographic South Pole are the top and bottom of the Earth, respectively.

Magnetic poles are located in the two spots where Earth's magnetic field becomes perfectly vertical. The magnetic North Pole is located in the geographic north, but it is actually the south pole of Earth's magnetic field if one looks at the directionality of the imaginary field lines. The same applies to the magnetic South Pole: It is located in the south, but it is the north pole of the magnetic field. This has important implications for the workings of compasses. Magnetic poles are not stationary poles. Because the magnetic field is generated by the turbulent convection of Earth's outer core, the field is always moving, and the poles follow.

Geomagnetic poles can be thought of as hypothetical versions of the magnetic poles. Earth's magnetic field approximates the shape of a field created by a dipole (such as an ordinary bar magnet). The geomagnetic poles are located at the two hypothetical points where the imaginary axis of the dipole-like field intersects with the earth's surface. Because the earth's field is not caused by a true dipole, there are anomalies in the field that cause the magnetic poles to vary from the hypothetical geomagnetic pole locations.

The magnetic and geomagnetic poles wander, sometimes substantially, with the turbulent and ever-changing motion of the magnetic field. In recent years, for example, the magnetic North Pole has been moving in the range of 40 kilometers (25 miles) per year, while the magnetic South Pole has been moving about 15 kilometers (9 miles) per year. On rare occasions (meaning every hundreds of thousands of years), Earth's magnetic field, and thus the pole orientations, spontaneously reverses in a process that takes several thousand years.

The last pole flip (dubbed the Brunhes-Matuyama reversal) occurred about 780,000 years ago, and geologic research shows evidence of 171 such reversals during the past 71 million years. These reversals are thought to simply indicate the chaos of the earth's churning outer core, which can cause a virtual tangled mess in the magnetic field. Some scientists theorize instead that reversals are caused by external triggers such as the impact of large objects, such as a meteorite, which could potentially disrupt the core's dynamo. The sun's magnetic field has been observed to reverse on a much faster scale, approximately every nine to twelve years.

Standard magnetic compasses have been in common use since about 250 B.C.E in ancient China. These compasses indicate the direction of the magnetic North Pole, which is typically near, but not at, "true" or geographic north. The compass needle is a magnetized bar, and its north pole is attracted to the south pole of the earth's magnetic field, which, as previously mentioned, is located in the earth's geographic north region. Magnetic compasses become virtually useless when one is near a magnetic pole; the error between the magnetic pole and corresponding geographic pole is too great.

ANIMALS AND EARTH'S MAGNETIC FIELD

It has long been assumed that humans do not innately sense the presence or effects of the earth's magnetic field, although research now suggests that this assumption might be worth reconsidering. Cryptochrome, a plant and animal protein that plays a role in circadian rhythms, has been demonstrated to have something to do with some animals' sense of the earth's magnetic field. Researchers were able to create transgenic *Drosophila* (fruit flies) that express the human version of the protein, human cryptochrome 2, rather than their own native cryptochrome. (Human cryptochrome 2 is found in human retinas.) The transgenic flies were able to detect and respond to a magnetic field. While this suggests that humans have a protein that can act as a magnetic sensor, it is still a long way from showing that human bodies can actually make use of this sensor.

There is, however, ample evidence that a variety of animals sense and use the magnetic field, particularly for migratory purposes. Turtles, pigeons, and cows are just a few species that exhibit this "extra" sense. The process is still somewhat mysterious, although several theories dominate. It is possible that some or all of these animals have receptors in their heads or elsewhere in their bodies containing a mineral called magnetite, which aligns itself with the earth's magnetic field. Magnetite has been found in the noses of some migratory fish, such as rainbow trout and salmon. Another possibility involves cryptochrome protein, a photopigment present in animals' eyes, which could react chemically with the magnetic field to provide a visual map to the animal. Research does indicate a possible connection between cryptochrome and magnetoception in some migratory birds.

Cattle and deer also apparently sense the earth's magnetic field. One study observed that when grazing or resting, they tend to align themselves with the magnetic field, facing either the magnetic North Pole or the magnetic South Pole. Two of science's most oft-studied animals, fruit flies and zebrafish, have been found to have magnetoception, so the door to future research in this area is wide open.

Rachel Leah Blumenthal

FURTHER READING

Almeida, J. Sánchez, and Mari Paz Miralles. *The Sun, the Solar Wind, and the Heliosphere.* New York: Springer, 2011. Focused on current research, this text reviews findings and interpretations of all things related to the sun and its atmosphere, and to solar wind and its effect on planets.

Basavaiah, Nathani. *Geomagnetism: Solid Earth and Upper Atmosphere Perspectives.* New York: Springer, 2011. This textbook provides a complete overview of the history of geomagnetism and of recent findings and advances. Topics include the nature of interactions between the sun and the earth, space weather, and magnetic field anomalies.

Hulot, G., et al. eds. *Terrestrial Magnetism.* New York: Springer, 2011. This book provides a detailed look at the past and future of the earth's magnetic field, including discussions of movement and reversal of the poles and the magnetic field, observations from field studies, and measurements of the field from space.

Lanza, Roberto, and Antonio Meloni. *The Earth's Magnetism: An Introduction for Geologists.* London: Springer, 2011. Geared to graduate students and professionals, *The Earth's Magnetism* examines a range of geomagnetism-related topics. Well supplemented with case studies and illustrations.

Merrill, Ronald T. *Our Magnetic Earth: The Science of Geomagnetism.* Chicago: University of Chicago Press, 2010. Provides an interesting alternative to a textbook. Includes personal anecdotes to comprehensively educate readers about the mysteries and discoveries of geomagnetism.

Schrijver, Carolus J., and George L. Siscoe, eds. *Heliophysics: Space Storms and Radiation—Causes and Effects.* New York: Cambridge University Press, 2011. Focusing on the sun, this textbook covers several topics relevant to the earth's magnetic field, most notably solar wind, mass coronal ejections, and space weather. Includes a comprehensive overview of heliophysics as a field of research.

Turner, Gillian M. *North Pole, South Pole: The Epic Quest to Solve the Great Mystery of Earth's Magnetism.* New York: Experiment, 2011. Aimed at general readers with or without a science background, this book looks at topics such as the differences between magnetic poles and geographic poles, why the magnetic field occasionally reverses itself, how animals use magnetism for migration, and how the earth's magnetic field was formed.

See also: Earth's Core; Earth's Differentiation; Earth's Interior Structure; Earth's Mantle; Geobiomagnetism; Magnetic Reversals; Mantle Dynamics and Convection; Polar Wander; Solar Wind Interactions.

EARTH'S MANTLE

Aside from making up the vast majority of Earth's volume, the mantle contributed to the formation of Earth's atmosphere and is a driving factor in the movement of tectonic plates, which in turn are responsible for earthquakes, volcanic activity, mountain building, and other processes. Direct exploration of the mantle is difficult, if not impossible, but drilling projects, computer simulations, and other technologies are allowing researchers to continue to learn more.

PRINCIPAL TERMS

- **asthenosphere:** the second layer (from the top) of the earth's interior when Earth is divided rheologically; corresponds with part of the upper mantle
- **convection:** the transfer of heat or matter in a fluid medium
- **differentiation:** the separation of interior layers of a planetary body caused by chemical and physical differences
- **discontinuity:** a zone where seismic wave velocity changes abruptly from the adjacent zone
- **lithosphere:** the uppermost layer of Earth's interior when Earth is divided rheologically; a hard, rigid, rocky layer that includes the crust and the top of the mantle
- **mesosphere:** the third layer (from the top) of the earth's interior when Earth is divided rheologically; a dense, rigid layer that corresponds with most of the mantle
- **plastic:** describes a solid material with some fluid-like properties; deformable
- **rheology:** the study of the flow of matter; focuses on liquids and soft solids that behave like plastic
- **seismic wave:** a moving, energy-transferring disturbance that occurs because of an event, such as an earthquake, which releases low-frequency acoustic energy
- **subduction:** a process that can occur when tectonic plates collide; one plate slides underneath the other plate
- **tectonic plate:** a slowly moving chunk of Earth's uppermost layer, the lithosphere

FORMATION OF THE EARTH'S MANTLE

About 4.6 billion years ago, Earth began to form from the debris of a solar nebula. About 10 million years later, Earth's interior began to differentiate into layers, particularly a core, but also a primitive mantle and crust. This event is known as the iron catastrophe because rising temperatures allowed iron to settle out of Earth's molten interior emulsion, sinking toward the center to form the core. Meanwhile, the remaining material closer to the surface began to cool and harden into the primitive mantle and crust.

The mantle and crust that formed at this point did not exist for very long. According to the giant impact hypothesis, a Mars-sized protoplanet crashed into the young Earth about 4.52 billion years ago. The impact was shallow enough that the forming core remained unperturbed, but much of the primitive mantle and crust were ejected into Earth's orbit, along with Earth's early atmosphere. Some of this debris eventually accreted to form Earth's moon.

The collision released vast amounts of heat, so a large portion of Earth's material that remained became molten. During the next 150 million years, the molten material cooled and hardened to form Earth's new rocky mantle and crust. (The differentiation of Earth's layers did not really end here. The mantle and crust continue to differentiate to this day through the movement of tectonic plates.)

STRUCTURE AND CHARACTERISTICS OF THE EARTH'S MANTLE

The interior of the earth is generally described by two sets of divisions: chemical and rheological (physical). The mantle is the earth's middle layer according to the main chemical divisions: core, mantle, crust. It is further divided into an upper mantle and a lower mantle. The rheological divisions are made based on physical characteristics, particularly the way matter flows (or how "elastically" it behaves). The mantle spans several of these rheological layers: the very bottom of the lithosphere, the entire asthenosphere, and the entire mesosphere (not to be confused with one of Earth's atmospheric layers, also called the mesosphere).

The mantle begins at a depth of about 35 kilometers (km), or 22 miles (mi) below Earth's surface and stretches down to about 2,890 km (1,790 mi), and its temperature ranges from 500 degrees Celsius (C), or 932 degrees Fahrenheit (F) near Earth's surface to 4,000 degrees C (7,230 degrees F) at its boundary with Earth's outer core. (For comparison, one can consider the surface of the sun, where the temperature is an estimated 5,500 degrees C or 9,932 degrees F.) By the core-mantle boundary, pressure is approximately 1.4 million standard atmospheres.

The mantle makes up quite a substantial portion of Earth's interior, accounting for about 84 percent of Earth's total volume. While it is referred to as a solid layer, the mantle's high temperature and its composition (mainly silicates and peridotite, an igneous rock containing high levels of magnesium and iron) allow the upper mantle to behave in a plastic manner, flowing very slowly, up to about one centimeter each year. This provides a rocky sea upon which tectonic plates ride and also serves to conduct heat from the inner layers of Earth. Under more pressure and temperature because of its depth, the lower mantle is denser and more rigid.

The mantle's characteristics vary widely throughout the layer; structural subdivisions are important to keep in mind, and many of these subdivisions refer to areas where seismic waves behave differently than they do in the surrounding areas because of physical differences in the matter. Beginning closest to the surface, the subdivisions are as follows: At a depth of about 35 km (22 mi) under continents and just 5 to 10 km (3-6 mi) under oceans, the Mohorovičić discontinuity (usually referred to as the Moho) separates the mantle from the crust, except at mid-ocean ridges. The Moho is actually the border between the lithosphere and the asthenosphere. This discontinuity was discovered by Croatian seismologist Andrija Mohorovičić in the early twentieth century, who observed unexpected behavior of seismic waves in the region.

The upper mantle includes approximately the bottom 65 km (40 mi) of the rigid, rocky lithosphere (which is broken up into tectonic plates) and all of the asthenosphere, a viscous area that spans the region from 100 to 200 km (62-125 mi) below Earth's surface; parts of it may extend to a much greater depth, nearly 700 km (435 mi) below the surface.

Within the upper mantle, at a depth of about 220 km (137 mi), there is another zone where seismic activity changes abruptly: the Lehmann discontinuity. The velocities of two types of seismic waves, P waves and S waves, increase suddenly at this area, as observed by Danish seismologist Inge Lehmann in 1958.

The transition zone serves as the border between the upper and lower mantle and between the asthenosphere and the mesosphere. It lies between 410 km (255 mi) and 660 km (410 mi) below Earth's surface, and it is marked off by seismic discontinuities at 410 km, at 660 km, and at several other depths within this range. The zone is formed by the changing structure of a substance called olivine within one of the mantle's main components, peridotite. Under increased pressure and temperature as depth increases, olivine's crystalline structure is altered significantly enough to affect the seismic wave paths and velocities.

The lower mantle corresponds to the mesosphere; it is denser and less plastic than the upper mantle. The final 200 km (124 mi) of the mantle is known as the D zone, or the Gutenberg discontinuity, a region marked by an abrupt decrease in seismic wave velocity. This region leads into the core-mantle boundary.

Mantle Convection and Its Effect on Tectonic Plates

Heat transfer by convection within the mantle is the main driving factor of tectonic plate movements (and thus seismic and volcanic activity). Tectonic plates are large chunks of the lithosphere, and they ride slowly along the plastic asthenosphere much like items on a conveyor belt; convection provides the energy that fuels this movement.

In general terms, the process of convection requires a fluid medium such as a liquid or a gas, rather than a solid. While the mantle is generally referred to as solid, the upper layer is plastic enough to support diffusion and advection, two necessary processes that contribute to convection. Diffusion refers to the random movements (Brownian motion) of particles within the medium, while advection refers to the larger-scale moving currents of heat or mass within the medium.

Within the mantle, convection manifests as hot material pushing to the surface while colder material travels down to the core. This movement of matter leads to the large-scale slow movement of the upper mantle as a whole, providing the vehicle for tectonic plates to slowly move as well. One can picture the effect on one tectonic plate: As convection drives hot mantle material up, the material adds to the edge of a nearby tectonic plate by accretion and begins to cool by convection and conduction (direct heat transfer). Meanwhile, the cooler edge of the plate is subducting because it is cooler and denser than the edge, where hot material is being added.

Tectonic activity represents the continuing differentiation of Earth's layers in that the parts of the mantle and crust are continually being broken down and replaced with new material through tectonic-induced seismic and volcanic activity and through the formation of mountains and ocean trenches. Tectonic plates interact at three major types of boundaries: convergent boundaries, divergent boundaries, and transform faults.

At convergent boundaries (also known as collision or destructive plate boundaries), plates move toward each other and meet by colliding or by one plate sliding under the other (subduction). These interactions cause friction, high pressure, and melting, leading to volcanic activity, earthquakes, and mountain formation.

At divergent boundaries (also known as extensional or constructive boundaries), plates move away from each other, allowing convecting plumes of molten magma to flow up into the space from the mantle. Between continental plates, this interaction forms rift valleys. Between oceanic plates, this interaction forms volcanic islands and mid-ocean ridges (underwater mountains).

At transform faults (also known as conservative plate boundaries), material is neither created nor destroyed. These faults are generally located in mid-ocean ridges and between continents. They are typically zigzag-shaped and relieve stress caused by interactions at nearby convergent and divergent boundaries.

THE MANTLE'S EFFECT ON THE EVOLUTION OF EARTH'S ATMOSPHERE

Earth's first atmosphere was likely blown away with the primitive mantle and crust during the impact described by the giant impact hypothesis. From volcanic evidence, scientists infer that this earliest atmosphere was a poisonous mix containing about 60 percent hydrogen; 20 percent oxygen, which includes water vapor; 10 percent carbon dioxide; 5 percent hydrogen sulfide; and a variety of other gases.

The mantle had a significant effect on the formation of Earth's next atmosphere (still not quite like today's atmosphere). After the giant impact, what was left of Earth's mantle began to convect violently because of the heat transferred by the collision. The mantle needed to cool and partially harden to allow for more differentiation—particularly the formation of the crust—but the overall mantle temperature was much higher than it is now, so a higher percentage of the mantle was molten rather than solid. This caused heat to be pushed upward by a process called outgassing: Steam and gases were released through cracks in the crust and expelled from volcanoes, contributing to the formation of Earth's new atmosphere.

The mantle was just one factor of several, though. The creation of Earth's atmosphere also was shaped by the effects of solar radiation, early life forms, and, to a large extent, impacting comets, meteorites, and protoplanets, which delivered ice and water into the atmosphere and onto Earth's surface. Some early life forms contributed oxygen to the atmosphere through photosynthesis.

About 3.5 billion years ago, Earth's magnetic field formed, protecting the atmosphere from being stripped away again by solar wind. Over time, other factors altered the atmosphere into its current state, a stratified set of gaseous layers that protect Earth from solar radiation and that heat the surface by the greenhouse effect. Earth's atmosphere now is approximately 78 percent nitrogen, 21 percent oxygen, and 1 percent argon, carbon dioxide, and other gases; it also includes water vapor.

EXPLORATION OF THE MANTLE

Direct exploration of the mantle can be extremely difficult because most of it is buried under kilometers of crust, well beyond the reach of even the most modern drilling technology. Exploration is usually done in the sea, as the crust is thinner there, but drilling under the sea has complications of its own.

The first major attempt to directly explore the mantle was called Project Mohole; the goal was to drill through the crust and into the Moho discontinuity to learn more about Earth's composition, age, and interior processes. The project involved drilling through the sea floor. Phase I, the experimental drilling of five holes, was largely successful and suggested that the more ambitious second and third phases could be attempted. However, the project was abandoned in 1966 as costs rose and after the organizing research group, the American Miscellaneous Society, dissolved in 1964. From the holes drilled during Project

Mohole's first phase, the deepest reached about 183 meters (200 yards) below the ocean floor.

In 2007, scientists aboard the RRS *James Cook* had a chance to explore exposed mantle at a spot between Cape Verde and the Caribbean Sea, where a large hole exists in the crust. Later that same year, a ship called *Chikyu* set out from Japan to begin a project dubbed Chikyu Hakken; the goal was to drill 7 km under the seabed to reach the mantle—a deeper hole than any previously dug under the ocean. The project was initially expected to be completed in 2012, but the vessel sustained damages in March 2011, during the Tohoku earthquake and resulting tsunami.

An alternative to drilling has been proposed. This alternative is a self-sinking probe filled with radionuclides, whose decay would melt rock around the probe, allowing it to continue sinking. Hypothetically, these probes could reach the Moho underneath oceanic crust in about six months.

Computer simulations provide an easier, yet less direct, approach to exploring the mantle. In 2009, for example, scientists used a supercomputer to model the distribution of various iron isotopes throughout the mantle and the rest of Earth's interior during Earth's differentiation 4.5 billion years ago. Until technology provides an easier way to explore the mantle directly, much can be learned about Earth's interior structure from seismological data.

Rachel Leah Blumenthal

FURTHER READING

Blakey, Ronald C., Wolfgang Frisch, and Martin Meschede. *Plate Tectonics: Continental Drift and Mountain Building*. New York: Springer, 2011. This book provides an introduction to plate tectonics, covering Earth's early history to the present. Topics include subduction zones, mid-ocean ridges, and the formation of mountains.

Dickey, John S. *On the Rocks: Earth Science for Everyone*. New York: Wiley, 1996. From stardust to Earth's formation to Earth's planetary neighbors, this book is ideal for Earth science novices. Easy to understand but comprehensive.

Gilmour, Iain, and Neil McBride. *An Introduction to the Solar System*. New York: Cambridge University Press, 2004. A textbook targeted for introductory college courses, this is an excellent resource for exploring the formation of the earth and other planets. Includes summaries and a glossary.

Lutgens, Frederick K., and Edward J. Tarbuck. *Earth Science*. 13th ed. Upper Saddle River, N.J.: Prentice Hall/Pearson, 2012. Originally published in 1976, this introductory textbook is geared to undergraduates, including those without a science background. Examines geology, astronomy, and other Earth science topics.

Williams, Linda D. *Earth Science Demystified*. New York: McGraw-Hill, 2004. A quick and easy self-teaching guide aimed at readers without formal science training. Covers the basic range of Earth science topics and includes summaries, questions, and a sample "final exam."

Wu, Chun-Chieh. *Solid Earth*. Vol. 26 in *Advances in Geosciences*. London: World Scientific, 2011. This book highlights the results of research papers published in seismology, planetary exploration, the solar system, and other topics relevant to Earth science.

See also: Continental Drift; Creep; Cross-Borehole Seismology; Deep-Earth Drilling; Deep-Focus Earthquakes; Discontinuities; Earthquakes; Earth's Age; Earth's Core; Earth's Differentiation; Earth's Interior Structure; Earth's Lithosphere; Earth's Magnetic Field; Faults: Normal; Faults: Strike-Slip; Faults: Thrust; Faults: Transform; Geodynamics; Lithospheric Plates; Mantle Dynamics and Convection; Metamorphism and Crustal Thickening; Mountain Building; Plate Motions; Plate Tectonics; Seismic Wave Studies; Slow Earthquakes; Stress and Strain; Volcanism.

EARTH'S OLDEST ROCKS

The oldest-known rocks on Earth have absolute (radiometric) ages approaching 3.8 billion years. Although Earth apparently has no rocks resulting from the first 760 million years of its history, rocks with ages ranging back to the earliest age for the terrestrial planets, about 4.56 billion years, occur for many meteorites and are closely approached in age by some rocks from the moon.

PRINCIPAL TERMS

- **absolute date/age:** the numerical timing, in years or millions of years, of a geologic event, as contrasted with relative (stratigraphic) timing
- **geochronology:** the study of the absolute ages of geologic samples and events
- **half-life:** the time required for a radioactive isotope to decay by one half of its original weight
- **isotopes:** species of an element that have the same numbers of protons but differing numbers of neutrons and, therefore, different atomic weights
- **mass spectrometry:** the measurement of isotope abundances of elements, commonly separated by mass and charge in an evacuated electromagnetic field
- **nuclide:** any observable association of protons and neutrons
- **radioactive decay:** a natural process by which an unstable (radioactive) isotope transforms into a stable (radiogenic) isotope, yielding energy and subatomic particles
- **radiogenic isotope:** an isotope resulting from radioactive decay of a radioactive isotope

STRATIGRAPHIC TIME SCALE

Present knowledge of the oldest rocks on Earth developed slowly and descriptively until the 1950's, when it became possible to measure the absolute (quantitative) ages of minerals and rocks by radiometric means. These means involve the instrumental (commonly, mass spectrometric) measurement of unstable (radioactive) and stable (radiogenic) isotopes, or species of elements that differ only in their masses.

Prior to the ability of physicists, chemists, and geologists to make absolute age determinations, the oldest rocks on Earth were known only through field relations. The main field relation used is stratigraphic sequence, which involves an application of the principle of superposition: In a sequence of undisturbed layered rocks such as sedimentary layers and lava flows, the oldest rock unit—that is, the first to be deposited—is at the bottom of the sequence. Another important field principle is the manner in which one rock is cut or cuts another rock unit. The obvious chronological conclusion is that the structure or rock that transects must be younger than the structure or rock that is transected. Through a combination of these stratigraphic and cross-cutting relationships, rock units studied and mapped can be assigned to a relative chronologic order and the geologic history of the mapped area worked out.

Accompanying the development of classical geologic principles was the understanding of the time dependence of biological evolutionary characteristics displayed by fossils found in the enclosing—primarily sedimentary—layers. Although it was early understood that fossil morphology changed through time from simpler to more complex forms, the time required for such evolutionary change could only be guessed. It is a tribute to early geologists and paleontologists that, before a quantitative measure of evolutionary scale was available, it was realized that enormous amounts of time probably were required between the deposition of rocks containing, for example, fossil collections of extinct marine animals such as trilobites and the deposition of those of containing fossils of horses.

The geologist's most important document, the stratigraphic column (the geologic time scale), was developed over the past several hundred years through the cumulative observations of field relations, paleontologic studies, and absolute dating methods. Its refinement will continue to be an important result of geologic endeavor. Correlation, the principal activity of the geologic study of stratigraphy, whereby rock units are related through their temporal and physical characteristics, was pioneered by William Smith in 1815. It enabled scientists to have some sense of the earth's oldest rocks, long before numbers of years could be assigned to paleontologic and physical geologic phenomena. Nevertheless, the Precambrian era—the vast period of geologic time that encompasses more than 85 percent of the known age of the earth—was not known to have harbored life or to have provided fossils until the past few decades. The discovery of widespread bacterial and stromatolitic fossils in Precambrian rocks has reversed this conclusion. Prior to these discoveries, the ages of the earliest rocks thus were surmised only through field relations and not through fossils.

RADIOMETRIC AGES

A misunderstanding of old rocks on Earth occurred because of the reasoning that the older the rock, the more opportunity for it to have altered, such as by weathering, tectonism (as in mountain building), or especially metamorphism. Thus, it was expected that the oldest rocks should be highly metamorphosed, as in much of the Precambrian terrain of Canada and other, commonly central continental areas of Precambrian rock (cratons or shields), and that essentially unaltered sediments and sedimentary rocks must be geologically young. Miscalculations in geologic age of billions of years occurred, owing to the incorrect correlation of rocks of similar petrology and metamorphic grade. Absolute age determinations, while not negating the essential premise of this theory, have nevertheless shown that some of the oldest rocks on Earth are not highly altered and that many young rocks may be highest-grade metamorphic and tectonized.

A major advance in geochronology has developed since radiometric ages were attached to points of the stratigraphic time scale and a quantitative framework for major fossil assemblages was established. Once the ages of characteristic, representative fossils are quantified through absolute age determinations, the ages of sedimentary rocks containing chronologically diagnostic fossils can be assigned by comparison of these "guide" fossils with points on the stratigraphic time scale. Thus, the field geologist may establish the approximate age of sediments or sedimentary rocks in his or her area of interest (and, through field relations, the qualitative ages of associated igneous and metamorphic rocks and of geologic structures), simply through fossil identification. Although fossils, especially diagnostic fossils, are rare in many sedimentary rocks (especially those sedimentary rocks that formed prior to about 600 million years) and are absent in most igneous and metamorphic rocks, the use of paleontology as a chronologic tool is routine and in most cases quicker and less expensive than are geochemical (radiometric) age determinations.

RADIOACTIVE DECAY

Antoine-Henri Becquerel presented his discovery of the phenomenon of radioactivity to the scientific community in Paris in 1895, laying the cornerstone for scientists' present understanding of Earth's oldest rocks. The finding was followed rapidly by the seminal work of Marie Curie in radioactivity, a term she was the first to use. Her discovery of the intensely radioactive radium as well as plutonium led Ernest Rutherford to distinguish three kinds of radioactivity—alpha, beta, and gamma—and, in 1910, with Frederick Soddy, to develop a theory of radioactive decay. Soddy later proposed the probability of isotopes, the existence of which was demonstrated on early mass spectrographs and mass spectrometers.

Rutherford and Soddy's theory of the time dependence of radioactive decay, followed by breakthroughs in instrumentation for the measurement of these unstable species and their radiogenic daughter nuclides by Francis William Aston, Arthur Jeffrey Dempster, and Alfred Otto Carl Nier, among others, caught the rapt attention of early geochronologists and had a revolutionary effect on the study of geology. In 1904, Rutherford proposed that geologic time might be measured by the breakdown of uranium in U-bearing minerals and, a few years later, Bertram Boltwood published the "absolute" ages of three samples of uranium minerals. The ages, of about half a billion years, indicated the antiquity of some earth materials, a finding enthusiastically developed by Arthur Holmes in his classic *The Age of the Earth*. Holmes's early time scale for Earth and his enthusiasm for the developing study of radioactive decay, although not met with immediate acceptance by geologists of his era, helped to set the stage for the acceptance of absolute age as the prime quantitative component in the study of geology and its many subdisciplines.

After the early study of the isotopes of uranium came the discovery of other unstable isotopes and the formulation of the radioactive decay schemes that have become the workhorses of geochronology, such as the rubidium-strontium, samarium-neodymium, potassium-argon, uranium-thorium-lead, and fission track methods. The formulation of the theory of radioactive decay of the parent, unstable nuclide (or the growth of the stable daughter nuclide), developed in the early 1900's, is the basis for the measurement of time, including geologic time, for any of these parent-daughter schemes used by geochronologists. Although each of the dating techniques is based on the formulation, differences occur in the kind of measurement and in the geochemical behavior of the several parent and daughter species. Thus, the geological interpretation of the data obtained is very different for the several chronometric schemes. These techniques for establishing absolute ages for minerals and rocks have been applied to the study of Earth's oldest rocks since the early 1900's and, with the development of mass spectrometry, more intensely since the 1950's.

DISTRIBUTION OF EARTH'S OLDEST ROCKS

Although not indigenous to Earth, the oldest rocks found on Earth (and also seen to fall to Earth) are the meteorites, many of which yield radiometric ages near 4.56 billion years, the accepted time of formation of many solar system materials. With respect to the oldest rocks indigenous to Earth, these rocks have the highest probability of being destroyed by ongoing geologic processes such as erosion, metamorphism, and subduction. It is no surprise that fewer and fewer outcrops are found as ages become older, deeper into Precambrian time. The oldest rocks are most commonly found in continental, cratonic regions because of geologic preservative features such as their protective superjacent rocks, their location in tectonically stable continental interiors, and their low density and thus lower propensity for subduction than the more common basaltic rocks.

Histograms of rock ages thus show fewer and fewer data the further back in time they go. Such figures also show a feature whose significance has not been immediately apparent: the clustering of ages in rather discrete groupings. These groupings correlate with regionally defined rock/tectonic units such as those of the Grenville and Superior provinces of the Canadian Shield and indicate the intense geologic activity that resulted in these Precambrian rocks. Many scientists believe that the "magic numbers" that mark the groupings represent geologic periodicity, perhaps a result of major, discrete plate tectonic episodes. Others, however, point out that many radiometric dates fall outside these groupings and that the picture is incomplete and thus misleading. Although certainly incomplete, the available data indicate to many that there is some patterning in both the chemical and chronologic analyses of these rocks.

Lewisian gneiss exposed in a quarry in South Uist, Scotland. Lewisian gneisses are among the oldest rocks in the world, having formed about 3 billion years ago. (Leo Batten/FLPA/Photo Researchers, Inc.)

Early radiometric results showed some ages far back in time, near 3 billion years, and further analyses confirmed their antiquity. The oldest rock was thought to be the Morton gneiss in Minnesota, at about 3.2 billion years and questionably older, until several cratons yielded rocks with ages near 3.5 billion years. One such exposure, at North Pole, Australia, is of special significance, because of the concurrence on its age by several chronometric schemes (3.5 billion years) and especially because of its well-preserved bacterial and stromatolitic fossil assemblage, the earliest known. (Equivocal chemical evidence for organic life in even older rocks has been described; the existence of well-developed life at 3.5 billion years presupposes the existence of earlier life.) The oldest rocks, however, appear to be the well-studied Amîtsoq gneiss and contiguous, related rocks in the Godthaab area of western Greenland. Although there is incomplete agreement as to the exact range and significance of these earliest ages, several are close to or perhaps slightly greater than 3.8 billion years. In 2008 the oldest rock on earth was discovered in the Nuvvuagittuq greenstone belt on the coast of Hudson Bay, in northern Quebec, and is dated from 4.28 billion years old. Some of the disagreement with respect to these rocks, as well as for similar rocks around the world, results from incompletely known and undoubtedly variable diffusive and "freezing-in" behavior of the parent/daughter nuclides of the several chronometric systems. This varying behavior commonly results in different "ages" (dates) for the same analyzed rock specimen. A further uncertainty is whether the several isotopic systems can be completely reset, on the whole-rock scale, in metamorphic terrains that have been metamorphosed to lower physicochemical conditions.

Although not all scientists may agree, minerals of even older ages have been analyzed from Archean sandstones of Australia. Zircons (residual mineral phases from the final stages of crystallization of igneous rocks, especially granites) were separated from this stratigraphic unit and analyzed by uranium-thorium-lead dating using an innovative technique, the ion probe mass spectrometer. Although many of these zircons have been analyzed, only a few have exceptionally old ages. However, these ages, ranging back to almost 4.3 billion years, are especially important. Because they are detrital (fragmental) in their host sandstone, they must have eroded from even more ancient rocks, perhaps granites or granitic gneisses, whose age, composition, and petrologic features are important to an understanding of the development of Earth's earliest crust. So far, their provenance (parental rocks) has not been found; apparently they have been completely eroded, altered, or buried by younger rock. If one accepts these earliest ages, crustal rocks existed on Earth less than 300 million years after Earth accreted from the solar nebula.

Analogy with Extraterrestrial Materials

It is useful to place Earth's oldest rocks within the framework of the ages of other available solar system materials, especially meteorites. Although the formation of the Earth—that is, Earth's time of accretion—is accepted by most scientists as having occurred about 4.56 billion years ago, it is obvious from the discussion above that no terrestrial rocks have ages this old. Earth's absolute age, therefore, as well as that of other solid materials of the solar system except for the sun, is known only by analogy with meteorites. Many of the meteorites have been dated by the techniques discussed earlier and give formational ages near 4.56 billion years; some are thought to represent the oldest and most primitive material in the solar system with the possible exception of cometary material and cosmic dust. A few apparently unprocessed (primitive) meteorites yield radiometric age and initial isotopic composition data that suggest formational ages slightly older than 4.56 billion years. The terrestrial planets (Earth, Mercury, Venus, and Mars) are thought to have originated at the same time as did the meteorites.

A few meteorites have ages significantly younger than 4.56 billion years. These rocks are considered to have originated from parent bodies that were large enough to have maintained internal heat and, therefore, igneous processes significantly after 4.56 billion years, as did the earth, with its continuing volcanism and other geologic processes that have resulted in rocks of all ages from 4.56 billion years to the present. Several of these exotic rocks, collected from ice fields in Antarctica, were recognized almost immediately as pieces of the moon, owing to scientists' familiarity with the Apollo missions' lunar rock collections. Even more spectacularly, a small collection of meteorites, long known to be different from the main collection of meteorites, were found to have crystallization ages of about 1.3 billion years, much younger than the accepted accretion age for solar system materials. These rocks must have originated from a body large enough to have maintained geologic processes between 4.56 and 1.3 billion years, unlike the moon, which has so far yielded no rocks younger than about 3.0 billion years. This parent body is widely assumed by scientists to be Mars, a theory that is much strengthened by the compositional similarity of gases dissolved in glass from these meteorites and atmospheric compositions of present Mars, as measured from the Viking lander in the 1970's.

Rocks returned from the moon by U.S. and Soviet space programs yield ages from 0.8 to 4.54 billion years. Although the moon is thought to have originated at the same time as the earth, it is not massive enough to have provided a continuing internal heat source to drive volcanic or tectonic processes to the present time. Instead, significant igneous activity decreased substantially after 3.2 billion years ago until approximately 1 billion years ago. A current and popular theory is that the moon originated by accretion in Earth's orbit from material ejected from Earth after a grazing impact with a Mars-size object. This hypothesis explains why the moon has a composition similar to the Earth's mantle but is poor in volatile elements, and why the moon's core is so small. Such an impact would be responsible for resetting the radiometric dates on the moon and perhaps the earth as well.

A widespread though not fully accepted theory for the early moon is that it underwent a massive, perhaps global melting not long after formation (whether by nebular accretion or Earth impact). Upon cooling, plagioclase feldspar crystallized, floated, and formed the earliest lunar crust (anorthosite), which thus dates from some time after moon accretion. If this theory is correct, no rocks older than the anorthosite will be found; this rock has yielded ages of 4.44 billion years and, arguably, somewhat older. If the moon underwent significant or complete melting, it is possible or likely that the earth experienced the same event, in which case there also will be no Earth rocks representing its earliest history. Finally, owing to Earth's continuing history of constructive and destructive geologic processes, it seems unlikely that significant amounts of rock will be found that date from Earth's first 200 million years.

GEOLOGIC APPLICATIONS

The absolute dating of geologic materials and events has had unprecedented influence on the evolution and understanding of geologic events on Earth, including Earth's origin and its oldest rocks, as well as other ancient minerals and rocks of the solar system. The ability of scientists to establish events in terms of actual years, rather than in relative terms such as "older than" or "younger than," has led to a realistic knowledge of Earth's origin and its oldest rocks and has led to calibrated time scales for major geologic and biologic processes such as organic evolution. Owing to their usefulness in the precise determination of the ages of very old rocks, dating methods such as uranium-thorium-lead, rubidium-strontium, and samarium-neodymium continue to be of major use in refining the sequence and meaning of Earth's oldest rocks and extraterrestrial materials.

E. Julius Dasch

FURTHER READING

Ashwal, L. D., ed. *Workshop on the Growth of Continental Crust.* Technical Report 88-02. Houston, Tex.: Lunar and Planetary Institute, 1988. A technical but interesting series of articles that bear directly on Earth's oldest rocks and related material. Suitable for college-level readers.

Blatt, Harvey, and Robert J. Tracy. *Petrology: Igneous, Sedimentary, and Metamorphic.* 3rd ed. New York: W. H. Freeman, 2005. Undergraduate text in elementary petrology for readers with some familiarity with minerals and chemistry. Thorough, readable discussion of most aspects of Earth's rocks. Abundant illustrations and diagrams, good bibliography, and thorough indices.

Faure, Gunter. *Isotopes: Principles and Applications.* 3rd ed. New York: John Wiley & Sons, 2004. This textbook, originally titled *Principles of Isotope Geology,* is an excellent though technical introduction to geochronology and the use of radioactive isotopes in geology and includes a thorough treatment of several dating techniques. Well illustrated and indexed. Suitable for college-level readers.

Grotzinger, John, et al. *Understanding Earth.* 5th ed. New York: W. H. Freeman, 2006. An excellent general text on all aspects of geology, including the formation of igneous and metamorphic rocks. Contains some discussion of the structure and composition of the common rock-forming minerals. The relationship of igneous and metamorphic petrology to the general principles that form the basis of modern plate tectonic theory is discussed. Suitable for advanced high school and college students.

Hall, Anthony. *Igneous Petrology.* 2d ed. Harlow: Longman, 1996. This introductory book provides a good understanding of igneous rocks and their geophysical phases. There are sections devoted to the study of petrology and magmatic processes. Well illustrated with plenty of diagrams and charts to reinforce concepts. This is a good resource for the layperson.

Keer, Richard A. "Geologists Find Vestige of Early Earth—Maybe World's Oldest Rock." *Science* 321 (2008): 1755-1755. A short article discussing the discovery of what is considered the world's oldest rock, found in Canada.

Tarbuck, Edward J., Frederick K. Lutgens, and Dennis Tasa. *Earth: An Introduction to Physical Geology.* 10th ed. Upper Saddle River, N.J.: Prentice Hall, 2010. This college text provides a clear picture of the earth's systems and processes that is suitable for the high school or college reader. It has excellent illustrations and graphics. Bibliography and index.

Taylor, S. R., and S. M. McLennan. *The Continental Crust: Its Composition and Evolution.* Reissued, Oxford, England: Blackwell Scientific, 1991. A technical review of processes contributing to the formation of Earth's oldest rocks. Suitable for college-level readers.

van Kranendonk, Martin, Hugh Smithies, and Vickie Bennett, eds. *Earth's Oldest Rocks: Developments in Precambrian Geology.* Amsterdam: Elsevier, 2007. A compilation of articles discussing early Earth. Provides an overview of the earth's formation, age, and tectonics. Best suited for researchers, graduate students, and advanced undergraduates.

Walker, Mike. *Quaternary Dating Methods.* New York: Wiley, 2005. This text provides a detailed description of current dating methods, followed by content on the instrumentation, limitations, and applications of geological dating. Written for readers with some science background, but clear enough for those with no prior knowledge of dating methods.

York, Derek, and Ronald M. Farquhar. *The Earth's Age and Geochronology.* Reprint. Oxford, England: Pergamon Press, 1975. Contains good accounts of the chronologic techniques required to date rocks and Earth's age but does not include the more recent work on the oldest rocks. Technical but suitable for college-level readers.

See also: Earth's Age; Experimental Petrology; Experimental Rock Deformation; Mass Spectrometry; Metamorphism and Crustal Thickening; Petrographic Microscopes; Potassium-Argon Dating; Radioactive Decay; Radiocarbon Dating; Relative Dating of Strata; Rock Magnetism; Rubidium-Strontium Dating; Samarium-Neodymium Dating; Uranium-Thorium-Lead Dating; Volcanism; Water-Rock Interactions.

EARTH TIDES

Earth tides are deformations of the crust of the earth as a result of gravitational interaction with the moon and the sun. Knowledge of the effects of these tidal forces is important to earth scientists who search for natural resources.

PRINCIPAL TERMS

- **deformation:** the alteration of an object from its normal shape by a force
- **gravimeter:** a device that measures the attraction of gravity
- **homogeneous:** having uniform properties throughout
- **oblate spheroid:** a spherically shaped body that is flattened at the polar regions
- **oscillate:** to fluctuate or to swing back and forth
- **pendulum:** a mass suspended in such a way that it can swing freely
- **perturb:** to change the path of an orbiting body by a gravitational force
- **synchronized rotation-revolution:** a situation in which the rotation rate of a body is equal to its rate of revolution

GRAVITATIONAL ATTRACTION

Earth tides are the deformation of the solid portion of the earth by the combined gravitational forces of the moon and the sun. Although other bodies within and beyond the solar system gravitationally attract the earth, the distances are great enough to make their tidal effect upon the earth negligible. Consider the Earth-moon system. According to Sir Isaac Newton's law of gravity, every particle of mass in the universe is attracted to every other particle of mass by a force that is directly proportional to the product of the masses and inversely proportional to the square of the distance between them. This means that gravity is always an attractive force, but its magnitude depends to a great extent upon the distance between the two bodies in question.

Since gravity is an inverse square law, the following relationship holds true: If the distance between two bodies is doubled, the attraction of gravity becomes one-fourth as great. If the distance between the bodies is tripled, the attraction becomes one-ninth as great, and so on. According to this law, each particle of the moon attracts each particle of the earth. Because these particles are not all equidistant from one another, the force of gravity varies in intensity.

Gravitational attraction is greatest between the particles that are closest. Therefore, the surface of the earth nearest the position of the moon is subjected to more attraction than is the surface of the earth opposite the moon. It is this difference in relative position that causes the tidal force and thus the deformation of the earth.

Albert Michelson measured the earth tides in 1913 by observing water tides in long horizontal pipes. He had assumed that the earth was rigid, but he did not observe the tidal values that the theory indicated he should. He used two 500-foot pipes at right angles, with 6-inch diameters and half-filled with water. They were buried in 6-foot deep trenches with concrete-lined viewing pits at the ends. As expected, there was more deformation due to tides in the north-south pipe than in the east-west pipe. The difference could be accounted for when the earth was assigned a rigidity so that it was able to respond to lunar gravitational forces by raising crustal tides to a height of several centimeters.

OCEAN TIDES

The ocean tides may be considered as being analogous to the earth tides. Like earth tides, ocean tides are caused by the gravitational forces of both the sun and the moon. Because of its relative closeness, the moon is the greater factor. Its gravitation causes the water in the oceans to bulge outward a distance of one meter or so. There are two water bulges on the surface of the earth: one in the direction of the moon and one in the direction opposite the direction of the moon. This latter bulge forms because of the reduced amount of gravity at that position on the earth's surface. Another way of looking at it might be as follows: The earth is being attracted toward the moon or, in a sense, is falling toward the moon. Therefore, the water on the lunar side is falling toward the moon and is actually ahead of the earth's surface. The water on the opposite side of the earth is also falling toward the moon but cannot quite keep up with the earth's surface and so forms a bulge. Theoretically, as the earth rotates with respect to the moon, the water level rises and falls as these bulges of water are swept around the earth. In reality, the height and timing of tides may vary considerably. In some bays, the tidal water may accumulate to heights of 10 meters and greater. Because there are two tidal bulges, there are two high tides per day.

The sun also exerts a tidal force on the earth, but because of its greater distance, its influence is only about one-half as great as the moon's. Extremely large high tides are generated when the sun, the moon, and the earth lie along a straight line. The tidal forces of the sun and the moon then act in the same direction. These tides are known as spring tides, though they have nothing to do with the season. The nature of the ocean tides provides an immediate observation and a fairly simple observation of the nature of tidal forces.

EARTH'S SHAPE

Tidal forces also have an effect on gravity, as does the shape of the planet. The ancient Greeks taught that the earth is a sphere. The philosopher Plato reasoned that all heavenly bodies are perfect and therefore must be spherical; because the earth was a heavenly body, its shape was thus spherical. In about the year 230 B.C.E., Eratosthenes calculated the circumference of the earth to be 12,560 kilometers, which is only 112 kilometers less than the current estimate. During the seventeenth century, several measurements were made on the earth's surface. The size of one degree of arc in the Northern Hemisphere proved to be somewhat smaller than a degree of arc farther south. It was concluded from these studies that the earth is flattened toward the poles and thus is not spherical. The shape of the earth is rather an oblate spheroid, as explained by Newton in his famous work of 1687, *Principia*.

If the earth were a perfect sphere and homogeneous in composition, the gravity measurements at all points on the surface would be identical and the orbits of earth satellites would be perfectly circular or elliptical. Because the earth's gravitational field is uneven, resulting from the fact that the earth is neither perfectly spherical nor homogeneous, the orbits of satellites are somewhat perturbed. The paths of satellites can be observed and plotted with a high degree of precision. The data indicate that the earth is an oblate spheroid, its radius 21 kilometers longer at the equator than at the poles. It behaves as though it were a fluid balanced between gravitational forces, which tend to make it spherical, and centrifugal forces resulting from its rotation, which tend to flatten it.

ACCELERATION OF GRAVITY

The acceleration of gravity near the earth's surface is measured in gals, in honor of Galileo. A gal is the amount of force that will accelerate a mass 1 centimeter per second per second, or 1 centimeter per second squared. The total value for the acceleration of gravity is 980 gals, which is equivalent to the more familiar value of 9.8 meters per second squared. It is known that when the moon is directly overhead, at a position known as the zenith, the value for the acceleration of gravity at that point on the earth's surface is slightly less than if the moon were in any other position. This phenomenon is a result of the gravitational influence or tidal force that the moon exerts on the earth. The attraction of the moon's gravity causes a point on the earth's surface to be distended slightly. Values for the amount of distension have been found to be about 0.073 meter. The fact that this point on the earth's surface has been gravitationally pulled away from the center of the earth will result in a slightly reduced value in the acceleration of gravity toward the center of the earth. These values have been found to be in the vicinity of 0.2 milligal. (A milligal is one thousandth of a gal.)

SYNCHRONIZED ROTATION-REVOLUTION

Subtle effects of tidal forces on the earth exist. When the earth and the oceans are subjected to tide-raising forces, energy caused by friction is dissipated. The result of this friction is the reduction in the period of the earth's rotation. In the case of a binary system such as the earth and the moon, the result of tidal forces produces a synchronized state of rotation-revolution. In other words, the rate that the moon rotates on its axis is the same as the rate at which it revolves around the earth in its orbit—which is the reason that the same face of the moon always points toward the earth. This particular phenomenon occurs elsewhere in the solar system; for example, the sun and Mercury, as well as Pluto and its moon Charon, form other such binary systems.

There is a law in physics that states that angular momentum is conserved. If the rotation rates of the earth and the moon are slowing but their masses stay the same, the distance between them must be increasing. Evidence from paleontological studies indicates that at one time, the earth had a faster rotation rate and the moon was much closer than it is today. It is now known that the moon is moving away from the earth 3.2 centimeters per year.

Vertical and Linear Deformation Studies

At the beginning of the nineteenth century, the concept that the earth was not perfectly rigid but in fact was somewhat deformable began to be accepted. The first studies of the deformation of the earth's crust were conducted in France in the early 1830's. These early studies were accomplished by using containers of mercury and comparing the motion of the liquid metal with the rise and fall of the ocean tides. The horizontal pendulum was the first instrument to record the effect of earth tides with scientific precision. It consisted of a rigid bracket whose base contained three leveling screws. At both the top and the bottom of the bracket (which resembled a C-clamp), two metal wires were attached. These wires were all attached to a metal arm in such a way as to suspend it in position. At the end of the arm was attached a small mass. The slightest vibration caused by changes of the ground would cause the pendulum arm to begin oscillating back and forth. This instrument was but the first of many types and variations of the pendulum.

In the 1900's, the gravimeter came into use in the field of exploration geophysics and was later used to detect the minute changes in gravity brought about by earth tides. Gravimeters are designed to measure the differences in the acceleration of gravity. There are several different types of these instruments, most of which consist of a mass suspended by springs. The greater the force, such as gravity, pulling on the mass, the more the spring stretches. The upward force is a function of the strength of the spring, or the spring constant. When the mass is in balance (not oscillating) the spring constant is equal to the force of gravity. Any change in gravity will then produce a corresponding change in the stretch of the spring. During a period of a maximum earth tide, gravity will be slightly reduced, resulting in a slight upward drift of the mass.

The pendulums and the gravimeter are used to study the vertical deformation of the earth's surface. The linear deformation may be measured by means of a device called an extensometer. The first results from the use of this device were reported in the early 1950's. The extensometer consists of a wire 1.6 millimeters in diameter that is held nearly horizontal between two fixed supports about 20 meters apart. A mass of 350 grams is suspended from the center of the wire by a smaller wire with a diameter of 0.2 millimeter. Variations in the 20-meter distance between the two fixed supports as a result of linear deformations of the earth's surface can cause variations in the tension of the main wire. These variations cause the suspended mass to oscillate vertically. By methods of calibration, the oscillation can be translated into values of linear deformation.

Economic and Geologic Applications

The knowledge of how earth tides function is necessary for an understanding of the deformable nature of the earth and of the earth's gravitational interaction with the moon and the sun. This knowledge is important to those who explore for the oil, gas, groundwater, and minerals that are necessary for life in the modern world. To the geophysicists who use the technique of gravity surveying, it is necessary to know whether the change in the value of gravity indicated by their instruments is caused by a subsurface geological structure or by the gravity of the moon or the sun.

For this reason, gravity surveyors must make what is known as a tide correction, which accounts for the time-varying gravitational attraction of the sun and the moon. The attraction is cyclic because the positions of the sun and moon are constantly changing with regard to a fixed position on the surface of the earth. To those earth scientists who use the technique of searching for magnetic anomalies, or areas where the earth's magnetism is greater or less than expected, the sun's effect on the earth's magnetic field is very important. The sun's tidal force produces wind currents in the earth's ionosphere in the same way that it produces ocean tides. Since these winds in the ionosphere consist of waves of charged particles, there is an associated electric current. With this current comes a fluctuating magnetic field. The geophysicist, therefore, needs to know if the sun's tidal force is causing deviations in the equipment being used.

David W. Maguire

FURTHER READING

Baugher, Joseph F. *The Space-Age Solar System.* New York: John Wiley & Sons. 1988. A well-illustrated, very readable volume on the planets, moons, and other bodies that make up our solar system. Suitable for the layperson.

Davidson, Jon P., Walter E. Reed, and Paul M. Davis. *Exploring Earth: An Introduction to Physical Geology.* 2d ed. Upper Saddle River, N.J.: Prentice Hall, 2001. An excellent introduction to physical geology, this book explains the composition of the earth, its history, and its state of constant change. Intended for high-school-level readers, it is filled with colorful illustrations and maps.

Hamblin, William K., and Eric H. Christiansen. *Earth's Dynamic Systems.* 10th ed. Upper Saddle River, N.J.: Prentice Hall, 2003. This geology textbook offers an integrated view of the earth's interior not common in books of this type. The text is well organized into four easily accessible parts. The illustrations, diagrams, and charts are superb. Includes a glossary and laboratory guide. Suitable for high school readers.

Howell, Benjamin F. *Introduction to Geophysics.* New York: McGraw-Hill, 1959. A technical volume dealing extensively with various areas in the study of geophysics. Topics such as seismology and seismic waves, gravity, isostasy, tectonics, continental drift, and geomagnetism are covered. The reader should have a working knowledge of differential and integral calculus. Suitable for college students of physics or geophysics.

McCully, James Greig. *Beyond the Moon: A Conversational, Common Sense Guide to Understanding the Tides.* Hackensack, World Scientific Publishing, 2006. Written in a manner that can easily be understood by the layperson, this text still covers the physics concepts behind tidal motions. The author gradually guides the reader through the topics to reach a strong understanding by the end.

Melchior, Paul. *The Earth Tides.* Elmsford, N.Y.: Pergamon Press, 1966. A highly detailed, highly technical volume on the discovery and the observation of earth tides. Goes into great detail on the evolution of the instrumentation used for earth tide detection. Suitable for college students of geophysics or engineering.

Robinson, Edwin S., and Cahit Coruh. *Basic Exploration Geophysics.* New York: John Wiley & Sons, 1988. A well-illustrated volume dealing with the science of geophysics both in theory and in applications. Contains well-developed chapters on seismic, gravity, and magnetic exploration techniques. The reader should have a working knowledge of algebra and trigonometry. Suitable for college students of geology, geophysics, or physics.

Spencer, Edgar W. *Dynamics of the Earth.* New York: Thomas Y. Crowell, 1972. An introduction to the principles of physical geology. Covers all aspects of geology, from introductory mineralogy through a study of the agents that shape the planet's surface. Concludes with units on global tectonics and geophysics. These later chapters tend to be somewhat technical, requiring the use of algebra. Suitable for college-level geology students.

Tarbuck, Edward J., Frederick K. Lutgens, and Dennis Tasa. *Earth: An Introduction to Physical Geology.* 10th ed. Upper Saddle River, N.J.: Prentice Hall, 2010. This college text provides a clear picture of the earth's systems and processes that is suitable for the high school or college reader. It has excellent illustrations and graphics. Bibliography and index.

Wilhelm, Helmut, Walter Zuern, Hans-Georg Wenzel, et al., eds. *Tidal Phenomena.* Berlin: Springer, 1997. A collection of lectures from leaders in the fields of earth sciences and oceanography, *Tidal Phenomena* examines Earth's tides and atmospheric circulation. Complete with illustrations and bibliographical references, this book can be understood by someone without a strong knowledge of the earth sciences.

Zeilik, Michael, and Elske Smith. *Introductory Astronomy and Astrophysics.* New York: Saunders College Publishing, 1987. A technical volume having to do with such topics as celestial mechanics, interactions of gravitational bodies, the planets, the origin of the solar system and the universe, stars, and cosmology. Some advanced mathematics is used. Suitable for college students of astronomy or astrophysics.

See also: Earth-Moon Interactions; Earth's Interior Structure; Experimental Rock Deformation; Geodynamics; The Geoid; Gravity Anomalies; Importance of the Moon for Earth Life; Lunar Origin Theories; Metamorphosis and Crustal Thickening; Plate Motions; Rock Magnetism; Solar Wind Interactions; Stress and Strain.

ELASTIC WAVES

The vibrations of the earth, felt as earthquakes, are elastic waves in soil and solid rock. These waves are similar to sound waves, which travel through the air, and sonic waves, which travel through water.

PRINCIPAL TERMS

- **body wave:** a seismic wave that propagates interior to a body; there are two kinds, P waves and S waves, that travel through the earth, reflecting and refracting off of the several layered boundaries within the earth
- **elastic material:** a substance that, when compressed, bent, stretched, or deformed in any way, undergoes a degree of deformation that is proportional to the applied force and returns back to its original shape as soon as the force is removed
- **homogeneous:** having the same properties at every point; if elastic waves propagate in exactly the same way at every point, they are homogeneous
- **ideal solid:** a theoretical solid that is isotropic, is homogeneous, and responds elastically under applied forces, stresses, compressions, tensions, or shears
- **isotropic:** having properties the same in all directions; if elastic waves propagate at the same velocity in all directions, they are isotropic
- **reflection:** when an elastic wave strikes a boundary between two substances or between two rock layers of different seismic velocities, part of the incident ray bounces back (reflects)
- **refraction:** when an elastic wave passes through a boundary between two rock layers of different seismic velocities, the rays passing through are bent (refracted) in another direction
- **surface wave:** a seismic wave that propagates parallel to a free surface and whose amplitudes disappear at depth; there are two kinds—"Rayleigh waves" (first described in 1885) and "Love waves" (first described in 1911), that travel at the surface around the earth

ELASTIC BEHAVIOR

Elastic waves are experienced frequently every day: Everything heard is an elastic wave in the air; every vibration felt, in the ground as a truck passes or in the floor from vibrations in a building, is from elastic waves in solid matter. Although the experience of elastic wave energy is familiar, the exact nature of this phenomenon is not something visible to the eye or easily described in a visual way. Ripples resulting from dropping an object in a still body of water move in ever-increasing circles away from the splash; such waves are not elastic, but rather are gravity waves. Yet, elastic waves are analogous to this example in that they originate from a disturbance and propagate outward and away in concentric circles or spheres.

An elastic wave moves in an elastic medium, which can be a solid, liquid, or gas. A substance is said to respond elastically if when it is compressed, stretched, bent, or submitted to shear forces, it deforms in proportion to the applied force and then immediately returns to its original unstressed state when the force is removed. A good illustration of this property is a spring scale. A 1-kilogram weight placed on the scale will cause the spring inside the scale to be stretched (deformed) into a longer length, such as dropping 1 centimeter. A 2-kilogram weight on the scale would cause it to move down 2 centimeters. Hence, the displacement of the spring is proportional to the force applied. When the weights are removed from the scale, it immediately returns to zero, its original unstressed length and shape. This is elastic behavior. If the spring of the scale stretched 1 centimeter for 1 kilogram and then stretched more or less than a centimeter for the second kilogram, it would not be elastic because it would not be a proportional response. Also, if the spring did not return to zero after the weight was removed but retained some permanent deformation, it would not be elastic.

In the case of wave propagation in the earth, consider the effect of striking the ground with a sledgehammer. When struck, the ground would be suddenly compressed, which would be transmitted to the neighboring soil and rock around and beneath the strike. Except in the immediate vicinity of the blow, where permanent deformation (non-elastic) may occur, the response of the neighboring soil and rock would be elastic. It would be temporarily compressed by the force of the blow and then immediately relax back into the former condition. A compression wave would irradiate spherically away from the blow, traveling across the surface like the ripples on a pond and down into the earth. In the passing of an elastic wave, the medium passing the wave is restored to its original unstressed state as if no wave had ever come through at all.

ELASTIC WAVE VELOCITIES

Elastic waves travel at certain velocities depending on the density and elastic stiffness or compressibility of the medium. If a substance is soft, elastic waves move more slowly; if it is very stiff, elastic waves move rapidly. In air, sound waves move at approximately 300 meters per second; in water, sonic waves move at roughly 1,200 meters per second. In rock, compressional waves move at a rate of from 3,000 to more than 10,000 meters per second, depending on the rock's hardness and how deeply it is buried.

If within a medium through which elastic waves can move the velocity is the same everywhere, that medium is called "homogeneous." If, in addition, at any given point in that medium the velocities are the same in all directions, the substance is termed "isotropic." In most rocks of the earth, which occur in layers, the deeper below the surface, the higher the velocity becomes. The increasing weight of the overburden acting on rocks found deeper in the earth causes their density and stiffness to change, and, generally, in the vertical direction the velocity of a seismic wave is different from its velocity in horizontal directions. Thus, many rocks of the earth are not isotropic; neither are they homogeneous. By analyzing seismograms from earthquakes, quarry blasts, and underground nuclear explosions, the inhomogeneities and anisotropies of the earth have been described to give a picture of what the earth's interior is like.

ELASTIC WAVE TYPES

There are two basic kinds of elastic wave: P waves, or compressional waves, and S waves, or shear waves. P waves are sometimes called "push-pull" waves because they consist of a series of pushes (compressions) and pulls (rarefactions), where the motion of a particle of matter as the wave passes by is parallel to the direction the wave passed. S waves are sometimes called "shake" or "shear" waves, because they consist of shearing or shaking motions where the movement of a particle of matter as the wave passes by is transverse, or perpendicular, to the direction the wave passed. A "Slinky" toy spring, held in two hands, can provide an illustration of sending waves back and forth. The alternate stretched and compressed parts of the spring move from one end to the other. If a long rope is tied to a post and the end is shaken up and down, a wave will move from the shaken end to the post, but the motion of the particles of the rope are up and down, transverse to the wave motion. P waves can move in all substances, solid, liquid, or gas. S waves can move only in solids. Compressional and shear waves are the only types that can propagate anywhere interior to a solid, like the rocks of the earth. These are called "body waves." Earthquakes generate both compressional and shear waves at the source where the fault moves.

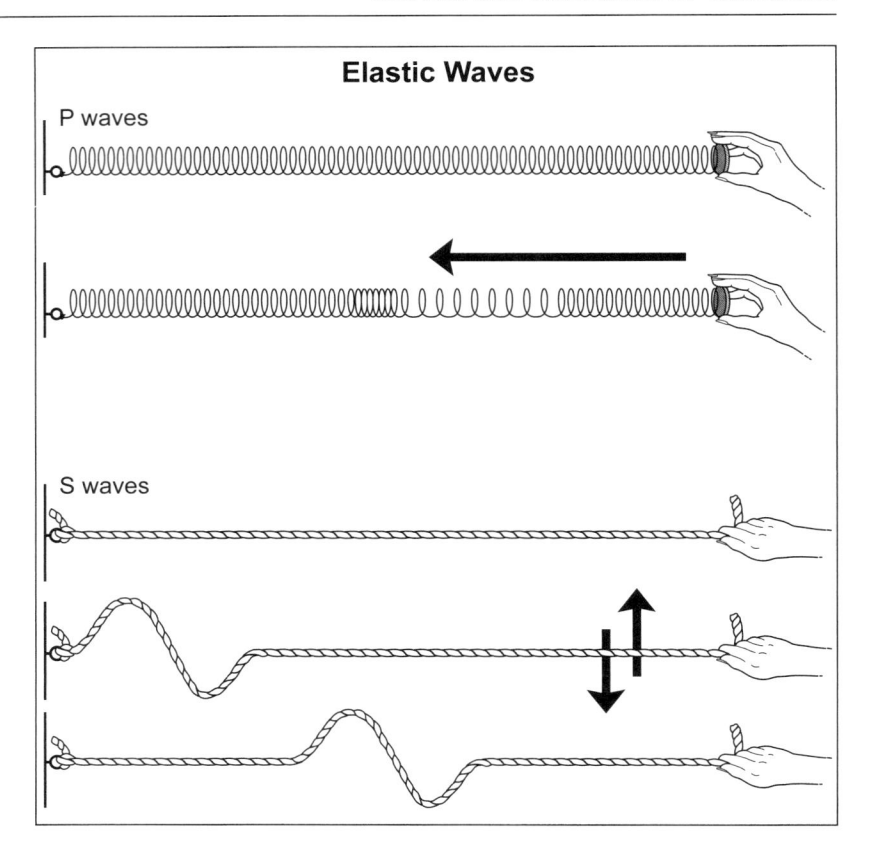

There are two other important kinds of elastic waves, but these travel only parallel to free surfaces, like the surface of the earth, and have amplitudes that decay with depth. They are called "surface waves." The two kinds are "Rayleigh waves" and "Love waves," each named after the scientist who discovered and described it.

When a Rayleigh wave passes by on the surface of the earth, a particle of soil or rock is first moved forward, then up, then backward, and then down to its starting point in an elliptical path. For Rayleigh waves, when the displaced particle is at the top of its elliptical motion, it is moving in the opposite direction of the Rayleigh wave front. This is called an "elliptic retrograde" motion. When a Love wave passes by on the surface of the earth, a particle of soil or rock is moved from side to side perpendicular to the direction of the wave front. Love waves are horizontally polarized shear waves traveling parallel with the surface.

With regard to velocity, compressional waves are the fastest; next are shear waves, which move at roughly six-tenths the speed of the compressional wave; slowest are the surface waves, which move at approximately nine-tenths the speed of shear waves.

ELASTIC WAVE FORMS

One final aspect needs to be described in talking about elastic waves, and this is the form of the wave. Regardless of the type of wave (P, S, Rayleigh, or Love), they all consist of trains of disturbances that move through the earth. A wave that has only one vibration is a pulse. In elastic waves, even ones that sound like sharp pops or explosions, there is a train of several cycles of vibration—sometimes a few seconds in duration and sometimes for many minutes or even hours. (Rayleigh and Love waves can be recorded for an hour or more on seismographs when generated by a very strong earthquake.)

The form of a wave is described by its frequency and its amplitude, as well as by its particle motion. Frequency is merely the number of times per second that the vibrations occur as the wave passes. Earthquake waves have frequencies of from several cycles per second down to several seconds per cycle. In addition to the time between peaks of an elastic wave's passage, there is a distance that can be measured between peaks. This is called the "wavelength." For waves in the earth, the wavelengths can vary from a few meters (for high-frequency P and S waves) to a kilometer or more (for low-frequency surface waves).

Hence, in an earthquake one part of a railroad track can be under compression, being sheared to the left, while a few hundred meters away another part is under tension, being sheared to the right, all at the same instant.

ELASTIC WAVE PROPAGATION COMPLEXITY

Even though a source of elastic waves may be simple, generating only one kind of wave, as soon as boundaries between differing layers are encountered, other kinds of waves result, reflecting and refracting in many directions. Those that eventually find themselves back at the surface can be recorded. When P and S waves arrive at the surface, it is their complex interaction at the surface that produces the Rayleigh and Love waves. With regard to earthquake-generated waves, not only do waves reflect from the source back to the surface off the boundaries of crust, mantle, and core, but some waves can pass completely through the earth and be recorded on the other side. During an exceptionally strong earthquake, waves can refract through to the other side and then return again through the core and mantle to be recorded again on the original side. Surface waves generated by large earthquakes have also been known to circumnavigate the globe, sometimes circling several times before their amplitudes become too small to detect. In very large earthquakes (those of 8.6 or more on the Richter scale), these waves have been measured to complete as many as ten or more passages around the world, taking approximately 3 hours for each trip.

A wave train of a single type can change from one type to another repeatedly along its ray path. This is of great interest to seismologists. For example, a P wave may start from where it is generated at an earthquake fault and travel down until it hits the *Mohorovičić*, or Moho, discontinuity, the boundary between the earth's upper crust and mantle below. There it can refract through, turning into an S wave, bending its direction of travel slightly, and taking on a new velocity. As it propagates farther and farther downward, it speeds up until it hits the boundary of the outer core, where it must either change again or reflect back toward the surface. If it passes through the boundary, it must transform back into a P wave because the outer core of the earth acts like a plastic liquid and will not permit the passage of S waves. Continuing on past the center of the earth, it would strike the boundary between core and mantle on the other side and, again, it could change back into an S wave. As it traveled up toward the other side of the earth, it would gradually slow in speed until it hit the Moho on the other side. There it could turn into a P wave again and move through the crust until it emerged at the ground surface. There it would be reflected back toward the earth's interior or, perhaps, follow a curved ray path that would skip back to the surface at another location.

During this long and varied path, the ray would travel at P-wave velocities when in a compressional phase and at S-wave velocities (roughly 40 percent slower) when in a shear phase.

Seismographs and Seismic Stations

Elastic waves in the earth are measured and recorded by various kinds of seismographs. Some measure vertical motions only, some horizontal; some measure compressional waves only, such as those that move through water. To describe the particle motion of a train of passing waves requires a set of three seismographs: one for vertical motion and two for horizontal—one for east-west motion and one for north-south.

Seismic stations permanently installed to monitor earthquakes are built in a variety of ways, depending on what is to be measured. A given seismic sensor can detect only a given band of frequencies; outside that band it is insensitive. Since earthquakes generate a wide range of frequencies from high to ultralow, some stations measure high frequencies (also called "short periods") while others measure low frequencies (called "long periods"). Moreover, seismic sensors respond only to a given range of amplitudes. Hence, some earthquake observatories have extremely sensitive instruments that detect and magnify even the tiniest vibrations 100,000 times or more. Other stations also have so-called strong motion instruments that do not record at all, unless a real jolt passes through. This diversity in equipment is necessary, because when strong high-amplitude seismic waves hit a high-magnification instrument, the readings go off the scale and cannot be deciphered. On the other end, strong-motion equipment is insensitive to smaller tremors.

Reflection Seismic Profiling

Because the interaction of seismic waves with the details (inhomogeneities) within the earth's interior enables what is there to be described, even though it is buried out of sight, artificially generated seismic waves are sometimes used to find oil and other things of interest belowground. "Reflection seismic profiling," a method used by oil companies the world over, usually entails the use of an explosive to send elastic waves into the ground. Then, by an array of seismic sensors called geophones, deployed to catch the reflections at the land surface, geophysicists can deduce the structures of the subsurface. Some seismic profiling methods employ only P waves, while some have been successful with artificial S-wave sources.

When P and S waves hit a boundary between two rock layers, they each split into four parts. A P wave, for example, will reflect back both a P wave and an S wave but will transmit a portion of its energy through the boundary into the next layer down that also splits into a P wave and an S wave. The wave bouncing back is the "reflected" portion of the incident wave, while the part that passes through is the "refracted" portion. An incident S wave similarly splits into four parts, a reflected P and S and a refracted P and S.

Practical Applications

Understanding elastic wave propagation within the earth not only has been the means by which seismologists have been able to define the inner structure of the earth but also has enabled the discovery of almost all of the oil deposits found since the mid-twentieth century. Other practical applications include the monitoring of underground nuclear testing to verify that countries are living up to their treaties. Also, because submarine earthquakes are the cause of seismic sea waves, and because the seismic waves passing through the earth travel several times faster than the seismic sea waves (or tsunamis) do through water, these destructive waves from the sea can be predicted, sometimes hours before they strike a coastline, thus saving many lives.

David Stewart

Further Reading

Aki, Keiiti, and Paul G. Richards. *Quantitative Seismology: Theory and Methods.* 2d ed. 2 vols. Sausalito: University Science Books, 2002. This is an advanced text on elastic waves and earthquake seismology. A modern version of Ewing, Jardetzky, and Press (cited below) and the treatise by Love (also cited below), it is written at the graduate university level.

Bedford, A., and D. S. Drumheller. *Introduction to Elastic Wave Propagation.* New York: John Wiley & Sons, 1996. This text is an excellent introduction to seismology and the behavior of elastic waves within the earth.

Bolt, Bruce A. *Inside the Earth.* Fairfax, Va. Techbooks, 1991. This book is an elementary but thorough treatment of seismic waves in the earth, useful to lower-level college or advanced high school students.

Bullen, K. E., and Bruce A. Bolt. *Introduction to Theory of Seismology.* 4th ed. Cambridge, England: Cambridge University Press, 1985. A thorough treatment of elastic wave theory, it is readable at the undergraduate science-major level.

Dahlen, F. A., and Jeroen Tromp. *Theoretical Global Seismology.* Princeton: Princeton University Press, 1998. Intended for the college-level reader, this book describes seismology processes and theories in great detail. The book contains many illustrations and maps. Bibliography and index.

Ekstrom, Goran, Meredith Nettles, and Victor C. Tsai. "Seasonality and Increasing Frequency of Greenland Glacial Earthquakes." *Science* 311 (2006): 1756-1758. This article provides examples of how elastic waves can form, where they register, and what changes they represent in the earth.

Ewing, W. Maurice, Wenceslas S. Jardetzky, and Frank Press. *Elastic Waves in Layered Media.* New York: McGraw-Hill, 1957. This is a definitive text on elastic waves in layered media as they occur within the earth. Mathematical, it is written for graduate-school-level or advanced physics students.

Graff, Karl F. *Wave Motion in Elastic Solids.* New York: Dover, 1991. This is an advanced text on elastic waves and earthquake seismology. Some sections may be suitable for the general reader, but it is written at the college level. Bibliographical references and index included.

Love, A. E. H. *A Treatise on the Mathematical Theory of Elasticity.* 4th ed. Mineola, N.Y.: Dover, 2011. First published in 1884 in Cambridge, England, this comprehensive work remains important. The Love waves of earthquake seismology were first described and discovered by the author. Best suited for graduate students.

Mollhoff, M., C. J. Bean, and P. G. Meredith. "Rock Fracture Compliance Derived from Time Delays of Elastic Waves." *Geophysical Prospecting* 58 (2010): 1111-1122. An article demonstrating the use of elastic wave measurements as a tool for estimating rock fracture compliance. An excellent article providing numerical evidence and experimental evidence to support concepts. Some mathematical background may be helpful.

Richter, Charles F. *Elementary Seismology.* San Francisco: W. H. Freeman, 1958. This is a classic work by the seismologist who devised the Richter scale. While dated in some subject areas, it is very readable and thorough. The chapter on elastic waves and their propagation through the earth is excellent.

See also: Creep; Cross-Borehole Seismology; Discontinuities; Earthquake Distribution; Earthquake Engineering; Earthquake Hazards; Earthquake Locating; Earthquake Magnitudes and Intensities; Earthquake Prediction; Earthquakes; Earth's Lithosphere; Experimental Rock Deformation; Faults: Normal; Faults: Strike-Slip; Faults: Thrust; Faults: Transform; Lithospheric Plates; Notable Earthquakes; Plate Motions; Plate Tectonics; San Andreas Fault; Seismic Observatories; Seismic Reflection Profiling; Seismic Tomography; Seismic Wave Studies; Seismometers; Soil Liquefaction; Stress and Strain.

ELECTRON MICROPROBES

The electron microprobe is an analytical tool used to determine the elemental composition of earth materials. The instrument is valuable to geologists because it can analyze very small samples that are approximately 5 microns in diameter.

PRINCIPAL TERMS

- **diffraction:** the bending of waves around obstacles, a process that allows photons of a specific wavelength to be analyzed
- **electron:** one of the fundamental particles of which all atoms are composed; it has an electrical charge of -1
- **electron shell:** a region around the nucleus of an atom that contains electrons; each electron in each shell will have a specific energy associated with it
- **photon:** a form of energy that has the properties of both particles and waves; electromagnetic (light) radiation

Development of the Electron Microprobe

Geologists often want to determine the concentration of elements in minerals and glasses. Prior to the invention of the electron microprobe, elemental analysis of geological samples was difficult and time-consuming. Furthermore, the analysis of very small samples, less than 0.1 gram, was virtually impossible. The electron microprobe provides geochemists with a rapid means of determining the elemental composition of geological samples even as small as about 5 microns in diameter. This ability to analyze extremely small samples has provided geochemists with a new understanding of the processes that form the rocks found within the earth.

In 1947, Canadian scientist James Hillier made the first patent application for an apparatus that would use a focused beam of electrons as a source of energy to analyze solid materials. In 1949, French scientists Raimon Castaing and André Guinier presented a paper in Delft, Netherlands, on the application of the electron microscope to analysis of samples, and they drew heavily on Hillier's earlier work. In 1951, Castaing developed the first usable electron microprobe as part of his dissertation research at the University of Paris. The microprobe introduced a whole new era of analytical research for geochemists. Although crude by today's standards, the instrument that Castaing and Guinier built contained all the fundamental elements of modern, more sophisticated electron microprobes.

Fundamental Process of Analysis

The electron microprobe is a complex machine, but the theory behind the analysis is relatively simple. All materials are composed of atoms. These atoms contain clouds of electrons that surround the nucleus, which is composed of neutrons and protons. These clouds of electrons are called shells and have a variety of geometric shapes. Although the exact positions of the electrons within each shell cannot be predicted, each has associated with it a discrete and known energy. These energy values are different for each element in the periodic table. Shells that are closer to the nucleus are called inner shells, and those farther away are called outer shells.

If an electron can be removed from an inner shell, then an electron from an outer shell will drop into the empty position within the inner shell. As the electron falls from an outer shell into an inner shell, it releases some energy in the form of photons. Each photon will have a characteristic wavelength associated with it, which is a function of this change in energy as the electron moves from one shell to another. Because the change in energies for all the electrons in all the elements is known, one can determine which element was responsible for the production of the photon simply by knowing the wavelength of the photon. The relative concentration of each element can be determined by counting the number of photons of a characteristic wavelength that have been generated. Thus, the type and concentration of each element in a geological sample can be determined.

Electrons are very small but can be removed by other electrons if they are moving fast enough. By accelerating a beam of electrons to a high velocity onto the sample surface, scientists can randomly "knock out" electrons of the shells surrounding the nucleus. That is the fundamental process whereby the electron microprobe performs an elemental analysis of a mineral or gas.

Parts and Process

All electron microprobes contain the following fundamental parts: a filament that acts as a source of electrons, an anode used to accelerate the electrons, a series of electromagnets that focus the beam of electrons, a sample holder, a crystal spectrometer used to determine the wavelength of the emitted photons, a photomultiplier tube used to determine the number of emitted photons, a vacuum chamber that contains all the previous parts, and a computer system for reporting findings.

The filament is a very thin wire that carries an electrical current that causes electrons to be emitted. These electrons are attracted to a positively charged metal plate with a hole in it through which the electrons pass. Because electrons have a negative charge, their path can be bent by electromagnets. Electron microprobes use a series of electromagnets that can adjust the focus of the electron beam from a large surface area (about 5 millimeters) to a very fine point (about 5 microns) on the surface of the sample.

The sample holder on modern instruments usually contains slots for up to ten samples and standards. The standards are used to calibrate the instrument for the material that is to be analyzed and have been subjected to rigorous elemental determinations by independent laboratories. The sample holder can be moved so that many different determinations can be made on each sample. As the sample is bombarded by electrons, photons are emitted in all directions, some of which strike a series of crystal spectrometers. These spectrometers are composed of special crystals whose structures act as diffraction gratings for the photons. Thus, they can be tuned to select specific photons of the characteristic wavelength desired for the element under analysis.

Those photons that pass through the crystal spectrometers are counted by photomultiplier tubes, also called the detectors. Each detector sends a signal to a computer that then converts the number of counts to relative concentrations for the elements being analyzed. The entire system is contained in a vacuum because electrons and photons can be absorbed by the air, which will reduce the lowest concentrations that can be analyzed.

Elemental Mapping Techniques

Within the geological sciences, the electron microprobe has become one of the most valuable tools to modern geochemists. It is used to find the composition of virtually all naturally occurring minerals and glasses with an accuracy previously unattainable. The electron microprobe can thus be used to understand the variation of the composition of naturally occurring materials over very small distances. The electron microprobe has been used to understand the origin and evolution of rocks, to determine the composition of new minerals, to develop new theories for the formation of ore deposits, and to help study how the earth has been affected by human activities.

Prior to the invention of the electron microprobe, the best that a geochemist could hope for was the bulk composition of single large crystals found within rocks. Furthermore, older analytical techniques were, for the most part, highly destructive to the sample under investigation. Thus, once the analysis was complete, the sample was usually lost and no check on accuracy was possible. With the invention of the electron microprobe, these problems were solved. The sample is prepared by making a highly polished surface. By virtue of the very fine focus available on the electron microprobe, many points on the sample can be analyzed. This process is called elemental mapping of samples.

By using elemental mapping techniques, geochemists can determine the variation of the composition from the core to the rim of minerals. Much information on the origin and evolution of rocks can be deduced using this method of research. The composition of igneous rocks (those rocks that have solidified from a molten magma) often changes during the crystallization history. By determining the chemical variation in minerals that formed during the crystallization of the magma, geologists can understand better how and in which sequence these minerals formed. Quite often, the composition of certain minerals found in rocks is a function of the temperatures and pressures of the environment in which they formed. Thus, these two parameters can be determined using an electron microprobe. That is especially valuable for metamorphic rocks (those rocks that formed from previously existing rocks as a result of a change in temperature and pressure) as well as for igneous rocks. Sedimentary rocks (those rocks that form at or near the earth's surface) are also good candidates for study using the electron microprobe. Many sedimentary rocks have been subjected to a variety of processes during their formation, which include erosion, transportation, and deposition. The very small changes in the composition of the minerals found in sedimentary rocks often reflect processes that formed them.

ANALYZING RARE AND VALUABLE MATERIALS

The electron microprobe, used in the years following its creation to investigate samples collected from the earth, was central to scientific study of the origin and evolution of the rocks that were collected on the moon. Thus, for example, the modification of the moon's surface by meteorite impacts has been understood, in part, by using the electron microprobe on samples collected by the Apollo mission astronauts. Furthermore, geologists have been able to determine the composition of the materials collected on the moon, which, in turn, has helped them in understanding the origin of the earth.

Geologists are continuously finding minerals that have never been described. Although the crystal structure of these new minerals can be determined using an X-ray diffractometer, the composition is best found using the electron microprobe. Most minerals that are discovered today are rare, and it is imperative that they be able to be preserved. For this reason, the electron microprobe is an ideal instrument to use.

Economic ore deposits are occurrences of rocks or minerals that may be extracted from the earth at a profit. Exploration geologists need to understand how known ores formed so that they can find new deposits. One key part of this research is understanding which processes cause a variation in the mineral composition found in such occurrences. Thus, large mining companies often own electron microprobes as analytical tools.

APPLICATION TO ENVIRONMENTAL GEOLOGY

The earth has been affected by the industrial activity of humankind. In an effort to repair the damage caused by this activity, geologists study how certain activities have affected earth materials and how to prevent further damage to the earth.

The application of the electron microprobe is also widespread in the area of environmental geology. One example involves the use of clay minerals to act as a barrier to toxic materials by absorbing them onto the clay's surface. By finding the amount of a given toxic material that has been absorbed by a particular clay, geologists can decide whether the clay is a suitable barrier in that case. Environmental scientists have needed the aid of the electron microprobe to provide quality microscopic-scale analyses of geological materials to aid in resolving other issues as well. These include the amount of hazardous material released by mining operations into the environment, and the migration of toxic substances in groundwaters.

A. Kem Fronabarger

FURTHER READING

Birks, L. S. *Electron Probe Microanalysis.* 2d ed. New York: Wiley-Interscience, 1971. This classic book explains the history of the development of the electron microprobe, its theory, and its use. Written at a technical level, it requires a fundamental understanding of chemistry and geology. Several excellent technical articles on the subject are listed. Figures within the text are good.

Elion, Herbert A., and D. C. Stewart. *A Handbook of X-Ray and Microprobe Data.* Elmsford, N.Y.: Pergamon Press, 1968. This handbook is used by geochemists to interpret data obtained from the microprobe. Highly technical and intended for the advanced student. Tables and lists are extremely complete.

Goldstein, Joseph, et al. *Scanning Electron Microscopy and X-Ray Microanalysis.* 3rd ed. New York: Springer, 2003. An excellent resource for anyone working in a SEM-EMPA lab. Provides technique and methodologies along with theory.

Heinrich, Kurt F. J., ed. *Quantitative Electron Probe Microanalysis.* U.S. National Bureau of Standards Special Publication 298. Washington, D.C.: Government Printing Office, 1968. This book is designed to introduce the student to the methods that are acceptable for obtaining quantitative results from the electron microprobe. Provides some understanding of the theory of the instrument and offers additional sources of information.

Ionescu, Corina, Volker Hoeck, and Lucretia Ghergari. "Electron Microprobe Analysis of Ancient Ceramics: A Case Study from Romania." *Applied Clay Science* 53 (2011): 466-475. A unique application of electron microprobe analysis. The authors assume readers have some background in geology.

McKinley, Theodore D., Kurt F. J. Heinrich, and D. B. Wittry, eds. *The Electron Microprobe.* New York: Wiley, 1966. This series of articles discusses the state of the art of the electron microprobe during the 1960's. Although somewhat dated concerning advances in automation and standardization, it does provide a remarkably complete discussion of the theory and use of the microprobe. Some of the articles can be understood by the freshman college student with limited background. Each article has an excellent bibliography.

Murr, Lawrence Eugene. *Electron and Ion Microscopy and Microanalysis: Principles and Applications.* 2d ed. New York: M. Decker, 1991. Murr's book describes in great detail the principles, theories, and applications of modern microscopy. Sections are devoted to the field study of electrons and ions, as well as advancements in optical engineering. Written for the student with a background in the sciences. Illustrations, bibliography, and index.

Potts, Phillip J., et al., eds. *Microprobe Techniques in the Earth Sciences.* London: Chapman and Hall, 1995. This book describes and illustrates many techniques and the theories concerning microprobe analysis in relation to the earth sciences. With a strong focus on analytical geochemistry, the book can be technical at times. Best for advanced undergraduates and graduate students.

Reed, S. J. B. *Electron Microprobe Analysis and Scanning Electron Microscopy in Geology.* 2d ed. Cambridge: Cambridge University Press, 2010. A thorough look into electron and petrofabric microprobe analysis, this book examines the techniques, tools, and procedures involved in scanning electron microscopy. Illustrations, index, and bibliographic references.

See also: Earth Resources; Electron Microscopy; Experimental Petrology; Geologic and Topographic Maps; Geothermometry and Geobarometry; Infrared Spectra; Mass Spectrometry; Neutron Activation Analysis; Petrographic Microscopes; X-ray Fluorescence; X-ray Powder Diffraction.

ELECTRON MICROSCOPY

Electron microscopy uses a "bombardment" of electron particles rather than light beams to obtain magnified images of specimen material. The process depends on carefully placed electromagnetic fields to focus the negatively charged electron particles. Much greater detail in magnification is obtained because electron particle waves are approximately four times shorter than light waves.

PRINCIPAL TERMS

- **anode:** in early cathode-ray tubes, this positively charged plate attracted negatively charged electrons that, passing through a tiny aperture in the plate, formed an electron beam
- **cathode-ray tube:** a tubular device, the interior chambers of which are vacuum-sealed, through which an electron beam passes, recording a dot-like image on a fluorescent screen
- **condenser lens:** the first stage of electromagnetic focusing, which "bends" the electron beam into a tightly concentrated focal point before it passes through the specimen
- **electronic lens:** electromagnetic fields inside the electron microscope that interact with the electron beam, bending the trajectory of its particles
- **objective lens:** the second, magnifying stage of electronic lens focusing, which occurs after the electron beam passes through the specimen
- **scanning electron microscopy (SEM):** a later (mid-1960's) development in electron microscopy, in which surface structure images are obtained by a process that records patterns of electron emissions off the surface of the specimen as it is subjected to the impact of the microscope beam
- **transmission electron microscopy (TEM):** electron microscopy in which all elements of electron activity involved in contact with the specimen are transferred simultaneously to the image

Development of the Technology

Electron microscopy involves the substitution of a beam of electrons and a series of electric or electromagnetic fields in the place of light beams and optical glass lenses in the common microscope. Because electron particles exhibit the same—but much shorter—wavelike movements found in light, electron beam focusing allows a much higher degree of resolution, or detailed contrast, when images of the effects of their passage through specimens are recovered, after magnified refocusing, on the fluorescent screen of an electron microscope.

A first step in developing what would later become standard technology came in 1895, when German physicist Wilhelm Röntgen noted that solid materials released invisible rays in the first stage of what was then called a cathode-ray device (later known in the field of electron microscopy as an electron "gun"). In time, scientists discovered that these atomic-particle rays not only shared the electromagnetic characteristics of light but also moved in wavelengths that were about 100,000 times shorter than those of light waves.

Among the earliest scientists to pioneer the field of electron microscopy was Englishman Sir Joseph John Thomson. In 1897, Thomson performed an experiment with electron rays that would yield the basic technology for the first television tube and, with essential adaptations which were developed by German physicist Ernst Ruska in the 1930's, the first electron microscope. In essence, Thomson's cathode-ray device, which was not yet conceived of in terms of magnifying possibilities, created a beam of tiny negatively charged particles, or cathode rays—that is, electrons. This beam resulted from a glow discharge in a gas and an incandescent metallic filament in a vacuum. These rays (in reality, negatively charged electron particles) were caused to accelerate toward and pass through a tiny hole in a positively charged plate that served as an anode. The result was a beam of high-velocity rays, which could be recorded on a fluorescent screen. Thus was established the basic technology that would later produce the first television screen, when multiple electronic "pinpoints" would be used to create a complex image. Use of this process to obtain magnification of objects placed in the path of a ray, or electron beam, became possible only after the development of the "electronic lens."

Stages of Magnification

Starting with the basic principle of a cathode-ray tube, the pioneers of electron microscopy experimented with several technological adaptations that aimed at focusing electron beams to obtain magnifying effects when focused rays are refocused and projected onto a fluorescent screen. In the simplest of terms, the electron lens magnifies by several stages. Each stage must take place in a vacuum.

After passing through the tiny aperture in the positively charged anode (the first phase in Thomson's 1897 cathode-ray experiment), the electron beam is acted upon by a first electronic lens, called the condenser lens. This process results from the effect of an electric or electromagnetic field, located within the tube device, which interacts with the "descending" electron beam. The effect of such an electronic lens is to "bend" electron paths toward an axis, just as converging glass lenses bend light rays toward a focal axis. Once focused by the condenser lens, the highly concentrated electron beam strikes, in a separate chamber sealed off by O-rings, a carefully prepared specimen. The beam actually passes through the object to be magnified. Because the beam is made up of particles, however, the physical effect of particle bombardment will vary according to varying levels of material density in the specimen itself. This is the key to the lighter/darker image produced in the final stage of image projection.

As the electron beam emerges (on the downward side of the specimen), two other electronic lenses (electromagnetic fields) influence its path. First, a minutely focused objective lens spreads the concentrated beam created by the condenser lens. Then, a projector lens "captures" the resultant image, which is recorded on a fluorescent screen as a picture. Details revealed in this image—as in images associated with common photography—appear lighter when highly exposed (that is, when large numbers of electrons passed directly through less dense areas of the specimen) and darker when underexposed (a substantial number of electrons were

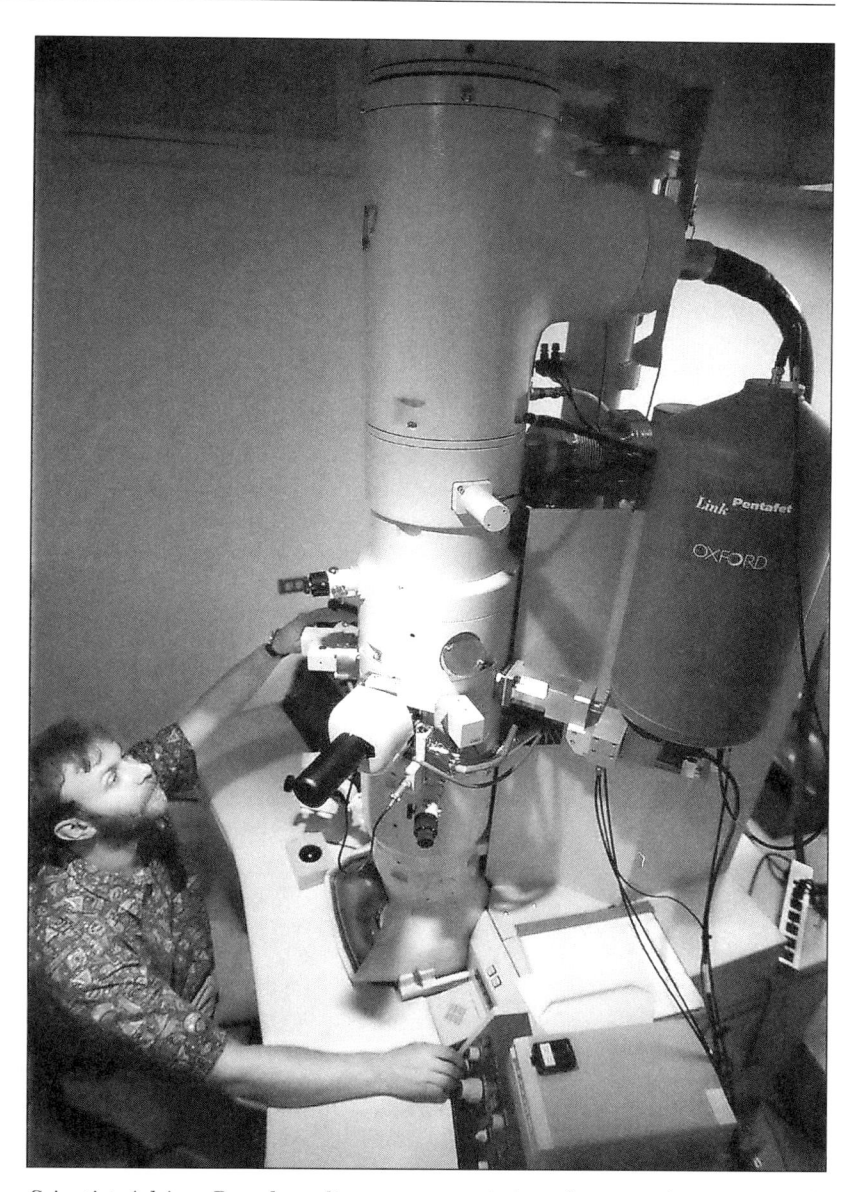

Scientist Adrian Brearley adjusts a transmission electron microscope in the Department of Earth and Planetary Sciences at the University of New Mexico. Brearley published a paper in a 1999 issue of Science *suggesting that a meteorite that fell in New Mexico in 1969 may have once had water in its composition.* (AP/Wide World Photos)

"scattered" upon impact with denser areas within the specimen). Because electron waves are approximately four times shorter than are light waves, the amount of minute detail recorded in this image (that is, the high degree of resolution obtained) promised to thrust electron microscopy to the forefront of laboratory research, especially following first stages of commercialization of the technology pioneered by Ernst Ruska in the 1930's. Ruska, along with two other scientists involved in the advanced technology of scanning tunneling microscopy, would be awarded the Nobel Prize in Physics in 1986.

SOLUTION OF TECHNICAL PROBLEMS

Eventual success in promoting electron microscopy originally hinged on the solution of several technical problems associated not with the phenomenon of electromagnetic lenses (which proved to be surprisingly easily adjustable, merely by varying the intensity of current) but with the effects of electron "bombardment" on different types of specimens. Effective solutions to several such problems were not found until several years after World War II, when not only German but also other firms producing scientific instruments would compete for commercialization of electron microscopy.

The most obvious technical problem was connected with the heat created in the process of concentrating a stream of electrons. This heat enters the specimen even if the time of exposure is very limited. Absorption of externally induced heat into the specimen caused automatic image distortion because of increased molecular agitation—a state that would not have existed prior to exposure to an electron beam. One way this problem was reduced was by use of condenser lenses to create what is called "small region radiation" techniques. Double-stage focusing reduced the area of electron bombardment to an absolute minimum and, therefore, reduced the number of electrons (that is, the intensity of the bombardment and thus the origin of heat) that were needed to obtain a detailed image. These developments increased the possibilities of using electron microscopy in research involving organic specimens.

A second technical problem involved the condensation of minor residual gases, mainly hydrocarbons, inside the chambers within which the various stages of electromagnetic focusing take place. Such condensation was particularly bothersome if it gathered on the specimen itself, causing a general darkening, and even distorting of the image produced. Research directed by the electron microscope's original inventor in Berlin, Ernst Ruska, solved this problem by "bathing" all surfaces surrounding the specimen with superchilled liquid air at the time of each experiment, which kept the specimen warmer than any of the other elements in its environment. Thus, any condensation that might occur gathers on surrounding elements rather than on the specimen itself.

Resolutions limited by aberration problems have been partially resolved with lens correctors. These have increased resolution to achieve image production at atomic dimensions, with magnifications above 50 million times.

SCANNING ELECTRON MICROSCOPY

What has been described up to this point applies to what is known as transmission electron microscopy (TEM), in which all elements of electron activity involved in contact with the specimen are transferred simultaneously to the image. In addition, researchers, again working mainly in Germany beginning in the mid-1930's, extended the technology of electron microscopy into a somewhat more complex domain—that of scanning electron microscopy (SEM). SEM technology relates to an entire subfield known as surface studies. Here, the image is built up in time sequence as a far tinier electron beam than that used in TEM moves across ("scans") the specimen. The image is derived from secondary electron currents that are released from a very thin surface layer in the atomic structure of the specimen being examined. The technology needed to detect, and then to "capture," the image of such surface-layer emissions accurately took a number of years to develop. Even following a breakthrough in secondary electron retrieval by British researchers at the University of Cambridge in 1948, the first commercial production of an effective SEM did not come until 1965. Since that date, SEMs have been used primarily to create images that resemble three-dimensional photography of the "topography" (meaning, cross sections of molecular structures) at the surface of specimen materials.

SCIENTIFIC AND INDUSTRIAL APPLICATIONS

A number of special fields of research relating to the earth sciences depend on diverse applications either of basic transmission electron microscopy or of scanning electron microscopy. Among them should be mentioned dark-field electron microscopy, in which the angle of the bombarding electron beam is "tilted" to produce bright spots in a darker field—a method that is particularly useful in determining whether there are crystalline structures in the specimen. The importance of scanning electron microscopy, together with computerized simulation in three dimensions, is particularly apparent in the field of crystallography and in the detailed analysis of minerals.

By far the most widely recognized application of electron microscopy in earth sciences relates to the petroleum industry and its by-products, either petrochemicals or related synthetic materials. In such areas, use of analytical electron microscopic techniques enables researchers to observe variations in molecular linkages that characterize "families" of synthetic products derived from petroleum. The results of such findings are critical in the search for new synthetic materials that are often much better suited to the specialized needs of modern industry and aerospace science than are natural metals and their alloys.

It is important to note that electron microscopy involves applied uses as well as investigative uses. Some of the former enable alteration of the molecular or atomic structures of natural substances, especially metals and their alloys, in ways that can make them more useful for technological purposes. An example is to be found in amorphization technology, which utilizes the electron irradiation process in the electron microscope to cause a break in crystalline periodicity (regular, latticelike chains in the atomic structure of substances), creating what specialists call a "zigzag" atomic chain that alters the basic electronic and/or chemical nature of the substance in question. Without necessarily being conscious of the fact, humans are surrounded by materials, many of them originally of natural mineral or metallic origin (especially those necessary for use as conductors in high-energy-intensity situations), that have been "transformed" for a number of special functions by processes connected with electron microscope technology.

Byron D. Cannon

FURTHER READING

Beyer, George L., et al., eds. *Microscopy*. New York: Wiley, 1991. This reference tool is a wonderful introduction to microscopy. It contains descriptions of microscopes, the history of their use, and research techniques. The authors analyze microscopy procedure and protocol in a clear and understandable manner. Includes illustrations, diagrams, index, and bibliography.

Chandler, Douglas, and Robert W. Roberson. *Bioimaging: Current Techniques in Light and Electron Microscopy*. Sudbury, Mass. Jones and Bartlett Publishers, 2009. Presents a full background in microscopy along with modern techniques for both electron and light microscopes. Provides useful information on basic cell components. Discusses microscopy principles such as refraction, reflection and polarization.

Egerton, R. F. *Physical Principles of Electron Microscopy: An Introduction to TEM, SEM, and AEM*. New York: Springer, 2010. The text discusses current practices in electron microscopy. A useful resource for graduate students, professors, researchers, or any professional using an electron microscope. Written in a manner understandable to beginners but detailed enough to be of use to all.

Fujita, Hiroshi. "The Process of Amorphization Induced by Electron Irradiation in Alloys." *Journal of Electron Microscopy Technique* 3 (1986): 245-256. This article investigates one area in which electron microscopy proved to be more useful to metallurgists than any other method in altering the atomic structure of key alloys. What is involved here is the careful calculated removal of crystalline periodicity (regular "latticelike" chains of atoms) in key substances through irradiation.

Goldstein, Joseph, et al. *Scanning Electron Microscopy and X-Ray Microanalysis*. 3rd ed. New York: Springer, 2003. An excellent resource for anyone working in a SEM-EMPA lab. Provides technique and methodologies, as well as theory.

Hunter, Elaine Evelyn. *Practical Electron Microscopy: A Beginner's Illustrated Guide*. 2d ed. New York: Cambridge University Press, 1993. As the title suggests, this is an excellent guide book for the beginner. Includes step-by-step descriptions of common techniques and practices within the electron microscopy field.

Ionescu, Corina, Volker Hoeck, and Lucretia Ghergari. "Electron Microprobe Analysis of Ancient Ceramics: A Case Study from Romania." *Applied Clay Science* 53 (2011): 466-475. A unique application of electron microprobe analysis. The authors assume readers have some background in geology.

Johnson, John E., Jr. "The Electron Microscope: Emerging Technologies." *Journal of Electron Microscopy Technique* 1 (1984): 1-7. The maiden article in the specialized journal that provides up-to-date terminology and changing methodologies for the entire field of electron microscopy. Of particular interest in this survey of the "state of the art" in 1984 is a discussion of analytical electron microscopy, an emerging subfield dedicated to the identification of types of atoms present in a given specimen and their probable arrangement.

Kuo, John, ed. *Electron Microscopy Methods and Protocols*. 2d ed. Totowa, N.J.: Humana Press, 2007. This guidebook discusses advancements and protocols in molecular and electron microscopy. Many of the procedures discussed are accompanied by an illustration. Includes bibliographic references and an index.

Lee, W. E., and K. P. D. Lagerhof. "Structural and Electron Diffraction Data for Sapphire." *Journal of Electron Microscopy Technique* 2 (1985): 247-258. This article discusses methods for using electron microscopy and computerized simulation to obtain three-dimensional images of crystalline structures. The discussion focuses specifically on sapphires, but the technology applies to a wider field of crystal structures.

Reed, S. J. B. *Electron Microprobe Analysis and Scanning Electron Microscopy in Geology*. 2d ed. Cambridge: Cambridge University Press, 2006. A thorough look into electron and petrofabric microprobe analysis, this book examines the techniques, tools, and procedures involved in scanning electron microscopy. Illustrations, index, and bibliographic references.

Ruska, Ernst. *The Early Development of Electron Lenses and Electron Microscopy*. Translated by Thomas Mulvey. Stuttgart, West Germany: S. Hirzel Verlag, 1980. An autobiographical account by (1986) Nobel laureate Ernst Ruska. Concentrates not only on Ruska's own contributions to the mid-1930's invention of the first successful electron microscope but also on other scientists' work in the field before and after.

Swift, J. A. *Electron Microscopes*. New York: Barnes & Noble Books, 1970. An account of the past techniques and technology of electron microscopy. Although outdated, this text is probably the most accessible and easily understood guide to electron microscopy. Different techniques of specialized work with electron microscopes are surveyed, with appropriate bibliographical references for more detailed coverage.

Watt, Ian M. *The Principles and Practice of Electron Microscopy*. 2d ed. New York: Cambridge University Press, 1997. A bit outdated, but still a good introduction. This college-level textbook is a good survey of electron microscopy. Goes beyond mere general treatment of a number of specialized areas, especially where scanning methods—usually the object of brief discussion—are concerned.

Williams, David B., and C. Barry Carter. *Transmission Electron Microscopy: A Textbook for Materials Science*. 2d ed. New York: Plentum Press, 2009. A readable, yet thorough text covering the apparatus, methodology, and problem solving of TEM. Each chapter has a summary and reference. This textbook is written for college students; some background in science is necessary.

See also: Electron Microprobes; Experimental Petrology; Geologic and Topographic Maps; Infrared Spectra; Mass Spectrometry; Neutron Activation Analysis; Petrographic Microscopes; X-ray Fluorescence; X-ray Powder Diffraction.

ELEMENTAL DISTRIBUTION

Ocean floors are composed of a dark, fine-grained rock called basalt that is more depleted in silicon and potassium and is richer in magnesium and iron than are the abundant light-colored granitic rocks on the continents. Igneous rocks that form where one oceanic plate is being thrust below another are generally intermediate in composition. Certain ore deposits occur only where specific plate tectonic processes take place, thereby enabling a geologist to focus the search for these deposits.

PRINCIPAL TERMS

- **andesite:** a volcanic rock that is lighter in color than basalt, containing plagioclase feldspar and often hornblende or biotite
- **basalt:** a dark-colored, volcanic rock containing the minerals plagioclase feldspar, pyroxene, and olivine
- **granitic rock:** a light-colored, intrusive rock containing large grains of quartz, plagioclase feldspar, and alkali feldspar
- **limestone:** a sedimentary rock composed mostly of calcium carbonate formed from organisms or by chemical precipitation in oceans
- **P waves:** the first waves from earthquakes to arrive at a seismic station; because they travel at different speeds through different types of rock, they may be used to deduce the rock types below the surface
- **peridotite:** a dark-colored rock composing much of the earth below the crust; it usually contains olivine, pyroxene, and garnet
- **plate tectonics:** the theory that assumes that the earth's crust is divided into large, moving plates that are formed and shifted by volcanic activity
- **sandstone:** a sedimentary rock composed of larger mineral grains than those forming shales, such that they are deposited from faster-moving waters
- **sedimentary rock:** a flat-lying, layered rock formed by the accumulation of minerals from air or water
- **shale:** the most abundant sedimentary rock, composed of very tiny minerals that settled out of slowly moving water to form a mud

Earth's Oceanic and Continental Crusts

The surface of the earth may be broadly divided into the oceanic crust and the continental crust. The oceanic crust is on the average "heavier," or denser, than the continental crust. Both the continental and oceanic crusts are less dense than the underlying rocks in the earth's mantle. The continental and oceanic crusts can thus be considered a lower-density "scum" floating on the denser mantle, somewhat analogous to an iceberg floating in water. Because the denser oceanic crust sinks lower into the mantle than the continental crust, much of the oceanic crust is covered by the oceans, but the less dense continental crust is mostly above the level of the oceans. Also, seismic waves from earthquakes indicate that the oceanic crust is much thinner (about 6 to 8 kilometers) than the continental crust (about 35 to 50 kilometers). The density difference between the oceanic and continental crusts is related to the kinds of minerals composing them. The oceanic crust contains more of the denser iron- and magnesium-rich minerals, olivine (iron and magnesium silicate) and pyroxene (calcium, iron, and magnesium silicate), than does the continental crust. The continental crust contains much more of the less dense minerals, quartz (silica) and alkali feldspar (potassium, sodium, and aluminum silicate), than does the oceanic crust. In addition, the oceanic crust contains much of the feldspar called calcium-rich plagioclase (calcium, sodium, and aluminum silicate) than does the continental crust.

This difference in mineralogy between the oceanic and continental crusts is reflected in their average elemental composition. The oceanic crust is enriched in elements concentrated in olivine, pyroxene, and calcium-rich plagioclase, and the continental crust is enriched in those elements concentrated in quartz and alkali feldspar. Thus, the continental crust contains larger concentrations of silicon dioxide (60 weight percent in the continental crust versus 49 weight percent in the oceanic crust) and potassium oxide (2.9 versus 0.4 weight percent) and lower concentrations of titanium dioxide (0.7 versus 1.4 weight percent), iron oxide (6.2 versus 8.5 weight percent), manganese oxide (0.1 versus 0.2 weight percent), magnesium oxide (3 versus 6.8 weight percent), and calcium oxide (5.5 versus 12.3 weight percent) than does the oceanic crust. The other major elements, aluminum and sodium, are fairly similar in concentration in both the oceanic and continental crusts.

The mantle is even denser than the crust, since it contains the dense minerals olivine, pyroxene, and garnet (magnesium and aluminum silicate) in the rock called peridotite. It does not contain the less dense minerals, quartz and feldspar. Thus, the mantle is even more enriched in iron oxide and magnesium oxide and more depleted in potassium oxide, sodium oxide, and silicon dioxide than are the crustal rocks.

COMPOSITION OF OCEANIC CRUST

Oceanic and continental crusts also vary substantially in composition. The continental crust is considerably more heterogeneous than is the oceanic crust. The oceanic crust consists of an upper sediment layer (about 0.3 kilometer thick), a middle basaltic layer (about 1.5 kilometers thick), and a lower gabbroic layer (about 4 to 6 kilometers thick). Basalts and gabbros both contain olivine, pyroxene, and calcium-rich plagioclase. They differ only in grain size; the basalts contain considerably finer minerals than do the gabbros. The basaltic and gabbroic layers are thus very similar in composition. They are also of fairly constant thickness across the oceanic floors. The gabbroic layers disappear over oceanic rises, or linear mountain chains on the oceanic floors. The basaltic rocks are believed to form at the rises by about 20 to 30 percent melting of the underlying peridotite in the upper mantle. The newly formed oceanic crust and part of the upper mantle are believed to be slowly transported across oceanic floors, at rates of about 5 to 10 centimeters per year, to where this material is eventually subducted or thrust underneath another plate.

The thickness of sediment on ocean floors varies considerably. It is nearly absent over the newly formed basalts at oceanic rises. It is thickest in basins adjacent to continents where weathering and transportation processes carry large amounts of weathered sediment into the basins. The composition of oceanic floor sediment varies as well. It contains varied amounts of calcite or aragonite (calcium carbonate minerals), silica (silicon dioxide), clay minerals (fine, aluminum silicate minerals derived from weathering), volcanic ash, volcanic rock fragments, and ferromagnesian nodules.

Finally, a few volcanoes composed of basalt form linear chains on the ocean floor, away from the rises or subduction zones such as the Hawaiian Islands. These ocean-floor basalts are similar in composition to those at oceanic rises, except that they contain greater amounts of potassium. The amount of basaltic rocks produced by these ocean-floor volcanoes is insignificant, however, compared to the vast amounts of basalt produced at oceanic rises.

COMPOSITION OF CONTINENTAL CRUST

In contrast to the oceanic crust, the continental crust is quite heterogeneous in mineralogy and chemical composition. About 75 percent of the surface of the continents is covered by great piles of layered rocks called sedimentary rocks. The average thickness of these sedimentary rocks on the continental crust is only about 1.8 kilometers, although they may locally range up to 20 kilometers in thickness. The main kinds of sedimentary rocks on the continents are the very fine-grained shales or mudrocks (about 60 percent of the total sedimentary rocks), the coarser-grained sandstones (about 20 percent of the total), and limestones or dolostones (about 20 percent of the total).

The shales or mudrocks are composed of very small grains of mostly clay minerals and quartz. The resultant composition of the shales is often high in the immobile elements, aluminum and potassium, and low in the mobile elements, sodium and calcium.

Sandstones vary in composition depending on which rocks weather to form the sandstone, the distance of the sandstone from the source, and the intensity of weathering. Sandstones formed close to a source of granitic rocks may have a composition similar to that of the granitic rock: high in silicon and potassium and low in magnesium, iron, and calcium compared to basaltic rocks. Sandstones formed a long distance from the source have more time to be weathered. Thus, these sandstones may have most of the unstable minerals weathered away to clays or soluble products in water (for example, sodium), and they may be enriched in silicon because of the abundance of the stable mineral quartz.

Limestones typically form in warm, shallow seas by the action of organisms to produce most of the calcium carbonate in these rocks. Thus, limestones are enriched in calcium and depleted in most other elements. The dolostones are enriched in magnesium as well as calcium.

Some places, such as the Great Plains in the United States, consist mostly of alternating limestones and shales formed in ancient, shallow seas. (Thus, the average composition of the surface rocks in these areas may be considered an average of that of shale and limestone in whatever proportion they occur.) The average composition of sedimentary rocks on the continents is significantly different from that of the granitic rocks that weathered to form them. The average sedimentary rocks on continents are much more enriched in calcium (because of carbonate rocks), carbon dioxide (also because of carbonate rocks), and water (because of incorporation in clay minerals), and they are depleted in sodium (because of its solubility).

The thickness of these sedimentary rocks is still small compared to the 35- to 50-kilometer thickness of most of the continental crust. Only about 5 percent of the continental crust by volume is composed of sedimentary rocks. Most crustal rocks are igneous rocks or their metamorphic equivalents. Metamorphic rocks form in the solid state at high temperatures and pressures because of their deep burial in the earth. A substantial percentage of these igneous rocks of the upper continental crust are either granitic rocks (quartz and alkali feldspar rock) or andesitic rocks (plagioclase-rich rock). Basaltic rocks probably compose only about 15 percent of the upper continental crust.

CONTINENTAL MARGINS AND RIFT ZONES

Most of the granitic rocks and andesites originally formed along subduction zones, where oceanic crust is being thrust or subducted below either oceanic or continental crust. There also may be some basalts formed along these subducted plates. These basalts, andesites, and granitic rocks that formed along continental margins may eventually be plastered along the edges of the continents, resulting in the gradual growth of the continents. Other basalts are formed in portions of continents, called continental rifts, that are being stretched apart much like taffy. These basalts are considerably more enriched in potassium than basalts formed on ocean floors. For example, a large fraction of the states of Washington, Oregon, and Idaho is covered with these rift basalts extruded as lavas since about 20 million years ago. The total volume of about 180,000 cubic kilometers for these basalts is still comparatively insignificant; therefore,

Basalt, an igneous rock formed by the rapid cooling of lava rich in magnesium and iron. (Doug Martin/Photo Researchers, Inc.)

basalts make only a small contribution to the composition of the average upper continental crust.

COMPOSITION OF THE MIDDLE AND LOWER CRUST

The composition of the lower continental crust is much more difficult to determine than that of the upper continental crust because the rocks forming the lower crust are not exposed at the surface. Estimates of about 50 percent granitic and 50 percent gabbroic rocks in the lower crust have been reached. Thus, the lower continental crust is more enriched in the basaltic components, calcium, magnesium, iron, and titanium, and depleted in the granitic components, potassium and silicon, than is the upper continental crust.

The average compositions of the middle and lower oceanic and continental crusts are difficult to determine because they cannot be directly sampled. Much of the information about the nature of the crust below the surface comes from the behavior of seismic waves given off by earthquakes, from heat-flow measurements, and from the composition of rock fragments brought up by magma passing through much of the crust. In addition, there are places in the crust where rocks from the lower crust have been uplifted to the surface, so their composition can be examined in detail.

The speed of the earthquake waves through the oceanic crust is consistent with the crust being composed of a thin upper layer of sediment (indicated by P-wave velocities of 2 kilometers per second), a thicker middle layer of basalt (P-wave velocities of

5 kilometers per second), and a thick lower layer of mostly gabbro (P-wave velocities of 6.7 kilometers per second). The thicker continental crust, however, has P-wave velocities (6.1 kilometers per second) consistent with mostly granitic rocks below the overlying sedimentary rock veneer (2 to 4 kilometers per second). The lower continental crust has P-wave velocities (6.7 kilometers per second) similar to those expected for lower-silica rocks like gabbro, so there is probably more gabbro mixed with granitic rocks in the lower crust.

HEAT-FLOW MEASUREMENTS

How fast heat flows out of the earth may also be used to estimate the composition of crustal rocks. Variation in heat flow at the surface depends on how much heat is flowing out of the earth below the crust; the distribution of radioactive elements in the crust, such as uranium, thorium, and potassium, that give off heat; and how close magmas are to the surface. Oceanic ridges and continental rift zones, for example, have high heat flow, suggesting that magmas are close to the surface. In contrast, the heat loss from much of the ocean floor and over much of the continents with old Precambrian rocks (older than about 600 million years) is considerably lower because of the lack of magma close to the surface. It is surprising, however, that the oceanic floor and continents with old Precambrian rocks have similar low heat flow, as the abundant granitic rocks in the continents ought to be more enriched in the heat-producing radioactive elements than is the oceanic crust. That suggests that many of the granitic rocks at depth in these parts of the continental crust are depleted in radioactive elements, perhaps because of melting processes carrying away the radioactive elements in the magmas during the Precambrian. Also, this finding is consistent with the presence of abundant basaltic rocks depleted in radioactive elements in the lower crust.

GLIMPSES INTO EARTH'S INTERIOR

There are places on the earth, such as the island of Cyprus in the Mediterranean Sea, that appear to be ruptured portions of the entire oceanic crust and part of the upper mantle. In Cyprus, the lower zone is composed of peridotite, olivine-rich rocks, or pyroxene-rich rocks, as are predicted to occur in the mantle. These rocks correspond to the P-wave seismic velocities of 8 kilometers per second. There is a rather abrupt change to the next overlying layer of mostly gabbros that correspond to the sharp decrease in P-wave velocities to about 6.7 kilometers per second. These rocks grade upward into basalt corresponding to the upper igneous rock layers of the oceanic crust with P-wave velocities of about 5 kilometers per second. The basalt and gabbros are also penetrated by a multitude of tabular igneous dikes that were feeders of magma to the overlying basalt at the surface. Finally, there are overlying sedimentary rocks corresponding to the upper oceanic layers with P-wave velocities of about 2 kilometers per second.

Deep drill holes provide scant information about the composition of the crust at depth. Drill cores provide mostly information about sedimentary rocks; they also give some information about the first igneous rocks just below the sedimentary cover. Unfortunately, deep drilling is costly and limited in depth and distribution. Generally, wells are never drilled deep enough to obtain samples from the intermediate and lower crust, and none reach the mantle.

Some volcanoes derive their magma from the upper part of the mantle. These volcanoes often bring up fragments of mantle material and random samples of crustal rocks thrown from the volcanic conduits' walls. Although volcanic vents of this kind are not common and sample an extremely small distribution of lower continent and upper mantle material, they are extremely important.

Finally, meteorites are used as analogies of the interior of the earth. Iron meteorites, rich in iron and nickel, are thought to approximate the composition of the core of the earth. Stony meteorites with compositions close to that of peridotite are thought to match the composition of the mantle of the earth.

GUIDE TO ORE DEPOSITS

A knowledge of the overall distribution of rock types and the corresponding elemental compositions of these rocks over the earth can give geologists a guide to where to look for certain kinds of ore deposits, as certain ores occur in certain kinds of rocks. The most generalized pattern is the association of certain types of ores with certain tectonic environments. Both oceanic rises and subduction zones tend to heat waters and drive the resultant metal-rich waters toward the surface. Oceanic rises often contain sulfide-rich, copper and zinc hot-water deposits.

These hot-water deposits at subduction zones are often enriched in copper, gold, silver, tin, lead, mercury, or molybdenum.

One example of subduction zone deposits is the copper porphyry deposits. These important ore deposits are formed in granitic rocks that crystallized at shallow depths below the surface in areas where an oceanic plate is being subducted, or thrust below a second plate. They are especially abundant around the rim of the Pacific Ocean. The copper ores contain low copper concentrations (0.25 to 2 percent) and have some associated molybdenum and gold. These low-grade ores are often profitable to mine because of the large volume of ore (over a billion tons in some places) that can be rapidly extracted from the rock. A geologist looking for such ores designs an exploration campaign to search out only areas with active or inactive subduction zones. Also, the geologist looks for certain compositions of granitic rocks intruded at fairly shallow depths below the surface that have been exposed to erosion near the top of the intrusion, as these are the places where the copper porphyries form. Hundreds of these copper porphyry deposits have been discovered, accounting for about half of the copper ores of the world. Copper is used in wires to transmit electricity and in bronze and brass.

Robert L. Cullers

FURTHER READING

Ahrens, L. H. *Distribution of the Elements in Our Planet.* New York: McGraw-Hill, 1965. This book provides a clear summary of the composition of the solar system and the earth. The elements are grouped in a geochemical classification. Directed to the nonspecialist.

Craig, J. R., D. J. Vaughan, and B. J. Skinner. *Resources of the Earth.* 3rd ed. Englewood Cliffs, N.J.: Prentice-Hall, 2001. This is an excellent book describing the distribution of ore deposits on the earth. Information is provided on the history and use of the elements, geologic occurrence, and reserves. For a nonscience major in college or interested layperson. There is a glossary of technical terms.

Emsley, John. *The Elements.* 3rd ed. Oxford: Oxford University Press, 1998. Emsley discusses the properties of elements and minerals, as well as their distribution in the earth. Although some background in chemistry would be helpful, the book is easily understood by the high school student.

Greenwood, Norman Neill, and A. Earnshaw. *Chemistry of Elements.* 2d ed. Oxford: Butterworth-Heinemann, 1997. An excellent resource for a complete description of the elements and their properties. The book is filled with charts and diagrams to illustrate chemical processes and concepts. Bibliography and index.

Krebs, Robert E. *The History and Use of Our Earth's Chemical Elements: A Reference Guide.* 2d ed. Westport, Conn.: Greenwood Press, 2006. This book defines geochemistry and examines its principles and applications. A good resource for the layperson interested in the field of geochemistry and in the earth's elements. Accessible to high school readers. Illustration, charts, and bibliography.

Skinner, B. J., et al. *The Dynamic Earth: An Introduction to Physical Geology.* 5th ed. New York: John Wiley & Sons, 2006. This is one of many introductory geology textbooks for college students that has a chapter on mineral and energy resources in the earth. The interested reader with some understanding of geology can find information here about the major ore deposits and their distribution within the earth.

Smith, D. G., ed. *The Cambridge Encyclopedia of the Earth Sciences.* New York: Crown, 1981. This reference is written for the reader with some background in science who needs to locate information on a specific earth science topic. Chapters 4 ("Chemistry of the Earth"), 5 ("Earth Materials"), and 10 ("Crust of the Earth") might be most appropriate for further reading related to elemental distribution. There are also chapters on plate tectonics.

Szefer, P., and Grembecka, M. "Chemometric Assessment of Chemical Element Distribution in Bottom Sediments of the Southern Baltic Sea Including Vistula and Szczecin Lagoons: An Overview." *Polish Journal of Environmental Studies* 18 (2009): 25-34. This article presents current methodology and analysis of element distribution in sediments. The authors use data to determine the origin of some chemical elements.

Utgard, R. O., and G. D. McKenzie. *Man's Finite Earth.* Minneapolis, Minn.: Burgess, 1974. This book is written as supplementary reading for college geology courses. A section on earth resources that gives some insight on ore distribution and how it relates to public policy is suitable for a layperson.

Wedepohl, Karl Hans. *Geochemistry*. Translated by Egon Althaus. New York: Holt, Rinehart and Winston, 1971. This book gives nontechnical descriptions of the elemental distributions within the solar system and the earth. Some knowledge of chemistry and geology is necessary for full use of the book. Chapter 7 gives specific information on the distribution of elements in the earth's crust.

Xuejing, Xie, et al. "Digital Element Earth." *Acta Geologica Sinica* (English Edition) 85 (2011): 1-16. A mapping project aimed to compile the geospatial data on chemical element distribution across the earth's surface. Thirty years of work by the authors resulted in the Digital Element Earth (DEE) collection of geochemical information available to industry, agriculture, government, and the general public. The article presents a summary of the multiple projects completed as part of this collection and the future of the DEE.

See also: Continental Drift; Creep; Experimental Rock Deformation; Faults: Transform; Freshwater Chemistry; Geochemical Cycle; Geothermometry and Geobarometry; Lithospheric Plates; Oxygen, Hydrogen, and Carbon Ratios; Phase Changes; Phase Equilibria; Plate Tectonics; Subduction and Orogeny; Water-Rock Interactions.

ENGINEERING GEOPHYSICS

Engineering geophysics involves the application of earth science to problems of interest to the engineering and groundwater geologist, including site evaluation, resource exploration, and pollution monitoring. Research methods in the field are focused on discovering shallow targets by seismic, electrical, magnetic, electromagnetic, gravity, and radar surveys.

PRINCIPAL TERMS

- **geophysical survey array:** a description of the orientation and spacing of sensors and energy sources relative to one another for a geophysical survey
- **geophysical target:** the object or surface that one wishes to detect by means of a geophysical survey; knowledge of the target is essential to selection of a survey type
- **physical property contrast:** the difference in a characteristic (density, velocity) between the object of interest and its surroundings
- **site evaluation:** a process whereby a site is selected or rejected as a location for a particular use such as construction or mining
- **survey line:** a usually straight line along which points are located where geophysical measurements will be taken

SEISMIC REFRACTION AND REFLECTION

The methods of engineering geophysics include a number of surveying techniques that detect the physical characteristics of materials and objects in the earth's subsurface. The technique may detect such phenomena as rock and soil layers, the thickness of glaciers, disturbed or disrupted soils, buried walls or foundations of archaeological interest, buried metal drums, buried sand and gravel deposits, and aquifers. Each of these targets has a distinct set of physical properties that may be sensed from the surface with the right kind of survey. The selection of a particular geophysical surveying instrument and method depends upon the size, depth, and other characteristics of the target, as well as upon the nature of the surrounding materials. The most commonly used methods are seismic refraction and reflection, ground-penetrating radar, magnetic and gravity surveys, and electrical and electromagnetic techniques.

The seismic refraction method relies on the different wave velocities of rocks and soils and the principles of refraction, which govern all wave motion. A seismic survey is accomplished using a number of seismic sensors, or "geophones," that are arrayed in a straight line and are usually spaced at regular intervals. Geophones consist of a spring-mounted magnet moving within a wire coil to generate an electrical signal. Recent designs use microelectromechanical systems technology to generate an electrical response to ground motion. These phones will detect the arrival of seismic waves from an artificial energy source such as a hammer blow, a dropped weight, or an explosive charge. The additional equipment includes a digital recording system that measures the interval between the time that the energy is put into the ground and the time that it is picked up at the geophones. The velocities of the seismic waves are then calculated from the distance between the energy source and a particular geophone and the time elapsed for the wave to reach that geophone. The data are simplest to interpret where layers are flat and of constant thickness; more complex equations allow interpretation where layers are tilted and of variable thickness.

Seismic reflection profiling is used to detect surfaces at depth. In general, the seismic array of the energy source and geophones is kept under 10 meters in total length because the waves that are detected are traveling along nearly vertical paths. The elapsed time for the wave to reach a geophone is a function of the velocities of the subsurface layers and their depths. The applications of this method for shallow targets are more limited than seismic refraction because of surface noise and the difficulties in identifying individual reflections.

GROUND-PENETRATING RADAR

Ground-penetrating radar, or GPR, works on the same principles as seismic reflection but does not suffer from the same noise problems for shallow investigations. This technique measures the time necessary for a controlled pulse of radar energy to return from a reflecting surface. The equipment involves a transmitter and receiver that may be towed across an area to detect buried objects and soil layers.

While radar waves can travel great distances in air or space, they are rapidly absorbed by rocks, soils, and especially water. For this reason, GPR is used for depths of 10-15 meters. The depths of radar penetration are even less in areas underlain by clayey soils or shale, both of which contain large amounts of water in the crystal structure, or where the water table is high. For most engineering purposes, GPR is a fast and efficient survey method in relatively smooth terrain.

MAGNETIC AND GRAVITY SURVEYS

Magnetic and gravity methods involve the detection of variations in the natural magnetic and gravity fields of the earth. Changes in the magnetic field are caused by the presence of materials that have a high magnetic susceptibility. These substances include very magnetic materials, such as metallic iron and steel, but also may include rocks and soils that have elevated quantities of the mineral magnetite. Either the very magnetic metals or the slightly magnetic rocks and soils may be detected by surveying the area of interest using a magnetometer. Magnetic surveys are often used to locate metallic targets that are quite small.

A gravity survey detects variations in the gravitational field as a result of changes in density of subsurface materials. Denser materials will cause a slight increase in gravity, while less dense objects cause a small decrease in gravity. A small object may be detected only if it has large contrast in density. If the density contrast of the target is small, as it is in most engineering applications, the object must be large to be detectable. Gravity surveys are usually employed to find large bodies of rock or soil as opposed to smaller objects. The survey techniques perhaps used most in engineering geophysics detect the electrical properties of subsurface materials. The contrasts in electrical properties of natural and human-made materials are very large, so surface surveys can detect anomalous regions easily.

ELECTRICAL AND ELECTROMAGNETIC TECHNIQUES

Electrical-resistivity surveys involve an array of electrodes planted in the ground in one of a number of patterns. Generally, two electrodes are used to introduce a current into the ground, while other electrodes measure the voltage drop between two points in the array. The pattern of the array and the spacing of the electrodes are chosen based on the character of the target and whether the investigator wishes to see a single depth or different depths. Measurements are interpreted by a series of complex equations that model the depths, thicknesses, and resistivities of the soil and rock layers.

Terrain conductivity is an electromagnetic method that measures the electrical properties in the subsurface without the necessity of placing electrodes. The method uses a pair of electrical coils, one to "broadcast" and the second to receive. The first coil is energized with an alternating current, creating an alternating magnetic field that penetrates into the earth. Where this magnetic field encounters a good conductor in the subsurface, a secondary electrical current is generated or induced. The secondary electrical current is accompanied by a secondary magnetic field that is detected at the surface by the receiving coil. The depths of penetration of the conductivity method are related to the spacing of the two coils and to the power of the electrical system.

All the methods used in engineering geophysics involve time and expense. Nevertheless, they are used because they may provide additional data on the continuity of conditions discovered from outcrops, drill holes, or excavations. In addition, anomalous regions within the site may become targets for more detailed study by drilling or coring.

SITE EVALUATION

Geophysical surveys are used to detect potential problems and provide additional information on a site undergoing evaluation. Engineering geology involves, for example, site evaluations for buildings, highways, well fields, dams, and sanitary or hazardous waste sites. Engineering geologists also deal with questions of slope stability, where data on the thickness of the unstable or moving part of the slide may be needed to assess the feasibility of construction or to develop plans for slope stabilization. In all these cases, information on the depth to bedrock, the lateral continuity of soil or rock layers, and the existence and orientation of surfaces within the rocks may be needed. Additionally, the geologist may have to detect human-made objects within an area. These include buried or abandoned gas and power conduits, storage tanks, and metal drums. One of the major problems requiring engineering geology today is the monitoring and correction of contaminant leakage from hazardous waste disposal sites across the country. In these situations, data are required to define geological conditions around the site or to map the extent and location of the plume of the contaminants that may be moving off the site.

For each of these needs, a geophysical survey may be useful. The first step in the selection of a specific technique depends on the identification of the target. Those physical properties of the target that contrast most with the properties of the surrounding rock and soil will indicate which geophysical technique may be most valuable. Next, the expected size, shape, orientation, and depth of the target will be guides in the selection of the best survey methods. Finally, the character of the terrain (whether it is dry or wet and boggy, topographically smooth or irregular) will be considered. Three examples may serve to elucidate this kind of planning.

LANDFILL APPLICATION

One hypothetical application of engineering geophysics and landfills involves a site at which a large number of 55-gallon steel drums containing toxic chemicals were buried many years ago. In this case, the drums need to be located and removed from the site. The task must be accomplished carefully, for it is essential that none of the drums be split or broken. Knowledge of the exact location of the drums would greatly lessen the chances of spilling toxic waste during the process of removal. A number of geophysical methods might apply, but the geologist wishes to choose the surest method. The geologist begins by considering the physical properties and sizes of the bodies in an effort to eliminate some methods. While the density of the drums is likely to be slightly different from that of the surroundings, the drums are too small to result in a measurable gravity anomaly. Similarly, a seismic refraction survey is not likely to detect individual objects as small as the drums even if there is a large contrast in velocity between the metal drums and the other material on the site.

Each of the remaining methods reviewed earlier might be useful. Seismic reflection and GPR would be able to detect reflections off the drums. In this situation, however, the drums are probably too shallow to use the seismic method, since surface-wave noise will be high. GPR, in contrast, would do the job very well and has worked in other areas. The only weakness of GPR is that the survey line must cross directly over a drum to detect it. To ensure finding all the drums, a very close, and therefore expensive, set of GPR survey lines would be needed.

Electrical methods could detect the large contrast in the resistivity of the metal drums. Either a resistivity or a terrain-conductivity survey could sense the drums. Because the electrical or electromagnetic field extends out beyond the direct survey line, either method could locate drums slightly off the direct line of the survey. A drawback to the method, however, would be the possible existence of salty water or other electrolytes in the landfill. If these fluids are present, their high electrical conductivities could mask the presence of the drums. Because the drums are steel, they will cause a magnetic anomaly, and a magnetic survey will be able to detect even those drums off of the survey line. Although the magnetometer will detect other iron objects in the landfill, the method may be the fastest and least expensive method to survey the site for this particular target and should result in locating all the drums.

LANDSLIDE APPLICATION

In the second example, the geologist needs to determine the depths to the slip-surface, or base, of a landslide. In this case, the target is too deep for GPR, and the variations in density and magnetic susceptibility are expected to be too small to produce a significant anomaly. Previous studies, however, have shown that seismic refraction and electrical surveys can be quite successful in solving this kind of problem. The upper, mobile, and generally disrupted part of the landslide will have a different seismic velocity from that of the lower, undisturbed rocks or sediments. While the thickness of the upper layer is likely to be variable and the slip-surface complex, the seismic method has been useful. Electrical methods will depend either on a difference in conductivity of the disrupted upper layer and the undisturbed rock beneath or on the conductivity of the slip-surface itself. The exact nature of the conductivity contrast will depend on the local geology, the soil moisture, and the groundwater conditions.

WATER SUPPLY APPLICATION

The third example involves a common problem with water supplies in coastal areas where fresh groundwater floats typically above a deeper zone of salt water. In this case, withdrawal of fresh water from wells leads to a drawdown of the water table and also to intrusion from the sides and from below of salt water from the ocean. In many places, the supply of potable water is very limited by the occurrence of salty water beneath the thin lens of fresh water. The depth to and migration of saline water can be monitored by wells at critical places, but to obtain a more continuous set of data across a region, a geophysical survey is desirable. The greatest contrast in physical property will be the high electrical conductivity of the salt water versus the low values of the fresh water. Surface electrical surveys have been successfully used to estimate the thickness of the freshwater layer above salt water and to monitor the extent of the migration of salt water laterally into freshwater aquifers. Similar applications of electrical techniques have been used to detect acid mine drainage or to monitor movement of electrically conductive pollutant plumes from industrial or waste-disposal sites.

Data Interpretation

Engineering geophysics involves the application of general geophysical techniques for shallow investigations. The methods are appropriate for depths from a few meters to a few hundred meters. Applications of geophysics to shallow targets often involve greater complexity in interpretation than for deep targets because the roughness and irregularities of buried objects or surfaces are large compared to the depth. Similarly, many of the targets of interest to the engineer are not simple in shape. Surfaces such as bedrock buried beneath soil or recent sediments or the fracture patterns in bedrock are inherently complex. Furthermore, the transitional character of physical properties in near-surface environments may be difficult to detect or model. Examples of such transitions include velocity changes caused by gradations from unweathered to weathered bedrock and density and electrical property changes caused by variation in the percentages of clay or sand in surficial gravels.

While geophysics is often used in a qualitative fashion, in many areas, mathematical modeling allows a quantitative measure of the depth, size, shape, and composition of the target. Seismic refraction and reflection and GPR generally give the most unequivocal information about the subsurface. Magnetic and gravity surveys yield information on the depths, sizes, and character of subsurface objects, but usually more than one interpretation will fit the data. Electrical and electromagnetic methods also supply data that can be interpreted in a number of ways. Thus, some geophysical methods are less exact than others, and all involve uncertainties in the interpretation. For these reasons, it is important to gather geophysical information in areas where other kinds of geological data are available from surface mapping, drilling, or excavation. The combination of geophysical and geological data provides an excellent basis for the evaluation of sites of engineering interest, for estimation of water or other resources, and for the monitoring of hazards.

Geological and Civil Engineering Value

The power of geophysics lies in its use to detect anomalous regions and to evaluate the lateral continuity from one point to another across an area of interest. In many cases, the existence of an anomaly can guide the engineering geologist to the critical area where drilling or excavation will be used to acquire data. In other cases, extensive drilling and sampling in an area may tell much about the region, but information is needed between the sample sites. In these cases, geophysical surveys may be used to show the extent to which the regions between sample sites are the same as the sample sites themselves. The nature of the array will allow the scientist to look to different depths in a single area or to improve the accuracy and precision of the data.

The use of geophysics to solve problems in geological and civil engineering has become more common as the need for more detailed information on site assessment and evaluation has become apparent. While nothing can match the factual nature of soil and rock sampling by drilling, excavation, or examination of outcrops, the need to test the continuity from one sample site to another requires additional data. Geophysical methods can supply that additional information through detection of the bulk physical properties of materials in the subsurface or the interfaces between the different materials.

The bulk physical properties that can be detected include density, magnetic susceptibility, seismic-wave velocity, and electrical conductivity or resistivity. The surfaces separating different bodies in the subsurface may be thick enough to have their own bulk properties, but generally they act as reflectors of seismic or radar wave energy.

Donald F. Palmer

FURTHER READING

Armstrong, Amit, Roger Surdahl, and H. Gabriella Armstrong. "Peering Into the Unknown." *Public Roads.* 72 (2009): 3. This article presents applications of geophysics methodology to road construction. The authors discuss techniques used by transportation engineers when planning to build or repair a road, including integrity testing.

Beck, A. E. *Physical Principles of Exploration Methods: An Introductory Text for Geology and Geophysics Students.* New York: Halsted Press, 1981. This text offers a mathematically rigorous, theoretically oriented introduction to geophysical prospecting of field data are derived and used in examples. The treatment of electrical and electromagnetic methods is especially good. An index is supplied. Most references in the bibliography are from geophysical journals.

Das, Braja, M. *Principles of Geotechnical Engineering.* 7th ed. Pacific Grove: CL Engineering, 2009. This book explores and clarifies the principles, policies, and applications of geotechnical engineering. Graphs, charts, and maps reinforce concepts throughout the chapters.

Dennen, William H., and Bruce R. Moore. *Geology and Engineering.* Dubuque, Iowa: Wm. C. Brown, 1986. The authors have provided a nonmathematical introduction to the field of engineering geology with excellent diagrams and an easily understood text. The treatment of geophysical methods is qualitative but useful. The discussion of the accuracy and precision of analytic data is an excellent and meaningful addition. A short glossary defines many terms used in engineering geophysics.

Dobrin, Milton B., and Carl H. Savit. *Introduction to Geophysical Prospecting.* 4th ed. New York: McGraw-Hill, 1988. One of the standard texts in geophysical exploration, this reference is complete on all topics except ground-penetrating radar. Derivation of equations from first principles and evaluation of problems that arise in data reduction and interpretation make this one of the best books for the serious student. Although the book does not focus on shallow targets, it offers a sound background with many applied examples.

Fetter, C. W. *Applied Hydrogeology.* 4th ed. Westerville, Ohio: Charles E. Merrill, 2000. The text is a superb introduction to all aspects of groundwater studies. The mathematical treatment is good, and graphs and figures are excellent. A chapter on field methods includes a brief review of geophysical methods of investigation with informative examples and mathematically treated solutions. A short section on ground-penetrating radar includes examples of its use. An excellent glossary defines virtually all terms used in groundwater studies.

Griffiths, D. H., and R. F. King. *Applied Geophysics for Engineers and Geologists.* Elmsford, N.Y.: Pergamon Press, 1976. Although outdated, this text considers application of classic geophysical techniques for shallow targets, with good treatments of problems of data reduction and drift corrections. Interpretation of simple and complicated geological settings is discussed. The bibliography provides many references to engineering or groundwater uses of geophysics.

Jusoh, Zuriati, M. N. M. Nawawi, and Rosli Saad. "Application of Geophysical Method in Engineering and Environmental Problems." *AIP Conference Proceedings* 1250 (2010): 181-184. This article provides an example of the use of geophysics methodology and techniques for the practical purpose of surveying a building site.

Legget, Robert F. *Cities and Geology.* New York: McGraw-Hill, 1973. This book presents a complete view of the importance of geology in urban areas. The text covers historical examples of engineering and planning problems and forms a good background for consideration of geophysical methods applied to geology. The references are compiled by chapter to aid in finding further readings on a particular subject. A well-annotated section offers suggestions for further reading.

Lowrie, William. *Fundamentals of Geophysics.* 2d ed. New York, Cambridge University Press. 2007. Excellent overview of geophysics topics written for a student with strong physics background. Lowrie presents the mathematics at a level understood by mid-level university students.

Maund, Julian G., and Malcolm Eddleston. *Geohazards in Engineering Geology.* London: Geological Society, 1998. An in-depth look at the policies and practices of geophysics engineering, as well as the dangers associated with the profession. Suitable for college-level readers.

McCann, D. M. *Modern Geophysics in Engineering Geology*. London: Geological Society, 1997. For the careful reader with a technical background, McCann offers a look at the science and practical applications of geophysics. Illustrations, bibliography, and index.

Plummer, Charles C., and Diane Carlson. *Physical Geology*. 12th ed. Boston: McGraw-Hill, 2007. A college-level introductory geology textbook that is clearly written and wonderfully illustrated. An excellent sourcebook of basic information on geologic terminology and fundamentals of geologic processes. An excellent glossary.

Rahn, Perry H. *Engineering Geology: An Environmental Approach*. 2d ed. Upper Saddle River, N.J.: Prentice Hall, 1996. This book offers a good introduction to the field of engineering geology, with an emphasis on environmentally sustainable practices. There are plenty of maps and graphs to illustrate key engineering concepts. References and index.

Sharma, Vallabh P. *Environmental and Engineering Geophysics*. Cambridge, England: Cambridge University Press, 1997. In a somewhat technical manner, Sharma describes the developments and policies within the field of engineering geophysics. A close look is also given to the role the environment plays in engineering decisions. Bibliography and index.

See also: Deep-Earth Drilling Projects; Freshwater Chemistry; Gravity Anomalies; Ocean Drilling Program; Ocean-Floor Drilling Programs; Remote Sensing Satellites; Rock Magnetism; Seismic Reflection Profiling.

ENVIRONMENTAL CHEMISTRY

Environmental chemistry is the study of the chemical interactions that take place in a natural environment, how to control them when necessary, and what happens when unnatural chemicals are introduced into a natural environment. Air, water, and soil chemistry are all affected by materials injected into the corresponding environmental systems by way of human activity. All natural environments function according to the same chemical principles that apply in a chemistry laboratory. Regular testing and monitoring of environmental systems determines how those materials affect and are affected by environmental system processes.

PRINCIPAL TERMS

- **absorption:** the capture of light energy of a specific wavelength by electrons in a specific atom or molecule
- **bioaccumulation:** the process whereby a pollutant material becomes concentrated in the body tissues of organisms as they consume other organisms
- **bioactive:** those materials that are capable of biological or biochemical activity when present in living systems
- **chlorofluorocarbons:** carbon-based compounds to which various numbers of fluorine and chlorine atoms are bonded instead of the hydrogen atoms of the parent hydrocarbon compounds
- **colorimetric:** relating to the measurement of the intensity of absorption of a specific color, or the light of a specific visible wavelength
- **eutrophication:** the depletion of dissolved oxygen in water by the growth of algae and aquatic plants, and the subsequent decomposition of vegetable matter
- **interface:** the surface between two materials at their point of contact with each other—as, for example, air and water, liquid and solid
- **partitioning:** a physical process by which components of a solution are extracted from one solvent directly into a second solvent immiscible with the first
- **spectrophometric:** relating to the measurement of the intensity of absorption of light in a specific wavelength that may or may not be in the visible range, including infrared and ultraviolet wavelengths
- **turbidity:** the degree to which a fluid scatters or diffuses light, caused by solid microparticles or liquid microglobules suspended in the fluid medium

THE CHEMICAL NATURE OF THE ENVIRONMENT

Modern chemical science is based on the quantum mechanical model of the atom, the underlying tenet being that all material in the universe is matter composed of atoms. Accordingly, anything and everything in the physical universe is chemical in nature. Matter interacts with matter and energy, as the atoms of matter interact with each other and with energy. The quantum mechanical model of the atom requires that the atoms interact in specific ways, however, and according to strict mathematical rules that are neither entirely known nor entirely understood.

The quantum mechanical model describes atoms of matter as consisting of a small, dense nucleus. This nucleus contains almost the entire mass of the atom in a specific number of discrete particles called protons and neutrons. The protons each bear a single positive charge, while the electrically neutral neutrons fulfill the role of a sort of "nuclear glue," countering the mutual repulsion of the protonic charges. For the atom itself to be electrically neutral, it contains a number of negatively charged electrons equal to the number of protons. The electrons occupy a space around the nucleus that is about 100,000 times larger in diameter than the nucleus.

According to the mathematical rules of quantum mechanics, electrons are allowed only to have specific energies depending on their position in the space around the nucleus; each specific energy-position combination is referred to as an orbital. The lowest-energy electrons, and hence the most tightly bound electrons, are those situated closest to the nucleus, while any other electrons in the atom range outward in order of energy; therefore, the outermost electrons are the most energetic and the least tightly bound by the nucleus. Interaction between atoms takes place only at the level of the outermost electrons involved. Within that restriction, any interaction that can take place will take place. This means that atoms of different elements can combine in an infinite number of ways, producing stable combinations of atoms with specific spatial arrangements called molecules. Electrons in molecules also must adhere to stringent energy levels.

The interaction of matter, as atoms, with energy is somewhat more complicated, but must nevertheless obey the same rule system of quantum mechanics as applies to atomic structure. The most common interaction of atoms occurs through the direct action of light energy on the electrons within the atoms and molecules. Absorption of light energy by an electron in one atomic or molecular orbital raises the level of the orbital that it must occupy. This is often sufficient to break the molecule apart and, in so doing, drive uncontrolled reactions that are usually undesirable. The fading of color and the embrittlement of plastics left in bright sunlight are examples of matter interacting with light energy. In other cases, the interaction of light energy with specific molecules also may be highly desirable, as is the case when chlorophyll is activated by sunlight to produce glucose and oxygen from carbon dioxide and water.

The Physical Interface

Substances exist normally in one of three states: solid, liquid, or gas. Earth is an agglomeration of substances that are found in any and all of these states. As determined by ancient philosophy, the four basic elemental forms of water, air, earth, and fire represent the three basic states of matter plus energy, all of which are the material medium of environmental chemistry. The physical environment of Earth can thus also be considered in the context of physical interaction between matter in all of these states, and in the context of chemical interactions.

It is an unavoidable reality that the physical and chemical contexts affect each other directly in various ways. The rates of chemical reactions, for example, are affected by the extent to which different chemical materials come into contact with each other at a physical interface, such that the larger the interface area, the faster an overall chemical reaction occurs. Similarly, the more intimately mixed two gases become, such as ozone and fluorocarbon compounds, the more rapidly and completely they react.

The concept of the physical interface between materials is complex. In the case of gases, no definitive interfacial surface can be defined as existing between two masses. Diffusion occurs where the two masses come into contact; the only real definable interface for interaction between them is the "surface" of the individual atoms or molecules as they interact. At the other end of the interface surface spectrum are the massive interfaces between air and water, air and land surface, and water and the solid earth. In the overall structure of Earth, one could also include the more diffuse divisions between the crust and the underlying mantle of the planet. As the source of volcanic and seismic activity, these also have significant roles in the environmental chemistry of the planet.

Environmental Chemistry and Human Activities

While there is nothing humans can do to control or even affect the natural geochemical processes associated with volcanic or seismic activity, the practical field of environmental chemistry focuses much more closely on those aspects of human society that enact change gradually. It has become increasingly apparent that human activities since the Industrial Revolution have had a significant impact on the natural workings of the planetary environment, especially with regard to air and water quality. Agriculture, industry, and transportation are the components of human activity that affect the environment most significantly. In all cases, the principal means of these changes is the ejection of waste and unnatural materials into the environment.

Since the early nineteenth century, the main gaseous component has been gaseous carbon dioxide from the combustion of fuels. Additionally, significant environmental effects have been observed as a result of other materials produced through combustion, unnatural quantities of naturally occurring materials (such as methane), and the indiscriminate use and release of chlorofluorocarbon (CFC) compounds as propellants and refrigerants. Unnatural materials and unnatural quantities of normal materials can now be found in the environment everywhere in the world because of the release of liquids and dissolved materials into the environment—as effluents from production and processing facilities, mining operations, industrial and transportation accidents, warfare, and many other human activities.

Solid materials produced as a result of human activities are also problematic with regard to environmental chemistry. Not the least of these materials is the mass quantity of solids placed in landfills every year. Even more insidious environmental effects have been wrought through the nanoparticulate matter deposited in the environment by combustion fuels—namely, the millions of tons of lead and manganese oxides that have been produced by internal combustion engines.

Human energy use and production also plays a role in environmental chemistry. Excess city light and heat have affected air and water quality, while the production of electrical energy through nuclear processes has led to an entirely separate environmental chemistry problem of its own: the management of waste and spent materials and the release of radioactive materials in quantity into the environment.

ENVIRONMENTAL CHEMISTRY OF THE ATMOSPHERE

Perhaps the most recognizable of atmospheric chemical effects is the generation of smog: an accumulation of noxious gases, smoke, and nanoparticulate matter that forms throughout cities. Gas-phase chemical reactions occurring between components of smog and normal atmospheric gases are known to contribute to an assortment of undesirable effects, including acid rain and asthma. Nitrogen oxides and sulfur dioxide react with water and oxygen to produce nitrous, nitric, sulfurous, and sulfuric acids, all of which become dissolved in raindrops as water vapor condenses in the atmosphere. The resulting acidic solutions, often appearing long distances from their sources, react chemically with carbonate stone such as limestone, marble, and concrete. This results in the slow chemical erosion of infrastructure, directly bringing about destruction and occasioning structural upkeep that costs billions of dollars.

Acidic solutions also dissolve solid materials on the surface, percolating through soils. These materials are then carried into waterways and into other sources of drinking water. In more extreme or prolonged cases, acidic precipitation has been identified as the agent responsible for the ecological deaths of otherwise pristine bodies of water, rendering them unable to support an aquatic ecology.

Julie Masura, right, a researcher with the University of Washington-Tacoma environmental science program, removes a collection container from a filter that was pulled through the Thea Foss Waterway, in Tacoma, Wash. Masura and other scientists are developing methods to measure the level of microplastics—tiny plastic particles no larger than a ladybug—in seawater and sediments by sampling the waters of Puget Sound. At left, Joyce Dinglasan-Panlilio, a UWT assistant professor of environmental chemistry, gathers a water sample. (AP Photo)

Other causes of smog include chlorine, bromine, and ground-level ozone, all of which react with oxygen and with each other in a multitude of ways by radical mechanisms. These interactions produce various oxyacids and other chemical compounds that have inimical environmental effects. Breathing air that contains these reactive materials causes damage to lung tissue and can lead to emphysema, asthma, and other respiratory difficulties. In some cities around the world, smog has become a serious health issue, requiring constant monitoring of air quality. Government environment regulatory agencies now issue daily statements of air quality as a regular feature of official meteorological reports.

While ground-level ozone is considered a health hazard, upper atmosphere ozone is essential to the survival of life on Earth. The diffuse layer of ozone at very high altitudes acts as a shield for the planet against the continuous influx of ultraviolet radiation. Absorption of the energy of ultraviolet light by the ozone molecule results in the formation of an electronically excited species that re-emits that energy at a different wavelength that is harmless with regard to living tissue.

In the twentieth century, the chemical compound known as CFC (chlorofluorocarbon) was commonly used as a propellant and refrigerant. The low boiling point and ready compressibility in the gaseous state of CFCs made them ideal for that application, but at the same time also made them easy to dispose of as they dissipated readily into the atmosphere. Ignorance of their environmental effects allowed them to be used for many years, until satellite-based monitoring of the upper atmosphere revealed that the accumulation of CFCs had seriously depleted the ozone shield by way of chemical reactions that produced oxygen dichloride and other exotic chemical species that did not interact with ultraviolet radiation.

Air quality and gaseous effluent monitoring is carried out through a number of methods. Direct sampling for instrumental analysis, using tandem gas chromatography-mass spectrometry (GC-MS), can detect airborne materials in the range of parts per billion. This is especially important when the materials of interest are highly carcinogenic materials (such as dioxins) that do not exist naturally in the environment. Other monitoring methods include ground-based and satellite-based infrared detection of composition and temperature and the use of monitoring tags for exposure to specific chemical compounds. Such tags use the occurrence of a specific chemical reaction to bring about a color change in the surface of the tag.

ENVIRONMENTAL CHEMISTRY OF WATER

Airborne materials can become dissolved in condensate water to produce acidic precipitation. Both acidic and nonacidic precipitation, including agricultural irrigation, dissolve materials in the soil and carry that material into the nearest waterways or into the water table as the liquid percolates through the soil. The leachates so formed can include every compound and element in the soil. For naturally occurring materials in their normal natural proportions, this is not problematic. However, the injection into the environment, through human activities, of many materials makes it essential that the chemical effects of those materials in the environment be monitored and understood.

For example, the use of tetraethyl lead in gasoline over a period of fifty years (before its use in North America was banned) had deposited some nine billion kilograms of lead in the environment as nanoparticulate lead oxides from vehicle exhaust gases. This resulted in a sudden, drastic increase in the amount of lead detectable in drinking water. In Europe, where estate wineries keep meticulous records of production, the amount of lead detectable in wine could be traced through analysis of vintage wine samples to the exact year that the use of leaded gasoline began in that region. In North America, the uptick in lead accumulation coincided with an increase in the levels of birth defects and disorders, and in mental health issues in children.

Lead is a known neurotoxin, as is the manganese that was used to replace the lead in gasoline fuels. Another neurotoxin is mercury, a common waterborne effluent of the mining and pulp and paper industries. Mercury has the insidious property of accumulating in the fatty tissue of organisms, as do many other heavy metals such as cadmium. The process of bioaccumulation works to concentrate those materials in organisms as it progresses up the food chain. By this process, creatures that consume smaller creatures as food, such as fish, may contain quantities of bioactive heavy metals that greatly exceed the baseline amount of those materials in the environment. Chemical compounds such as pesticides (for example, dichloro-diphenyl-trichloroethane, or DDT), industrial chemicals (such as polychlorinated biphenyls, or PCBs) and carcinogenic chemicals (dioxin) also can be concentrated by bioaccumulation.

In agriculture, both organic and mineral fertilizers are used to augment the nutritive value of soils to increase or maintain crop yields. The primary components of mineral fertilizers, and to a lesser extent the organic fertilizers, are nitrogen (as nitrate) and phosphorus (as phosphate). Without this addition, mineral nutrients essential for plant growth in soils can become depleted as successive crops remove them faster than natural processes can replenish them.

Irrigation also works toward the depletion of minerals and other nutrients from the soil by dissolving and carrying them away as the water percolates downward to the water table. Coupled with the overuse of inorganic fertilizers, this can and has led to elevated amounts of nitrates and phosphates in waterways and water tables. The presence of these extra nutrients contributes to the excess growth of algae and aquatic plants, and to the eutrophication of bodies of water.

To an extent, this problem is alleviated by the practice of crop rotation, in which different crops are grown on the same plots of land in successive years.

Typically, a crop that draws nutrients from the soil in one year will be succeeded in the following year by a crop that incorporates nutrients naturally by nitrogen fixation or deep-root mineral extraction. For example, corn is often rotated with beans on heavy soils with a high clay component, and tobacco is generally rotated with rye grain on light, sandy soils. Additionally, the likelihood of crop disease increases when the same crop is grown on the same fields year after year without rotation through other crops.

WATER TESTING

Water testing is carried out both regularly and upon request using a much broader variety of techniques. Because materials are dissolved in water, numerous instrumental analytical techniques are available to measure contents and properties of samples brought in for testing.

Spectrophotometric methods detect specific ions and organic compounds both qualitatively and quantitatively. Samples are prepared for analysis in different ways. In some cases, specific compounds are added that combine with dissolved ions or organic compounds to form specific complexes. Such complexes are often colored, whereas the dissolved ions and organic compounds are not. Colorimetric analysis can then be used to determine the concentration of the complexed material in the prepared sample, and hence the original concentration of the ions and organic compounds of interest.

Spectrophotometric analysis can be used to determine the presence and concentration of specific materials directly. In each case, analysis depends on the measurement of the absorption of light at characteristic wavelengths that are uniquely associated with the specific materials for which testing is being carried out. The Beer-Lambert law describes the relationship between the absorption of light energy and the concentration of the material that is absorbing the light.

Methods such as high-performance liquid chromatography (HPLC) and tandem GC-MS can determine concentrations as low as only a few parts per billion. Chromatographic methods are extremely versatile. They are based on the equilibrium between the differential adsorption of dissolved materials on a solid surface and the solubility of those materials in a fluid phase (either gas or liquid).

In chromatographic separations, the fluid phase containing the dissolved materials passes through a second, solid phase. Components of the gas or liquid solution equilibrate between being dissolved in the fluid medium and being adsorbed on the solid medium. As they progress through the column of the solid medium, they become increasingly separated from each other. Effective separation brings each component of the solution out of the column at different times.

When chromatographic methods such as gas chromatography or HPLC are connected in tandem with a second procedure, such as mass spectrometry, it becomes possible to identify and measure the components of complex mixtures such as environmental water samples in a single step. In tandem processes the efflux from the first stage becomes the input material for the second stage without any intermediate isolation or purification step. In some cases, such as the analysis of organic compounds in aqueous solutions, it is expedient to extract the organic compounds by partitioning to eliminate physical interference by water in the analysis.

As a general rule, the presence of water makes organic analysis very difficult simply because of the physical properties of that material. Organic solvents, being much more volatile than water, are also much more easily removed than water and so make analysis much easier.

Richard M. Renneboog

FURTHER READING

Baird, Colin, and Michael Cann. *Environmental Chemistry.* Toronto, Ont.: W. H. Freeman, 2008. This introductory, college-level textbook presents the field of environmental chemistry in a balanced discussion of soil, water, and air chemistry as it relates to current issues such as global warming, renewable energy, and hazardous waste.

Girard, James E. *Principles of Environmental Chemistry.* 2d ed. Sudbury, Mass.: Jones and Bartlett, 2010. This book presents the study and practice of environmental chemistry in a thorough manner, with each chapter devoted to one specific aspect of the field.

Hites, Ronald A. *Elements of Environmental Chemistry.* Hoboken, N.J.: John Wiley & Sons, 2007. An elementary introduction to basic chemical principles as they are applicable to some of the basic aspects of environmental chemistry. Features practical problems that demonstrate the methods to readers as they are solved.

Trimm, Harold H., and William Hunter, III. *Environmental Chemistry: New Techniques and Data* Candor. New York: Apple Academic Press, 2011. Discusses environmental chemistry as an interdisciplinary field of study that continually changes, addressing new analytical methods and results and new environmental issues.

Van Loon, Gary W., and Stephen J. Duffy. *Environmental Chemistry: A Global Perspective.* New York: Oxford University Press, 2010. Examines environmental chemistry in the context of the chemical nature of the three physical-foundation environments of the planet. Also analyzes the role of human activities.

Weiner, Eugene. *Applications of Environmental Aquatic Chemistry: A Practical Guide.* 2d ed. Boca Raton, Fla.: CRC Press, 2008. This book provides detailed information about the nature of environmental aquatic systems and the nature and behavior of material pollutants within them.

See also: Biogeochemistry; Climate Change: Causes; Experimental Petrology; Fluid Inclusion; Freshwater Chemistry; Geobiomagnetism; Glaciation and Azolla Event; Mass Extinction Theories; Milankovitch Hypothesis; Nucleosynthesis; Oxygen, Hydrogen, and Carbon Ratios; Radioactive Decay; Water-Rock Interactions.

EXPERIMENTAL PETROLOGY

Experimental petrology is the laboratory simulation of chemical and physical conditions within and at the surface of the earth. A wide variety of apparatuses are used to routinely obtain and control precisely the range of temperature and pressure conditions known to occur up to 150 kilometers in depth. Other apparatuses allow access to the much greater temperatures and pressures of the transition zone of the earth's mantle. Experimental data allow petrologists to interpret quantitatively the evolution of natural rocks and the earth as a planet.

PRINCIPAL TERMS

- **component:** a chemical entity used to describe the compositional variation of some phase
- **igneous rock:** any rock that forms by the solidification of molten material, usually a silicate liquid
- **metamorphic rock:** any rock whose mineralogy, mineral chemistry, or texture has been altered by heat, pressure, or changes in composition; metamorphic rocks may have igneous, sedimentary, or other, older metamorphic rocks as their precursors
- **mineral:** a naturally occurring solid compound that has a specific chemical formula or range of composition; a mineral normally has regular crystal structures such that its internal arrangement of atoms is predictable
- **phase:** a chemical entity that is generally homogeneous and distinct from other entities in the system under investigation; compositional variation within phases is described in terms of components
- **phase equilibria:** the investigation and description of chemical systems in terms of classical thermodynamics; systems of specified composition are generally investigated as a function of temperature and pressure
- **thermodynamics:** the area of science that deals with the transformation of energy and the laws that govern these changes; equilibrium thermodynamics is especially concerned with the reversible conversion of heat into other forms of energy

TEMPERATURE AND PRESSURE CONDITIONS

A wide range of chemical and physical conditions exist within and at the surface of the earth. Experimental petrology is the simulation of these conditions in a carefully controlled laboratory environment. Temperature and pressure are the two primary physical parameters that change during geological processes. On or near the earth's surface, temperatures ranging between 0 and 1,300 degrees Celsius are observed. Such temperatures are easily attained at low pressure in the laboratory. Much higher temperatures occur deep within the earth and other planets. Pressure increases proportionally with increasing depth below the earth's surface as a direct result of the greater mass of material overlying the material below. The pressure at the earth's surface is that exerted solely by the overburden of the atmosphere, which defines the low pressure limit attained in geological processes.

At a depth of 3,000 kilometers below the earth's surface, near the boundary between the solid silicate mantle and the liquid outer core, the pressure approaches 1.2 million times that exerted by the atmosphere on the surface. As the pressure within the earth increases, so does the temperature, such that at the core-mantle boundary, the temperature approaches nearly 3,000 degrees Celsius. Experimental petrologists have apparatuses that will routinely obtain pressures of 50,000 times that of the atmosphere and temperatures up to 2,000 degrees Celsius, which corresponds to the physical conditions that occur at approximately 150 kilometers below the earth's surface. Other devices available to petrologists since 1980 allow access to temperatures and pressures as great as those found at 1,000 kilometers deep.

NATURAL- AND SYNTHETIC-SYSTEM EXPERIMENTS

The style of experiments done by petrologists to quantify the conditions of formation of a particular suite of rocks or some widely occurring rock type is nearly as varied as is the number of active investigators. These experiments, however, may be broadly grouped into two major categories: The first would include all experiments done on natural rock or mineral systems, while the second would include experiments done on simpler synthetic systems that are analogous to the much more complex natural systems in some way.

Most experimental petrologists tend to work within one of these categories almost exclusively. Experiments on natural systems are necessarily more complicated and difficult to interpret because even the simplest rocks contain three to four chemical components (the simplest chemical entity that may be used to describe the system under consideration), while most rocks contain ten major and several minor components. In contrast, experiments on synthetic systems can be designed to isolate and study an individual chemical component. In such experiments it is considerably less difficult to demonstrate attainment of equilibrium and interpret the finding in terms of classical thermodynamics.

QUENCH AND GAS-MIXING FURNACES

Experimental petrologists have developed many different apparatuses that allow them to achieve conditions in the laboratory that mimic those found in the earth. The simplest apparatuses are quench and gas-mixing furnaces. Typical working conditions for these apparatuses are pressures that range from moderate vacuums to 1 atmosphere and temperatures from 0 to 1,600 degrees Celsius. Slight variations in pressure on the order of 1 percent normally occur during the operation of these furnaces. Temperatures may be controlled and measured with a precision of several to tens of degrees depending on the particular setup.

Standard 1-atmosphere quench furnaces are set up with a vertical ceramic tube around which some resistive heating elements are either wound or placed in close proximity. The elements are then heavily insulated to prevent large heat losses to the laboratory atmosphere. The vertical geometry is required to enable rapid quenching of material held in the hot zone of the furnace to water at 25 degrees Celsius (or some other suitable quench medium) at the lower end of the vertical tube. Samples are generally held at the end of a ceramic rod by thin platinum loops. At the end of an experiment, the platinum loops are melted by passing a small electrical current through them, which allows the samples to fall by gravity directly into the water, where they are cooled rapidly.

Standard vertical tube furnaces may be modified to include the capability to have a gas mixture of known composition flow through the tube, replacing the static air environment. Such gas mixtures are commonly composed of species of carbon and hydrogen that, when mixed in known proportions, will fix the oxygen activity of the furnace atmosphere. The control of oxygen activity allows experimental petrologists to investigate chemical systems that contain transition metal cations, whose valence state would otherwise change in an uncontrolled and perhaps unwanted way. Samples range in size from 0.1 to 50 milliliters. One-atmosphere furnaces also may be mounted horizontally, if rapid quenching and gas mixing are not required. These apparatuses are found in virtually every experimental petrology laboratory around the world and have been used to investigate a wide variety of petrological problems.

COLD-SEAL VESSELS

Another common apparatus found in experimental petrology laboratories is the cold-seal vessel designed by O. F. Tuttle in the late 1940's. These apparatuses have been enormously important in the investigation of metamorphic and igneous processes that occur at middle to lower crustal levels. Cold-seal vessels typically are operated at pressures of several hundred to several thousand atmospheres and between 25 and 900 degrees Celsius. Modifications of the original design have allowed petrologists to obtain pressures up to 12,000 atmospheres and slightly higher temperatures. The pressure vessel is fabricated from a superalloy rod usually composed of nickel and chromium with smaller amounts of other metals. The rod has a small-diameter hole drilled into it to yield a container that is similar in shape to a test tube. The walls of the vessel are kept thick to support the high pressures and temperatures that occur during the course of an experiment. A pressure seal is formed by a cone-in-cone fitting at the open end of the vessel. High temperatures are obtained by placing the pressure vessel inside a simple muffle furnace (kiln), which is usually mounted vertically. The end of the vessel with the pressure seal remains outside the furnace and thus remains cold throughout the experiment. The pressure medium in most experiments is water, but to obtain pressures of more than 8,000 atmospheres, argon gas is used.

PISTON CYLINDERS

The last common apparatus found in experimental petrology laboratories is called a piston cylinder. These apparatuses were originally designed and built in the late 1950's and early 1960's. Much of the current understanding of melting relations in basaltic and ultramafic systems has been gained by using the piston cylinder. The typical operating conditions range from pressures of 5,000 to 60,000 atmospheres and temperatures between 25 and 1,800 degrees Celsius. These conditions are similar to those of the earth's deep crust and shallow upper mantle. The apparatus consists of a small piston pressing into a cylinder, which compresses the solid materials of the furnace assembly. One end of the cylinder usually abuts a massive end load. The piston is pressed against the furnace assembly by the use of a hydraulic ram. The ratio of the areas of the piston and the ram allows one to calculate the pressure obtained inside the cylinder. Furnace assemblies consist of small graphite cylinders inside Pyrex or salt outer sleeves with inner sleeves of similar materials. Sample sizes typically range between 0.01 and 0.1 milliliter, which is an order of magnitude smaller than those used in 1-atmosphere experiments. Noble metal capsules that are welded shut are commonly used to contain the sample materials and to isolate them from the rest of the furnace assembly. Temperature is generated by passing a current through the graphite furnace as a result of its finite resistance. Although this apparatus is generally quite easy to operate, careful calibration of experimental temperatures and pressures is necessary prior to its use. Large pressure corrections as a result of frictional forces may arise depending on the materials and design of the furnace assembly.

STUDYING BASALTIC ROCKS

The pioneer of modern igneous petrology, N. L. Bowen, used 1-atmosphere quench furnaces as described earlier to study a simple analog system for the evolution of basaltic rocks. Bowen investigated a synthetic iron-free diopside-albite-anorthite system as a function of composition and temperature. Because the system was iron-free, there was no need to control the oxygen activity during the course of the run by a gas-mixing apparatus. The system does contain sodium, however, which is notoriously volatile at high temperatures. To combat this problem, the sample charges were enclosed in platinum foil. At each bulk composition, a series of experiments was conducted to determine the onset and completion of melting. The starting materials were previously synthesized crystalline pyroxene and plagioclase.

Although this method could yield erroneous results because the melting point was only approached from the low-temperature side and not reversed from the high-temperature side, the analysis of the run products was extremely sensitive to small amounts of melting, recorded as glass. The run products were rapidly quenched and then ground to a fine powder. Gain mounts immersed in oil allowed Bowen to detect minute amounts of crystalline material and thus to determine precisely the location of the liquids in temperature-composition space. To determine the liquid composition, experiments were conducted to yield only quenched liquid (glass), which was then analyzed by conventional wet chemical techniques. Today, with highly developed electron microbeam capabilities, the liquid and two solid phase compositions could all be determined simultaneously. Bowen was able to apply his experimental results to the petrogenesis of basaltic rocks. The experiments clearly demonstrated that with decreasing temperature, plagioclase composition will become more sodic and less calcic when coexisting with a diopsidic pyroxene and liquid of approximately basaltic composition. Modern experiments on basalt and peridotite systems incorporating controlled amounts of volatiles began in earnest in the 1960's and continued through the 1980's. By the 1990's a reasonably self-consistent model for the evolution of terrestrial basalt magmas was available.

STUDYING GRANITIC SYSTEMS

Granitic systems were studied extensively by Tuttle and Bowen in the late 1950's. Until the time of their experiments, many geologists did not believe that granite batholiths (large intrusive igneous rocks) were the products of crystallization from silicate liquids at moderate pressure. The experiments of Tuttle and Bowen were conducted in cold-seal pressure vessels at temperatures below 700 degrees Celsius and pressures between 500 and 4,000 atmospheres. Their experiments were some of the first to use the apparatus designed by Tuttle. The solubility of water in synthetic granitic melts was determined as a function of temperature and pressure. Water was added directly to the experimental charges, which were welded shut inside platinum capsules. These experiments demonstrated conclusively that many granitic batholiths crystallized from water-bearing silicate melts.

STUDYING MAGMAS

A final example of experimental techniques for solving petrological problems is an investigation of the solubility of carbon dioxide in basaltic liquids at high pressure. In the late 1970's, D. Eggler, using a piston-cylinder apparatus similar to the one described earlier, demonstrated that carbon dioxide had significant solubility at pressures of 30,000 atmospheres. To study the effect of carbon dioxide in the melting relations of synthetic systems whose compositions approximated the earth's upper mantle, silver oxalate was added to the experimental charges, which were then sealed by welding in platinum capsules. At high temperature, the silver oxalate decomposes to produce carbon dioxide or other carbon bearing species, depending on the composition of the liquid present and the oxygen activity during the course of the experiments. The addition of carbon dioxide decreases the solidus to lower temperatures at constant pressure and tends to favor the formation of orthopyroxene over olivine as a result of changes in the melt structure. These experiments, and others like them, are useful in helping to constrain the genesis and evolution of alkali-rich, silica-poor magmas.

STUDYING VOLCANOES AND EARTHQUAKES

Experimental petrology has provided geologists with quantitative data that allow them to understand many complex geologic processes. The process of magma genesis deep in the earth's upper mantle, and its subsequent migration from depth to the surface is a process that would not be as well understood today if not for experimental petrologists and their work. The eruption of volcanoes at the surface of the earth is merely one example of the type of process upon which experimental petrology bears. One such eruption occurred in 1980 at Mount St. Helens, Washington. Studies of volcanic rocks, which generally integrate experimental data, geochemistry, and field geology, have revealed that the processes that governed the eruption of Mount St. Helens are still operating there and at other sites worldwide where oceanic crust is subducted below continental crust. In the Cascade province of the western United States, for example, Mount Shasta, Mount Bachelor, and Mount Rainier have many characteristics in common with Mount St. Helens, and these volcanoes might be expected to erupt in a similar manner in the near future. Such an eruption of Mount Rainier or Mount Bachelor could prove to be extremely dangerous to the large population centers of Seattle and Portland.

The techniques of experimental petrology also allow geologists to develop geothermometers and geobarometers, which may be applied to pieces of the earth that are entrained in magmatic eruptions worldwide. The compositions of the coexisting phases in these fragments of rock have permitted the petrologist to determine temperature-depth profiles for the upper 200 kilometers of the earth, gaining valuable information on the composition and state of regions of the earth that are not accessible to direct observation. This information is crucial to the understanding of how and why rocks deform during earthquakes. For rocks of fixed composition, ambient temperature is the primary variable that determines whether rocks will be able to deform in such a way as to produce an earthquake.

Glen S. Mattioli

FURTHER READING

Blatt, Harvey, and Robert J. Tracy. *Petrology: Igneous, Sedimentary, and Metamorphic.* 3rd ed. New York: W. H. Freeman, 2005. Undergraduate text in elementary petrology for readers with some familiarity with minerals and chemistry. Thorough, readable discussion of most aspects of petrology. Abundant illustrations and diagrams, good bibliography, and thorough indices.

Campos, Cristina, et al. "Enhancement of Magma Mixing Efficiency by Chaotic Dynamics: An Experimental Study." *Contributions to Mineralogy & Petrology* 161 (2011): 863-881. This article introduces a new apparatus used in experimental petrology. The authors discuss the need for the new apparatus in studying petrogenesis, as well as results from their experiments. A good example of experimentation in petrology.

Carmichael, Ian S. E., Francis J. Turner, and John Verhoogen. *Igneous Petrology.* New York: McGraw-Hill, 1974. Many colleges use this classic text for a first course in igneous petrology. It includes extensive discussions on all aspects of the formation of igneous rocks. The text is highly technical but quite readable. Suitable for college-level students.

Edgar, Alan D. *Experimental Petrology: Basic Principles and Techniques.* Oxford, England: Clarendon Press, 1973. The most complete book on experimental petrology presently available. All aspects, from basic thermodynamic treatment of data to starting materials and apparatus, are covered in detail with excellent illustration. The material is suitable for the undergraduate or graduate student whose interest or specialty is in experimental petrology.

Ernst, W. G. *Petrologic Phase Equilibria*. San Francisco: W. H. Freeman, 1976. This text outlines the elements of classical thermodynamics and experimental approaches to acquiring the thermodynamic data necessary for geothermometry and geobarometry. A short but well-illustrated section on experimental petrology is included. Primarily concerned with the application of data acquired by experiment to the interpretation of igneous and metamorphic rocks.

Fyfe, W. S., and Mackenzie, W. S. "Some Aspects of Experimental Petrology." *Earth-Science Reviews* 5 (1969): 185-215. Though outdated, this article provides a good account of the then current research methods in petrology. Highlights topics which benefit from experimentation in petrology.

Gregory, Snyder A., Clive R. Neal, and W. Gary Ernst, eds. *Planetary Petrology and Geochemistry*. Columbia, Md.: Geological Society of North America, 1999. A compilation of essays written by scientific experts, this book provides an excellent overview of the field of geochemistry and its principles and applications. The essays can get technical at times and are intended for college students.

Grotzinger, John, et al. *Understanding Earth*. 5th ed. New York: W. H. Freeman, 2006. An excellent general text on all aspects of geology. Discussions on the formation of igneous and metamorphic rocks and on the structure and composition of the common rock-forming minerals are included. Also discussed is the relationship of igneous and metamorphic petrology to the general principles that form the basis of modern plate tectonic theory.

Hall, Anthony. *Igneous Petrology*. 2d ed. Harlow: Longman, 1996. This introductory book provides a good understanding of igneous rocks and their geophysical phases. There are sections devoted to the study of petrology and magmatic processes. Well illustrated and with plenty of diagrams and charts to reinforce concepts, this is a good resource for the layperson.

Holloway, J. R., and B. J. Wood. *Simulating the Earth: Experimental Geochemistry*. Winchester, Mass.: Unwin Hyman, 1988. A book that describes the different philosophies, apparatus, and applications of experimental geochemistry. Although far from comprehensive, the text will give the college-level student a good idea about the nuts and bolts of experimental work.

Mortimer, Charles E. *Chemistry: A Conceptual Approach*. 3rd ed. New York: D. Van Nostrand, 1975. This book is aimed at advanced high school and beginning college students. An excellent basic chemistry text designed to accompany a first course in general chemistry. Extensive descriptions of all basic chemical phenomena are included. Problems follow each chapter, and a separate answer book is available.

Newton, Robert C. "The Three Partners of Metamorphic Petrology." *American Mineralogist* 96 (2011): 457-469. This article discusses field observations, experimental petrology, and theoretical analysis of the earth's crust. It highlights how each of these contributes to studies, and that collaboration is necessary to move forward in petrology.

Ulmer, G. C. *Research Techniques for High Pressure and High Temperature*. New York: Springer-Verlag, 1971. An edited volume of research papers that deal specifically with the finer points of how to conduct different types of experimental work. The book, although highly technical, remains quite readable and should be suitable for college-level students and above.

Uyeda, Seiya. *The New View of the Earth: Moving Continents and Moving Oceans*. San Francisco: W. H. Freeman, 1971. This college-level text provides an in-depth outline of the modern theory of plate tectonics. The author has placed all the relevant observations in their historical context, which allows the reader to become familiar with the people involved in the development of the central paradigm of the earth sciences.

See also: Earth's Age; Electron Microprobes; Electron Microscopy; Experimental Rock Deformation; Geologic and Topographic Maps; Geothermometry and Geobarometry; Infrared Spectra; Mass Spectrometry; Neutron Activation Analysis; Petrographic Microscopes; Phase Equilibria; Plate Tectonics; Radioactive Decay; Water-Rock Interactions; X-ray Fluorescence; X-ray Powder Diffraction.

EXPERIMENTAL ROCK DEFORMATION

To understand how and why rocks deform, experiments are done using laboratory apparatus that simulate some of the conditions found in the earth's crust and mantle. These experiments have shown that the mechanical behavior of rocks is complex, but can be deciphered. The results help to develop an intuition which leads to meaningful interpretations of field situations.

PRINCIPAL TERMS

- **brittle behavior:** the sudden failure of a sample by catastrophic loss of cohesion
- **confining pressure:** pressure acting in a direction perpendicular to the major applied stress in a rock deformation experiment
- **dislocation:** a defect in a crystal caused by misalignment of the crystal lattice; the presence of dislocations greatly reduces the stress necessary to produce permanent deformation
- **ductile behavior:** permanent, gradual, nonrecoverable deformation of a solid; sometimes called plastic deformation
- **elastic behavior:** recoverable deformation where the strain is proportional to the stress
- **pore pressure:** the pressure in the fluid within the pores of a rock
- **strain:** a measure of deformation including translation, rotation, dilatation, and distortion; it is usually measured as a percentage or ratio and results from stress
- **stress:** the intensity of forces (force per unit area) acting within a body; may refer to a particular stress acting in a particular direction on a particular plane or to the collection of all stresses acting on all planes at that point

BRITTLE VS. DUCTILE BEHAVIOR

Experiments in rock deformation aim to develop an understanding of how rocks behave mechanically. Under conditions found at the surface of the earth, most rocks behave as brittle solids, but when subjected to the conditions found at depth within the earth, those same rocks behave in a very ductile manner. If a chalk-sized piece of rock were squeezed between the jaws of a vise, it would fail in a sudden fashion, almost like an explosion. The pieces of rock recovered from such an experiment would be sharp shards similar to the remnants of a rock that was smashed with a hammer. The breaking strength, pore pressure, and internal angle of friction determine the failure criterion for a given type of rock. The failure criterion encompasses the combinations of factors that can cause the rock to fail. Engineers can apply such criteria to determine the stability of a slope, for example, or the spacing required for pillars in an underground mine.

Often a rock will deform without breaking. Rock layers are sometimes buckled into folds that can vary from angular kink bands to smoothly undulating surfaces. Other rocks seen in roadcuts and outcrops show, by the patterns of their banding and textures, that they have deformed in a very ductile way, flowing much like a fluid. To simulate this behavior in a laboratory experiment, the conditions under which deformation occurred must be considered.

DEFORMATION CONDITIONS

The most striking examples of ductile deformation come from deep crustal or upper mantle depths. What are the pressure, temperature, and strain rate like at depths of 20 kilometers? The pressure produced by 20 kilometers of rock with a density of approximately 2.7 grams per cubic centimeter is about 5.2 kilobars, which is approximately 75,000 pounds per square inch, or 5,000 atmospheres of pressure. By studying minerals found in volcanic rocks that come from deep sources, scientists have learned that the temperatures at a depth of 20 kilometers are 250-500 degrees Celsius. If an object deforms so that its length changes by 1 percent in a second, it is undergoing a deformation with a strain rate of 0.01 per second. Geologists have found examples of deformation that incorporate datable features, and from these have learned that strain rates of 10^{-13} to 10^{-14} per second are typical for geological processes. At such rates, the length of an object would change by about one part per million per year. This is too slow to study in the laboratory, so experiments are run at strain rates on the order of 10^{-5} per second. The results are extrapolated to estimate how rocks would behave at very low strain rates. An extrapolation over nine orders of magnitude is risky but is supported by theoretical considerations.

Experiments have been designed to study the effects of pressure, temperature, and strain rate, as well as pore pressure, anisotropy, and water content. The emerging picture reveals that rocks exhibit a complex mechanical behavior. At low confining pressures, low temperatures, and high strain rates, they behave as elastic solids when subjected to stresses up to their breaking strength, then fail in a brittle fashion. At high confining pressures, high temperatures, and low strain rates, they deform in a ductile fashion, with yield strengths and viscosities that are functions of temperature and strain rate. One result of this behavior is that when subjected to the very low strain rates associated with convection, mantle rocks flow easily; when subjected to the high strain rates resulting from the passage of those seismic waves called shear waves, mantle rocks respond like elastic solids.

DISLOCATIONS

Results from flow experiments indicate that at high temperatures and pressures and low strain rates, rocks deform by the movement of offsets in crystal lattices called dislocations. This is similar to the way a caterpillar moves forward—only a few of its legs are in motion at any one time, but the movements propagate along as waves, and the whole animal moves forward. As dislocations move through a crystal, only a few bonds are broken at a time, but the entire crystal deforms as a result.

The study of dislocations and how they move has resulted in a better understanding of ductile deformation of rocks and other materials. Several different mechanisms, such as power law creep and diffusion creep, have been found to be active in different substances under different conditions. Flow laws have been formulated, and maps have been constructed which show, for a given mineral, which flow law will dominate the deformation for a given stress difference and temperature. Because rocks are aggregates of different minerals, their behavior is more complex than that of any single mineral. Progress is being made, however, and eventually the behavior of the material of the crust and upper mantle will be better understood.

STUDYING LOWER MANTLE AND CORE BEHAVIOR

To study lower mantle and core behavior, experiments have been designed using diamonds as platens (flat plates that exert or receive pressure). Diamonds can withstand very high pressures and are transparent to visible light. This transparency permits visual observations of phase changes and allows samples to be heated to very high temperatures using lasers. Conditions similar to those within the earth's core can be simulated in such experiments, but the measurements that are possible are limited by the need to use small samples.

STUDYING FRACTURE AND FLOW

The methods used to study rock deformation in the laboratory depend on whether fracture or flow is the subject of investigation. Many studies of fracture are motivated by the desire to understand how earthquakes occur, and, if possible, to develop means of predicting them. Because damaging earthquakes frequently occur in rock near the surface, these experiments are done at low temperatures and confining pressures. Studying the flow of rock at high temperatures and high confining pressures develops insights into mantle convection and plate tectonics. The general procedure is to prepare a sample of the rock to be studied; attach strain gauges to the sample to monitor changes in strain during the course of the experiment; insert the sample between the platens of the press; adjust confining pressure, temperature, and pore pressure to the conditions of interest; and then squeeze the sample while recording data from the sensors.

An experiment with no confining pressure is called a "uniaxial" test, because all the stress is applied along one axis. A hydrostatic confining pressure can be applied to the sample by immersing it within a medium and then compressing that medium. Although the terminology is not quite correct, this kind of experiment is usually called a "triaxial" experiment. The confining medium can be a solid (such as talc), a fluid (commonly kerosene), or a gas.

During experiments exploring the fracturing process, the sample may be instrumented by gluing small microphones to it. Just before the rock fails catastrophically, several small acoustic events often can be located within the sample by triangulation from a number of microphones. These noises are thought to be produced by the extension of small, naturally occurring fractures as they grow in response to the increasing stress. Determining the relationship between these events and the fracture that finally forms offers a promising means of earthquake prediction: Foreshocks are commonly recorded in the vicinity where a large earthquake is about to occur and may be similar to the acoustic events observed in the laboratory.

High confining pressures, high temperatures, and low strain rates are needed to study the flow of rock. Such experiments are technically difficult, particularly if they run for long periods of time. Commonly, the temperature is increased above that typically expected to occur within the earth at the pressures being studied, so as to increase the rate of deformation. At the completion of the experiment, the sample may be recovered, sliced into thin sections, and studied under a microscope. If the textures produced in the laboratory resemble those observed in samples from the crust and mantle, it is likely that similar processes are active. The flow laws operative during the experiment can be determined and then adjusted for differences in temperature and strain rate, which permits the extrapolation to the conditions present within the earth to be conducted with more confidence.

UNDERSTANDING FORMATION OF GIANT STRUCTURES

A topographic map from the Valley and Ridge Province of the Appalachian Mountains shows sinuous ridges tracing out elaborate folds in a coherent pattern extending for hundreds of miles. The landscape south of San Francisco is dominated by long, linear valleys parallel to the San Andreas fault. Roadcuts near the Thousand Islands reveal swirling, flowing patterns which appear to have formed as if the marble there behaved like a fluid.

Each of these phenomena is a striking demonstration of how rocks deform when subjected to the mammoth stresses involved in mountain building and plate tectonics, yet they are all very different from one another. To understand how such giant structures are formed, scientists have performed experiments in the laboratory on small samples of the rocks from which the structures are made. They have learned that the behavior of rock is a function of its environment at the time it is being deformed, the size of the stresses applied to it, and the rate at which those stresses are applied.

FRACTURE ORIENTATIONS AND PATTERNS

Some of this behavior can be compared to that of three familiar materials: modeling clay, beeswax, and Silly Putty. Modeling clay shows a behavior that varies with its environment, particularly temperature. A piece of cold modeling clay is difficult to work with. Most people spend a few minutes kneading it in their warm hands; its behavior changes noticeably as it warms. A piece of very cold modeling clay may shatter if it is dropped on the floor, unlike a piece that has been warmed. If the pieces were reassembled, there would be a pattern of fractures related to the orientation of the rock within the vise. Much larger, but similar, fractures occur within the crust of the earth, which are called joints or faults. Theoretical considerations and data from laboratory experiments are used to interpret the orientations and patterns produced by these brittle fractures.

If a tennis ball were put in the vise, it would shorten in the direction it was squeezed and would get fatter in the plane of the jaws of the vise. A rock deforms elastically, just like a tennis ball, but at a much smaller scale. Sensors, called strain gauges, attached to the rock sample will record these tiny changes in shape. Careful monitoring of the stress applied by the vise and the strain experienced by the rock would help to determine the elastic constants that describe the behavior of the rock before failure begins. These constants, called Young's modulus for compressional stresses and Poisson's ratio for shearing stresses, can be used to calculate seismic wave velocities. As failure occurs, fractures grow across the sample. In the rock-in-a-vise example, these fractures will usually form perpendicular to the jaws of the vise, corresponding to what are called extension joints. If the sample were enclosed in a jacket that provided pressure on its sides, the experiment would be conducted with a confining pressure present. Under these conditions, many fractures might form at an angle to the jaws of the vise, producing a set of what are called conjugate shear joints. Alternatively, one fracture might develop, and the sample might slip in opposite directions on both sides of this fracture. Such a fracture corresponds to a fault in the field. Measuring the angles at which these fractures form would show that they are somewhat constant for fractures produced in the same material. By increasing the confining pressure, the stress needed to break the sample also increases. Graphing the results permits the determination of another material constant, called the internal angle of friction. A comparison of this angle with the angle at which conjugate fractures and faults form shows that they are simply related.

Pore Pressure

These results characterize some of the mechanical behavior of the rock from which the sample was taken. Young's modulus, Poisson's ratio, breaking strength, and the internal angle of friction are material constants that vary little among different samples from the same rock. Different types of rocks have different elastic constants and strengths, just as they have different densities.

Fluids within the pores of a rock play a significant role in its brittle behavior. Experiments that control the pressure of such pore fluids show that the strength of the rock decreases as the pore pressure increases. Some of a rock's resistance to failure is provided by the pressure of one grain against the next. Pore pressures reduce this pressure and so weaken the rock, which helps to explain why most catastrophic landslides have occurred after heavy rainfall. Slopes that are stable when dry can weaken as the pore pressure within them increases to become unstable and to fail. High pore pressure may also facilitate movements on thrust faults deep within the earth.

Size and Rate of Stresses

The behavior of beeswax varies with the size of the stresses applied to it. A chunk of beeswax feels hard and makes a sharp, rapping sound when struck against a table. The fact that hives and statues in wax museums maintain their shape for years attests to the ability of beeswax to resist the forces of gravity over long periods of time. Yet it is easy to stick a thumbnail into a chunk of beeswax. The stress produced by the edge of a nail is greater than the strength of the beeswax, and it deforms, whereas the stresses produced by gravitational force are less than the strength of the beeswax and are unable to cause it to deform.

Silly Putty shows a behavior which varies with the rate at which stresses are applied to it. Throw a sphere of Silly Putty onto the floor, and it bounces. But pull on it slowly, and it will stretch. In response to a rapid application of stress, Silly Putty behaves like a brittle, elastic solid. But when subjected to a slowly applied force, its behavior is much more like that of a fluid.

Otto H. Muller

Further Reading

Billings, Marland P. *Structural Geology*. 3rd ed. Englewood Cliffs, N.J.: Prentice-Hall, 1972. Chapter 2, "Mechanical Principles," includes an elementary review of experimental rock deformation. Descriptive, with little prior knowledge assumed, this book is suitable for the general reader.

Dahlen, F. A., and Jeroen Tromp. *Theoretical Global Seismology*. Princeton: Princeton University Press, 1998. Intended for the college-level reader, this book describes seismology processes and theories in great detail. The book contains many illustrations and maps. Bibliography and index.

Davis, George H. *Structural Geology of Rocks and Regions*. 2d ed. New York: John Wiley & Sons, 1996. Chapter 3, "Dynamic Analysis," discusses experimental rock deformation in a manner that is suitable for the general reader. It takes the reader step by step through an experiment involving the compression of a limestone, with the procedures, results, and interpretations of those results carefully described. The treatment is the most descriptive and least technical of the references listed here.

Doyle, Hugh A. *Seismology*. New York: John Wiley, 1995. A good introduction to the study of earthquakes and the earth's lithosphere. Written for the layperson, the book contains many useful illustrations.

Fossen, Haakon. *Structural Geology*. New York: Cambridge University Press, 2010. This text is well written and easy to understand. An excellent text for geology students or resource for geologists. Provides many links between structural geology theory and application. Photos and illustrations add great value to the text. Contains a glossary, references, an appendix of photo captions, and indexing.

Hobbs, Bruce E., Winthrop D. Means, and Paul F. Williams. *An Outline of Structural Geology*. New York: John Wiley & Sons, 1976. Section 1.4, "The Response of Rocks to Stress," provides a nice summary of the results obtained from experiments in rock deformation. Chapter 2, "Microfabric," presents excellent discussions of the development of microfabric and crystallographic preferred orientations in deformed rocks. This treatment includes many striking microphotographs, line drawings, and stereonet plots. Suitable for technically oriented college students.

Jackson, Ian, ed. *The Earth's Mantle: Composition, Structure, and Evolution*. Cambridge: Cambridge University Press, 2000. Intended for the college student, *The Earth's Mantle* provides a clear and complete description of the elements that make up the earth's mantle and the process of change that it has undergone since its formation. Includes bibliography and index.

Jaeger, John Conrad, Neville George Wood Cook, and Robert Zimmerman. *Fundamentals of Rock Mechanics*. 4th ed. New York: John Wiley & Sons, 2007. This book is a standard text in the field of experimental rock deformation. Although parts of it become so technical that they are suitable only for experts in the field, much of it is of interest to technically minded college students. Chapter 6, "Laboratory Testing," is thorough, lucid, and contains fewer complex equations than most of the book.

Mancktelow, N. S. "Fracture and Flow in Natural Rock Deformation." *Trabajos de Geologia* 29 (2009): 29-35. Mancktelow writes a clear and concise description, comparing brittle fractures to ductile flows. The article is well organized and has some excellent photographs providing examples of rock deformation. This is an excellent article for those with a general interest in geology to gain exposure to current research and peer-reviewed work.

Marshak, Stephen, and Gautam Mitra. *Basic Methods of Structural Geology*. Englewood Cliffs, N.J.: Prentice-Hall, 1988. Chapter 10, "Analysis of Data from Rock-Deformation Experiments," by Terry Engelder and Stephen Marshak, begins by providing a good overview of the subject, followed by results from several experiments which the reader is invited to interpret. This technique probably results in a better understanding and a firmer intuitive grasp of what is entailed in doing experiments on the mechanical behavior of rocks than any other reference listed here. Suitable for the general reader.

Paterson, Mervyn S., and Teng-fong Wong. *Experimental Rock Deformation: The Brittle Field*. 2d ed. New York: Springer, 2005. The book begins with experimental procedures, techniques and methodology. Along with the strong content in brittle rock deformation, this text is a good resource for geologists and upper-level geology students.

Suppe, John. *Principles of Structural Geology*. Englewood Cliffs, N.J.: Prentice-Hall, 1985. Chapter 4, "Deformation Mechanisms," and Chapter 5, "Fracture and Brittle Behavior," discuss the results of experiments in rock deformation. The emphasis is on flow laws and failure criteria and how these are related to the microstructures and chemical bonds within the rock. Suitable for college students, but a background in physical chemistry or material science would be helpful.

See also: Creep; Cross-Borehole Seismology; Discontinuities; Earthquake Engineering; Earthquake Prediction; Earthquakes; Earth's Crust; Earth's Mantle; Earth's Oldest Rocks; Elastic Waves; Elemental Distribution; Experimental Petrology; Faults: Normal; Faults: Strike-Slip; Faults: Thrust; Faults: Transform; Geothermometry and Geobarometry; Heat Sources and Heat Flow; Phase Changes; Rock Magnetism; San Andreas Fault; Seismic Observatories; Seismic Tomography; Seismometers; Stress and Strain.

F

FAULTS: NORMAL

Normal faults are common features that occur when the earth's crust is subjected to tensional forces. The sense of movement is primarily vertical and results in an extension of the crust. These faults are generally associated with broad-flexed or uplifted areas and are an integral part of the modern concept of plate tectonics.

PRINCIPAL TERMS

- **dip:** the angle of inclination of a fault, measured from a horizontal surface; dip direction is perpendicular to strike direction
- **fault:** a break in the earth's crust that is characterized by movement parallel to the surface of the fracture
- **fault drag:** the bending of rocks adjacent to a fault
- **footwall:** the crustal block underlying the fault
- **graben:** a long, narrow depressed crustal block bounded by normal faults that may form a rift valley
- **hanging wall:** the crustal block that overlies the fault
- **horst:** a long, narrow elevated crustal block bounded by normal faults that may result in a fault-block mountain
- **slickensides:** fine lines or grooves along a faulted body that usually indicate the direction of latest movement
- **stress:** the forces acting on a solid rock body within a specified surface area
- **throw:** the vertical displacement of a rock sequence or key horizon measured across a fault

NORMAL FAULT FORMATION

A normal fault is a fracture that separates two crustal blocks, one of which has been displaced downward along the fractured surface. Some workers use the term "gravity fault" to indicate an apparent normal fault if genesis, rather than geometry, is implied. Crustal blocks overlying or underlying a normal or reverse fault are commonly designated as the hanging wall and footwall blocks, respectively. In a normal fault, the hanging wall moves downward relative to the footwall. In a reverse fault, the hanging wall moves upward relative to the footwall, with a dip greater than 45 degrees. These are old descriptive terms that were used in the early English coal-mining districts. Faults that were inclined toward the down-dropped side were common in the area and the term "normal fault" was applied. At places where the movement was in the opposite direction, the breaks in the rock were designated as reverse faults. The displacement of normal faults, which can be intermittent, ranges from less than a meter up to thousands of meters. The inclination or dip of the fault can be from nearly horizontal to vertical but generally ranges from 45 to 60 degrees. In some areas, the angle of dip decreases with depth and results in a curved surface that is concave upward. This curved surface is termed a listric normal fault and is a common type of fracture in the Gulf coast region of the United States.

Normal faults are the product of a dynamic process that results in conditions of changing stress (force per unit area) along a plane of weakness in the earth's crust. The fault develops from a point along this plane. According to Lamoraal de Sitter, the stress is at a minimum at the starting point along the surface and at a maximum at the edges. Because of the edge conditions, the plane steepens and splits into several divergent smaller faults, or splays. These small segments may join to form a larger normal fault with a scalloped trace. Subparallel normal faults with smaller displacements generally accompany the large faults. At places, the adjacent beds are systematically fractured without significant displacement. These fractures are termed joints and generally have a high density (close spacing) near the fault.

The deforming forces can be related to three mutually perpendicular but unequal axes designated as maximum principal stress (σ_1), intermediate principal stress (σ_2), and minimum principal stress (σ_3). In the case of normal faults, the primary deforming force (σ_1) is vertical or nearly vertical. The least stress (σ_3) is horizontal. The normal faults are actually steeply inclined shear fractures that formed in response to forces promoting the sliding of adjacent blocks past each other. These fractures generally form at an angle of 30 degrees from the maximum principal stress. The orientation of the maximum principal stress is horizontal for thrust faults and for wrench (transform) faults.

NORMAL FAULT CLASSIFICATION

Normal faults are classified according to the type of displacement of fault blocks relative to a known point. Based on the slip or actual movement of formerly adjacent points on opposing fault blocks, three types are commonly designated: strike-slip, or movement along the trend of the fault; dip-slip, or movement directly down the fault surface; and oblique-slip, or diagonal movement down the fault surface. The movement along all these examples is translational. Consequently, no rotation of the blocks in respect to each other has occurred outside a disturbed zone adjacent to the fault. If the actual displacement is not known, the term "separation" is used by most geologists to indicate the apparent movement on a map or cross section. Heave and throw are the horizontal and vertical components, respectively, of the dip separation as measured along a vertical profile that is at right angles to the trend of the fault.

There are several varieties of normal faults. Detachment (denudation) faults have a low angle of dip (usually less than 30 degrees) and are common features in the western United States. Growth or contemporaneous faults are listric normal faults that are active during sediment accumulation. Layered rocks on the downthrown side of the fault are thicker than equivalent beds on the upthrown side. Smaller subsidiary or antithetic faults commonly form on the downthrown side of the main fault but dip in a direction opposite to the master fault. These are common features along the Gulf coast of the United States.

Some special faults may result from the same stress orientation as normal faults; that is, the maximum principal stress is vertical. These closely related faults, however, are characterized by rotational movement between blocks. For example, hinge faults increase in displacement along the length of the fault; linear features that were parallel before faulting are not all parallel after faulting. A pivotal (scissor) fault is another example of a rotational fault. In this type, the fault

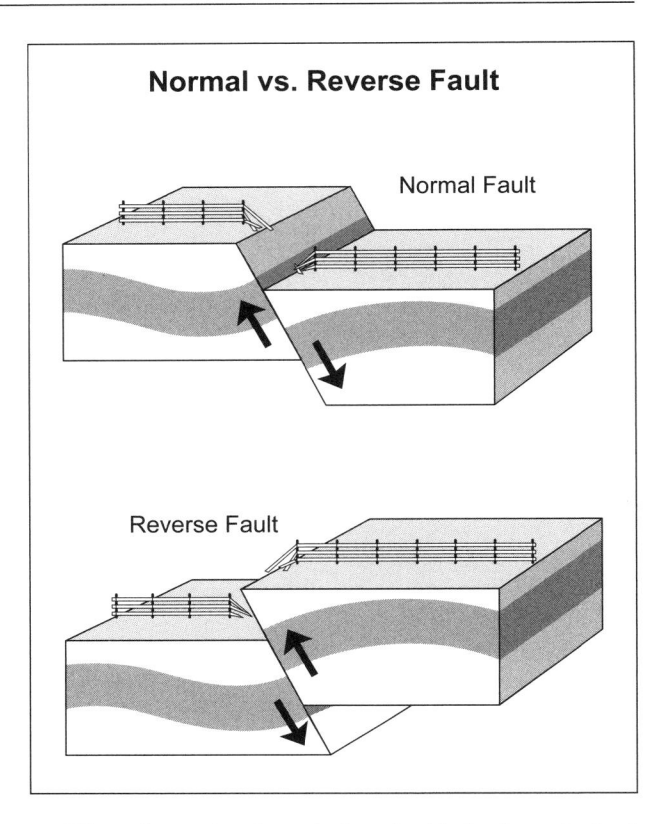

When the motion is such that the block above the fault plane slides down the fault, it is a "normal" fault. When the motion is such that the block above the fault plane slides up the fault, it is a "reverse" fault.

blocks pivot about an axis that is at right angles to the fault surface. The movement on the downthrown side is in opposite directions (up and down) along the length of the fault.

ASSOCIATED FEATURES

Major steeply dipping normal faults occur in the Colorado Plateau (Arizona and New Mexico) where these features are closely associated with regional flexures called monoclines. The western part of the plateau along the Colorado River is divided into large structural blocks by three north-trending faults. One of these faults, the Hurricane fault of Arizona and Utah, dips to the west and has a maximum displacement of 3,048 meters.

At some places, normal faults bound narrow blocks that have been displaced up or down. An uplifted or elevated block is called a horst; the depressed block is termed a graben. Topographically, these structural features may be represented by a series of mountain ranges and intervening valleys, respectively. The Basin and Range Province of the western United States is a good example of this horst-and-graben type complex. In this region, both low- and high-angle normal faults have shaped an area that extends from southern Oregon southward to northern Mexico; the area has been broadly uplifted and the crust stretched in an east-west direction by the normal faulting. Some estimates of the total extension across the region are more than 100 percent. The displacement along these large faults ranges up to 5,486 meters. In Europe, the Rhine graben is a classic example of a well-developed rift system. This narrow structural trough trends northward for nearly 300 kilometers through West Germany and controls the path of the upper Rhine River.

CLUES TO IDENTIFICATION

Normal faults are recognized in the field or on vertical aerial photographs and satellite images by identifying features characteristic of this type of fault. On the earth's surface, these faults occur as geological lines that are revealed by a sharp, curvilinear line in the bedrock that is usually accentuated by vegetation, a sharp contact in adjacent rock types in section or map view, a marked change in structural style, an abrupt change in topography, or an anomalous drainage pattern. Normal faults are usually recognized in vertical drill holes by an omission of rock layers; comparison of rock samples (drill cuttings) or mechanical logs from adjacent borings will generally reveal the part of the rock column that is missing. Caution, however, must be exercised to make sure that the strata have not been eroded, have not thinned, or have not changed character laterally.

There are many distinctive geometric, mineralogic, or physiographic features that are associated with large normal faults. Some of these features, however, are also characteristic of other types of faults and should thus merely be considered as "clues" in recognizing normal faults. Many normal faults are expressed at the surface by low cliffs or scarps that reflect the minimum displacement along the fault; however, these straight slopes are usually modified by erosion to form faultline scarps. The scarps may be notched by streams crossing the upthrown block at a high angle to the fault trend; continued erosion by these side streams may result in triangular-shaped bedrock facets on the footwall with fan-shaped stream deposits on the downthrown hanging wall block. Movement along the fault usually disturbs rocks adjacent to the break and results in beds along the fault being bent up or down; in other words, the rocks bend before rupture takes place. This phenomenon is called fault drag. Normal drag occurs when rocks on the upthrown side are bent down into the fault and rocks on the downthrown side are bent upward. Reverse drag occurs when beds on both sides of the fracture are bent down.

Because movement along the fault produces an irregular surface between the fault blocks at some places, subsurface fluids are provided an avenue to the surface. Both hot- and cold-water springs are common occurrences along large normal faults. Solutions moving along the fault may also deposit minerals such as calcite or quartz between the blocks. These fillings are usually stained yellow or reddish brown by iron oxide.

Movement along the fault is usually recorded by fine lines or by narrow grooves on the fault surface called slickensides. These features, however, may indicate only the latest movement along the fault. Impressions of the slicken lines are sometimes preserved on the outer surfaces of the mineral fillings. A series of larger scale (several centimeters or more of relief), parallel grooves and ridges may produce an undulating fault surface. The movement of large blocks along the fault usually produces low, steplike irregularities on the surface that are steeply inclined in the direction of movement. These features can be used to identify the direction of movement when a fault surface is poorly exposed.

As the fault blocks slide past each other, angular rock fragments are dislodged and may accumulate to form a tectonic breccia; a microscopic breccia, or mylonite, may also result from movement. In some cases, the dislodged rock may be ground to a pliable, claylike substance called gouge. At places, a large fragment of bedrock, called a horse, is caught along a normal fault.

CHARACTERISTIC MAP PATTERNS

Normal faults generally occur in definite region patterns that are easily represented on geologic maps. Zones of overlapping, or *en echelon*, normal faults are common in the Gulf coast region of the United States. In Texas, individual faults within the

Balcones and Mexia-Luling fault systems are not continuous along strike but overlap with adjacent faults that have a similar trend. These fault zones generally follow the path of the buried Ouachita fold belt and mark the boundary between the geologically stable area of Texas and the less stable Gulf coast region. Parallel or subparallel faults in an area also form a distinctive map pattern; if most of the faults are downthrown in the same direction, these structural features are designated as step faults. Radial fault patterns are common over or around central uplifts or domed areas of the crust. These faults are generally associated with local stretching of the crust that results from the emplacement of salt masses (plugs) or igneous intrusions.

Some normal faults are also closely related to the development of plunging (inclined) folds and form characteristic map patterns. According to de Sitter, steeply dipping normal cross faults may form nearly at right angles to the trend of concentric (formed of parallel layers) folds. These faults are parallel to the principal deforming force and occur during the folding process. The maximum displacement occurs along the highest part (crest) of the fold; these faults usually die out along the flanks. Longitudinal crest faults may occur parallel to the trend of the folds. These normal faults are perpendicular to the principal deforming force and probably form as the compressional forces diminish.

ROLE IN SHAPING EARTH'S SURFACE

Faults have played a significant role in shaping the earth's surface throughout geologic time. The occurrence of normal faults is closely tied to modern plate tectonics, a unifying concept for the geological sciences. The faults are generally associated with modern and ancient divergent lithospheric plate boundaries, both on continents and in the ocean basins. The regions adjacent to modern plate margins, which are characterized by high heat flow and shallow-focus earthquake activity, are places where new oceanic-type crust is generated. In modern ocean basins, inferred normal faults bound narrow down-dropped blocks (grabens) along the axis of the mid-ocean ridge system, the longest continuous geologic feature on earth. Topographically, these structural troughs form deep valleys along the ridge crest. Individual troughs range up to 30 kilometers wide and are filled or partly filled with sediment. The Mid-Atlantic Ridge, a mountain range along the midline of the Atlantic basin, extends northward from Antarctica to Iceland. In Iceland, measurements across the boundary between the North American and Eurasian plates indicate that the crustal blocks are currently moving apart at the rate of a few centimeters per year.

In the Middle East, along the Red Sea, steeply dipping normal faults are associated with a large dome or uplift over a plumelike hot spot in the earth's mantle. Near the Afar region of Ethiopia, the uplift has been subdivided by three radial fault systems that intersect at angles of about 120 degrees. These systems are characterized by large, high-angle normal faults that initially formed a series of down-dropped blocks, or grabens. The three-pronged structural feature represents a "triple junction" that separates the African, Arabian, and Indian-Australian lithospheric plates. The East African rift system, which consists of both east and west zones, forms the second prong; it trends northward for nearly 5,000 kilometers and is marked by a series of elongate lakes. The maximum displacement on the bordering faults is nearly 2,500 meters at some places. The east-trending third arm of this large feature is a rift that is partly occupied by the Gulf of Aden.

Earth scientists have also been able to identify historical divergent lithospheric plate boundaries from regional geologic and geophysical (application of physics to geological problems) studies. In the modern Appalachian mountain chain in the eastern part of the United States, a series of elongated structural troughs (grabens and half-grabens) occur along the axis of the range. These structural features, which extend from Nova Scotia in Canada southwestward to North Carolina, contain thick deposits of Triassic (period of geologic time ranging from about 200 to 245 million years ago) sedimentary rocks with associated igneous rocks. Internally, steeply dipping normal faults divide the troughs into narrow tilted blocks that range up to 10 kilometers wide. Some of the border faults were active during Triassic deposition and have a cumulative displacement of nearly 4,000 meters. The formation of these troughs probably marked the separation of North America and Europe about 200 million years ago.

ECONOMIC IMPORTANCE

Normal faults are also economically important. These faults serve as traps for hydrocarbons at many places. Migrating oil and gas moving updip from a place of origin, usually a sedimentary basin, are trapped against the fault, which acts as an impermeable barrier or seal. If the fault is not completely sealed, however, it may serve as an avenue for fluids to move to a higher level. Most commercial hydrocarbon deposits occur on the upthrown side of the fault where "rollover" of the rock layers has provided a suitable site for the accumulation of hydrocarbons. The faults are also the locus of metallic mineral deposits. Mineralization may occur in the openings along the fault or in the adjacent fractured rock. Drag ore, related to fault drag, occurs at some places. Also, rich ore bodies are moved downward along younger normal faults. A classic example is at the United Verde extension mine near Jerome, Arizona. There, a rich copper deposit was displaced more than 500 meters vertically.

Donald F. Reaser

FURTHER READING

Billings, M. P. *Structural Geology*. 3rd ed. Englewood Cliffs, N.J.: Prentice-Hall, 1972. A popular college textbook that presents basic concepts and structural features in a clear, concise, and understandable manner. Emphasizes a field approach to recognizing and solving geological problems. Suitable for upper-level undergraduate geology students.

Buck, Roger, et al, eds. *Faulting and Magmatism at Ocean Ridges*. Washington, D.C.: American Geophysical Union, 1998. This collection of essays covers topics including seismology, magmatism, active faults, and seafloor spreading. The articles lean toward the technical but are illustrated with charts, maps, and graphs. Bibliography and index.

Buck, W. Roger, Luc L. Lavier, and Alexei N. B. Poliakov. "Modes of Faulting at Mid-ocean Ridges." *Nature* 434 (2005): 719-723. The article presents new information on plate separation at ridges. Understandable to the layperson; the article has few, yet well-explained, equations. This provides useful information for identifying processes which produce particular characteristics in faults.

Cloos, E. "Experimental Analysis of Gulf Coast Fracture Patterns." *American Association of Petroleum Geologists Bulletin* 52 (1968). This journal article describes the results of experimental work with clay and dry sand models. Model grabens bounded by normal faults as well as single normal faults accompanied by fault drag are produced by applying tensional forces in a pressure box. These model fractures are compared to the Texas Gulf coast fracture pattern. Suitable for high-school-level readers and college students who are interested in geological models.

Coble, Charles R., E. C. Murray, and D. R. Rice. *Earth Science*. Englewood Cliffs, N.J.: Prentice-Hall, 1986. A general textbook designed for junior high school students and interested laypersons. In the structural section of the text, challenging scientific questions are presented for the reader. Also, the activities of specialists in the structural career field are summarized.

Davis, G. H. *Structural Geology of Rocks and Regions*. 2d ed. New York: John Wiley & Sons, 1996. This book is very readable and takes a practical approach to regional tectonics. Basic concepts and principles of structural geology are emphasized. Suitable for upper-level undergraduate geology students.

De Sitter, L. U. *Structural Geology*. New York: McGraw-Hill, 1959. This book effectively relates geological theory with practice; it compares similar geological phenomena, both on a small and large scale, from different parts of the world. An advanced book designed primarily for students with a good background in geology.

Doyle, Hugh A. *Seismology*. New York: John Wiley, 1995. A good introduction to the study of earthquakes and the earth's lithosphere. Written for the layperson, the book contains many useful illustrations.

Fossen, Haakon. *Structural Geology*. New York: Cambridge University Press. 2010. This text is well written and easy to understand. An excellent text for geology students or resource for geologists. Provides many links between structural geology theory and application. Photos and illustrations add great value to the text. Contains a glossary, references, an appendix of photo captions, and indexing.

Jacobs, J. A. *Deep Interior of the Earth*. London: Chapman and Hall, 1992. Deals in detail with all aspects of the earth's inner and outer core. The origin of the core, its constitution, and its thermal and magnetic properties are discussed in detail. Well suited to the serious science student.

Judson, Sheldon, and Marvin E. Kauffman. *Physical Geology*. 8th ed. Englewood Cliffs, N.J.: Prentice-Hall, 1990. An interesting and well-written book on physical geology designed for introductory college students. Includes a section on geotectonics that details features associated with normal faults. An excellent glossary of technical terms.

McClay, Kenneth R. *Thrust Tectonics*. London: Chapman and Hall, 1992. This collection of papers was presented as part of the Thrust Tectonics Conference held at Royal Holloway and Bedford New College, University of London, in 1990. The advanced nature of the collection makes this most useful for the college student.

Mitra, Shankar, et al., eds. *Structural Geology of Fold and Thrust Belts*. Baltimore: Johns Hopkins University Press, 1992. A good discussion of physical geology focusing on the structure and processes of thrust faults and folds. Suitable for the college reader. Illustrations, bibliography, and index.

Palmer, Donald F., and I. S. Allison. *Geology: The Science of a Changing Earth*. 7th ed. New York: McGraw-Hill, 1980. An interesting and well-written book on physical geology that is designed for high school or introductory college students. An excellent section on geotectonics that details features associated with normal faults.

Park, R. G. *Foundations of Structural Geology*. 3rd ed. New York: Routledge, 2004. An excellent reference for high school or college students specifically interested in structural geology. Good coverage of the relationship between geologic structures and plate tectonics.

Shelton, J. W. "Listric Normal Faults: An Illustrated Summary." *American Association of Petroleum Geologists Bulletin* 68 (1984). This article provides the reader with details about the characteristic features of a specific type of normal fault. The paper also discusses the general geometry, causes, and occurrences of most types of normal faults. A number of illustrations, mostly cross sections, are included in the text as an aid to understanding the concepts presented.

Wibberley, Christopher A. J., and Shipton, Zoe K. "Fault Zones: A Complex Issue." *Journal of Structural Geology* 32 (2010): 1554-1556. Provides a brief account of the geology of normal faults.

See also: Creep; Cross-Borehole Seismology; Discontinuities; Earthquake Hazards; Earthquake Magnitudes and Intensities; Earthquake Prediction; Earthquakes; Elastic Waves; Experimental Rock Deformation; Faults: Strike-Slip; Faults: Thrust; Faults: Transform; Mountain Building; Notable Earthquakes; Plate Tectonics; San Andreas Fault; Seismic Observatories; Seismic Reflection Profiling; Seismic Tomography; Seismometers; Stress and Strain.

FAULTS: STRIKE-SLIP

Strike-slip faults separate portions of the earth's crust that have moved horizontally past each other. They can be thousands of kilometers in length, with offsets of hundreds of kilometers across them. Many of the most devastating earthquakes occur along strike-slip faults.

PRINCIPAL TERMS

- **dip:** a measure of slope; the angle between a plane and the horizontal, measured in the vertical plane perpendicular to the strike of the plane
- **fault:** a fracture in the earth's crust across which there has been measurable movement
- **plate tectonics:** a theory that holds that the surface of the earth is divided into about twelve rigid plates that move relative to one another, producing earthquakes, volcanoes, mountain belts, and trenches
- **slip:** the relative motion across the surface of a fault
- **strike:** the orientation of a horizontal line on a plane; it is measured using compass directions and represents the angle between the horizontal line on the plane and a horizontal line in the north direction

CHARACTERISTICS OF STRIKE-SLIP FAULTS

A fault is a surface within the earth's crust across which displacement has occurred. A surface can be curved, flat, tilted, vertical, or horizontal. A curved surface may be subdivided into smaller pieces, each of which can be considered to be a plane. At any location, therefore, a fault can be thought of as a planar element. The orientation of this element is referred to by two angles: its "strike" and its "dip." The strike is the orientation of a horizontal line on the plane, given as a compass direction. The dip is the angle at which the plane tilts down into the ground.

Stipulating that a fault occurs within the earth's crust implies something about the scale of the process. Although some may consider fractures on a centimeter or meter scale to be faults, in general faults are expected to be tens of meters or kilometers in size, and some can be hundreds of kilometers long. Faults seldom consist of a single, clean fracture, so the term "fault zone" is used when referring to the region of complex deformation associated with the fault plane.

Displacement across the fault, or "slip," refers to the amount and direction of relative motion of the blocks of rock on opposite sides of the fault. If this motion is entirely horizontal and neither block has moved up or down, the slip will be in the direction of the strike on the fault surface; such faults are called "strike-slip" faults. Conversely, if the motion is in the direction of the slope on the fault plane, it is called a "dip-slip" fault. There are two kinds of dip-slip faults. When the motion is such that the block above the fault plane slides down the fault, it is a "normal" fault. When the motion is such that the block above the fault plane slides up the fault, it is a "reverse" fault.

For many reasons, some mechanical and some geological, most faults on earth are one of these three types. Less common are faults with significant components of both dip-slip and strike-slip motions, which are called "oblique-slip" faults. In addition, a surface may be a strike-slip fault over most of its length, but portions may be dip-slip or oblique-slip. If the relative motion between the blocks of rock on either side is predominately horizontal, it is considered to be a strike-slip fault.

TYPES OF STRIKE-SLIP FAULTS

Because their motion is restricted to being horizontal, only two kinds of strike-slip faults are possible. These are defined on the basis of how the block of rock that is across the fault from the observer appears to move. It is not necessary to know which block actually moved; only their relative motion is important.

Consider two buses parked next to each other with their drivers' sides adjacent and facing in opposite directions. If either bus were to move forward, passengers in one bus would see the other bus appear to move forward. They might not even be certain whether it was their bus moving or the other one. To passengers looking out the side windows, the view would be of a bus moving to the left, past the windows. This relative motion could be called "left-lateral" or "sinistral." Similarly, if either bus were to move in reverse, passengers in either bus, looking out the side windows, would see the other one move to their right. The relative motion between the buses could be called "right-lateral" or "dextral." Such relative motion has nothing to do with the absolute motions (movement across the ground) of either bus. If one bus were to move in reverse and the other bus were to move forward somewhat more rapidly, relative motion would be left-lateral or sinistral, even though both buses were moving across the ground in the same direction.

Geologists can study features that are offset across a strike-slip fault to determine the sense of relative motion and then name the fault either right-lateral or left-lateral. As an example, if a fence is offset across an east-west-trending fault such that the fence north of the fault has moved horizontally to the east (or the fence south of the fault has moved horizontally to the west), then it is a right-lateral strike-slip fault.

This concept was successfully applied for decades but needed modification when seafloor spreading was discovered. Offsets along mid-ocean ridges, where new ocean crust is being manufactured, appear to have a sense of relative motion that is opposite the sense of motion observed along the strike-slip faults connecting them. This can be demonstrated by considering twelve fast-growing trees planted in pots. Imagine that they are all in two parallel north-south lines initially. Next, imagine that the six at the north end of the lines (three in each line) are moved a few meters to the east. This produces an offset in the lines with a right-lateral sense. Finally, imagine that all of the trees are knocked down: Those on the western lines fall to the west, while those on the eastern lines fall to the east. Temporarily ignoring some tenets of biology, imagine that all the trees continue to grow taller, but because they are lying down, this growth is horizontal. Along the offset, trees to the north will be growing to the west, and trees to the south will be growing to the east. This motion is left-lateral, opposite the offset in the pots. Mid-ocean ridge segments are thus not good indicators of the sense of motion on the strike-slip faults connecting them because new crust is created along them. The term "transform" fault refers to this situation and to similar manifestations of strike-slip faults.

RECOGNIZING STRIKE-SLIP FAULTS

Two approaches are used to find and delineate strike-slip faults: identifying features that have been juxtaposed or offset by the fault, and detecting features in the landscape that are known to have been produced by the faulting process. If an area has experienced fault displacements after human development, obvious offsets of anthropogenic structures are generally easy to find, measure, and date. Highways, railroads, fence lines, pipelines, and buildings have been studied over the years to decipher recent displacement histories on a great number of faults. Other, less obvious features include surveys of real estate boundaries, such as town and village borders. Although less tangible, these surveys are usually done with high precision and provide good estimates of regional deformation associated with faulting.

Faulting is episodic by nature, and movement on a particular fault may recur on a time scale of centuries. Such lengthy recurrence intervals are too long to be reflected in offsets of most cultural features, so topographic characteristics are also examined. Offsets in the courses of rivers and streams, interruptions in hills and valleys, and other topographic changes can reveal relative displacements along a fault. Unlike anthropogenic structures, such alterations of a natural topography can be difficult to date with great precision. By extending the time scale back thousands of years, however, they often provide important data.

Interruptions in the deposition in lakes and swamps near strike-slip faults can reveal considerable detail about the timing and intensity of former earthquakes. In an area called Pallet Creek, along the San Andreas Fault in California, evidence for eight earthquakes has been dated using radiocarbon techniques. The earliest occurred at about 750 C.E., and the most recent in 1857. Unfortunately, the slip associated with each of these earthquakes is not easily ascertained from the sedimentary strata.

During an earthquake, a fault may slip several meters in a matter of seconds, but, averaged out over millions of years, most faults have displacement rates on the order of a few centimeters per year. Over the course of 1 million years or so, the displacement across most faults will be on the order of a few tens of kilometers. The San Andreas Fault is thought to have been in existence for about 29 million years. Hundreds of kilometers of displacement are possible over this vast period of time. Topographic features persist for thousands, perhaps even hundreds of thousands of years, but few could be expected to exist after millions of years. Here, the juxtaposition of rock types and the records left in the magnetic minerals of the ocean floor can be used to delimit the time of inception and the subsequent movement history of a fault.

A high-angle reverse fault in Woburn, Quebec. (Geological Survey of Canada)

The faulting process itself produces clues, distinct from offsets or juxtapositions, that can reveal the presence of a fault. In general, the broken rock in the vicinity of the fault will weather and erode more easily than sound rock some distance away. This results in the formation of long, linear valleys along fault lines.

By grinding up the rock in the immediate vicinity of the fault, faulting often modifies the groundwater system. Sometimes rock that was impermeable has its permeability increased by the new fractures produced. Other times, a permeable rock is rendered less permeable because the conduits that permitted water to flow through it are disrupted by the fracturing process. These changes may result in the formation of springs or lakes called "sag ponds" directly above the fault.

Finally, some faults are revealed by the earthquakes that occur on them. By studying the seismographic data obtained from an earthquake, geophysicists can determine the location of the earthquake and the direction in which the blocks of rock moved, even if the fault involved lies kilometers beneath the surface.

Occurrence

The surface of the earth is made up of twelve or so tectonic plates that are roughly 100 kilometers thick and persist with little deformation within them for hundreds of millions of years. Plates diverge from each other along ridges (generally beneath the oceans but occasionally running through a continent, such as the African rift valley). Connecting ridge segments, which may be separated by hundreds of kilometers, are strike-slip faults. Plates converge, with one plate moving beneath the other, along subduction zones. The transform fault is a special class of strike-slip faults, where such faults form a plate boundary. When, as is often the case, the direction of convergence is not perpendicular to the subduction zone (oblique convergence), a strike-slip fault may develop to accommodate the horizontal component of relative motion. Sometimes the horizontal stresses produced by plate convergence are sufficient to extrude a wedge-shaped piece of a plate to the side, which results in strike-slip faults. If plates move past each other without diverging or converging, the motion is horizontal and is accomplished along strike-slip faults.

The majority of strike-slip faults connect ends of ridge segments. In the ocean floor, such ridge segments are often called offset spreading centers (OSCs), and the reason for their existence is not well understood. Active faulting occurs along the transform faults between ridge segments, resulting in an age difference across a line that extends the fault beyond the offset region. Because the ocean floor cools, contracts, and sinks as it gets older, this age difference is often expressed by significant topographic relief, with cliffs stretching out for hundreds of kilometers from the transform fault as an oceanic fracture zone.

Oblique convergence is frequently accommodated by a partitioning of relative motion between a dip-slip subduction zone and a strike-slip fault parallel to the subduction zone. Some examples of strike-slip faults that are parallel to subduction zones are the Great Sumatran fault on the island of Sumatra and the Denali fault in Alaska. In 1995, the

Hyogoken-Nanbu earthquake near Kobe, Japan, occurred on one of these faults, killing more than 5,000 people and causing more than $200 billion of damage.

Horizontal stresses perpendicular to the trend of a subduction zone can also produce strike-slip faults. Sometimes called "watermelon seed" or "horizontal extrusion" tectonics, this process occurs when a segment of a plate is wedged off to the side by plate convergence. The motion of this plate segment is similar to that of a watermelon seed when it is squeezed between the thumb and forefinger. Mechanical engineers say that this type of deformation is caused by a "rigid indenter," and the faults produced may be called "indent-linked" strike-slip faults.

The subcontinent of India has acted as a rigid indenter as it has pushed northward into the Eurasian plate. Subduction produced the Himalaya Mountains and the Tibetan plateau; at the same time, huge, wedge-shaped pieces consisting of most of Southeast Asia have been moved off to the side. As much as one-half of the convergence has been accommodated by this eastward extrusion along the Altyn Tagh, Haiyuan, and related strike-slip fault systems, resulting in many devastating earthquakes. The North Anatolian fault in Turkey is another indent-linked strike-slip fault. In this case, much of Turkey is being extruded to the west as the Arabian and Eurasian plates converge. Movement on this fault brought Turkey 1.2 meters closer to Europe during a devastating 1999 earthquake.

If the relative motion between plates has little convergence or divergence, it may be taken up almost entirely along strike-slip faults. In California, motion between the North American plate and the Pacific plate occurs largely across the San Andreas fault, for example. Although it is presently a variety of transform fault that connects ridge segments in the Gulf of California to the Mendocino triple junction (where it meets a trench and another strike-slip fault), the San Andreas fault has such a complex history that it is often best to consider it as a plate boundary where most of the relative motion has a right-lateral strike-slip sense.

Formation and Secondary Features

There are often long periods of time between episodes of motion on strike-slip faults. During these periods, soil and other unconsolidated sediments can accumulate over the fault region. When motion again occurs on the fault, the offset in these new layers at the surface can be seen to develop in a complex but systematic way. Similar processes have been observed in laboratory models involving clay cakes being offset above moving plates.

Initially, many small offsets develop above the fault in a parallel, offset geometry resembling the slats on a venetian blind. These are called Riedel shears, conjugate Riedel shears, or P shears, depending on their angular relationship to the underlying fault. Complex technical issues are involved in their formation, but of particular interest is the fact that many minor faults form initially, and only later do the principal displacement shears develop. These are the strike-slip faults across which most of the movement occurs. Study of strike-slip faults in bedrock often reveal the complexities introduced by the early shears.

The complicated geometry of strike-slip faults means that motion across some parts of them will not occur in a simple, strike-slip sense. In some places, the fault surface will bend, or be offset, resulting in either extension or compression. Imagine an east-west-trending right-lateral strike-slip fault with an offset across which the eastern side has been offset to the south relative to the western side. In the vicinity of the offset, the crust will be extended. Sometimes called "transtension," this stretching may result in normal faults bounding a down-dropped block of crust, producing a "pull-apart" basin. The Salton Sea in Southern California and the Dead Sea in Israel are examples of these features.

Using the same geometry, but this time considering the offset to be of a left-lateral strike-slip fault, the vicinity of the offset would be in compression. This "transpression" might be expected to produce buckling and mountain ranges. The Transverse Ranges of California, occurring near the "big bend" of the San Andreas fault, probably owe their existence, in part, to these compressive stresses.

Significance

Many of the horizontal movements on the surface of the earth occur along strike-slip faults. If a complete set of slip data for all of the strike-slip faults on the planet could be constructed, it would reveal most of what is known about tectonics. Most strike-slip faults extend to the surface, providing exposures where they can be studied in detail. Knowledge of the geometry, offset history, and earthquake recurrence intervals on strike-slip faults is therefore more developed, and based on better data, than similar knowledge for normal or reverse faults.

Because they often cut through the continental crust, strike-slip faults are likely to traverse populated regions. Their effects on topography may even encourage development of the most earthquake-prone areas. Disastrous earthquakes are common on strike-slip faults today, as they have been in the past, and will certainly be in the future. Learning more about these faults and the earthquakes that occur along them is likely to help in predicting those earthquakes and in mitigating their negative consequences.

Otto H. Muller

FURTHER READING

Cunningham, W. D., and Mann, P. *Tectonics of Strike-Slip Restraining and Releasing Bends,* Special Publication no. 290. Geological Society of London, 2008. Highly technical, but thorough look at strike-slip tectonics. It explores the mechanics and distribution of bends.

Davidson, Jon P., Walter E. Reed, and Paul M. Davis. *Exploring Earth: An Introduction to Physical Geology.* 2d ed. Upper Saddle River, N.J.: Prentice Hall, 2001. Chapter 10, "The Conservative Boundary: Transform Plate Margins," provides an easily understood treatment of strike-slip faulting that covers the common transform types. However, it does not deal with indent-linked strike-slip faults. Profusely illustrated with colored maps and diagrams, it is suitable for high school readers.

Davis, George H., and Stephen J. Reynolds. *Structural Geology of Rocks and Regions.* 2d ed. New York: John Wiley & Sons, 1996. Provides a thorough, comprehensive treatment of faults and faulting, as well as a great deal of information about the strength of rock, the accumulation of strain, and other related aspects of geology. More technical than the other references cited.

Fossen, Haakon. *Structural Geology.* New York: Cambridge University Press. 2010. This text is well written and easy to understand. An excellent text for geology students or resource for geologists. Provides many links between structural geology theory and application. Photos and illustrations add great value to the text. Contains a glossary, references, an appendix of photo captions, and indexing.

Fowler, Christine Mary Rutherford. *The Solid Earth: An Introduction to Global Geophysics.* 2d ed. Cambridge: Cambridge University Press, 2004. This book provides an outstanding treatment of how offsets between plates can be determined, excellent descriptions of the detailed structure of oceanic transform faults, and a useful discussion of the extrusion tectonics associated with the formation of the Himalaya Mountains. Although not suited for high school readers, its treatment of these topics requires no mathematics beyond algebra and trigonometry.

Wyld, Sandra J., and James E. Wright. "New Evidence for Cretaceous Strike-Slip Faulting in the United States Cordillera and Implications for Terrane-Displacement, Deformation Patterns, and Plutonism." *American Journal of Science* 301 (2001): 150-181. Discusses the relationship between strike-slip faulting and terrane displacement. Highly technical.

Yeats, Robert S., Kerry Sieh, and Clarence R. Allen. *The Geology of Earthquakes.* New York: Oxford University Press, 1997. A thorough and detailed exploration of all aspects of earthquakes, with emphasis on the geological evidence used to study them, this book is an excellent resource. Although some of the treatment may be too detailed for a beginner or casual reader, these areas can be skimmed over easily. Concepts are explained well, and great care is taken to keep terminology concise and understandable. Profusely illustrated with black-and-white maps and diagrams, it also has a very useful index.

See also: Creep; Cross-Borehole Seismology; Discontinuities; Earthquake Distribution; Earthquake Hazards; Earthquake Magnitudes and Intensities; Earthquake Prediction; Earthquakes; Earth's Mantle; Elastic Waves; Experimental Rock Deformation; Faults: Normal; Faults: Thrust; Faults: Transform; Lithospheric Plates; Metamorphism and Crustal Thickening; Notable Earthquakes; Plate Motions; Plate Tectonics; San Andreas Fault; Seismic Observatories; Seismic Tomography; Seismometers; Stress and Strain.

FAULTS: THRUST

Thrust faults are the result of compressional forces that exceed the natural strength of rocks and cause them to break and move. They can trigger earthquakes, create mountain ranges, and serve as natural traps for gas and oil deposits.

PRINCIPAL TERMS

- **dip:** the angle between a fault plane and a horizontal surface
- **fault:** a fracture or zone of breakage in a mass of rock that shows evidence of displacement or offset
- **footwall:** the block of rock that lies directly below the plane of a fault
- **head wall:** the block of rock that lies directly above the plane of a fault; it is also known as a hanging wall
- **reverse fault:** the same thing as a thrust fault, except that its fault plane dips at more than 45 degrees below the horizontal
- **scarp:** a steep cliff or slope created by rapid movement along a fault
- **slip:** amount of offset or displacement across the plane of the fault, relative to either the dip or the strike
- **strike:** the orientation of a fault plane on the surface of the ground measured relative to north

THRUST FAULT PRODUCTION

A mass of rock below the surface of the earth usually cracks and fractures when it loses its resistance to an applied force. Rocks break when their ability to store energy is exceeded. When a rock shows some evidence of movement or displacement along the zone of breakage, a fault is created. Thrust faults are commonly the result of strong compressional (squeezing) forces acting on relatively brittle, older subsurface rock that has moved upward and over or on top of a mass of younger, adjacent rock. It is a particular kind of fault and one of many types that exist.

The zone of breakage between the once-united masses of rock is known as the fault plane. The motion of the rocks on either side of this plane and the plane itself are usually parallel to each other. The blocks of rock on both sides of a fault plane are known as walls, a term that comes from the days of the early prospectors, who were really the first field geologists. Because the presence of a fault marks a zone of weakness in the ground, either mineral-rich groundwater or hot fluid magmas will eventually find and follow this path of least resistance toward the surface and deposit ores, minerals, or gemstones. Prospectors would seek out faults, as they knew that a fault was likely to be the home of some valuable material. Once a fault was located, a mine shaft would be dug to follow the trace of the fault below ground.

The head wall, or hanging wall, was the wall above the miner's head; the footwall was the wall below his feet. Head walls and footwalls exist only in faults that are not vertical. In terms of the overall structure of the fault, the hanging wall is the block that occurs above the fault plane and the footwall is the rock below the fault plane. In thrust faults, the head wall always moves relatively upward and the footwall moves relatively downward. The term "relatively" is used because it is usually very difficult for a geologist to determine exactly which block has moved. For example, both blocks could have moved upward but the hanging wall moved farther; both blocks could have dropped but the footwall dropped farther; the hanging wall could have remained stationary while the footwall dropped; or the footwall could have remained stationary while the hanging wall moved upward.

FAULT ORIENTATION

The orientation of the fault relative to the earth's surface is of great importance. It allows the fault to be located and mapped as a place to avoid during construction, especially if it is an active fault or one with the potential for continued movement. A fault's orientation, or strike, is measured by the trace of the fault plane as it would appear on a horizontal plane and is measured in degrees from the magnetic North Pole. The plane's angle of tilt, or dip, is measured from a horizontal position down to the fault plane. The dip direction is always perpendicular to the strike direction.

A fault can be straight in form and consist of one sharp, clean break, or it can have a highly irregular form and be composed of multiple breaks. Thrust faults of the latter type may be so closely spaced as to form a highly complex zone that may be hundreds of meters wide. Fault planes can also be curved, adding to the complexity. Geologists have located and mapped many small thrust faults at very shallow depths below the surface and have found some large thrust faults that extend down to a depth of 700 kilometers.

In general, the total displacement, or offset, in a rock along a thrust fault may be large or small, and horizontal, vertical, or oblique, depending on the strength of the compressional force and the rock type involved. Typical thrust faults have low dip angles in which lower strata are often pushed above higher strata. They often place older rocks above younger rocks. An important factor in determining this displacement is the angle of the fault plane. Steeply dipping planes will show a small vertical uplift on the order of a few meters or less; shallowly dipping planes may exhibit a long horizontal displacement extending for many kilometers. Displacements are also described in terms of their relative motion. A "dip-slip" occurs when the movement of the rock is parallel to the dip direction of the fault plane, a "strike-slip" indicates motion parallel to the strike, and an "oblique slip" occurs somewhere between the strike and dip.

FAULT VARIETIES AND AGES

Three types of faults involve movement of the hanging wall upward with respect to the footwall: Thrust faults are characterized by fault planes that dip at angles less than 30 degrees; overthrust faults are thrust faults with very large displacement; and reverse faults dip at angles greater than 45 degrees from horizontal. In large displacement thrust faults, the fault surface may be quite irregular; therefore, the fault plane may approach horizontal or even reverse direction. Overthrusts are very common in areas of intense folding where there is shortening, and usually thickening, of the crust. The offset of rock along the path of the fault is usually greatest near its middle and decreases at either end, or the thrust may terminate in strike-slip faults.

The age of rocks may be used to determine the relative age of faults. Since a fault cuts through a block of rock, the fault must be younger than the rock. Therefore, if a rock is found to be 100 million years old and is cut by a fault, then the fault must be younger than 100 million years old.

LOCATING THRUST FAULTS

Structural geologists study faults to try to understand what was happening to the earth's crust at the time of faulting and to determine the origin of the force that caused the rocks to break and move. To study a thrust fault, the geologist must first accurately locate and measure the fault in the field. It is sometimes difficult to determine where a fault exists, especially if it is very old. The geologist, acting as a detective, must rely on direct physical evidence that he or she can gather from above and below the surface.

When a fault intersects the surface, it usually forms a fault line. The existence of this line is commonly indicated by some noticeable feature. Photographs of the land taken from high-altitude aircraft reveal these features as offsets or disruptions in rows of planted crops or trees; sharp breaks in the channels of streams; unusual linear alignments of springs flowing at the base of a mountain or along a valley; raised sections of land, such as beach and river terraces; and fault scarps. A fault scarp is a recent, sharp break in the surface of the ground that has a straight and very steep slope. The height of a scarp is directly related to the amount of upward motion along a fault. Unless the fault is still active, however, it usually is not exposed at the surface but is buried beneath the cover of more recently deposited sediments. In this case, a geologist must rely on subsurface evidence.

EVIDENCE OF THRUST FAULT MOTION

The subsurface location of a thrust fault is not always easy to find. The field geologist must rely on direct or indirect evidence that is not usually visible on the surface. Direct subsurface evidence of the existence of a thrust fault can be obtained from the examination of rocks within mine shafts, highway tunnels, or excavation sites dug for building foundations. Similarly, "roadcuts" (highway excavations that run through mountains and valleys) and natural outcroppings of rock at the surface may indicate the presence of a thrust fault. In these situations, a geologist would look for any evidence that suggests that massive blocks of rock have moved relative to one another.

When huge blocks of rock are broken and continue to rub against one another, certain physical features are produced as a result of the friction between these moving blocks. Sure indicators of faulting are slickensides, which are recognized as highly polished and finely scratched, or striated, surfaces of rock along the fault plane. The direction of the striations is parallel to the direction of the last movement along the fault. Depending on the amount of friction generated between these blocks, their inward-facing surfaces may be crushed into a fine, soft, claylike powder known as "gouge" or may become fault breccias, rocks consisting of small, angular fragments. Microbreccias are formed when the crushed fragments along the fault plane are microscopic. Mylonites are a special type of solid microbreccia; they have a streaked appearance in the direction of motion. Pseudotachylites are a kind of microbreccia that does not appear streaked but exists as a thin glassy film because of the melting of rock from frictional heat. Other evidence of thrust fault motion would be the overlapping or repetition of the same rock units (like overlapping shingles on a roof) and an abrupt termination of a rock unit along its trend. One or more of these faulting criteria may be present, or evidence may be completely missing.

DRILLING FOR EVIDENCE

If there are no accessible outcrops or underground viewpoints, the field geologist must turn to the expensive direct method of evidence-gathering: drilling. Drilling into the ground with specialized drilling rigs allows access to the subsurface. The examination of small broken rock samples brought up by a diamond drill bit or solid "rock cores" brought up by hollow drill bits is a direct means of studying rocks that do not exist at the surface. These samples can be compared with those taken from well-known nearby regions, and any disruption or missing units of rock may indicate the presence of a fault. If a subsurface geologic map exists, the geologist can predict which rocks exist below the drill site and use this information in conjunction with the collected rock samples.

If the evidence from drilling does not match the regional geology, then one or more faults may have been at work shuffling the sequence of rocks below. Sometimes large blocks of rock get caught between faults when they move and become bent, folded, or twisted, creating a highly complex pattern that is not easy to understand using drill core data alone; the uncertainty of what lies below ground increases.

EVIDENCE FROM GEOLOGIC MAPS

Indirect evidence of thrust faulting can be had from a geologic map of an area. Geologic maps show the distribution, thickness, age, and orientation of the various types of rock that would be seen at the surface of the earth were all the soil covering removed. Any older rock formation that sits on top of a younger rock formation was most likely moved to its location by thrust faulting. Rock formations that appear out of place with the overall sequence of rocks probably suffered a similar fate.

Several notable large thrust faults exist in the United States. They can be seen in the Rocky Mountain region as a series of sharp, parallel ridges similar to the teeth on a saw—hence the term

A thrust fault showing fault drag in Atacam Province, Chile. (U.S. Geological Survey)

"sawtooth," as in the peaks of the Sawtooth Mountain Range in Montana. In the same state, a low-lying slab of rock known as the Lewis Overthrust shows a horizontal displacement of about 24 kilometers with a fault plane that dips at an angle of less than 3 degrees. Thrust faults occur in most mountain ranges on the earth such as the Appalachian Mountains, the European Alps, and the Himalaya.

ENGINEERING AND ECONOMIC APPLICATIONS

A fault is a zone of weakness in a rock; therefore, it may continue to move over a long period. Information about the rate of rock movement is very valuable, especially when the motion along a thrust fault (or any fault) is rapid enough to trigger the release of stored energy within a rock, causing an earthquake. Accurate mapping of thrust faults is also important, since the potential for future earthquakes must be carefully evaluated, especially in highly urbanized areas. A complete study of thrust faults is not easy, however, because there are many variables to consider: the fault's exact location and total horizontal extent, the orientation and strength of the rocks relative to the fault, the direction and amount of movement, and the fault's previous earthquake history. These factors are critical to decisions of where to construct nuclear power plants, dams, housing projects, and cities.

Thrust faults have great economic potential, since they have the ability to act as "traps," or reservoirs, for deposits of migrating oil and natural gas. In such a case, one impervious rock type is brought into fault contact with a petroleum-bearing rock. The impermeable rock now acts as a barrier to any further upward fluid migration and allows oil to accumulate beneath it. Similarly, in mineral exploration, large thrust faults have been known to harbor exploitable quantities of radioactive and other rare minerals, needed for use in industry and medicine, that were either deposited by igneous activity or precipitated by circulating, mineral-rich groundwater. Thrust faults also serve as natural underground pipelines, allowing circulating groundwater easier access to the surface of areas that otherwise might have been deserts.

Steven C. Okulewicz

FURTHER READING

Billings, Marland P. *Structural Geology*. 3rd ed. Englewood Cliffs, N.J.: Prentice-Hall, 1972. This is an introductory college-level textbook for all aspects of structural geology. Chapters 8 through 12 discuss the variation, classification, and recognition criteria for faults. Chapter 10 is devoted to thrust faults. The book is well written and clearly illustrated with many line drawings. Also included are black-and-white photographs of faults as they appear in the field and a useful end-of-chapter bibliography. The bible of structural geologists.

Cubas, N., et al. "Prediction of Thrusting Sequence Based on Maximum Rock Strength and Sandbox Validation." *Trabajos de Geologia* 29 (2009): 189-195. Development of procedures by which thrust faulting events can be identified, including experimentation with sandbox models and field research. Lots of diagrams are provided; the article is written in a manner that requires some scientific background.

Fossen, Haakon. *Structural Geology*. New York: Cambridge University Press. 2010. This text is well written and easy to understand. An excellent text for geology students or resource for geologists. Provides many links between structural geology theory and application. Photos and illustrations add great value to the text. Contains a glossary, references, an appendix of photo captions, and indexing.

Lahee, Frederic H. *Field Geology*. 6th ed. New York: McGraw-Hill, 1961. Within this thick book are described the various historical techniques for the measurement, mapping, relative age determination, and interpretation of rock formations in the field. A large part of Chapter 8 is devoted to faulting, with a description of applied field mapping techniques that are used in the construction of surface and subsurface maps. The chapter is easy to read, and the fault types are illustrated with either line drawings or block diagrams. Although dated, this work is still a classic and the field book carried by most practicing geologists.

McClay, Kenneth R. *Thrust Tectonics*. London: Chapman and Hall, 1992. This collection of papers was presented as part of the Thrust Tectonics Conference held at Royal Holloway and Bedford New College, University of London, in 1990. The advanced nature of the collection makes this most useful for the college student.

Mitra, Shankar, et al., eds. *Structural Geology of Fold and Thrust Belts*. Baltimore: The Johns Hopkins University Press, 1992. A good discussion of physical geology focusing on the structure and processes of thrust faults and folds. Suitable for the college reader. Illustrations, bibliography, and index.

Parker, Sybil P., ed. *McGraw-Hill Encyclopedia of Geological Sciences*. 2d ed. New York: McGraw-Hill, 1988. A source that covers every aspect of geology. Topics are arranged alphabetically. The entry on faults and fault structures discusses fault movement, procedures for locating faults, stress conditions, and examples of various types of fault. Includes references to other entries and a bibliography. Contains some equations, but they are not essential to the reader's understanding.

Poblet, J., and Lisle, R. J. *Kinematic Evolution and Structural Styles of Fold-and-Thrust Belts,* Special Publication 349. Geological Society of London, 2011. Provides an overview of current research of fold-and-thrust structures and processes. It covers theoretical modeling, methodology and experimentation.

Spencer, Edgar W. *Introduction to the Structure of the Earth*. 3rd ed. New York: McGraw-Hill, 1988. This college-level structural geology textbook was written from the "plate tectonics" point of view and covers the entire earth. Several chapters deal with various faults that were formed as a result of colliding continents and ocean basins and are indicators of intense rock deformation. Spencer describes the many types of thrust faults. Many regional examples are provided and clear line drawings are included. There is a thorough end-of-chapter bibliography.

Tarbuck, Edward J., Frederick K. Lutgens, and Dennis Tasa. *Earth: An Introduction to Physical Geology*. 10th ed. Upper Saddle River: Prentice-Hall, 2010. This basic geology textbook contains a section on rock deformation. It provides diagrams to illustrate the various types of fault and the concepts of strike and dip, and it includes a photograph of an overthrust formation. Review questions and a list of key terms conclude the chapter. For high school and college-level readers.

Thornbury, William D. *Principles of Geomorphology*. 2d ed. New York: John Wiley & Sons, 1968. Geomorphology is the study of landforms on the earth's surface. This basic college-level textbook describes the many types of landform that are produced by various geologic processes. Chapter 10 is an informative, highly readable, and well-illustrated chapter that discusses the landforms created by fault activity and how they are recognized.

See also: Creep; Cross-Borehole Seismology; Discontinuities; Earthquake Distribution; Earthquake Engineering; Earthquake Hazards; Earthquake Locating; Earthquake Magnitudes and Intensities; Elastic Waves; Experimental Rock Deformation; Faults: Normal; Faults: Strike-Slip; Faults: Transform; Lithospheric Plates; Mountain Building; Plate Tectonics; San Andreas Fault; Seismic Observatories; Seismic Reflection Profiling; Seismic Tomography; Seismometers; Stress and Strain; Subduction and Orogeny.

FAULTS: TRANSFORM

Transform faults occur along fracture zones found at the mid-oceanic ridges. The ridges are areas of erupting ultramafic lavas, which cause seafloor spreading, which, in turn, drives the moving lithospheric plates. Geologists have found that the offset ridges do not move in relation to each other and have been essentially fixed relative to the spreading sea floor.

PRINCIPAL TERMS

- **Curie temperature:** the temperature below which minerals can retain ferromagnetism
- **divergent boundary:** the boundary that results where two plates are moving apart from each other, as is the case along mid-oceanic ridges
- **ferromagnetic:** relating to substances with high magnetic permeability, definite saturation point, and measurable residual magnetism
- **fracture zone:** the entire length of the shear zone that cuts a generally perpendicular trend across a mid-oceanic ridge
- **hypocenter:** the initial point of rupture along a fault that causes an earthquake; also known as the focus
- **magnetic anomalies:** patterns of reversed polarity in the ferromagnetic minerals present in the earth's crust
- **mid-ocean ridge:** a long, broad, continuous ridge that lies approximately in the middle of most ocean basins
- **rift valley:** a region of extensional deformation in which the central block has dropped down with relation to the two adjacent blocks
- **seafloor spreading:** the hypothesis that oceanic crust is generated by convective upwelling of magma along mid-ocean ridges, causing the sea floor to spread laterally away from the ridge system
- **transcurrent fault:** a fault in which relative motion is parallel to the strike of the fault (that is, horizontal); also known as a strike-slip fault

TRANSFORM FAULT OCCURRENCE

Faults are regions of weakness or fractures in the earth's crust along which relative movement occurs. The simplest type of fault is the so-called normal fault, in which one block of crust is displaced vertically downward with respect to the other. A reverse or thrust fault is one in which the block is driven upward with respect to the other block. Yet another type of crustal displacement takes place horizontally, as one block slides laterally with respect to the other. These strike-slip or transcurrent faults are related to the most complex class of faults, known as transforms.

Because they are regions of crustal transformation, transform faults are found almost exclusively along the mid-oceanic ridges that nearly encircle the globe and along the boundaries of tectonic plates. The ridges are the sites of newly forming crustal materials composed of very dense magmas of relatively high iron and magnesium content. As the new crust forms, lava flows act to push the oceanic crust laterally away from the spreading ridge, as the sea floor spreads out at a rate of a few centimeters per year.

MID–OCEANIC RIDGE SYSTEM

If one were able to view the mid-oceanic ridge system from orbit, it would quickly become apparent that the spreading centers do not occur along a smoothly continuous line but rather are broken into scores of offset ridges. The offset is marked by a fracture zone, which serves as the border between two spreading centers. Because the ridges are displaced, it was first believed that these fracture zones were simply transcurrent faults, along which right or left lateral displacement would be observed from opposing sides. The ridges are fixed with relation to each other and appear to have been so for long periods of geologic time. Clearly, a new type of faulting was being observed.

If one were to voyage to the bottom of the mid-Atlantic Ocean to view one of these faults, its highly unusual nature would become clear only after one had traveled hundreds of miles along its entire length. Starting at the west end of the transform fault, one would find that the crust is slowly moving toward the west on either side of the fracture zone. Because the crust is essentially moving in the same direction, earthquakes are rare events in this region.

Most transform faults are oriented at right angles to the mid-oceanic ridges. Typically, they extend from ridge to ridge (the active part of the fault); if spreading has been taking place for long periods of geologic time, fracture zones extend out from the ridge systems. These transforms and fracture zones

199

are thought to be ancient areas of weakness in the crust that formed when the ocean basin (such as the Atlantic) began forming. The discontinuous nature of the ridge system is believed to be structurally ancient. When viewed globally, the spreading axis is offset from the earth's rotational axis, with the ridges corresponding to lines of longitude and the transforms roughly approximating latitudinal lines. Spreading rates are greatest at the equatorial regions of the globe.

VOLCANIC ERUPTIONS AND EARTHQUAKES

The underwater ridge mountains are marked by a distinctive rift valley, centered along the range. Volcanic eruptions emanating from the rift create pillow-shaped lavas. The ridge is a boundary between the earth's lithospheric plates; it is a divergent boundary, for the sea floor is spreading laterally in opposite directions. A view of the ridge along the transform fault would reveal that the line of mountains is broken and offset. An observer passing over the high ridge on the north side of the fault would suddenly notice that the fault had taken on the appearance of a transcurrent fault. Movement on the south side of the fault is right lateral, and although the crust is traveling west, on the north side of the fault, the movement is toward the east. An observer on either side of the fault would see right lateral displacement.

Because crustal movement is in opposing directions between the offset ridges, earthquakes occur frequently in this area. The crust's relative motion is horizontal so that the focus or hypocenter (the actual point of rupture in the rock) of transform earthquakes is shallow—typically less than 70 kilometers deep, whereas trench earthquakes can be up to 700 kilometers deep. Magmatic eruptions and earthquakes offer convincing evidence that the earth's crustal plates are far from stationary. Past the offset ridge, crustal motion is once again in the same direction, and no lateral motion would be observed on opposite sides of the fracture. This transformation of crustal displacement along the shear zone is the derivation of the term "transform fault."

While the Atlantic Ocean is a basin in which new crust is forming, causing the rifting of the continents and the growth of the ocean basin at a rate of a few centimeters per year, the opposite is the case in the Pacific Ocean basin. There, the transform faults are a bit more complex. An example would be the New Zealand fault, which terminates on both ends at subduction trenches rather than at ridges. New Zealand is seismically active, because the transform fault passes directly through both of the large islands.

SAN ANDREAS FAULT

The most famous of all transform faults forms a distinctive type of lithospheric plate boundary. Extending from the East Pacific Rise off the coast of western Mexico to the Mendocino fracture zone and the Juan de Fuca ridge system off the coast of Washington State is the 960-kilometer-long, 32-kilometer-deep system of strike-slip faults known as the San Andreas. The San Andreas fault forms a boundary between the northward-trending Pacific plate and the North American plate. Horizontal displacement along this huge fault system is estimated at 400 kilometers since its inception nearly 30 million years ago. The rate of displacement has been measured at an average of 3.8-6.4 centimeters annually.

How did this impressive plate boundary form, and how does it fit into the transform fault model? About 60 million years ago, the east Pacific was also home to a third plate, the Farallon, which eventually was subducted under the North American plate. The Farallon plate was pushed eastward into the North American plate by a spreading ridge system and trench that were eventually overridden by the continental plate. The remnants of this ridge system are found at the ends of the San Andreas fault. Because the Pacific plate's motion was northward, the ridge system was converted into a transform fault, characterized by its right lateral strike-slip displacement.

TRIPLE JUNCTIONS

The complexity of three plate interactions, or triple junctions, explains the complex nature of transform fault systems such as the San Andreas. The remnants of the doomed Farallon plate are presently found as the Juan de Fuca plate, which is bounded by spreading ridges and transform faults to the west and a subduction zone where the plate is being inexorably pushed into the earth's upper mantle, under the North American plate. Inland, the Cascade volcanoes, most notably including Mount St. Helens, have their volcanic fires fueled by a rising magma plume that is generated by the melting of the subducted oceanic plate.

North of the Juan de Fuca plate is another large transform system similar to the San Andreas fault, called the Queen Charlotte transform. An extension of the San Andreas system, the Queen Charlotte fault begins to the south at the Juan de Fuca ridge but becomes inactive at its northern end, whereas the San Andreas is a ridge-ridge transform that is bordered by the East Pacific Rise to the south and Gorda Ridge to the north.

The great ridge system that was overridden by the continental plate is responsible for the faulted structure of the Basin and Range Province, and the triple junction of the plates may have given rise to the magma plume or "hot spot" that is responsible for the Snake River plain volcanics and the thermal activity at Yellowstone National Park. Clearly, understanding the evolution of transform faults such as the San Andreas and the Queen Charlotte is pivotal to the understanding of the complex mountain scenery of western North American.

Evidence from the Sea Floor

A revolution in the earth sciences had its germination in the ideas of the German meteorologist Alfred Wegener, who argued that the continents had once been joined in one supercontinent and had since drifted apart. Wegener's theories of continental drift were not taken seriously until evidence from the sea floor forced a rethinking of modern geology in the 1960's. Oceanographic research in the 1950's led to a new picture of the ocean basins. Far from the featureless abyssal plains they were once thought to be, the basins proved to be marked by dramatic mountain ranges characterized by rift valleys. In 1960, a Princeton University scientist, H. H. Hess, suggested that convection currents in the earth's upper mantle were driving volcanic eruptions along the rifts, causing the sea floor to spread and the continents to move apart.

The real breakthrough in understanding the ocean floor came as a result of numerous cross-Atlantic voyages by research vessels towing submerged magnetometers in an effort to measure the strength of the crustal rock's residual magnetism. As the basaltic magma erupts at the spreading rift, magnetite freezes out of the melt at 578 degrees Celsius, and the earth's magnetic field orientation is frozen into it. This temperature is known as the Curie temperature.

Magnetic Anomalies

When the first magnetic maps of the sea floor were produced, scientists groped for an explanation of the alternating nature of the field's polarity, which changed in a random pattern over short distances. Two research students at the University of Cambridge, Fred Vine and Drummond Matthews, solved the riddle of the magnetic anomaly stripes by proposing that the stripes represented reversals in the earth's magnetic field over time. Because the anomaly patterns were symmetrical with respect to the axis of the spreading centers, this was proposed as evidence in favor of the Hess model of seafloor spreading.

The huge fracture zones that appeared to offset the mid-oceanic ridges were mapped with magnetic anomaly measurements, and seismic data indicated that seismic activity was concentrated between the offset ridges along the fracture zones. Networks of seismic instruments also enabled geophysicists to study the direction of the spreading crustal movement, through geometrical solutions known as first motion studies or fault-plane solutions.

In 1965, Canadian geophysicist J. Tuzo Wilson dubbed the faults "transforms" because of the changing of relative motion along the length of the fracture. He reasoned that the faults were not causing the offset of the spreading ridges, but that the offset is what caused the appearance of a transcurrent fault between the ridges and an inactive segment, in which the sea floor was spreading in the same direction beyond the zone of offset.

Geophysicists mapping the magnetic anomalies were aided by the work of terrestrial geologists, who were able to identify magnetic reversal patterns in terrestrial lavas that were identical to those found in oceanic rocks. Radiometric dating established a magnetic anomaly time scale that would allow scientists to determine the ages of rocks laterally out from the ridges and hence to deduce the rates of seafloor spreading. Armed with the paleomagnetic data and with Wilson's notion of transform faults as directional guides, scientists were able to show that the Atlantic Ocean had indeed been opening at a rate of a few centimeters per year and that the landmasses of western Europe and eastern North America were once essentially in contact.

IDENTIFYING TRANSFORM FAULTS

Wilson and others went on to study the more complex faults of the eastern Pacific. The San Andreas fault had been identified long before as a transcurrent fault. The new paradigm of plate tectonics placed it in a global tectonic scale; the fault is now accepted by most as a kind of transform fault that connects to spreading ridge centers to the north and south.

While faults such as the New Zealand Alpine and San Andreas are fairly accessible to scientists, most of the planet's transform faults lie below thousands of meters of ocean. Echo sounding, radar, sonar seismographic, and magnetic anomaly maps have helped scientists to locate the earth's transform faults. The 1978 Seasat mission radar mapped the ocean floor from space, producing detailed information on the globe's mid-oceanic ridge and fault system. In addition, deep-sea submersibles have been piloted to the rifts, where eruptions of lavas and fault displacements have been observed directly.

PLATE TECTONICS

The theory of plate tectonics in the 1960's caused a revolution in the study of geology, geophysics, and even paleontology. Post-World War II oceanographic research led to the discovery that the mid-oceanic ridges are sites of newly forming crust and resulted in Wilson's explanation of transform faults as ancient fracture zones offsetting the spreading ridges.

With a vast supply of paleomagnetic, seismic, geologic, and other evidence at their disposal, geologists have been able to reconstruct earth history and the relative motions of the continents, using transform faults as directional guides and magnetic reversals to determine the rates of plate movement. Like any successful theory, plate tectonics is elegantly simple; it explains nearly all the earth's diverse landforms and rock formations. Changes in continental distribution may also be linked to the extinction events that punctuate the geologic time scale.

Geologists' understanding of the interaction between ridge systems, subduction trenches, and transforms has led to a better understanding of the paleogeography of the American West and how that complex mountainous region of active faults and recent volcanic activity came into being. Aside from contributing to the fundamental understanding of earth processes, plate tectonics theory explains forces that can influence human lives. Active plate boundaries are the sites of earthquake and volcanic activity. In California, millions of people live and work astride one of the world's largest transform faults, the San Andreas. Residents of Vancouver and Seattle are similarly threatened by the offshore Queen Charlotte Islands fault. On the other side of the Pacific Ocean, New Zealand is nearly bisected by the New Zealand Alpine fault, making it a seismically active region. Whenever human beings decide to build their homes and cities near these moving regions of the earth's crust, the possibility of geologic catastrophe is very real.

David M. Schlom

FURTHER READING

Condie, Kent C. *Plate Tectonics and Crustal Evolution*. 4th ed. Oxford: Butterworth Heinemann, 1997. An excellent overview of modern plate tectonics theory that synthesizes data from geology, geochemistry, geophysics, and oceanography. Of special interest is Chapter 6, on seafloor spreading, and Chapter 9's treatment of the Cordilleran system, including a discussion of the evolution of the San Andreas fault. A very helpful tectonic map of the world is enclosed. The book is nontechnical and suitable for a college-level reader. Useful "suggestions for further reading" follow each chapter.

Cox, Allan, and R. B. Hart. *Plate Tectonics: How It Works*. Palo Alto, Calif.: Blackwell Scientific, 1986. A valuable treatment of the geometrical relationships and movements of the earth's lithospheric plates. Designed for the reader who has a basic qualitative knowledge of plate tectonics but who wishes to learn more, particularly about quantitative analysis of plate movements. Filled with easy-to-follow exercises that demonstrate plate motions, particularly those associated with transform faults.

Doyle, Hugh A. *Seismology*. New York: John Wiley, 1995. A good introduction to the study of earthquakes and the earth's lithosphere. Written for the layperson, the book contains many useful illustrations.

Kearey, Philip, Keith A. Klepeis, and Frederick J. Vine. *Global Tectonics*. 3rd ed. New York: Wiley-Blackwell, 2009. A great overview of tectonics, Chapter 4 has a section covering transform faults. Written for the college student.

Lambert, David, et al. *The Field Guide to Geology*. 2d ed. New York: Facts on File, 2007. For the beginning student of geology, this reference work is filled with marvelous diagrams that make the concepts easy to understand. Suitable for any level of reader.

McClay, Kenneth R. *Thrust Tectonics.* London: Chapman and Hall, 1992. This collection of papers was presented as part of the Thrust Tectonics Conference held at Royal Holloway and Bedford New College, University of London, in 1990. The advanced nature of the collection makes this most useful for the college student.

Mitra, Shankar, et al., eds. *Structural Geology of Fold and Thrust Belts.* Baltimore: Johns Hopkins University Press, 1992. A good discussion of physical geology focusing on the structure and processes of thrust faults and folds. Illustrations, bibliography, and index.

Redfren, Ron. *The Making of a Continent.* New York: Times Books, 1983. Richly illustrated with dramatic photographs, this book is a lucid discussion of plate tectonics with respect to the continent of North America. Contains excellent explanations of seafloor spreading and transforms, along with a section on the San Andreas fault.

Shea, James H., ed. *Plate Tectonics.* New York: Van Nostrand Reinhold, 1985. A collection of classic and key scientific papers, mostly from the 1960's, that together constitute a sweeping overview of plate tectonic geology. Of special interest are papers by Fred Vine and J. Tuzo Wilson on the magnetic anomalies off Vancouver Island and a paper by L. R. Sykes on transform faults at the mid-oceanic ridges. With chapter introductions by the editor, the work is suitable for a college-level reader with an interest in the history of plate tectonics theory.

Shepard, Francis P. *Geological Oceanography.* New York: Crane, Russak, 1977. Chapter 2 addresses seafloor spreading and faulting of the oceanic crust. Photographs, diagrams, and supplementary reading lists augment the text, which is suitable for a beginning geology or oceanography student.

Sullivan, Walter. *Continents in Motion.* 2d ed. New York: American Institute of Physics, 1993. Dedicated to Harry Hess and Maurice Ewing, two late pioneers of plate tectonics theory, this is the classic popular work on moving crustal plates. Well-written explanations of transform faults and their roles in seafloor spreading and a discussion of the San Andreas fault are included in the highly readable text.

Tarbuck, Edward J., Frederick K. Lutgens, and Dennis Tasa. *Earth: An Introduction to Physical Geology.* 10th ed. Upper Saddle River, N.J.: Prentice Hall, 2010. This basic geology textbook contains a section on rock deformation. It provides diagrams to illustrate the various types of fault and the concepts of strike and dip, and it includes a photograph of an overthrust formation. Review questions and a list of key terms conclude the chapter. For high school and college-level reader. It has excellent illustrations and graphics.

Wilson, J. Tuzo, ed. *Continents Adrift and Continents Aground.* San Francisco: W. H. Freeman, 1976. Selected, classic readings from *Scientific American* are introduced with commentary by Wilson, a leading figure in the history of plate tectonics theory. Chapter 2 deals with seafloor spreading and transform faults with a classic article by Don L. Anderson on the San Andreas fault. Suitable for a general audience. Provides a historical perspective of plate tectonics. Contains a bibliography.

Wyllie, Peter J. *The Way the Earth Works: An Introduction to the New Global Geology and Its Revolutionary Development.* New York: John Wiley & Sons, 1976. Wyllie's book has a very informative section on transform faults and earthquake studies. An extensive list of suggested readings augments the text, which is suitable for a college-level reader.

Young, Patrick. *Drifting Continents, Shifting Seas.* New York: Franklin Watts, 1976. A good entry-level discussion of plate tectonics theory, written by a journalist with a knack for simplifying complex concepts. Contains a brief glossary, indexed with a bibliography. Suitable for high school readers.

See also: Continental Drift; Creep; Cross-Borehole Seismology; Discontinuities; Earthquake Distribution; Earthquake Engineering; Earthquake Hazards; Earthquake Locating; Earthquake Magnitudes and Intensities; Elastic Waves; Experimental Rock Deformation; Faults: Normal; Faults: Strike-Slip; Faults: Thrust; Lithospheric Plates; Magnetic Reversals; Magnetic Stratigraphy; Notable Earthquakes; Plate Tectonics; Rock Magnetism; San Andreas Fault; Seismic Observatories; Seismic Tomography; Seismometers; Stress and Strain; Subduction and Orogeny; Volcanism.

FISSION TRACK DATING

When the isotope uranium-238 decays by fission in certain minerals, charged nuclei create a trail of damage, called a fission track. In transparent minerals, fission tracks can be enlarged by chemical etching until they are visible in an optical microscope. The age of the mineral can then be determined from the number of fission tracks and the uranium concentration in the mineral sample.

PRINCIPAL TERMS

- **crystal:** a solid having a regular periodic arrangement of atoms
- **fission fragment:** one of the lighter nuclei resulting from the fission of a heavier element
- **fission track:** the damage along the path of a fission fragment traveling through an insulating solid material
- **glass:** a solid that has no regular periodic arrangement of atoms
- **isotopes:** two atoms of the same element having different numbers of neutrons and thus different atomic weights
- **spontaneous fission:** the splitting of an unstable atomic nucleus into two smaller nuclei

NUCLEAR FISSION

The fission track dating technique is applicable to any type of rock containing minerals or glasses that record fission tracks and have sufficient uranium concentration to produce a detectable number of fission events within an appropriate-sized sample. Samples containing uranium at a concentration of one part per million can easily be dated by this technique if they are more than 100,000 years old. Correspondingly higher concentrations of uranium are required for younger samples, and lower concentrations are required for older ones. Because of its broad applicability, the fission track dating method has been applied to all three major terrestrial rock classes—sedimentary, metamorphic, and igneous.

Nuclear fission is a process by which an unstable atomic nucleus splits into two smaller nuclei, or fission fragments. The fission process releases a large amount of energy, causing the fission fragments to fly apart. In solid matter, each fission fragment can travel about 10-20 micrometers before coming to rest. Because of their high energy, these fission fragments do not carry along all the electrons from the original atom. Therefore, they have a positive electric charge during their passage through the surrounding matter. The passage of a charged particle through certain types of material gives rise to localized damage along the path of that particle. The damage caused by fission fragments can be observed in some materials, making it possible to determine the number and location of fission decays that have taken place since that material was formed.

SPONTANEOUS FISSION

Spontaneous fission is a random radioactive decay process. For many elements with atomic numbers higher than that of lead, although alpha decay is usually the dominant mode of decay, spontaneous fission decay is also observed. The half-life (the time for half of any initial amount of an isotope to decay) for spontaneous fission has been measured for those elements having significant spontaneous fission decay. The age of a sample containing uranium-238 depends on the number of fission tracks, the half-life for fission decay, and the number of uranium-238 atoms present. The number of atoms can be found from mass spectroscopy of a small part of the sample or by methods involving neutron irradiation of the sample.

In natural mineral samples, uranium is the only element currently present for which spontaneous fission is significant. Because uranium-235 has a much longer spontaneous fission half-life and a much lower abundance than uranium-238, the major contribution comes from the fission of the latter. In minerals that still survive from the very early era of solar system formation, plutonium-244 can have contributed many more fissions than did uranium because plutonium-244 has a very short spontaneous fission half-life. Minerals old enough to display plutonium-244 fission tracks occur in meteorites and in some ancient lunar samples, but no terrestrial rocks preserving a record of such fission have yet been found.

DETERMINING THE NUMBER OF FISSION DECAYS

The age of a natural mineral sample containing a significant amount of uranium can be calculated if the number of spontaneous fission decays since the formation of that mineral can be determined. The present abundance of uranium in the sample is first determined. The number of fission decays, along with the known spontaneous fission half-life of uranium-238 and the measured uranium abundance, is what is used to calculate the time that has elapsed since formation of the mineral.

The damage caused by the passage of each fission fragment is examined to determine the number of fission decays that have occurred in the mineral since its formation. In 1959, it was observed that the passage of fission fragments through natural silicate minerals gave rise to a population of short damage trails, which could be viewed in a transmission electron microscope. Because of the low uranium abundance in natural silicates and the high magnification required to observe these small damage trails, however, the use of the transmission electron microscope to count fission decays in such samples was not routinely practical. The major breakthrough permitting the routine fission track dating of natural minerals came in the 1960's, when a technique was developed to permit the damage trails to be observed at low magnification in an ordinary optical microscope. In these damage trails, chemical activity is greater than in the surrounding undamaged mineral. Certain chemical etches will attack the damage trail more rapidly than they will the surrounding crystal. Initially, the etch removes material along the damage trail. As the hole lengthens, however, the etch also attacks the walls, enlarging the diameter of the hole. These etched holes, or fission tracks, can then be easily counted using an optical microscope.

Ion Explosion Spike Model

The detailed mechanism by which a fission fragment interacts with the mineral structure to produce the damage trail has not been positively determined. The "ion explosion spike" model, however, is the generally accepted mechanism. In the ion explosion spike model, the positively charged fission fragment passes through a crystalline mineral, which consists of a periodic array of positively charged nuclei, each surrounded by orbiting electrons. The charged fission fragment removes electrons from some of the atoms along its path, leaving a line of positively charged ions in the crystal. If the electrical conductivity of the crystal is low, a significant time elapses before the ejected electrons can migrate back to the ionized nuclei, restoring local electric neutrality. During this time, the positively charged nuclei along the path of the fission fragment repel one another electrostatically, causing displacements in the crystal structure. Once electrical neutrality is restored, the displaced atoms remain. This damage trail is visible in a transmission electron microscope as a disruption of the periodic array structure or can be enlarged by chemical etching.

The damage to the crystal structure can, however, be removed (or "annealed") by heating the mineral. The temperature required to anneal fission tracks depends on both the type of mineral and the duration of exposure to heat. For time scales appropriate to most geological measurements (thousands of years or longer), the track annealing temperatures of common minerals range from less than 100 degrees Celsius to more than 600 degrees. Thus, the age actually measured by fission track dating is the time interval from the present back to the time when the mineral was last heated above its annealing temperature.

Chemical etches appropriate to reveal fission tracks have been found for more than one hundred minerals. Some etches are quite simple. For example, a boiling sodium hydroxide solution is appropriate for the common mineral feldspar. However, some other minerals, such as olivine, require etches that are mixtures of several chemicals. Despite the fact that the ion explosion model seems applicable only to crystalline solids, fission tracks can also be revealed in most glasses when a hydrofluoric acid etch is used.

Dating of Volcanic Material

The products of a volcanic eruption are frequently quite rich in uranium and volcanic glasses can be dated by the fission track method. While the major terrestrial volcanoes themselves can be dated with greater precision using other techniques, volcanic debris dated by fission track techniques has proved useful in establishing the time scale for sedimentary accumulation on the ocean bottom. Wind-blown debris from major volcanic eruptions frequently accumulates as discrete layers of ash in the deposited sediments. Tiny volcanic glass fragments from the ash layers of ocean bottom cores collected by the Deep Sea Drilling Project have been dated using fission tracks, providing a chronological framework for the sediment deposition.

Fission track dating has proved to be especially valuable in the investigation of ocean-bottom, or seafloor, spreading. A model for the evolution of the floors of the ocean basins proposed that the mid-ocean ridges were sites where fresh, hot lava intrusions were deposited. After cooling, the lava would be displaced horizontally, and a new deposition would occur at the ridge. Thus, age determinations for ocean-bottom rocks at various distances from a mid-ocean ridge would constitute a direct test of the sea-floor spreading hypothesis. Potassium-argon dating had been applied to such rocks, but the ages obtained were unreliable because of the way argon-40 reacts with lava. The fission track dating technique was applied to samples taken at the Mid-Atlantic Ridge and at various distances up to 140 kilometers from the ridge. The results showed material at the ridge to have an age of 10,000 years before the present and material from the most distant point to have an age of 16 million years. As the distance of the sample from the mid-ocean ridge increased, the fission track age also increased. Thus, the ocean-bottom spreading hypothesis was supported.

Dating of Meteorites and Lunar Material

The fission track technique has also allowed scientists to date meteoritic impact events. Such events produce impact glass, formed from the local rock and soil that was melted by the impact event. The age of the impact glass thus dates the impact event.

Additionally, fission tracks have proved useful in determining the ages of meteoritic and lunar samples. Generally, minerals extracted from meteorites give ages of about 4.5 billion years, consistent with the age of the solar system inferred by other radioactive dating techniques. Mineral grains extracted from some meteorites, and in rarer cases grains from lunar samples, however, exhibit far more fission tracks than would be produced in 4.5 billion years by the uranium in the samples. Once all other sources of tracks were excluded, the investigators attributed these tracks to the fission of now-extinct plutonium-244, which was present in the very early solar system. Given that the half-life of plutonium-244 is only 80 million years, minerals containing a substantial number of fission tracks from plutonium must have formed and cooled to below the annealing temperature within a few hundred million years of the last addition of fresh radioactive material to the solar system. Thus, grains exhibiting plutonium-244 fission tracks formed very early in the evolution of the solar system.

Plutonium-244 fission tracks have been used in the development of a technique to determine the rate at which the parent bodies of the meteorites cooled. These cooling rates then permit the sizes of the parent bodies to be inferred. Different minerals have different track annealing temperatures. The plutonium in meteorites is generally concentrated in phosphate minerals such as merrillite, which have very low track annealing temperatures (about 100 degrees Celsius for merrillite). These plutonium-rich minerals, however, occasionally occur adjacent to plutonium-poor silicate grains. As the fission fragments have ranges of 10-20 micrometers, some fission decays near the merrillite-silicate contact surface produce fission tracks in the silicate minerals. Typical silicates have track annealing temperatures of about 300 degrees Celsius. Thus, the plutonium-244 fission fragments from the merrillite begin to produce tracks in the adjacent silicate before the tracks were recorded in the merrillite itself.

Because the plutonium-244 decay rate is known, a comparison of the number of fission tracks in the silicates with the number in the merrillite gives the time it took for the meteorite to cool from the track annealing temperature of the silicate to that of the merrillite. When the ordinary chondrite meteorite St. Severin was examined by this technique, a cooling rate of about 1 degree per million years was found. Cooling-rate data obtained on a number of ordinary chondrites suggest that these meteorites come from below the surfaces of asteroidal-sized parent bodies, no more than about 300 kilometers in diameter.

Use in Archaeology

Because fission track dating measures the time interval since the last heating of the mineral above the track annealing temperature, it has proven to be particularly valuable in archaeology. Certain archaeological objects, such as pieces of pottery, are heated when they are manufactured. The age of manufacture of such an object can be determined if it or mineral grains within it record fission tracks.

The earliest application of fission track dating to an archaeological sample was to an obsidian knife blade found by L. S. B. Leakey in Kenya. The texture of the blade indicated that it had been heated after its manufacture. A small fragment of the knife blade, about one-tenth of a gram, was found to be about 3,700 years old. Fission track dating was subsequently applied to samples from the Olduvai Gorge beds, from which Leakey's team recovered the specimen of *Zinjanthropus*, a very early humanoid. Potassium-argon dating of volcanic material from this bed suggested that the actual age of *Zinjanthropus* was almost twice as great as had been inferred from the fossils associated with the bed. Fission track dating of volcanic pumice from the bed gave an age of 2 million years, consistent with the potassium-argon age. This confirmation of the age of *Zinjanthropus* resolved the controversy.

Many of the clays used in the manufacture of pottery contain crystals of zircon, a mineral rich in uranium. The high temperature the pottery reaches in the kiln erases all the tracks previously recorded in the zircons, and their high uranium concentration permits even short intervals since the heating to be established with reasonable precision. In one such study done in Japan, nine zircon-containing samples of pottery were dated, giving ages ranging from 700 to 2,300 years. This fission track technique has also been applied to many human-made glass samples, doped at high uranium concentrations (sometimes up to 1 percent) to color them. These uranium-rich glass samples can be dated by the fission track technique after only a few years of track accumulation.

Comparison with Other Techniques

The fission track dating technique has been applied to a wide variety of terrestrial and extraterrestrial materials. The main advantage of this technique is the large span of ages over which it can be employed, permitting the ages of objects from only tens of years old to more than 4.5 billion years old to be established. Although the ages obtained by this technique are generally not as precise as those available through radiocarbon and potassium-argon dating, fission track ages are frequently useful to confirm ages obtained by these other techniques when their applicability to the particular sample is questionable. Where such comparisons can be made, fission track dating has been shown to give correct ages ranging from less than a year to more than a billion years.

Because a large number of individual fission tracks must be counted if age is to be determined with a high degree of precision, fission track dating results are generally considered less reliable than those of techniques that use mass spectrometers to determine isotopic ratios. Fission track dating is particularly valuable for the range of ages from 40,000 years before the present, where radiocarbon dating ceases to be accurate, to about a billion years, where potassium-argon dating is relatively easy. Fission track dating is also applicable to very small samples. Individual mineral grains as small as one milligram in mass have been dated by this technique.

The fission track dating method has been adopted by many laboratories throughout the world because of its simplicity, broad applicability, and low cost. Analyses can be performed in laboratories equipped with only simple chemical etching facilities and optical microscopes. Fission tracks are recorded in a wide variety of crystals and glasses, and the technique is applicable over a wide range of sample ages.

George J. Flynn

Further Reading

Fleischer, R. L. *Nuclear Tracks in Solids: Principles and Applications*. 2d ed. Berkeley: University of California Press, 1980. This 605-page book describes all aspects of nuclear track formation and applications. Fission track dating is described in Chapter 4, along with the experimental procedures to apply this technique. The authors explain the limitations imposed by track fading caused by heat and explain the methods of correcting for such track loss. An extensive bibliography is included. This well-illustrated book is suitable for college-level readers.

_____. *Tracks to Innovation: Nuclear Tracks in Science and Technology*. New York: Springer, 1998. The author explains the method of fission track dating and describes experiments that have been conducted to compare it with other dating techniques. The book also emphasizes the mechanism of track formation and the use of solid-state track detectors to determine the charge and energy of each particle. Designed to acquaint geologists with the technique of fission track dating, this book is a suitable introduction for general readers.

Galbraith, Rex F. *Statistics for Fission Track Analysis*. New York: Chapman & Hall/CRC, 2005. Provides a strong mathematical foundation for fission track analysis. Excellent coverage of apatite fission tracks analysis. Recommended for students with a geology background. Individual chapter bibliographies.

Garver, J. I. "Fission-Track Dating." In *Encyclopedia of Paleoclimatology and Ancient Environments*, edited by V. Gornitz. New York: Kluwer Academic Press, 2008. An encyclopedia article providing foundational content on fission track dating. The encyclopedia contains over 200 articles discussing climate change over geological history.

Macdougall, J. D. "Fission-Track Dating." *Scientific American* 235 (December 1976) 114-122. This well-illustrated account of the fission track dating technique describes applications to terrestrial and meteorite samples. A series of diagrams illustrates the mechanism by which fission fragments are believed to produce damage trails in crystal structures. Accessible to general readers.

Wagemans, Cyriel. *The Nuclear Fission Process*. Boca Raton, Fla.: CRC Press, 1991. A clear description of the process and techniques involved with nuclear fission. Wagemans discusses the applications of such procedures in relation to fission track dating. Although the subject is complicated, this book presents information in a way that a college reader without much science background can follow. Bibliography and index.

Wagner, Geunther A., and Peter Van de Haute. *Fission Track Dating*. Boston: Kluwer, 1992. This is an excellent and thorough introduction to fission track dating. The authors discuss the history, protocols, techniques, and applications of fission track dating in a manner that is understandable by the person without a strong scientific background. Illustrations, diagrams, bibliography, and index.

Walker, Mike. *Quaternary Dating Methods*. New York: Wiley, 2005. This text provides a detailed description of current dating methods, followed by content on the instrumentation, limitations, and applications of geological dating. Written for readers with some science background, but clear enough for those with no prior knowledge of dating methods.

See also: Earth's Age; Earth's Oldest Rocks; Electron Microscopy; Environmental Chemistry; Geothermometry and Geobarometry; Nucleosynthesis; Plate Tectonics; Potassium-Argon Dating; Radioactive Decay; Radiocarbon Dating; Rubidium-Strontium Dating; Samarium-Neodymium Dating; Uranium-Thorium-Lead Dating.

FLUID INCLUSIONS

Fluid inclusions are small amounts of fluids trapped in minerals within rocks. They contain valuable clues regarding many geologic processes. The study of fluid inclusions also has a number of practical applications in the exploration for mineral and petroleum resources, the study of gemstones, and the search for a storage site for nuclear wastes.

PRINCIPAL TERMS

- **brine:** water with a higher content of dissolved salts than ordinary seawater
- **fluid:** a material capable of flowing and hence taking on the shape of its container; gases and liquids are both examples of fluids
- **glass:** a solid without a periodic ordered arrangement of atoms; it frequently forms when molten material is rapidly cooled
- **igneous rocks:** rocks formed by the crystallization of magma
- **immiscible fluids:** two fluids incapable of mixing to form a single homogeneous substance; oil and water are common examples
- **intrusive rocks:** igneous rocks formed from magmas that have cooled and crystallized underground
- **magma:** molten material capable of yielding a rock upon cooling
- **metamorphic rocks:** rocks that have transformed from their original condition as a result of changes in physical or chemical conditions within the earth
- **mineral:** a naturally occurring, inorganic substance with a regular periodic arrangement of atoms
- **ore:** any concentration of economically valuable minerals
- **sedimentary rocks:** rocks formed by the consolidation of material transported by and deposited from wind or water
- **volcanic rocks:** igneous rocks formed at the surface of the earth

PRIMARY AND SECONDARY FLUID INCLUSIONS

During its history, almost every rock will come in contact with at least one fluid. Geologists have long recognized that fluids play an important role in shaping and altering rocks and in determining the earth's geologic history. It has been said that the fluid phase is the critical "missing" phase in petrology (the study of rocks). Behind this statement is the widespread belief that the fluids eventually leave rock systems, so that only indirect evidence can be used to deduce their nature. There is a growing recognition, however, that small amounts of these fluids are often left behind, trapped in small cavities, as fluid inclusions. In most cases, these inclusions represent the only available direct samples of fluids active deep within the earth or in the distant past.

Fluid inclusions can form in a variety of ways, though two are most common. Fluids can be trapped during mineral growth, to yield primary fluid inclusions. Most of these inclusions probably form when fluid fills pits on the surface of a growing crystal. New material added to the crystal grows over the tops of these cavities, trapping the fluid. These inclusions provide information about the nature of the fluids during the growth of the host minerals. Fluids trapped after the growth of their host minerals are secondary fluid inclusions, which form when fluid enters a crack within a mineral. As the ends of the crack grow together, the fluid is trapped. Secondary fluid inclusions originally form a thin envelope with a high surface-area-to-volume ratio. With its very high surface energy, this envelope is very unstable. Therefore, the host mineral will frequently recrystallize around the inclusion, causing it to break up into a swarm of smaller but thicker inclusions with a smaller surface-area-to-volume ratio. Such recrystallization occurs when the fluid inside the inclusion dissolves material from some parts of the inclusion wall and precipitates new material on other parts. Secondary fluid inclusions are sources of information about the fluids that have interacted with the rock after the growth of the host mineral.

SIZE, SHAPE, AND APPEARANCE

Fluid inclusions vary greatly in their size, shape, and appearance. Those that have diameters larger than a few tenths of a millimeter can be seen with the naked eye, but they are rare. Fluid inclusions with diameters between 1 and 100 microns are common and can be studied with a microscope. Inclusions with diameters as small as 0.01 micron have been observed with electron microscopes.

Fluid inclusions occur in almost any shape, but particularly noteworthy are negative crystals—cavities shaped like a crystal of the host mineral. This is the shape with the lowest surface energy and in many cases appears to be the final result of the recrystallization of the host mineral around the inclusion. Many inclusions contain only one phase, liquid or gas. However, gas bubbles within a liquid are also common. If these bubbles are small enough, they will move vigorously back and forth. This motion is a consequence of their relatively high surface tension. Some fluid inclusions contain immiscible liquids such as oil and water. In this situation, one of the liquids generally lines the walls of the inclusion, and the other liquid forms a droplet inside it. Solids can precipitate out of the fluids trapped within an inclusion to form tiny crystals known as daughter minerals. Trapped magma will solidify upon cooling to form either a glass or a mass of tiny mineral grains. Although such inclusions may not be fluid now, the material was trapped as a fluid, and hence they are generally classified as fluid inclusions.

TYPES OF INCLUSIONS

The composition of fluid inclusions depends on the environment in which they are trapped. Many different kinds of inclusions have been found in materials formed at the surface or in the upper levels of the earth's crust. Glacial ice and amber (fossilized tree sap) contain gas inclusions, which serve as modified samples of the atmosphere. Studies of amber inclusions show that the oxygen content of the atmosphere fell from 35% to 20% about 65 million years ago at the time that the dinosaurs disappeared. Minerals in evaporite deposits (rocks formed by precipitation from evaporating water) may contain samples of the concentrated brines from which they formed. Inclusions of groundwater can be found in the mineral deposits formed on walls, ceilings, and floors of caves. Fluid inclusions can be found in the cements that hold grains together in sedimentary rocks, as well as in the minerals that line the walls of vugs (roughly spherical cavities) and fractures in these rocks. Two kinds of inclusions most frequently occur in these situations. Water-bearing inclusions generally contain fairly high amounts of dissolved salts. Hydrocarbon inclusions can contain natural gas (most commonly methane) or crude oil.

Fluid inclusions in gypsum crystals are tested for organic matter. (National Geographic/Getty Images)

Volcanic rocks can contain a number of different kinds of inclusions. Glass inclusions may represent trapped samples of the silicate liquid from which the rock crystallized. In some cases, inclusions represent immiscible liquids present as droplets in the main magma. These droplets may have consisted of another silicate-rich liquid now present as a glass or a fine-grained mixture of minerals, or they may have been a nonsilicate liquid rich in sulfur and iron, now represented by tiny sulfide mineral grains. Water-rich inclusions in volcanic rocks tend to be filled with either a concentrated brine or water vapor. Carbon dioxide is the other gas most commonly found in volcanic rocks.

COMPOSITION AT GREATER DEPTHS

Fluid inclusions are also found in rocks formed at greater depths in the earth. The rock portions of the earth that surround the iron core consist of the relatively dense mantle surrounded by the much thinner, less dense crust. Carbon dioxide is the most common fluid in inclusions in rocks from the upper mantle and in metamorphic rocks from the lower crust. Because these fluids are trapped at high pressures, they are often very dense, and the carbon dioxide is usually present as a liquid, which may or may not contain a gas bubble. Variable amounts of nitrogen and methane can be dissolved in these inclusions, and it appears that compositions range from pure carbon dioxide to pure nitrogen or methane. At low temperatures, dense liquid carbon dioxide and water form immiscible liquids. Thus, when water occurs in these dense carbon-dioxide-rich inclusions, it forms a separate liquid phase that generally lines the wall of the inclusions.

Water is the most abundant fluid in inclusions in metamorphic rocks formed in the upper levels of the crust, although carbon dioxide, methane, and nitrogen also occur. The kinds and amounts of dissolved solids vary greatly in these aqueous solutions and in many cases appear to reflect the nature of the rocks in which they were trapped. In the lower crust, the most abundant inclusions are rich in carbon dioxide. This change is coincident with a decrease in the abundance of water-bearing minerals in the deeper levels of the earth. Intrusive igneous rocks cool and solidify slowly within the earth. Under these conditions, inclusions of the silicate liquid from which the rock forms will become not glass but rather clusters of small mineral grains. Many other kinds of fluids can be trapped during or after the crystallization of an intrusive rock. Although inclusions rich in carbon dioxide, hydrocarbons, sulfide minerals, and nitrogen have been found, the most common inclusions in these kinds of rocks are water rich and in some cases contain large amounts of dissolved solids.

CHANGES UNDERGONE BY FLUID INCLUSIONS

Fluid inclusions can undergo a variety of changes after they are trapped. Most inclusions form at temperatures significantly above those normally found on the earth's surface. As an inclusion cools, the volume of liquid will decrease much more rapidly than will the volume of the surrounding solids. This differential shrinkage frequently results in the formation of a vapor bubble from the liquid. If the inclusion is heated, this bubble will disappear. For inclusions trapped near the earth's surface, the temperature at which the bubble disappears will be close to the temperature at which the inclusion was trapped. For fluids trapped at elevated pressures deeper within the earth, the temperature at which the bubble disappears (in the laboratory) will be lower than the initial temperature of trapping. As temperature decreases, the fluid may become saturated with one or more solid compounds; the result is the precipitation of daughter crystals.

The thermal expansion or contraction of solids is low enough that temperature changes have a negligible effect on the volume of most inclusions. Thus, inclusions can generally be treated as constant volume systems. Unless material leaks out of the inclusions, their density will usually not change after trapping. Dense fluids trapped at elevated pressures will exert pressure on the walls of the inclusion. If this pressure exceeds the strength of the host mineral, the inclusion will burst, and much of the fluid will leak out. Fluid inclusions may also leak slowly instead of catastrophically. One way this leakage can occur is by slow diffusion of molecules through the host mineral. Certain molecules have a greater tendency to diffuse through the host; leakage by diffusion, then, can change the composition of an inclusion. Recrystallization of the host mineral can cause changes in the shape of an inclusion, as well as cause a larger fluid inclusion to break into smaller ones. If the larger inclusion contains two or more fluids (for example, a liquid and a vapor), generally they will not split evenly between the new smaller inclusions. Thus, the compositions of the new inclusions will differ from each other and from the compositions of the original inclusion. The effects of these and many other possible secondary changes have to be carefully evaluated during any fluid inclusion study.

EXTRACTION OF FLUID FROM INCLUSIONS

Two major problems confront anyone trying to study the composition of a fluid inclusion: separating the fluid in the inclusion from the rest of the sample (including other inclusions) and obtaining an analysis of a very small sample. Most attempts to analyze fluid inclusions by conventional chemical methods have involved the extraction of fluid from many different inclusions in the same rock. One way to do this is to heat the rock. As temperature increases, so does the pressure exerted by the fluid on the walls of the inclusion. When the pressure exceeds the strength of the host mineral, the inclusion will burst open, and the fluid will escape—a process known as decrepitation. Another possibility is to crush the rock, thereby releasing the fluid. The fluid given off during crushing or decrepitation is collected, and its composition is determined by any of a number of different analytical techniques. Unfortunately, most samples contain different kinds of fluids trapped at different times. Analyses done by the methods above give the composition of a mixture of these fluids, which usually differs significantly from the composition of the fluid in any of the individual inclusions.

ANALYSIS OF INDIVIDUAL INCLUSIONS

A number of techniques have been developed which will permit a partial chemical analysis of a single inclusion. Most of these involve hitting the inclusion with a small, tightly focused beam of light—usually a laser. Under these conditions, radiation will be emitted or absorbed; its wavelengths will be characteristic of the kinds of molecules present, while the intensity of the radiation will be proportional to their concentration. Measuring the spectra, then, makes it theoretically possible to obtain the composition of the inclusion. Most attempts to do this have used the Raman spectrum, emitted when the inclusion is struck by a laser beam. Methods using the infrared spectrum have also been used.

Most information on the composition of individual inclusions has come from observations made under the microscope. For such studies, a wafer of rock is ground to a millimeter or less in thickness and then polished on both sides. Light can be transferred through many minerals at this thickness. Much information can be obtained by careful observation at room temperature.

HEATING-COOLING STAGES

Even more information can be obtained by observing changes in an inclusion as it is heated or cooled. For example, if a solid forms upon cooling and then melts at 0 degrees Celsius, when the sample is heated, the inclusion contains water with no dissolved solids (0 degrees is the melting point of pure water). To make these kinds of observations, the rock wafer is placed in the sample chamber of a heating-cooling stage, which is in turn placed onto the microscope stage. Most heating-cooling stages are capable of cooling a sample to about -180 degrees Celsius or heating it to 600 degrees while it is being observed under the microscope. Observations made over this temperature range usually allow scientists to identify the major fluid species present (for example, water, carbon dioxide, or methane) and put some limits on the amounts and kinds of dissolved solids. Moreover, heating-cooling stages are relatively cheap and easy to use and give results quickly. Finally, they provide important information on the density of inclusions. Thus, the heating-cooling microscope stage has been, and is likely to continue to be, the major tool of fluid inclusion studies.

At low temperatures, most fluid inclusions contain both a gas and a liquid. As the inclusion is heated, the fluid will homogenize; the volume of one of the fluids increases, and the other completely disappears. The temperature at which this homogenization occurs is a function of the density of the fluid inclusion. Thus, by observing the temperature at which homogenization occurs on the heating-cooling stage, researchers can often obtain the density of the inclusion. Such density data may supply important information about the temperatures and pressures at the time the fluid was trapped.

SEARCH FOR ECONOMIC RESOURCES

The principal use of fluid inclusions has been in the study of ancient fluids and their interaction with solid earth materials. Although this research has made important contributions to the understanding of the earth's geologic history, the interest has been largely academic. Nevertheless, a number of practical applications of fluid inclusion studies have also been found. Much of this application has involved the search for mineral resources and the study of petroleum migration.

Fluid inclusions have contributed to the search for mineral resources in many different ways. First, studies of these inclusions have contributed enormously to the understanding of the processes by which ore deposits form. Many ore deposits have been formed by hot, water-rich solutions (hydrothermal solutions), which circulate through the rocks, dissolving and removing elements from some areas and precipitating them as ore minerals in others. This understanding has in turn guided the selection of areas in which to search for ore deposits. Fluid inclusion studies can also contribute to knowledge of the geologic history of a specific area, including an understanding of the development of features most likely to control the emplacement of an ore body. Finally, fluid inclusions can be one of the telltale signs in the search for ore. The strategy is to increase the size of the "target"—the small area containing a valuable mineral—by looking for secondary effects that accompanied ore deposition but affected a larger area. The improvement in the chances of finding ore with this strategy can be compared to the increase in the chances of finding a nail rather than a needle in a haystack. Fluid inclusions surrounding an ore body can show special properties related to the development of the ore, and these anomalies can extend well beyond the deposits. A search for such anomalies has been used in ore exploration.

Petroleum originates in a fine-grained source rock, moves into a more permeable reservoir rock, and then migrates into a petroleum trap. Hydrocarbon inclusions in old reservoir rocks are potential clues to the process of petroleum migration. This is a relatively new area of research, but it shows good promise of increasing the understanding of the movement of oil underground, leading to improved strategies in the search for this oil.

OTHER PRACTICAL APPLICATIONS

The study of fluid inclusions has also been applied to the establishment of the authenticity of gemstones and investigation into long-term storage sites for nuclear waste. Establishing the authenticity of gemstones is one of the more important jobs of a gemologist. Many gemstones contain fluid inclusions; the nature of these inclusions often allows the expert to distinguish between synthetic and natural stones. In some cases, the fluid inclusions can be used to identify the source of the gem. Thus, for example, Colombian emeralds can often be recognized by fluid inclusions containing a concentrated brine and large daughter crystals of sodium chloride.

Because of its dangerous, highly radioactive nature, nuclear waste must be isolated for periods of time ranging from thousands to hundreds of thousands of years. Most proposals of ways to bring about such isolation involve the underground burial of the waste. The principal hazard with burial is that the nuclear waste could be dissolved in and carried away by fluids circulating through the local rocks. The study of fluid inclusions has been used to evaluate this danger. These studies are based on the realization that fluid inclusions represent a partial record of the fluids that moved through the rocks in the past and hence provide some basis for extrapolating into the future. Thus, for example, studies of fluid inclusions from salt beds have indicated that some of these beds probably have been penetrated by circulating groundwater. This finding is especially significant, since salt deposits have often been mentioned as possible repositories of nuclear waste.

Edward C. Hansen

FURTHER READING

Aharonov, Einat. *Solid-Fluid Interactions in Porous Media: Processes That Form Rocks.* Woods Hole: Massachusetts Institute of Technology, 1996. Aharonov examines the processes involved in rock formation. This is a technical book at times, but it can be understood by the careful reader.

Berner, Elizabeth K., and Robert A. Berner. *Global Environment: Water, Air, and Geochemical Cycles.* Upper Saddle River, N.J.: Prentice Hall, 1996. This book offers a clear and readable introduction to the processes that sustain life and effect change on the earth, including a useful section on aquatic geochemistry. Color illustrations and maps.

Correns, C. W. "Fluid Inclusions with Gas Bubbles as Geothermometers." In *Milestones in Geosciences: Selected Benchmark Papers Published in the Journal "Geologische Rondschau,"* edited by Wolf-Christian Dullo and Geologische Vereinigung e.V. New York, Springer-Verlag, 2010. This article was originally written in 1952, though it has been recognized multiple times since then as a significant paper in geoscience. A background in geology is helpful in understanding this article.

Emsley, John. *The Elements.* 3rd ed. Oxford: Oxford University Press, 1998. Emsley discusses the properties of elements and minerals, as well as their distribution in the earth. Although some background in chemistry would be helpful, the book is easily understood by the high school student.

Hollister, L. S., and M. L. Crawford, eds. *Mineralogical Association of Canada Short Course in Fluid Inclusions: Applications to Petrology.* Mineralogical Association of Canada, 1981. This book contains some good general information about fluid inclusions but concentrates on the information that fluid inclusions can give about rocks and rock-forming processes. Contains twelve separate articles by nine different authors. Some articles may be difficult for a reader with no previous knowledge of geology.

Lowenstein, Tim K., Brian A. Schubert, Michael N. Timofeeff. "Microbial Communities in Fluid Inclusions and Long-Term Survival in Halite." *GSA Today* 21 (2011): 4-9. An interdisciplinary look at fluid inclusions. This article presents a detailed analysis of community ecology within ancient rock, and proposes a hypothesis of how the species survived millennia.

Roedder, Edwin. "Ancient Fluids in Crystals." *Scientific American* 207 (October 1962) 38-47. One of the few articles on fluid inclusions intended specifically for a general audience. Written by the "dean" of American fluid inclusion studies: Roedder has done more to promote interest in fluid inclusion than any among his contemporaries in the English-speaking world. Very well illustrated and clearly written, it is an excellent introduction to the subject. Although it does not cover any of the many more recent discoveries, Roedder's description of the basic phenomenon remains valid.

_____. "Fluid Inclusion Studies of Hydrothermal Ore Deposits." In *Geochemistry of Hydrothermal Ore Deposits*, edited by L. B. Barnes. 3rd ed. New York: John Wiley & Sons, 1997. The bulk of this paper is an introduction to fluid inclusions and their interpretation, which should be comprehensible to someone without a technical background. The remaining portion of the article concentrates on fluid inclusions in ore deposits and is especially good for those with an interest in ore-forming solutions.

_____. *Reviews in Mineralogy*. Vol. 12, *Fluid Inclusions*. Washington, D.C.: Mineralogical Society of America, 1984. This book is certainly one of the best and one of the most complete works on fluid inclusions that is available in the English language. Written as an introduction to fluid inclusion research for the geologist, the text may be rough going in places for those with no previous geologic knowledge.

Schmatz, Joyce, Oliver Schenk, and Janos Urai. "The Interaction of Migrating Grain Boundaries with Fluid Inclusions in Rock Analogues: The Effect of Wetting Angle and Fluid Inclusion Velocity." *Contributions to Mineralogy & Petrology* 162 (2011): 193-208. A current article discussing fluid inclusion dynamics. The authors use many technical terms, so a knowledge of geology is beneficial.

See also: Carbon Sequestration; Deep-Earth Drilling Projects; Elemental Distribution; Environmental Chemistry; Experimental Petrology; Freshwater Chemistry; Geochemical Cycle; Geologic and Topographic Maps; Geothermometry and Geobarometry; Oxygen, Hydrogen, and Carbon Ratios; Phase Changes; Phase Equilibria; Radioactive Decay; Radiocarbon Dating; Relative Dating of Strata; Water-Rock Interactions.

FRESHWATER CHEMISTRY

The unique properties of water that are so important to geological processes depend upon the distinctive polar structure of the water molecule. The chemistry of fresh waters is highly variable. Throughout the earth, it is influenced by the atmosphere and locally influenced by reactions with the local soils and bedrocks. Changes through time in the chemistry of natural fresh waters are one of the best indicators of changes in the earth's surface environment.

PRINCIPAL TERMS

- **acid:** a substance that yields free hydrogen ions in solution
- **acidity:** the degree of a solution's being acidic as determined by the quantity of base needed to neutralize the solution
- **alkalinity:** the degree of a solution's being basic as determined by the quantity of acid needed to neutralize the solution
- **anion group:** a combination of ions that behaves as a single anion
- **colorimeter:** an instrument that measures the intensity of color produced when a reagent reacts with a substance in a solution; the intensity of color is used to quantify the amount of the substance in solution
- **covalent bonding:** a type of chemical bonding produced by sharing of electrons between overlapping orbitals of adjacent atoms; covalently bonded solids usually have low solubility in water
- **density:** a property of a substance expressed in units of weight per unit of volume, such as pounds per cubic foot or grams per cubic centimeter
- **eutrophication:** processes causing water bodies to receive excess nutrients that stimulate excessive plant growth
- **ionic bonding:** a type of chemical bonding that holds the constituents of a crystal together primarily by electrostatic attraction between oppositely charged ions
- **total dissolved solids (TDS):** a quantity of solids, expressed in weight percent, determined from the weight of dry residue left after evaporation of a known weight of water

Polar Molecule

The temperature of the earth lies within the range that permits water to exist as a liquid, a gas, and a solid. Earth is sometimes called the "blue planet" because of the brilliant blue color, seen from outer space, that results from more than three-quarters of the planet's surface being covered by water. Ultimately, this water comes from within the planet during magmatic differentiation and formation of the crust. The size of the earth provides a gravitational field that is sufficient to keep water vapor from escaping through geologic time. Steam, the gaseous form of water, is the dominant constituent of volcanic gases, and pressure generated by steam accounts for the power of most explosive eruptions. Ice in glaciers has been a major force in influencing the topography of vast areas of the continent.

The chemical symbol for water, H_2O, refers to a molecule composed of one hydrogen ion (H^+) and one hydroxyl ion (OH^-). When combined to form a water molecule, the two hydrogen ions are arranged to one side and have an angle of about 105 degrees between them. Because the positively charged hydrogen ions lie on one side of the molecule, they impart a positive charge to that side of the molecule while leaving a negative charge of the oxygen exposed at the other end. Therefore, each water molecule is like a tiny magnet, with a positive pole and a negative pole. The molecule has polar charge distributions and thus is termed a polar molecule.

Because each water molecule is like a magnet, as a substance, water will behave somewhat like a box full of very tiny magnets. The tendency for the magnets would be to line up positive pole to negative pole, and a similar tendency occurs in water, where the ionic attraction of the hydrogen of one water molecule for the oxygen of another water molecule actually draws water molecule to water molecule and creates a type of ordering. The cohesive ordering that results from this attraction is called hydrogen bonding, and it accounts for some remarkable physical properties that belong only to water.

Hydrogen Bonding

Water is one of the few liquid substances that do not simply become denser as they cool. The hydrogen bond allows liquid water molecules to pack themselves into a tighter pattern than would be possible if the molecules were not polar. Water achieves its maximum density of 1.00000 gram per cubic centimeter at 3.94 degrees Celsius. Liquid water molecules vibrate rapidly (about 10^{12} vibrations per second), but when the water is cooled to the freezing temperature of ice (0 degrees Celsius), these vibrations decrease to about 10^6 vibrations per second. At that point, new covalent bonds are able to overcome the force of the hydrogen bond. The slowing of vibrations as water cools permits some rearrangement and some creation of open space to begin at just below 3.94 degrees Celsius, even though ice crystals do not form until 0 degrees is reached. This accounts for the maximum density of water occurring at 3.94 degrees rather than at 0 degrees Celsius.

When actual freezing occurs, water molecules arrange themselves into a covalently bonded crystal structure that has open space not present in liquid water. Water goes from a density of 0.99987 gram per cubic centimeter as liquid to 0.917 gram per cubic centimeter as solid ice. The pronounced decrease in density causes ice cover to form on the surface of lakes and rivers rather than at the bottom. Without this property, aquatic life would probably not be possible, since ice on the surface insulates the water below, keeping it above 0 degrees Celsius and preventing complete freezing from bottom to top. The decrease in density is accompanied by an increase in volume; water expands as it freezes. When water seeps into cracks and pores of rock and soil and then freezes, the force generated by the expansion causes rock to break, soils to heave and swell, and small grains and crystals to spall away. Therefore, it is the hydrogen bond of water that ultimately permits water to become a powerful mechanical agent in weathering, as exemplified through frost heave in soils or frost wedging in rock.

Above 3.94 degrees Celsius, the density of water decreases with increasing temperature. At temperatures likely to be encountered in lakes and streams, the density of pure water can be closely approximated with a formula. The small changes in density that occur during heating, cooling, and freezing are responsible for seasonal circulation of water in freshwater lakes.

Water is unlike other substances in that its freezing point is lowered, rather than raised, by pressure, a result of a tendency at near-freezing temperatures for the covalent bonds of ice to collapse back to the

The hydrogen bond of water permits water to become a powerful mechanical agent in weathering, as exemplified through "frost heave," a process whereby frost in soils or wedged in rock lifts the rock, as shown here in the frost-heaved blocky boulders of Shepard Formation in the Boulder Pass area of Montana's Glacier National Park. (U.S. Geological Survey)

hydrogen bonds of liquid water. Simple pressure is sufficient to enact the transition. When an ice skater places pressure on the blades of the skate, the pressure melts the ice at the sharp edges of the blade, and the skater glides on a thin layer of water that is created momentarily by the pressure. Such skating is not possible on other substances such as solid carbon dioxide (dry ice). This same liquefaction under pressure permits glacial ice to flow over a thin layer of water created at the base of the glacier and for this water to flow into joints and cracks, refreeze, and "pluck" out sections of the bedrock as the glacier moves.

ELECTROSTATIC BONDING

Water is a powerful solvent because the water molecule is polar. The magnet-like quality of the water molecule strongly influences substances that are bonded primarily by electrostatic (ionic) bonds. Ions bound into many solid minerals by virtue of the ionic bond may therefore be dissolved into water. In this way, water picks up dissolved materials from the soils and rock. The dissolving power provided by the polar nature of the molecules is further assisted by their vibration. As the molecules vibrate faster, they act like high-speed jackhammers against solid particles, and those solids held together with ionic bonds are particularly affected. Heating is one way to increase the vibration of molecules. Therefore, fresh water tends to be a more effective solvent in warm climates.

Fresh waters contain dissolved solids and gases. The substances that are likely to occur in waters of this planet are those substances that are abundant and that occur in forms that are easily soluble. More than 99 percent of the earth's crust can be accounted for by only twelve elements (oxygen, silicon, aluminum, iron, calcium, magnesium, sodium, potassium, titanium, hydrogen, manganese, and phosphorus—in order of abundance from greatest to least), and nearly 100 percent of the atmosphere by five elements (nitrogen, oxygen, argon, hydrogen as water vapor, and carbon as carbon dioxide).

The most common dissolved substances in fresh water are bicarbonate, calcium, sodium, magnesium, potassium, fluorine, iron, phosphate, sulfate, chloride, and dissolved silica. These substances enter natural waters both from the atmosphere and from the rocks and soils contacted by the waters. Even rainwater is not completely pure and contains small amounts of silica, sulfate, calcium, chlorine, nitrate, carbonic acid, sodium, potassium, iron, and aluminum.

Some elements are crustally abundant but are not abundant as dissolved species in fresh waters because they form insoluble compounds. For example, aluminum and iron are abundant elements, but in most waters they form nearly insoluble hydroxides—aluminum hydroxide and iron hydroxide. Silicon is extremely abundant, but it forms strong covalent bonds with oxygen that are not possible for water to break. In nature, silicon forms minerals that include silica-oxygen structures with very low solubility in water.

ACIDIC AND ALKALINE WATERS

Although water is a good solvent because of the polar nature of the water molecule, it can be made an even more effective solvent if it can be rendered acidic. Chemists have devised the pH scale as a means for expressing the degree to which a solution is acid. The scale runs from 0 (most acid) to 14 (least acid), with a pH value of 7 termed neutral. Waters with a pH of less than 7 are termed acidic, and those with pH values greater than 7 are termed alkaline. Natural rainwater is acidic and has a pH of 5.6, because carbon dioxide in the atmosphere dissolves into the rainwater to make carbonic acid. (The atmosphere has a powerful effect on the composition of fresh water; the character of fresh water on this planet has changed through geologic time in accord with changes in the atmosphere.) Once rainwater strikes the ground, the weak acid is neutralized as it dissolves minerals in the rocks and soils. Therefore, even though natural rainwater is acidic, most fresh groundwater and surface waters are near neutral, with pH values between 6 and 8.5.

Unusually acidic waters occur around volcanic vents, geysers, and fumaroles, where gases such as sulfur dioxide react with water to produce sulfuric acid. Acidic waters also occur in bogs and marshes (pH 3.3-4.5), where carbon dioxide released during the decay of organic matter reacts with water to produce carbonic acid. The strongest acids occur where sulfide minerals such as pyrite and marcasite oxidize in the presence of air and water to produce sulfuric acid. Sulfuric acid is a strong acid, and its production through the weathering of sulfide minerals is so common around coal and metal mines that the waters released are termed acid mine drainage. Streams flowing from these sites often have pH values below 2. Thousands of miles of streams have become polluted by acid mine drainage. Alkaline waters occur in limestone terrains, where pH values between 8 and 9 are common. Extremely alkaline waters occur in playa lakes in contact with sodium carbonate or sodium borate, where pH values above 12 have been noted.

Extremely acidic and extremely alkaline waters permit substances that remain immobile in normal waters to be dissolved. For example, aluminum is normally insoluble but does dissolve in both very acidic and very alkaline waters. Acidic waters dissolve and transport metals such as iron and manganese and nonmetals such as phosphorus that would normally be present in most natural waters at low or undetectable levels.

ACIDITY VS. PH

A clear distinction needs to be made between pH and acidity, because these terms are too often misused interchangeably. In a large tank of pure distilled water at pH = 7, if a few drops of strong acid (a substance that releases free positively charged hydrogen ions into solution) are added, a low pH, perhaps pH = 3 throughout the tank, will soon be registered. If a few drops of strong base (a substance that releases free negatively charged hydroxide ions into solution) are added, however, the pH will easily rise back to 7. The measure of the amount of base required to get the tank to a pH of 7 is a measure of acidity. The tank had a low pH, but yet also a low acidity. Suppose another tank contains water with a considerable amount of dissolved solids and has a pH of 6. It may take many gallons of strong base to get the tank up to a pH of 7. In the second tank, then, there is water with a moderate pH but a very high acidity. In short, pH is the measure of the concentration of hydrogen ion present at a given time, but acidity is a measure of the amount of base needed to change the pH back to neutral. Natural waters with high acidity are usually high in dissolved solids and have a low pH. Acid mine drainage is an example.

AQUEOUS GEOCHEMISTRY

Freshwater chemistry is a part of the field of aqueous geochemistry, or low-temperature geochemistry, which is a specialty field of many geologists. Proper collection, analyses, and interpretation of data from natural waters require more than knowledge of chemistry. Success also requires knowledge about the natural environmental system, and this is why hydrology and aqueous geochemistry fall more properly within the province of geology than that of pure chemistry.

A typical water analysis includes determination of the levels of silica, aluminum, iron, manganese, copper, calcium, magnesium, strontium, sodium, potassium, dissolved oxygen, carbonate, nitrate, bicarbonate, sulfate, chloride, fluoride, phosphate, arsenic, selenium, boron, total dissolved solids, pH, acidity or alkalinity, temperature, and conductivity. If waters are polluted with unusual substances such as pesticides, sewage, or industrial wastes, specialized tests must be undertaken for these.

As soon as a sample is taken, the water is removed from its actual environment, and changes start to occur. Dissolved gases may leave, precipitates may form, and bacteria and microscopic algae may metabolize substances or die and release substances that were formerly solids. Because some parameters change so quickly and easily, some tests must be done immediately in the field. The temperature, conductance, pH, dissolved oxygen, and sometimes acidity or alkalinity is measured in the field. Small battery-operated instruments such as specific ion meters (for nitrate, sulfate, carbonate, and a number of elements), colorimeters, and conductivity meters are made by several manufacturers to be used in field analyses.

Pure water is a poor electrical conductor, but water's ability to conduct electricity increases markedly with the amount of dissolved solids. Therefore, a conductance test in the field is a rapid method to use to make a rough estimate of the amount of total dissolved solids. Sometimes rapid colorimetric tests for sulfate, phosphate, and some metals are done in the field, but more often the analyses of those substances are done in the laboratory.

Samples to be taken to the laboratory are filtered in the field to remove any microscopic suspended solids, because it is important that the water analyses show only dissolved substances. After the water is filtered, it is acidified with a few milliliters of high-purity acid to ensure that all dissolved species remain dissolved. The water is then put into a plastic container and filled to the top so as not to admit any air. The samples are usually placed in a dark cooler and kept refrigerated until they are analyzed in the laboratory.

DANGER OF CONTAMINANTS AND POLLUTANTS

Of the vast amount of water present on the planet's surface, only about 2.5 percent is fresh water; the remainder is in oceans and inland seas. Of that 2.5 percent, 2.15 percent is locked away in glaciers and polar ice caps. About another 0.3 percent of the total is accounted for by fresh groundwater. All the surface water commonly seen in freshwater lakes and streams amounts to only .009 percent of the earth's total water.

The small percentages do not reveal the true importance of fresh water. Fresh water is the major sculpting agent of the planet's land surface, and it is an essential substance for all terrestrial life-forms. Changes in the chemical composition of fresh water constitute one of the most sensitive indicators of changes in the environment. Changes may be local, as exemplified by the eutrophication of a small lake that receives phosphorus from a few local septic tanks, or they may be global, as exemplified by the acidification of sensitive lakes in northern Europe and Canada as a result of the burning of fossil fuels, which releases carbon dioxide and oxides of nitrogen and sulfur into the atmosphere.

The category "fresh water" includes a group of waters from wells, lakes, and streams with diverse chemistry. Even natural substances in excessive amounts pose problems for water consumption or industrial use. In addition to the toxic content of acid mine drainage, excessive amounts of nitrate, sodium, and fluoride present health hazards. Excessive amounts of calcium and magnesium promote buildup of scale in water tanks and boilers, and waters that are high in these elements (hard waters) must have these constituents chemically removed with a water softener prior to domestic and industrial use. High phosphate contents of waters promote algal blooms and eutrophication. High boron content is important in irrigation, because boron is toxic to many crops. Of all natural resources, fresh water ranks as one of the most essential to human survival. Continual monitoring of the chemistry of fresh waters is essential to ensure that these valuable resources are not rendered unsuitable by human-made contaminants and pollutants.

Edward B. Nuhfer

Further Reading

Berner, Elizabeth K., and Robert A. Berner. *Global Environment: Water, Air, and Geochemical Cycles.* Upper Saddle River, N.J.: Prentice Hall, 1996. This book offers a clear and readable introduction to the processes that sustain life and effect change on the earth, including a useful section on aquatic geochemistry. Color illustrations and maps.

Dodds, Walter K., and Matt R. Whiles. *Freshwater Ecology: Concepts and Environmental Applications of Limnology.* 2d ed. Burlington: Academic Press, 2010. Covers physical and chemical properties of water, the hydrologic cycle, nutrient cycling in water, as well as biological aspects. Written by two of the leading scientists in freshwater ecology, the text is an excellent resource for college students. Each chapter has a short summary of main topics

Drever, James I. *The Geochemistry of Natural Waters.* 3rd ed. Englewood Cliffs, N.J.: Prentice-Hall, 1997. The book is intended for students in advanced courses in geochemistry or water chemistry and also for professionals working in the area of water chemistry. Thermodynamic tables, references, and an index are included. A particular strength of the book is its inclusion of case study examples, particularly in the chapters on weathering and water chemistry.

Faust, Aly. *Chemistry of Natural Waters.* Stoneham, Mass.: Butterworth Publishers, 1981. Chapter 1, "Chemical Composition of Natural Waters," of this useful reference book requires little chemical or mathematical background. Later chapters cover thermodynamics, equilibria, reactions, and models. The final chapter deals with toxic metals in the aquatic environment.

Greenberg, Arnold, et al., eds. *Standard Methods for the Examination of Water and Wastewater.* 21st ed. Washington, D.C.: American Public Health Association, 2005. This reference has been the major reference for water chemists. It is the essential "cookbook" for water collection and water analysis laboratory procedures.

Hem, J. D. *Study and Interpretation of the Chemical Characteristics of Natural Water.* U.S. Geological Survey Water Supply Paper 2254. Honolulu: University Press of the Pacific, 2005. Written by one of the world's foremost water chemists, the usefulness of this reference has proved itself since the first edition appeared in 1959. In addition to including a condensed discussion of chemical thermodynamics, the writer provides many analyses of unusual water types and discusses, component by component, the common constituents of natural water. This section is virtually a water chemist's tour of the periodic table. The concluding sections deal with interpretation and presentation of analytical data and with the relationship of water quality to water use. Well indexed and contains an excellent list of references.

Krauskopf, Konrad B. *Introduction to Geochemistry.* 3rd ed. New York: McGraw-Hill, 2003. One of the best-written texts produced for students. Provides an excellent introduction to water chemistry, particularly with respect to equilibrium and chemical thermodynamics. The book is designed as a text for undergraduate students in geochemistry but may be understood by students with good high school chemistry and earth science courses.

Includes problem sets at the end of each chapter, good references, and tables of thermodynamic properties that are especially pertinent to water chemists.

Lam, Buuan, et al. "Major Structural Components in Freshwater Dissolved Organic Matter." *Environmental Science & Technology* 41 (2007): 8240-8247. Discusses the important aspects of dissolved organic matter in freshwater ecosystems, the global carbon cycle, and processes driving the cycle within Lake Ontario. A good example of water chemistry analysis for research.

Langmuir, Donald. *Aqueous Environmental Geochemistry*. Upper Saddle River, N.J.: Prentice Hall, 1997. A thorough look into aqueous geochemistry, this book examines the geochemical cycles and processes of water systems and their interactions with other cycles on the earth. Suitable for the careful high school reader or college student.

Nicholson, Keith. *Geothermal Fluids: Chemistry and Exploration Techniques*. Berlin: Springer-Verlag, 1993. Nicholson provides the reader with an examination of the behavior of geothermal fluids, as well as study and exploration practices. Slightly technical, this book is intended for the reader with some background in chemistry.

Sparks, Donald L., and Timothy J. Grundl, eds. *Mineral-Water Interfacial Reactions: Kinetics and Mechanisms*. Washington, D.C.: American Chemical Society, 1998. This collection of essays deals with the ongoing chemical reactions and processes that occur in aquatic systems. A technical piece intended for the person with a background in chemistry or earth sciences.

Stumm, Werner, and James J. Morgan. *Aquatic Chemistry: An Introduction Emphasizing Chemical Equilibria in Natural Waters*. 3rd ed. New York: John Wiley & Sons, 1996. This reference is indispensable for limnologists and aqueous geochemists. Thermodynamic principles are stressed from the outset, but a number of examples from natural systems are provided. References, a subject and author index, and a good set of thermodynamic tables make the book appropriate as a reference and as a college text.

van der Leeden, Fritz, Fred L. Troise, and D. K. Todd. *The Water Encyclopedia*. 2d ed. New York: CRC Press, LLC, 1990. A compilation of hundreds of pages of useful tables about water properties, water chemistry, and bodies of water of the world.

See also: Biogeochemistry; Elemental Distribution; Environmental Chemistry; Fluid Inclusions; Geochemical Cycle; Geothermometry and Geobarometry; Glaciation and Azolla Event; Nucleosynthesis; Oxygen, Hydrogen, and Carbon Ratios; Phase Changes; Phase Equilibria; Water-Rock Interactions.

G

GEOBIOMAGNETISM

Geobiomagnetism refers to the interaction of living organisms with the earth's magnetic field. Many animals, plants, and even bacteria have displayed in laboratory experiments the ability to sense and to use the earth's magnetic field in various ways, notably in navigation.

PRINCIPAL TERMS

- **biomagnetism:** the magnetic fields generated by living organisms
- **geomagnetism:** the magnetic field generated by the earth
- **magnetite:** an isometric mineral, an oxide that is sensitive to magnetic fields
- **magnetometer:** a device used to detect and measure magnetic fields
- **SQUID:** an extremely sensitive magnetometer capable of detecting and measuring very weak magnetic fields

NAVIGATION ABILITY OF LIVING ORGANISMS

The ability of living organisms to navigate accurately over great distances has long fascinated and baffled naturalists and life scientists. How are many species of birds able to migrate thousands of miles annually, often across open seas, and unerringly reach their destinations? How can homing pigeons find their way back to their coops after having been taken many miles from them in enclosed containers? How can honeybees, after having located a desirable food source miles away from their hives, not only return to the food source but also communicate its location to other honeybees? These are only three examples of the remarkable navigation abilities displayed by a variety of living organisms. Systematic research into methods by which animals navigate began in the late 1930's. In experiments in the 1940's and 1950's, researchers showed that living organisms use a variety of means to find directions. These means include celestial navigation (use of the sun and the stars to find directions), which is used by several species of migratory birds and some crustaceans, and navigation by sound reflection, which is used by bats and many forms of sea-dwelling mammals. Other methods include navigation by electricity, used by many species of fish, and navigation by using the earth's magnetic field. Life-forms as diverse as bacteria, butterflies, fish, and birds have built-in compasses, in the form of minute, magnetic, mineral grains, that enable them to orient to the earth's magnetic field.

BIRDS' USE OF EMF

German researcher Hans Fromme was conducting observations of several robins in a cage at the Frankfurt Zoological Institute in 1957 at a time that they were preparing for their annual migration to Spain. Fromme was not satisfied with the then-accepted theory that the robins found their way to Spain by celestial navigation, because radar data of the birds had shown that they flew straight toward their destination even when heavy cloud cover hindered the visibility of the sky. Fromme caged his birds in a windowless room. Nevertheless, when the robins outside began their southwestward migration, Fromme's birds became restless and fluttered up to the southwestern corner of their cage. Fromme reasoned that they were responding to some stimulus other than the stars or the sun. He guessed that this stimulus might be the earth's magnetic field. Scientists had long known that the earth acts in many ways as a giant electromagnet of considerable power. Until Fromme's experiments, however, few scientists suspected that geomagnetism affected living organisms or could be used by them for various purposes.

To test his theory, Fromme put his birds into a special steel chamber which reduced the power of the earth's magnetic field (EMF) to 0.14 gauss (a unit of measure for magnetic force). The average strength of the EMF at Frankfurt is 0.41 gauss. In this enclosure, the robins still became restless at their normal migration time, but their flutterings were random, no longer toward the southwest corner of the cage.

Further research showed that over a period of days the robins adjusted to the reduced magnetic field and once again flew toward the southwest corner of their cage. Fromme and his colleagues were able to "fool" the robins by creating an artificial magnetic field that created a false southwest. The robins rapidly adjusted to the artificial field and fluttered toward the southwest of the artificial field.

After Fromme's experiments proved that robins used the EMF to navigate, life scientists began investigating the effects of geomagnetism on a variety of life-forms, ranging from bacteria to higher vertebrates, including human beings. These experiments resulted in a series of dramatic and unexpected discoveries.

In addition to Fromme's experiments with robins, other experiments have demonstrated conclusively that many species of birds rely on the EMF to navigate. The homing pigeon provides perhaps the best example. Carefully conducted experiments showed that a simple bar magnet attached to the back of a homing pigeon's head completely disrupts its navigational ability. Other experiments showed that homing pigeons are remarkably sensitive to minute local fluctuations (anomalies) in the EMF, which may perhaps explain their remarkable homing ability.

Other Organisms' Use of EMF

One group of scientists observed anaerobic bacteria of the *Spirillum* type, which are usually found in aquatic mud. When taken into open water, the bacteria swim along magnetic field lines, natural or artificial, which take them toward the magnetic north pole in the Northern Hemisphere and the magnetic south pole in the Southern Hemisphere. This reaction takes them directly to their natural habitat, the mud of the sea floor.

Other scientists have shown that different species of insects use the EMF in a number of ways and for a variety of purposes. The common honeybee, for example, can communicate location of a food source to its hive mates by use of the EMF, without the hive mates having visited the site. The honeybee accomplishes this by performing a complicated series of movements (called a "waggle dance") on the honeycomb in which its movements are oriented by the EMF. The so-called compass termite of Australia uses the EMF in an entirely different way from the honeybee. Compass termites build large nests, sometimes 13 feet high and 10 feet long but only about 3 feet wide, the temperature of which is regulated by use of the EMF. The long axis of the nest always runs due north and south. This magnetic orientation has the advantage of exposing the long sides of the nest to the direct warming rays of the sun during the early morning and late afternoon. In the middle of the day, when the sun's rays might be too hot, however, only the relatively thin top edge is exposed to its direct rays.

Many species of fish also use the EMF. Sharks and rays, for example, are apparently able to detect changes in the EMF to locate potential prey. Scientists have shown that the fish interact with the EMF by introducing electrical fields into their environment, which they use to orient themselves in the EMF and to register fluctuations therein caused by magnetic anomalies or by other living creatures. The organ involved in the fishes' ability to interact with the EMF appears to be the electroreceptive ampullae of Lorenzini, which respond to very low electrical voltage gradients. Freshwater eels, both the European and the American varieties, apparently use the EMF to guide them from the rivers where they spend their adolescence to the Sargasso Sea, to which they migrate for purposes of reproduction once they have reached maturity.

Magnetic Properties of Higher Organisms

Scientists investigating the magnetic properties of higher organisms have also made spectacular and unexpected discoveries. Researchers in this area, called biomagnetism, have found that most organs in higher vertebrates (including humans) produce weak magnetic fields that can be detected and measured using the very sensitive instruments. The organs producing such magnetic fields include the liver, the brain, and the heart. Magnetic measurements of these organs provide information about them that no other sort of test, including X-rays and electroencephalograms, can yield. A number of researchers in the field of biomagnetism suspect that the magnetic fields produced by some living organisms allow them somehow to use the EMF for direction-finding. This relationship, however, has not yet been scientifically demonstrated.

Instruments for Studying Geobiomagnetism

The sensitive instruments necessary to study geobiomagnetism emerged from weapons research conducted during World War II. In their efforts to discover ever more efficient ways to detect enemy submarines and aircraft, scientists in both Allied and Axis countries investigated various applications of electromagnetism. Governments invested huge sums of money into scientific research projects that offered even a tenuous hope of producing revolutionary weapons. Some of the better-known results of these military-oriented scientific projects include radar, sonar, and nuclear fission. After the war, a part of the research conducted by military research projects led to the development of instruments capable of detecting the very weak magnetic fields produced by living organisms, or biomagnetism.

Biomagnetic fields are very faint, usually less than one-tenth that of the earth, and cannot be measured with the magnometers used to measure the EMF. Magnetic fields stronger than 1 gauss are measurable by a simple but sensitive magnetometer called a fluxgate. Measurement of weaker fields requires the use of an extremely sensitive cryogenic magnetometer called a SQUID (acronym for superconducting quantum interference device). No instrument yet devised, however, has been able to show how organisms interact with the EMF or which device or organ is involved in that interaction, although some clues have been discovered. Nevertheless, abundant evidence exists that such interaction does take place.

In the experiments with bacteria mentioned earlier, researchers introduced an artificial magnetic field pointing at right angles to the sea bottom. The bacteria invariably aligned themselves with the new field. When scientists cultured these bacteria in a largely iron-free medium, the bacteria lost their ability to orient themselves along the EMF. Upon examination, the researchers found the bacteria that were cultured in a natural environment contained 1.5 percent (dry weight) iron, which almost certainly is the agent which allows their interaction with the EMF. Exactly how this interaction occurs, however, is unknown.

THEORIES OF GEOBIOMAGNETISM

Geophysicists have proposed several theories about which mechanisms are at work in geobiomagnetism. The paramagnetic molecule theory states that molecules with unpaired magnetic spins are present in all living cells, which may line up with external magnetic fields, although this has yet to be demonstrated. Even if such alignment does occur, there is no evidence or even theory as to how an organism's nervous system could use the information to deduce the direction of the field. The electrodynamic theory states that if a force of electrically charged particles is introduced into a magnetic field, the field exerts a force which influences their direction of motion. Whether detectable effects can be produced in living organisms, allowing them to detect the weak EMF, continues to be debated and has yet to be demonstrated. According to the magnet hypothesis, the ingestion of magnetic material or the formation of magnetic material within specialized cells by living organisms allows them to sense the earth's magnetic field. Magnetotactic bacteria produce intercellular iron sulfide, greigite, which is magnetic. Magnetite, an iron oxide, in the trigeminal nerve cells of trout and other fish enables them to detect changes in the magnetic field. German scientist Dominik Heyers published evidence in 2007 that migratory birds use a visual link to the brain, allowing them to "see" magnetic fields.

APPLICATIONS BENEFITING HUMANKIND

Learning the methods by which living organisms sense and use the EMF for navigation could be beneficial. In theory, the EMF could be used to steer planes and ships to their destinations, forgoing the need for expensive and complicated navigational equipment.

The new field of biomagnetism has already yielded unexpected results in medical technology. It is not inconceivable that, as we learn more about the magnetic properties of living organisms and their interaction with the EMF, ways of treating malfunctions of bodily processes through manipulating these magnetic fields will evolve.

Paul Madden

FURTHER READING

Barnothy, Madeleine F., ed. *Biological Effects of Magnetic Fields.* 2 vols. New York: Plenum Press, 1964, 1969. The articles in this older but still valuable work cover the entire spectrum of research into the effects of the EMF on living organisms. Many of the articles use very technical language, but the average reader will nevertheless be able to gain insights into the scope of research being done and the possibilities presented by further investigation into the field of biophysics.

Blakemore, R. P., and R. B. Frankel. "Magnetic Navigation in Bacteria." *Scientific American* 245 (June 1981): 42-49. This article presents incontrovertible evidence that some forms of bacteria are able to use the EMF for purposes of navigation. The authors also make a compelling case that magnetite ingested by the bacteria is the agent that allows them to use this unique form of navigation. The article is written in such a way as to be intelligible to readers without advanced degrees in either physics or biology.

Dubrov, A. P. *The Geomagnetic Field and Life*. New York: Plenum Press, 1978. Dubrov surveys the entire spectrum of the effects of the EMF on living organisms, both proven and possible. Written for a general rather than a professional reading audience, the book is accessible to anyone with a moderate background in science.

Fenwick, Peter. "The Inverse Problem: A Medical Perspective." *Physics in Medicine and Biology* 32 (April 1987): 5-10. Valuable to anyone wishing an understanding of the new science of biomagnetism. The author explains how new techniques for measuring the magnetic fields produced by living organisms aid in solving perplexing problems in medical diagnoses.

Gulrajani, Ramesh M. *Bioelectricity and Biomagnetism*. New York: Wiley, 1998. A thorough look at the effects of biomagnetism on the earth and its lifeforms. Suitable for people with little scientific background, the book is clearly written and filled with helpful illustrations.

Hamblin, William K., and Eric H. Christiansen. *Earth's Dynamic Systems*. 10th ed. Upper Saddle River, N.J.: Prentice Hall, 2003. This geology textbook offers an integrated view of the earth's interior not common in books of this type. The text is well organized into four easily accessible parts. The illustrations, diagrams, and charts are superb. Includes a glossary and laboratory guide. Suitable for high school readers.

Ioannides, A. A. "Trends in Computational Tools for Biomagnetism: From Procedural Codes to Intelligent Scientific Models." *Physics in Medicine and Biology* 32 (January 1987): 77-84. Ioannides' article is conceptually relevant, in spite of the outdated technology. This article provides insight into the past problems in the seemingly unlimited applications of biomagnetic technology in medicine. The nonspecialist should read the article with a dictionary close at hand.

Jungreis, Susan A. "Biomagnetism: An Orientation Mechanism in Migrating Insects?" *Florida Entomologist* 70 (1987): 277-283. This article represents the start of a long list of studies conducted on the subject of EMF for insect navigation. Jungreis makes a convincing case that a number of insect species use the EMF as a navigational tool. The author also tested a number of migratory insects for significant levels of magnetic particles in their bodies that might help explain the mechanisms involved. Only one of five species tested displayed evidence of such particles. Readers with a moderate background in science will be able to follow this article with little difficulty.

Kholodov, E. A. *Magnetic Fields of Biological Objects*. Translated by A. N. Taruts. Moscow: Nauka, 1990. Although slightly technical, this book does provide great insight into the relationships among biological organisms, the earth, and magnetic fields.

Malmivuo, Jaakko. *Bioelectromagnetism: Principles and Applications of Bioelectric and Biomagnetic Fields*. New York: Oxford University Press, 1995. Malmivuo does a fine job describing the basic principles of bioelectromagnetism in terms that a person with little to no scientific background can grasp. Numerous charts and graphs help illustrate important points.

Markl, Hubert. "Geobiophysics: The Effect of Ambient Pressure, Gravity and of the Geomagnetic Field on Organisms." Translated by B. P. Winnewisser in *Biophysics*, edited by Walte Hoppe et al. New York: Springer-Verlag, 1983. Markl's article is written for the reader with substantial background in the sciences; nevertheless, it is worth the effort necessary to understand the author's arguments, because he addresses the problem involved in geobiomagnetism from a number of perspectives and offers insights not available elsewhere.

Plummer, Charles C., and Diane Carlson. *Physical Geology*. 12th ed. Boston: McGraw-Hill, 2007. A college-level introductory geology textbook that is clearly written and wonderfully illustrated. An excellent sourcebook of basic information on geologic terminology and fundamentals of geologic processes. An excellent glossary.

Reite, M., and J. Zimmerman. "Magnetic Phenomena of the Central Nervous System." *Annual Review of Biophysics and Bioengineering* 7 (1978): 167-188. This article suggests that the understanding of the functions of the central nervous systems of humans will be greatly enhanced as the study of biomagnetism proceeds. It gives one example of the vistas opened up by geobiophysical research.

Reppert, Steven M., Robert J. Gegear, and Christine Merlin. "Navigational Mechanisms of Migrating Monarch Butterflies." *Trends in Neurosciences*, 33 (2010): 399-406. A recent article addressing the Monarch butterfly's use of a magnetic compass in their yearly migrations. It suggests Monarchs provide the opportunity to research possible molecular and neural factors involved in the long-distance travel.

Street, Philip. *Animal Migration and Navigation*. New York: Charles Scribner's Sons, 1976. Street's book contains only one short section on geobiomagnetism, but there are strong suggestions in the sections examining animal navigation and migration that the navigational abilities of many species of life may be explained by their interaction in some manner with the EMF. Written for a general readership.

Walker, Michael M., et al. "Structure and Function of the Vertebrate Magnetic Sense." *Nature* 390 (November 1997): 371-376. This article describes the function of magnetic crystals within the various sensing organs of vertebrates.

Wiltschko, Wolfgang, and Roswitha Wiltschko. "Magnetic Orientation and Magnetoreception in Birds and Other Animals." *Journal of Comparative Physiology.* 191 (2008): 675-693. This article provides current research on magnetic reception in animals with a comprehensive overview of the subject and species known to possess the trait. Written in a manner that is easy to follow without a science background, yet detailed and technical enough to provide useful information.

See also: Biogeochemistry; Earth-Moon Interactions; Earth's Magnetic Field; Environmental Chemistry; Magnetic Reversals; Magnetic Stratigraphy; Polar Wander; Remote-Sensing Satellites; Rock Magnetism.

GEOCHEMICAL CYCLE

The geochemical cycle describes the movement of and changes in elements through the earth's atmosphere, surface waters, biosphere, crustal sediments and rocks, and upper mantle. Human activities have altered the earth-surface part of the geochemical cycles of many elements, such as carbon, sulfur, nitrogen, and phosphorus.

PRINCIPAL TERMS

- **biosphere:** all living organisms, including plants and animals
- **carbonate rocks:** sedimentary rocks composed mainly of carbonate minerals whose structure includes a carbon atom linked to three oxygen atoms
- **crust:** the outermost layer of the earth; the continental crust is between 30 and 40 kilometers thick, and made of dominantly silicon-rich igneous rocks, metamorphic rocks, and sedimentary rocks, while the oceanic crust is only 5 kilometers thick, and made of magnesium and iron-rich rocks such as basalt
- **flux:** the rate of transfer of an element from one reservoir to another
- **hydrosphere:** the waters of the earth, including rivers, lakes, and oceans
- **lithosphere:** the outer, rigid part of the earth, consisting of crustal rocks and the upper mantle
- **plate tectonics:** very slow movement of sections of the earth's lithosphere, called plates, away from one another in some areas and toward one another in other areas
- **reservoir:** a place on or in the earth where an element remains for a period of time
- **sedimentary rock:** a rock formed by the consolidation of loose sediments deposited at the earth's surface by water, air, or ice or precipitated from solution
- **silicate rocks:** rocks containing silicate minerals, whose structure contains silicon linked to four oxygen atoms
- **upper mantle:** the fairly rigid part of the earth's interior below the crust of the earth down to about 700 kilometers, composed of magnesium- and iron-rich rock

EARTH-SURFACE CYCLE

Geochemical cycles refer to the transfer of various chemical elements between different reservoirs on the earth. The largest reservoirs include the atmosphere, hydrosphere, biosphere, and lithosphere (crustal rocks and sediments and the upper mantle). The basic geochemical cycle, which is often referred to as the rock cycle, has a time scale of hundreds of millions to billions of years. Subcycles within the rock cycle involve only certain of these reservoirs or small parts of them and occur on a shorter time scale. The atmospheric-hydrologic-biological-sedimentary cycle which operates on the earth's surface will be referred to as the earth-surface cycle and has a time scale of 10 million to 100 million years. Within the earth-surface cycle are many smaller cycles, such as the oceanic cycle, which takes hundreds of years, or the biological cycle, which varies seasonally. The rock cycle is continuous, including an earth-surface cycle and a deeper, subsurface cycle. The earth-surface part of the cycle is driven by the atmospheric-hydrologic cycle, and elements are transported by water or as atmospheric gases. The biosphere, or plants and animals living on the land and in the oceans, is strongly involved in the chemical changes which occur on the earth's surface, and it is also greatly affected by them. Geochemical cycles which involve strong biogenic interaction are often referred to as biogeochemical cycles.

Rain falling on the earth surface erodes rock and soil and carries fragments of them to rivers. The rain, which contains atmospheric gases and dissolved chemicals, also chemically reacts with rocks and minerals in the soil in a process known as weathering. As a result, some elements become dissolved in water and new minerals are formed. Plants and bacteria living in the soil are also involved in weathering, and gases are exchanged with the atmosphere. Certain elements are taken up from the atmosphere and soil solutions and stored within the biota (the plants and animals of a region), and others are cycled by the biota and released to waters and the atmosphere. Rivers transport dissolved chemicals resulting from weathering and biological activity, as well as suspended solid particles of rocks, soils, and organic matter. Some of this material may be deposited on the land surface, but most of it is ultimately carried by rivers to the ocean.

The Geochemical Cycle

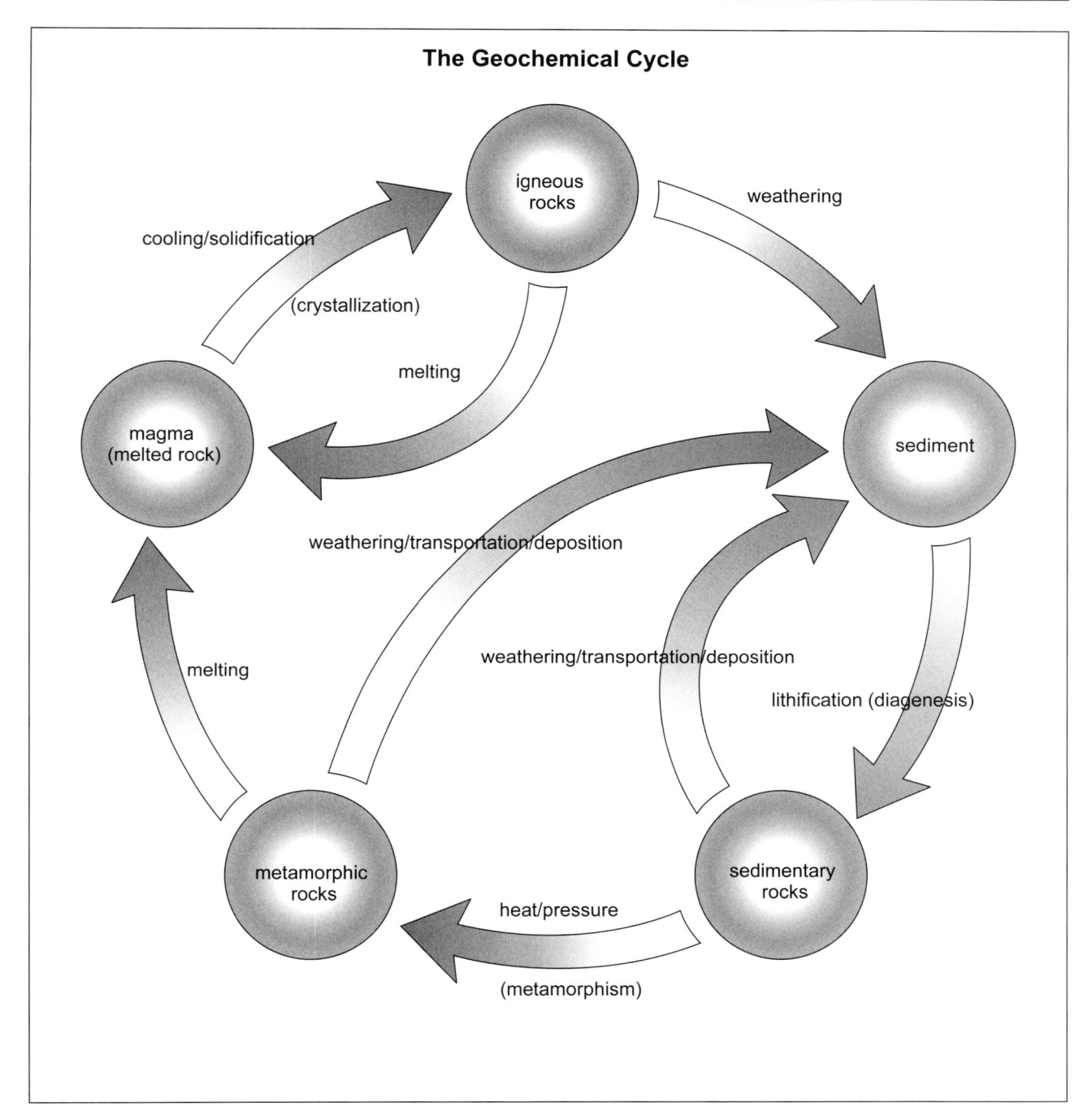

A diagram shows the geochemical cycle with the following components and transitions:
- **igneous rocks** → **sediment** (weathering)
- **sediment** → **sedimentary rocks** (lithification (diagenesis))
- **sedimentary rocks** → **metamorphic rocks** (heat/pressure, metamorphism)
- **metamorphic rocks** → **magma** (melting)
- **magma** → **igneous rocks** (cooling/solidification, crystallization)
- **igneous rocks** → **magma** (melting)
- **metamorphic rocks** → **sediment** (weathering/transportation/deposition)
- **igneous rocks** → **sediment** (weathering/transportation/deposition)

INTERNAL OCEANIC CYCLE

Upon reaching the ocean, inorganic and organic particles which have been suspended in water are deposited on the ocean floor. These particles make up most of the sediment on the bottom. Dissolved chemicals in the oceans along with atmospheric gases may be taken up by organisms living in the ocean surface waters, where there is light, to make cellular matter and shells. The shells accumulate on the sea floor and eventually form deposits of calcium carbonate or limestone. When the surface organisms die, most

227

of the cellular organic matter is broken down and rapidly recycled. A small part of the organic matter falls into the deeper oceans, however, which have been isolated from the ocean surface for thousands of years. Here, organic matter may be broken down into chemicals and gases and slowly recycled to the surface. Thus, there is an internal oceanic cycle between the bottom and surface ocean waters.

Part of the organic matter which reaches the bottom survives and is buried in oceanic sediments, with a small fraction of it ultimately transformed to petroleum upon burial. Some dissolved chemicals, such as sodium chloride salt, can precipitate directly out from shallow ocean water to form sediments. Water is evaporated and other gases are released from the ocean surface into the atmosphere, where they are transported back to land and rained out again. Thus, the earth-surface cycle involves rain, erosion, weathering, river transport, oceanic sedimentation, evaporation, and atmospheric transport of water and gases.

SUBSURFACE PART OF CYCLE

The subsurface part of the rock cycle begins when oceanic sediments are buried, become compressed, and lose water to form hard sedimentary rocks. These sedimentary rocks may then be uplifted to form part of the land, where they are eroded and begin the earth-surface cycle again. Other sedimentary rocks become involved in plate tectonics at the edges of continents, are buried to greater depths, and are subjected to heat and pressure to form metamorphic rocks. If the rocks are heated enough, they melt and are mixed with deeper rocks to form molten rock or magma, which then moves toward the earth's surface. Some magma erupts on the land or sea floor to form volcanic rocks. Other magma cools beneath the surface to form crystalline igneous rocks such as granite. Gases escape from the magma and are released to the atmosphere. Deeply buried igneous and metamorphic rocks may be uplifted in the rock cycle into mountain ranges, where they become part of the earth-surface cycle and are eroded. Thus, over the history of the earth, the chemical elements which make up rocks have been through many cycles of erosion and weathering, river transport, oceanic deposition, burial, metamorphism, melting, and uplift.

The rock cycle has apparently been operating for most of the age of the earth. Over a long period of time, the rock cycle is roughly in steady state; in other words, the total amount of chemicals within the system remains constant although the locations of these elements vary. Sedimentary rocks have been found in Greenland that are 3.8 billion years old and were formed by the same sort of geochemical cycling which occurs today. The major changes between that time and the present are the evolution of life from a very primitive form, the gradual cooling of the earth, an increase in the amount of oxygen in the atmosphere, and the growth in the size of the continents. These changes are unidirectional, and the rock cycle is essentially superimposed upon them.

CARBON CYCLE

Cycles of individual elements are included within the rock cycle—for example, the carbon cycle or the sulfur cycle. The importance and size of any reservoir, the size of the fluxes between reservoirs, and the amount of chemical change that occurs in a reservoir vary from element to element. The carbon cycle is a good example of an element cycle.

Considering the earth-surface carbon cycle first, the atmosphere contains carbon in the form of carbon dioxide gas. Although the amount of carbon contained in the atmosphere is not very great, atmospheric carbon dioxide is an important reservoir because it provides a means of exchanging carbon between the other reservoirs, affects climate via the "greenhouse effect," and is part of the air that we breathe. The forests and terrestrial plants and animals also make up a carbon reservoir. Plants use carbon dioxide from the atmosphere to produce organic matter through the process of photosynthesis. When organic matter is broken down in a yearly biological cycle, carbon dioxide is released back to the atmosphere. Deforestation, the cutting and burning of tropical forests by humans, releases excess stored plant carbon to the atmosphere as carbon dioxide. Another large terrestrial carbon reservoir is carbonate rocks, such as limestone, and shales, or rocks that contain organic carbon. When carbonate rocks and other rocks containing silicate minerals, such as feldspar, are weathered, atmospheric carbon dioxide is taken up, and dissolved carbon in the form of bicarbonate is released to rivers. Rivers transport dissolved bicarbonate and organic carbon, from plant and animal life which has not been decomposed, to the oceans. Terrestrial carbon also occurs in fossil fuels, such as coal, oil, and gas. When humans burn fossil fuel for energy, carbon dioxide is released to the atmosphere and the cycling of carbon is speeded up.

The oceans are another important earth-surface carbon reservoir. Carbon occurs as dissolved bicarbonate ion and as dissolved carbon dioxide gas. The upper part of the ocean rapidly exchanges carbon dioxide with the atmosphere. Small, floating marine plants living in the surface oceans produce their organic matter by taking up atmospheric carbon dioxide through photosynthesis. When they die, most of their remains are decomposed by bacteria, and carbon dioxide is recycled to the atmosphere. A small part of this organic matter, however, is not destroyed. It falls to the bottom of the ocean to be recycled into bottom waters or buried in oceanic sediments along with organic matter carried into the ocean from rivers. Marine organisms also secrete shells and hard parts, removing bicarbonate from ocean water to form calcium carbonate and releasing part of the carbon to the atmosphere as carbon dioxide. Corals are an example of calcium carbonate secretion. Much of this carbonate material is redissolved at depth, thereby recycling bicarbonate to ocean water again, but part of the calcium carbonate becomes buried in ocean sediments.

The long-term part of the carbon cycle occurs over millions of years, when the organic carbon and calcium carbonate buried in oceanic sediments are converted into rock through burial, compaction, and heating. Oceanic muds containing organic carbon are converted into shales. Sediments composed of calcium carbonate become limestone rock. When limestones and organic carbon are heated, metamorphosed, and melted at depth, carbon dioxide is released. Part of this carbon dioxide escapes to the atmosphere from volcanoes on the land, and some is released at midoceanic ridges. Ultimately, limestone rocks containing calcium carbonate and shales containing organic carbon are uplifted to the land surface to be weathered again, beginning the earth-surface cycle anew.

Long and Short Time Scale Processes

To study geochemical cycles, geochemists combine information from various specialties, such as geology, geochemistry, geophysics, oceanography, biology, hydrology, and atmospheric sciences. They must have a broad knowledge of the long and short time scale geologic processes involved in the cycling of an element and an idea of how these processes interact on the earth's surface with the atmosphere and biosphere. They must know the major reservoirs for various elements, the concentrations of elements within these reservoirs, processes that change the elements within the reservoirs, and the fluxes of elements between various reservoirs. The relative importance of various processes and fluxes depends on the time scale involved. Processes that are important on a short time scale are often unimportant in long-term cycling.

Computer Modeling

Computer modeling is one technique that is used in the study of geochemical cycles, particularly those that involve many reservoirs and fluxes. With the carbon cycle as an example, it can be seen which type of information is needed to model geochemical cycles.

The earth-surface carbon cycle has a time scale of decades to tens of thousands of years. Information about the present size of the atmospheric reservoir of carbon, in the form of carbon dioxide, comes from atmospheric science. Measurements are made of the average yearly concentration of carbon dioxide in the atmosphere and the way in which this concentration is increasing with time. Knowledge of the amount of atmospheric carbon dioxide over the last thousands to tens of thousands of years can be obtained from the analysis of air bubbles trapped in glacial ice, which is sampled from below the surface by ice cores. Estimates of the size of the reservoir of carbon stored in the form of forests and terrestrial biota and the amount of forests that are being destroyed, with their carbon recycled to the atmosphere, come from biologists. The amount of carbon stored as fossil fuel and the amount being burned each year, releasing carbon dioxide to the atmosphere, is estimated by economic geologists, particularly those who study petroleum and gas reserves and coal reserves. The oceans are an important reservoir in the carbon cycle. The amount of carbon dioxide being taken up from the atmosphere by the surface ocean and the rate at which it is being transferred to the deeper oceans represent important fluxes, which come from a knowledge of ocean chemistry provided by oceanographers. All this information is used by geochemists to study the short time scale carbon cycle and to predict increases in the concentration of atmospheric carbon dioxide.

MODELING LONG TIME SCALE CARBON CYCLE

In modeling the long time scale carbon cycle over millions of years, different reservoirs and fluxes become important. Geochemists use information about past conditions on the earth, preserved in sedimentary rocks. To estimate the amount of carbon stored in sedimentary rocks, a geochemist must know the average carbon content of different types of sedimentary rocks, such as carbonates and shales, and also the abundance of various sedimentary rock types as a function of age—both those rocks exposed at the earth's surface and those at depth in the subsurface. This information comes from compilations of vast amounts of data. The average chemical composition and amount of water being carried by world rivers gives a measure of how much weathering is occurring on land. Information about ancient oceans is preserved in the ratio of isotopes, or heavier and lighter forms of different elements, that are found in rocks formed from these oceans. For example, the ratio of the isotopes carbon–13 and carbon–12 in marine carbonates gives an indication of how much organic carbon was being buried in the ocean millions of years ago. Fossil records of past life preserved in the rocks give an idea of past climate conditions. For example, certain types of organisms, such as crocodiles, require a warm climate; therefore, their presence in polar regions in the past suggests higher levels of atmospheric carbon dioxide, which tends to raise earth temperatures especially at high latitudes.

From modeling the carbon cycle using information of the type discussed above and other data, it has been concluded that changes in the concentration of atmospheric carbon dioxide have probably occurred on the time scale of the rock cycle—that is, millions of years. The important fluxes involved are the uptake of atmospheric carbon dioxide in the weathering of silicate rocks and in the formation of organic carbon, the burial of limestone and organic carbon in sediments, and carbon dioxide release from volcanoes and midoceanic ridges.

HUMAN ALTERATION OF GEOCHEMICAL CYCLES

Human activities such as fossil fuel burning, industry, and agriculture have altered the earth-surface part of some of the more important geochemical cycles, particularly the carbon, sulfur, and phosphorus cycles. For example, there have been large changes in the transport rate of sediments in streams, and excess carbon dioxide and sulfur dioxide have been released to the atmosphere by human activity. It is important to know the natural fluxes within these element cycles to determine how important human changes are and what their overall effect may be. In trying to correct human changes in our environment, it is important to understand the connections between different parts of the geochemical cycle. By studying the long-term rock cycle we can see how large changes were in the geologic past and how they were stabilized within the cycle. This gives us an idea of what the range of fluxes has been over the earth's history and how the earth reacts to change.

The short-term geochemical cycling of carbon dioxide gas within the earth-surface carbon cycle on a time scale of years to hundreds of years has been of particular interest recently because of the observed increase in the concentrations of atmospheric carbon dioxide and concern about the "atmospheric greenhouse effect." Both burning of fossil fuel and deforestation by humans release carbon dioxide to the atmosphere. Some of this excess carbon dioxide is taken up by the oceans and stored there. To predict how fast the concentration of atmospheric carbon dioxide will change over time, it is necessary to know the fluxes of carbon between the various reservoirs. Increased carbon dioxide should lead to increased temperatures on the earth, the so-called atmospheric greenhouse effect. Long-term changes in the concentration of carbon dioxide have probably also occurred and affected past climates on the earth. These changes can be deduced from the long-term carbon cycle.

Humans have also greatly altered the geochemical phosphorus cycle. Phosphorus is an important nutrient, a food source for plants and animals. The release of phosphorus by humans in sewage and from agriculture and industry has increased plant growth in lakes and in estuaries. When these plants die and decay, oxygen in the water is often completely used up. Both the increased plant growth and the oxygen depletion make lakes and estuaries less usable by humans.

ACID RAIN

Another topic of current interest which involves geochemical cycles is acid rain. Acid rain speeds up the weathering part of the geochemical cycle, releasing more and different ions into solution. It also changes the chemistry of and alters the geochemical cycle in lakes. The primary gases which cause acid rain are sulfur dioxide, which forms sulfuric acid, and nitrogen dioxide, which produces nitric acid. Thus, a knowledge of the atmospheric-earth surface geochemical sulfur cycle and the nitrogen cycle are important in understanding acid rain and its effects. Sulfur dioxide gas is produced primarily by fossil fuel burning. From studying the sulfur cycle, it is known that the amount of sulfur released into the atmosphere from fossil fuels is about equal to that formed by natural processes. Thus, humans are greatly altering the sulfur cycle. Nitrogen dioxide is produced mainly by fossil fuel burning in cars and in power plants. Another important source is forest burning, particularly to clear land in the tropics. Human sources of nitrogen dioxide add up to nearly three-quarters of the total amount released into the atmosphere. Clearly, that indicates an alteration of the geochemical nitrogen cycle.

Elizabeth K. Berner

FURTHER READING

Albarede, Francis. *Geochemistry: An Introduction.* 2d ed. Boston: Cambridge University Press, 2009. A good introduction for students looking to gain some knowledge in geochemistry. Covers basic topics in physics and chemistry, isotopes, fractionation, geochemical cycles, and the geochemistry of select elements.

Berner, Elizabeth K., and Robert A. Berner. *Global Environment: Water, Air, and Geochemical Cycles.* Upper Saddle River, N.J.: Prentice Hall, 1996. This book discusses the processes that sustain life and affect change on the earth. Topics include the hydrologic cycle; the atmosphere; atmospheric carbon dioxide and the greenhouse effect; acid rain; the carbon, sulfur, nitrogen, and phosphorus cycles; weathering; lakes; rivers; and the oceans. It is understandable to the college-level reader.

Berner, Robert A., and Antonio C. Lasaga. "Modeling the Geochemical Carbon Cycle." *Scientific American* 260 (March 1989): 74-81. This article discusses the modeling of the carbon cycle over millions of years and should be understandable to the general reader. Changes that may have occurred in atmospheric carbon dioxide are discussed, along with a discussion of the factors important in regulating atmospheric carbon dioxide over geologic time. The illustrations, particularly of the carbon cycle, are very helpful.

Davidson, Jon P., Walter E. Reed, and Paul M. Davis. *Exploring Earth: An Introduction to Physical Geology.* 2d ed. Upper Saddle River, N.J.: Prentice Hall, 2001. An excellent introduction to physical geology, this book explains the composition of the earth, its history, and its state of constant change. Intended for high-school-level readers, it is filled with colorful illustrations and maps.

Garrels, Robert M., and Fred T. Mackenzie. *Evolution of Sedimentary Rocks.* New York: W. W. Norton, 1980. This book discusses the geochemical cycle of sedimentary rocks over a long time scale. It is a college-level textbook and though the data are dated, it has detailed information about many parts of the rock cycle, and summarizes the geochemical cycling of sedimentary rocks.

Gregor, C. Bryan, et al., eds. *Chemical Cycles in the Evolution of the Earth.* New York: John Wiley & Sons, 1988. The prologue gives a summary for the college-level reader of the historical development of the concepts of geochemical cycling. The rest of the book covers other aspects of geochemical cycling in more detail than would interest the nonspecialist.

Lerman, A. "Geochemical Cycles." In *The Oxford Companion to the Earth*, edited by Paul Hancock and Brian J. Skinner. New York: Oxford University Press, 2000. The article provides an excellent overview of geochemical cycles, focusing on the water cycle, sodium cycle, and carbon cycle. Accessible to the lay reader.

Nicholson, Keith. *Geothermal Fluids: Chemistry and Exploration Techniques.* Berlin: Springer-Verlag, 1993. Nicholson provides the reader with an examination of the behavior of geothermal fluids, as well as study and exploration practices. Slightly technical, this book is intended for the reader with some background in chemistry.

Siever, Raymond. "The Dynamic Earth." *Scientific American* 249 (September 1983): 46-55. This article discusses for the general reader the geochemical cycle in the context of the history of the earth. Topics covered include geologic time, the rock cycle, the earth-surface cycle, the carbon cycle, and atmospheric carbon dioxide over geologic time. There is also a discussion of evolutionary changes in the earth over time. Excellent illustrations.

Tarbuck, Edward J., Frederick K. Lutgens, and Dennis Tasa. *Earth: An Introduction to Physical Geology*. 10th ed. Upper Saddle River, N.J.: Prentice Hall, 2010. This college text provides a clear picture of the earth's systems and processes that is suitable for the high school or college reader. It has excellent illustrations and graphics. Bibliography and index.

Trabalka, J. R., ed. *Atmospheric Carbon Dioxide and the Global Carbon Cycle*. Honolulu: University Press of the Pacific, 2005. This reference provides a summary of the global carbon cycle and how it is related to changes in atmospheric carbon dioxide. For the college-level reader.

Woodwell, George M. "The Carbon Dioxide Question." *Scientific American* 238 (January 1978): 34-43. This article discusses deforestation as a source of atmospheric carbon dioxide and shows how this flux is estimated. It also summarizes the present global carbon cycle and how excess carbon dioxide produced by humans can be taken up. For the general reader. The illustrations are excellent.

See also: Carbon Sequestration; Climate Change: Causes; Earth's Mantle; Elemental Distribution; Environmental Chemistry; Fluid Inclusions; Freshwater Chemistry; Geothermometry and Geobarometry; Milankovitch Hypothesis; Nucleosynthesis; Oxygen, Hydrogen, and Carbon Ratios; Phase Changes; Phase Equilibria; Water-Rock Interactions.

GEODETIC REMOTE SENSING SATELLITES

Geodetic remote sensing (RS) satellites are orbital spacecraft equipped with a wide range of sensory technology. These systems conduct detailed surveys of the earth's topography, taking into account changes in land mass positioning and elevation, magnetic fields, and seismic and volcanic activity. The use of RS satellite technology in geodetic studies helps scientists generate invaluable data on variations in the earth's geophysical systems, including postglacial rebound, plate tectonics, and ocean circulation. Using the data collected from such technologies, geodesists can create working models of the earth's geophysical characteristics.

PRINCIPAL TERMS

- **geodesy:** the study of the earth's shape, topography, and physical features and forces
- **global positioning system (GPS):** the network established by satellites to navigate and map the earth
- **lithosphere:** layer of large plates believed to be floating on molten rock beneath the earth's outer crust
- **mantle:** superheated layer of molten rock located between the earth's core and the outer crust
- **plate tectonics:** a theory that states that beneath Earth's outer crust there is a series of plates in constant motion and through which magma flows
- **post-glacial rebound:** the process by which the earth's crust slowly returns to its original position after a glacier dissipates or moves away from the subduction zone
- **subduction:** geodynamic process whereby an extremely heavy object located on the outer crust pushes down on the crust and the tectonic plates beneath it

BASIC PRINCIPLES AND HISTORY

In the study of Earth's shape, topography, and physical features and forces (the field known as geodesy), it is often critical to obtain a comprehensive profile before study can begin. Geodesists have increasingly looked to satellite technology to assist them in analyzing a wide range of geodynamic processes and geological events (such as volcanism and earthquakes). The remote sensing (RS) technologies aboard these spacecraft utilize many different types of survey approaches. Among the RS technologies used are laser and radar altimeters, global positioning systems (GPSs), synthetic aperture radar, and satellite-to-satellite tracking systems.

The geodetic RS satellite is rapidly becoming an invaluable tool in the study of geophysics and geodynamics. For example, RS satellites can be used to study fault zones, the movement of Earth's subterranean plates (as identified by the theory of plate tectonics), and postglacial rebound (the outer crust's ability to return to its original position after being weighed down by glaciers). RS satellites use infrared, thermographic, and three-dimensional imaging systems to give detailed surveys that are virtually impossible to conduct from the ground. These systems can focus on even the tiniest changes in the earth's surface, changes that would otherwise go unnoticed by the naked eye.

BACKGROUND AND HISTORY

The size, shape, and profile of the earth have for millennia been points of interest and debate among scientists. The fourth century B.C.E philosopher Aristotle was the first to attempt to calculate the circumference of Earth. About one century later, Greek astronomer and mathematician Eratosthenes employed a triangulation method, using the sun's rays as a point of reference, to determine the planet's circumference. Eratosthenes' triangulation approach would be used through the centuries that followed, as scientists attempted to better understand not only the earth's size but also its shape.

In the sixteenth century, the Royal Academy of Sciences in Paris looked to end the debate over Earth's shape by sending expeditions to the border between Sweden and Finland and to what is now Ecuador in South America. Once on site, the teams studied the north-south curvature of the planet, concluding that the earth was predominantly spherical.

By the early nineteenth century, the United States and other countries began to establish scientific networks to map their respective coastlines and borders. These networks, however, were often disparate, their means for calculating distances unreliable and imprecise. With the advent of the satellite, however, such calculations were made with exceptional clarity and precision.

In the 1960's, the U.S. Army's Ballistic Research Laboratory used satellite positioning to triangulate an area spanning Maryland to Minnesota to Mississippi. Building on that test, researchers were able to connect each mapping network and generate a precise profile of the United States. Satellites are far more advanced now, and they can carry a wide array of technologies. These technologies can provide comprehensive, multidimensional profiles of the entire Earth and monitor changes and events that occur in specific locations.

GEODESY: ACTIVE AND PASSIVE SENSOR APPLICATIONS

Two general types of remote sensors are used in this arena. The first type of sensor is passive: It simply detects emissions from the target source. An example of this type of sensor is the radiometer, which detects naturally generated microwave energy radiating from within the earth. The second sensor type is active: It sends a signal to the target area and measures the energy based on the return signal. An example of an active remote sensor is radar, which sends a radio signal to the target area and captures the echo.

There is a wide range of applications for remote sensing technologies to the fields of geology, geophysics, and geodesy. One area involves the study of long-term changes to the earth's outer crust and surface (a process known as deformation). In some cases, researchers use onboard, gravitational field sensors to determine the degree of deformation in a given area over a specific time period.

For example, in 2010, scientists utilized the remote sensors aboard the Gravity Recovery and Climate Experiment (GRACE) satellites to study the thickness of the lithosphere (the layer of rock plates continually moving beneath the earth's outer crust). GRACE compiled more than seven years of data, monitoring fluctuations in the earth's gravitational field as it passed over one geographic location (the region comprising Scandinavia and parts of northern Europe and Eastern Europe known as Fennoscandia). Using these data, researchers created one- and three-dimensional models of the lithosphere and determined the approximate viscosity of the layer of molten rock located between the core and the lithosphere, called the mantle.

Researchers from the University of Chicago and the University of Wisconsin install weather and global positioning system (GPS) instruments on iceberg B-15A. It is the first time an iceberg has been monitored in this manner, and the data will allow an unprecedented understanding of how giant bergs make their way through the waters of Antarctica and beyond. (Science Source)

RS satellites are also useful for tracking crustal movement. According to the theory of plate tectonics, the massive plates that exist beneath the lithosphere are in constant motion, driven both by the heat radiated from the mantle and core and by the downward force of gravity, in tandem with the weight of oceans and glaciers. Under the weight of glaciers, for example, the lithosphere is pushed down (a process called subduction). When a glacier dissipates, the lithosphere slowly pushes back out, causing deformation in the outer crust as the lithosphere returns to its previous position. This is called postglacial rebound. Scientists use evidence of postglacial rebound to analyze the movement of the earth's crust and to understand the geological history of a given area.

In this arena, geodetic RS satellites have proved highly useful. Researchers have used GPS, coupled with altimeters (which use lasers and radar to calculate the elevation of a surface object) and space-based imaging systems, to calculate the speed at which the crust beneath the Antarctic ice sheet is moving.

In some cases, instead of causing the formation of mountains and ridges, crustal deformation causes rock to pull horizontally along the surface. Using GPS, scientists in the Baltic region calculated the rate of horizontal strain occurring along the crust in three separate areas. However, in each area horizontal deformation occurred at different rates. This information led scientists to conclude that horizontal strain in this region is being caused not by one but by several geodynamic processes.

Sometimes, the geodynamic changes occurring on the Earth's surface are accompanied by the emission of energy, such as energy in seismic waves or the release of heat. While it is possible to detect some form of energy emissions from the ground, geodetic remote sensing satellites can capture the sources and volumes with greater precision. For example, in the late 1990's, scientists used the infrared and other thermal imaging systems aboard the National Aeronautics and Space Administration's Moderate-Resolution Imaging Spectroradiometer satellites to classify the different types of land cover on the earth's surface (types including snow, ice forest, and rock). This study helps geodesists analyze the different forms of energy emissions from beneath the outer crust and how they travel through various types of surface cover.

Radar

Like their ground- and air-based counterparts, satellite radar RS systems are active sensors, sending their targets radio waves that in turn bounce back in the form of an echo. Radar has undergone a significant evolution recently, particularly with regard to its ability to provide a clear and precise image of its target. This characteristic is particularly important for geodetic RS satellite applications, because satellites must be both effective from a great distance and able to operate in all types of weather and at any time of day.

One of the most useful radar RS systems in geodesy is the Synthetic Aperture Radar (SAR). This type of radar emits microwaves at the target, illuminating the target in any weather condition and at any time of day. SAR also is beneficial because its signal coherently returns to the system uncorrupted by the sun's rays. Furthermore, SAR can retain all of the information it obtains during a flyover. With repeated flyovers, the satellite can therefore conduct a phase comparison. Such a comparison is useful in tracking even minute changes in the position of the earth's surface.

Thermal/Infrared Imaging

Geodesy is often concerned with the differences in radiation emitted from a target. For example, variations in heat along an ice pack indicate different levels of thickness in the ice pack or activity under the ice. In this regard, many satellites used for geodetic surveys will include thermal and infrared sensors alongside other technologies. Thermal imaging also helps scientists understand fluctuations in temperature in the oceans—fluctuations potentially stemming from seismic or volcanic activity or other types of geodynamic phenomena.

Since the 1970's, for example, scientists have used passive thermal and infrared sensors to map the ice pack in the Antarctic. Their studies have also enabled them to effectively map the various degrees of sea-surface temperatures. In addition, scientists use infrared and thermal imaging sensors to record and study temperature variations in the lithosphere, shedding light on changes in plate tectonics in a given area.

Global Positioning System

GPS has become one the most integral components of geodesy. GPS helps scientists pinpoint changes in the earth's crust down to a few millimeters. This level of accuracy has proved highly useful in monitoring crustal movements and seismic activity. For example, Turkey (a country with a long history of severe earthquakes) has a GPS network of nearly two thousand stations throughout greater Istanbul along a major fault line. This intricate network has helped scientists study with greater effectiveness the activity along that fault zone, calculating any horizontal stress that occurs during a seismic event.

The use of the GPS network in geodetic study is not limited to land-oriented applications. Because of its accuracy, GPS also is used by scientists to study the ocean floor. Such studies are difficult because they often use other forms of sensors, mostly those used for determining ocean depth. However, GPS has been known to provide a greater degree of accuracy. Scientists have combined the use of GPS technologies with acoustic positioning systems to create an effective survey system for mapping the ocean floor and monitoring noteworthy changes.

RELEVANT GROUPS AND ORGANIZATIONS

Because of the implications of using satellite-based RS systems to create detailed geodetic surveys, a number of groups and organizations are involved.

Government agencies frequently play an important role in organizing and funding geodetic surveys. The U.S. Geological Survey (USGS), for example, operates two RS satellites for geodetic surveys. The *Landsat 5* and *Landsat 7* crafts have helped the USGS develop one of the most extensive remote sensor data archives in the world. Also, in 2011 the National Oceanic and Atmospheric Administration (NOAA) launched its update of the extensive surveys and maps in its vast archives. Using GPS and other passive and active RS equipment based on satellites, the NOAA project updates data that were first compiled in the early 1980's.

With the ever-increasing availability of GPS technology, a large number of university geology, geography, and earth science departments are using geodetic RS satellite technologies. Based at such institutions, scientists are using the data collected by these satellite systems to study volcanic and seismic activity, climate change (such as rising water levels and shrinking ice packs), and plate tectonics.

Geodetic surveys using RS satellite technologies are not limited to academic or government arenas. Data recorded by satellite-borne RS systems are proving highly beneficial for energy companies in search of new oil, gas, and mineral deposits. These companies often use two- and three-dimensional imaging systems, GPS, and other RS systems aboard satellites like *Landsat 7* and *Spot-5* to locate new sites for exploratory drilling. These companies also use such technologies for risk assessment and management purposes, calling upon them to locate and mitigate oil spills.

IMPLICATIONS AND FUTURE PROSPECTS

Geodesy has seen tremendous evolution, thanks largely to the introduction of GPS and other RS technologies on low- and high-altitude satellites. Whereas approaches to geodesy were until the twentieth century somewhat limited, the advent of the satellite and its many uses have enabled cartographers and geodesists to fill in most of the gaps left by ground-based technologies and approaches.

RS satellite systems are far more precise and powerful than those developed even in the mid- to late twentieth century. According to the U.S. Naval Observatory, thirty-two operating GPS satellites along with hundreds of other orbiting spacecraft are now in orbit. Many of the onboard sensors are capable of penetrating thick cloud cover and hundreds of feet of seawater from high orbit. Hurdles for scientists and engineers still need to be overcome. For example, some GPS satellites are negatively influenced by the ionosphere (part of the earth's upper atmosphere), especially during magnetic storms and other disturbances. Scientists continue to seek ways around these hurdles so that geodetic RS satellites can continue to evolve.

GRACE and other satellite programs implemented during the first decade of the twenty-first century have provided a great deal of geodetic data in a relatively short time, inspiring scientists to build even more state-of-the-art RS satellite systems to examine more of the earth's unexplored areas, such as its lithosphere and other parts of the planet's interior. Indeed, the evolution of RS satellite systems and their application to the exploration of the earth is expected to continue into the long term.

Michael P. Auerbach

FURTHER READING

Leick, Alfred. *GPS Satellite Surveying*. 3rd ed. Hoboken, N.J.: Wiley, 2004. Presents an overview of the global positioning system and its uses. Discusses the systems utilized by this network and its applications to the fields of in geodetic and geographical surveying.

Lillesand, Thomas, Ralph W. Kiefer, and Jonathan Chapman. *Remote Sensing and Image Interpretation*. 6th ed. Hoboken, N.J.: Wiley, 2008. Provides a review of remote sensing and its applications. Outlines the different types of remote sensors and helps students understand how to use RS systems in their own respective fields of study.

Martin, Angel, et al. "Compact Integration of a GSM-19 Sensor with High-Precision Positioning Using VRS GNSS Technology." *Sensors* 9, no. 4 (2009): 2944-2950. Focuses on magnetic fields and the use of RS satellite technology to capture them. Discusses the positioning of these sensors (including GPS technologies) to record magnetic anomalies and to catalog the different magnetic fields that are radiated from the earth's subsurface regions.

Maus, S., et al. "Earth's Lithospheric Magnetic Field Determined by Spherical Harmonic Degree 90 from CHAMP Satellite Measurements." *Geophysical Journal International* 164, no. 2 (2006): 319-330. Examines the analysis of the lithosphere and the earth's magnetic field. The article focuses on information collected by the CHAMP satellite, which uses RS equipment to study the shape and dynamics of the lithosphere's magnetic emissions.

Peltier, W. R. "Global Glacial Isostatic Adjustment: Paleogeodetic and Space-Geodetic Tests of the ICE-4G." *Journal of Quaternary Science* 17, nos. 5/6 (2002): 491-510. Discusses the significance of postglacial rebound and the use of geodetic RS satellite technology to explore this field. The authors use the data collected to create models to understand postglacial rebound.

Seeber, Gahnter. *Satellite Geodesy*. Rev. ed. Berlin: Walter de Gruyter, 2008. An extensive review of the different applications of satellite geodesy. Discusses the different orbits of satellites, timing, and other aspects of satellite technology, providing an overview of the many satellites used in geodetics.

See also: Geodynamics; Geologic and Topographic Maps; Remote-Sensing Satellites.

GEODYNAMICS

Geodynamics focuses on the processes that cause the formation of and changes to Earth's crust and mantle (the superheated region between the planet's core and crust). Geodynamicists study Earth's plates and the processes by which they move atop the mantle, causing seismic and volcanic activity. Geodynamicists also study crustal deformation, subduction, and postglacial rebound. These processes and phenomena are studied by analyzing rock samples and observing environmental changes on Earth's surface from orbiting satellites.

PRINCIPAL TERMS

- **elastic deformation:** the process by which material beneath a heavy object, such as a glacier, is deformed and returned to its original form after the weight is lifted
- **last glacial maximum:** the prehistoric period (approximately 30,000 years ago) in which the glaciers covering Earth were at their thickest
- **lithosphere:** the layer of large plates believed to be floating on molten rock beneath the earth's outer crust
- **magma:** molten rock pushed outward from the earth's core
- **mantle:** the superheated layer of molten rock located between Earth's core and its outer crust
- **plate tectonics:** the theory stating that beneath the earth's outer crust there exists a series of plates in constant motion and through which magma flows
- **postglacial rebound:** process by which the earth's crust slowly returns to its original position after a glacier dissipates or moves away from the subduction zone
- **subduction:** geodynamic process whereby an extremely heavy object located on Earth's outer crust pushes down on the crust and the tectonic plates underneath

Basic Principles

The field of geodynamics entails the study of the deformation of Earth's crust as caused by the planet's internal processes and external elements. Geodynamicists study the movement of the massive plates at an area above the mantle known as the lithosphere (a superheated layer of molten rock between the core and the outer crust).

According to the theory of plate tectonics, these plates are in constant motion, occasionally coming into contact with each other and causing seismic activity. Scientists who focus on geodynamics also study the movement of molten rock (magma) from the mantle outward, through fissures in the tectonic plates and ultimately the outer crust, occasionally through volcanoes. Such activity creates new physical characteristics on the planet's surface, such as mountains and calderas (deep, bowl-shaped depressions that are formed after a volcano collapses into a depleted magma chamber following volcanic events).

In addition to the earth's internal geological processes, geodynamicists study the deformation of the planet's outer crust caused by glaciers and oceans. These massive bodies of water change the shape of the material beneath them and cause subduction (a process whereby the crust and the tectonic plate beneath it are pushed down under the weight of a glacier or ocean). When the weight either dissipates or moves from a subduction zone, the material returns to its original form (a concept known as elastic deformation).

Background and History

The scientific study of the Earth's geological, seismological, volcanic, and geodynamic activity is a relatively young field. For millennia, however, humanity has acknowledged such phenomena, taking into account the often catastrophic results of volcanoes, earthquakes, and floods. Not until the mid-eighteenth century, however, did scientists begin to consider the more gradual deformations that were taking place.

In 1785, for example, Scottish geologist James Hutton first suggested a connection between the major seismic and volcanic events that had previously been recorded and the changes in the earth's surface that occurred gradually. In time, scientists speculated that the earth's continents were moving from one another, although it was not until 1915 that German geophysicist and meteorologist Alfred Wegener formally introduced the theory of continental drift—the notion that in a period of hundreds of millions of years Earth's continents (once all conjoined) have been in motion.

Wegener's theory has since been disproved, but his concept shed light on why fossils of the same species were found thousands of miles away on other continents. His theory also formed the basis of the theory of plate tectonics. In the 1960's, scientists who placed seismographs near nuclear test sites noticed that volcanoes, earthquake zones, and other active geologic surface features seemed to be located along the fault lines between tectonic plates. Furthermore, scientists studying the crust beneath the ocean noticed that the magnetic material contained in the rock indicated historic reversals in the earth's polar magnetic field, providing further proof that the planet's lithosphere is always in motion. Plate tectonics rests at the heart of the field of geodynamics, helping create a framework for understanding the processes that continue to shape Earth's surface.

The Lithosphere and Magma Flow

Geodynamics focuses on Earth as a planet that is always developing. Because of the planet's internal and external forces, processes, and elements, the outer crust is subject to continual deformation. The central internal element of this deformation is the motion of tectonic plates.

As molten rock moves from Earth's interior, it cools and contracts. However, the core's heat gives the rock buoyancy, allowing the plates to float atop the mantle. These plates are always moving and, occasionally, contacting one another. Owing to its low density and gravitational forces, magma will move through these plates at fault zones on its way to and through the crust.

The lithosphere has varying degrees of density. In some areas, it is highly dense and thick, pulling down the outer crust and causing lower land elevations. Examples of such lower elevations include the plains of North America. In less dense areas of the lithosphere, however, the horizontal forces that draw rock outward, coupled with magma moving outward, makes this area of the lithosphere denser than the crust, leading to increased land elevations and mountain ranges. For example, scientists believe that the Andes and Himalayas were formed this way.

An important element in the deformation of Earth's crust is the amount of stress placed on the lithosphere. Factors contributing to lithospheric stress include the pressures caused by the flow of material from the mantle, the regional density of the lithosphere, gravity, and the level of friction placed on the lithosphere by mantle convection (the flow of hotter and less dense molten rock toward the lithosphere). This stress contributes significantly to the topography of the outer crust. The results are particularly evident along the plate boundaries, such as those found in North America and Western Europe, where larger ridges and mountains are formed.

Subduction and Postglacial Rebound

Another important element on which geodynamics focuses is subduction. In this process, the lithosphere beneath the earth's oceans is pushed down, causing plate displacement. The space left by the subducting plate, known as the subduction zone, is filled by outward-flowing magma. When two plates separate beneath the ocean, the ocean floor becomes deformed, creating new mountains and ridges. When ocean plates come into contact with continental plates, as commonly occurs along the boundaries of the Pacific Ocean (the region known as the Ring of Fire), volcanic and seismic activity increases. Magma that flows into these subduction fields commonly contains water and certain volatile minerals, which cause explosive volcanic activity and, ultimately, significant deformation of the outer crust.

Although subduction causes this downward movement of the lithosphere (and the tectonic plates within it), the effects of subduction are not permanent. In fact, when the pressure above is removed, the lithosphere and the outer crust slowly return to their original positions, a phenomenon known as postglacial rebound. Scientists study this aspect of geodynamics by analyzing the layers of rock after the dissipation of ancient glaciers and by studying the deformation caused by modern glaciers. Postglacial rebound explains why the earth's surface is moving north at 0.035 inch per year. Earth's center of mass is located at the North Pole, and before the glaciation associated with the Ice Age, the outer crust was positioned relative to this center. Glaciers pushed the surface down and away from North Pole, but when the glaciers dissipated, the elasticity of the crust caused a gradual return of Earth's center of mass to that region.

Scientists study postglacial rebound by observing folds in the rock that were influenced by glaciers. In the Italian, French, and Swiss Alpine ranges, for example, scientists have unearthed significant evidence of postglacial rebound. Such evidence is available because this region is among several in the world in which the postglacial rebound from the last glacial maximum (a period about 30,000 years ago, in which ancient glaciers were at their thickest) had the most pronounced deforming effects, forming high peaks and other formations. The accentuated evidence of postglacial rebound located in this and other mountainous regions has enabled geologists and geophysicists to formulate theories about the rates at which rebound occurs and the effects of the last glacial maximum on the earth's lithosphere.

COMPUTER MODELING

One of the most effective approaches to studying geodynamics is mathematical and computer modeling. Through the creation of general formulae, scientists can input data from a specific area to study such physical factors as temperature and pressure. In one 2011 study, for example, scientists sought to analyze the chemical and mineralogical composition of the Caribbean plate using an area in eastern Cuba as their point of reference. Using a basic pressure and temperature formula, and a software program to compile data on minerals, scientists were able to create a profile of the area's thermal history and, within this arena, the subduction that occurred there.

Modeling can compile data from samples around the world to create a general framework for understanding geodynamic processes. In the case of one study, information gleaned at fifty-six different subduction zones was compiled to create a more comprehensive illustration of plate geometry, ages, velocities, and other characteristics in general. Such models help scientists understand the temperatures and other forces that influence subduction zones.

SATELLITE TECHNOLOGY

In many cases, evidence of deformation, surface movement, and postglacial rebound is visible not from the ground but far above the earth. In 2009, the European Space Agency launched the Gravity Field and Steady-State Ocean Circulation Explorer (GOCE) to study the planet's gravitational field and the ocean floor's topography. GOCE has been utilized by geodynamicists to study the lithosphere's structure and densities. By recording gravitational data, scientists now are able to study subduction zones, postglacial rebound, and plate deformation more comprehensively.

GOCE is far from the only satellite used in the study of the deformation of the earth. National Aeronautics and Space Administration's (NASA) Gravity Recovery and Climate Experiment (GRACE) satellite, launched in 2002, provides time-lapsed data on postglacial recovery. Scientists also use the worldwide global positioning system to study tectonic motion. Furthermore, NASA and other organizations use satellites to study the planet's geodynamic activity using onboard radar, gradiometers, and other technologies to gather data about tectonic movement and deformation.

OTHER SCIENTIFIC DISCIPLINES

In many cases, geodynamics is greatly aided by the application of other scientific disciplines. Wegener's view of continental drift, for example, was born in part by paleontology, as Wegener (and others) observed that fossils of the same species could be found on opposite sides of the globe. His observations led to the claim that at one point the continents were connected. Wegener's theories led to the widely accepted geodynamic concept of plate tectonics.

Geophysicists frequently look to prehistoric eras for evidence of the earth's geodynamic properties. One example of this interdisciplinary approach is a 2006 study of the Tibetan plateau. Scientists used paleoclimatologic information to study the collision of the two plates in this region of East Asia. Based on an analysis of unearthed deposits (dating back to the Eocene period, about fifty million years ago), geophysicists determined that the plateau had been elevated by more than 4 kilometers (2.5 miles) as long as thirty-five million years ago.

RELEVANT GROUPS AND ORGANIZATIONS

One of the key government agencies involved in the study of the earth's geodynamic phenomena is the U.S. Geological Survey (USGS). USGS scientists conduct analytical studies of geodynamic activity in the United States and its territories around the world. The USGS frequently partners with private and public organizations and scientists on large-scale research projects.

In addition to supporting the USGS, the U.S. government provides funding for private research. The National Science Foundation offers a wide range of scientific grants, including funding for geodynamic research. The recipients of such grants are often affiliated with either a private research organization or a university.

The field of geodynamics requires a great deal of theory, mathematical computation, and extensive data collection, calling for the expertise of university researchers and faculties. At some educational institutions, geodynamics research is part of geology and earth science course curricula. However, some universities (such as Michigan State University) form groups of geodynamics-oriented researchers to collaborate and develop scientific papers, books, and other media on the subject. University professors and researchers commonly collaborate with peers at other institutions, sharing data and findings to support theories and research efforts.

Energy companies are always seeking new oil and gas deposits. Such findings are usually unearthed in subduction zones and other areas where seismic, geodynamic, and volcanic activity is prevalent. For this reason, the petroleum industry often calls upon full-time geophysicists and geodynamics consultants to help them find such deposits. These professionals largely use seismic equipment and data to survey a given area and locate new deposits.

Implications and Future Prospects

The study of geodynamics is critical to understanding volcanic and seismic activity, as well as the forces that created (and continue to reshape) the earth's surface profile. Some of the most significant developments in geodynamics research (such as plate tectonics) have been introduced only in the last few decades of the twentieth century. Much of this science has relied on the collection and analysis of minerals or the visual study of rock formations beneath the many strata (layers) of the earth's surface. Scientists studying geodynamics also record seismic waves, a process that helps estimations of crustal density and magma viscosity.

Although these scientific approaches remain important, the study of geodynamics has been greatly aided by the introduction of new technologies. Satellite systems, for example, have enabled geophysicists to survey the earth's deformations, magnetic fields, and other key characteristics. Updates to computer systems are enabling scientists to continually gather larger and larger amounts of data and create comprehensive models of the processes occurring far beneath Earth's surface.

Geodynamics has been further aided by the advent of the Internet. Scientists from around the world can now connect with one another through a vast array of networks to share theories, data, and other information about advances in geodynamics and changes in the earth's internal and surface areas. The Internet also enables scientists located anywhere in the world to access remote observatories and seismic stations in real time. The speed by which information is collated and shared and the volume of data compiled through the use of twenty-first century technologies continues to benefit the ever-evolving field of geodynamics.

Michael P. Auerbach

Further Reading

Dolphin, Glenn. "Evolution of the Theory of the Earth." *Science and Education* 18, nos. 3/4 (2009): 1-17. Offers recommendations on how to teach the various theories and concepts of geodynamics to high school students. Develops a curriculum for educating ninth graders in this field, including reading materials and teaching methods.

Freymueller, Jeffrey T. "Active Tectonics of Plate Boundary Zones and the Continuity of Plate Boundary Definition from Asia to North America." *Current Science* 99, no. 12 (2010): 1719-1732. Reviews the existing theories regarding tectonic movement and activity along plate boundaries. Focuses on plate boundary zones in Asia and western North America, analyzing the deformation that continues to occur in these regions.

Kearey, Philip, Keith A. Klepeis, and Frederick J. Vine. *Global Tectonics*. 3rd ed. Hoboken, N.J.: Wiley-Blackwell, 2009. Provides updated information about developments in the study of plate tectonics, including the implications of plate tectonics on the growing issue of climate change.

Lalleman, Serge, and Francesca Funicello. *Subduction Zone Geodynamics*. New York: Springer, 2009. This book discusses in detail the process of subduction. Describes how subduction occurs and how this concept affects Earth's outer crust and its internal composition and regions.

Norton, Kevin P., and Andrea Hampel. "Postglacial Rebound Promotes Glacial Re-advances." *Terra*

Nova 22, no. 4 (2010): 297-302. This article discusses the postglacial rebound that followed the last glacial maximum. Presents a three-dimensional model to illustrate the degree of rebound that has taken place, particularly in the Alps region of Europe.

Rollinson, Hugh. "When Did Plate Tectonics Begin?" *Geology Today* 23, no. 5 (2007): 186-191. Explores the question of when the earth's plates began to move. Discusses the two main theories about the starting point of plate movement and then theorizes that the plates began moving about 3.5 to 4 billion years ago.

Turcotte, Donald L., and Gerald Schubert. *Geodynamics*. New York: Cambridge University Press, 2002. This textbook offers a wide range of discussion points on geodynamics, geophysics, and geology. Among the topics the authors review are the chemical processes of geodynamics, seismology, subduction, and plate tectonics.

See also: Earthquakes; Earth's Interior Structure; Earth's Magnetic Field; Earth's Mantle; Geodetic Remote-Sensing Satellites; Isotope Geochemistry; Mantle Dynamics and Convection; Mountain Building; Plate Motions; Plate Tectonics; Remote-Sensing Satellites; Seismic Wave Studies; Slow Earthquakes; Stress and Strain; Subduction and Orogeny; Tectonic Plate Margins; Volcanism.

THE GEOID

The shape of the sea-level surface, over the oceans and under the continents, is given by the geoid. This shape differs from the best-fitting ellipsoid by amounts ranging up to approximately one hundred meters, and these variations provide valuable information concerning models of the convection and tectonics of the planet.

PRINCIPAL TERMS

- **ellipsoid of revolution:** a three-dimensional shape produced by rotating an ellipse around one of its axes
- **equipotential surface:** a surface on which every point is at the same potential, used here to include gravitational and rotational effects; no work is done when moving along an equipotential surface
- **global positioning system (GPS):** a group of satellites that go around Earth every twenty-four hours and that send out signals that can be used to locate places on Earth and in near-Earth orbits
- **Legendre polynomials:** mathematical functions used to describe equipotential surfaces on spheres
- **mantle convection:** thermally driven flow in the earth's mantle thought to be the driving force of plate tectonics

APPROXIMATING EARTH'S SHAPE

The geoid is an imaginary surface that is at sea level everywhere on the earth. Over the oceans, it is generally at mean sea level; under the continents, it is the elevation the sea would have if all of the continents were cut by narrow sea-level canals. It is usually represented as the difference in elevation between sea level and some ellipsoid representing the average shape of the earth, and its relief is on the order of one hundred meters. It is important in surveying and geodesy because elevations are measured above or below this surface, and it is important in geology and geophysics because its departures from a perfect ellipsoid reveal information about the earth's interior.

The shape of the earth can be approximated, with varying degrees of complexity and with different levels of success, by different mathematically defined shapes. If represented as a sphere with a radius of 6,371 kilometers, the shape is very simple, but it will have a radius that will be 7 kilometers too small at the equator and 15 kilometers too large at the poles. Nonetheless, this is adequate enough to be used for many scientific purposes.

For some purposes, however, a spherical shape is entirely inadequate. Gravity varies with the radius of the earth. Surveys seeking to detect density variations beneath the surface using sensitive measurements of gravity need a way to account for the gradual increase in radius from the poles to the equator. This change in radius, usually called the earth's "flattening," is obtained by dividing the difference between the equatorial radius and the polar radius by the equatorial radius. A modification to the spherical shape is obtained by letting the radius vary slightly with latitude, using a straightforward function that includes the value for flattening. This shape is known as the spheroid and sometimes called the "niveau spheroid"; and for many years it was used in gravity surveys. It permitted data reduction at a time when computers filled rooms, if not buildings. This simple formula is actually an approximation of a slightly more complex shape, the ellipsoid.

EQUIPOTENTIAL SURFACES

An ellipsoid of revolution is the shape of a solid produced by rotating an ellipse about one of its axes. If an ellipse with a minor axis equal to the polar diameter and a major axis equal to the equatorial diameter is rotated about the poles, the resulting shape is the earth's ellipsoid. This ellipsoid is used in studies when the sphere or spheroid is inadequate, and it forms the basis for the geoid.

The geoid is an example of an equipotential surface. If there was some way of sliding a mass around on its surface without any friction, no work would be done in moving that mass from place to place, because the mass would stay at the same potential. It is not difficult to calculate the shape of this surface for various idealized situations. It is also possible to measure the shape of the geoid. Much can be learned by comparing the observed shape with the shapes generated by the models.

If the earth were a stationary sphere of uniform density, the geoid would also be a sphere. If the earth were a rotating sphere of uniform density, the geoid would become an ellipsoid. This is because the rotation produces centrifugal force. Rotation involves an acceleration, which, when multiplied by a mass, must be balanced by a centripetal force—in this case, one supplied by gravity. For these purposes, using the non-Newtonian centrifugal force will prove simpler. This centrifugal force acts in a direction perpendicular to the rotation axis. At the equator, it would be directly opposed to the gravitational pull of the earth. An equipotential surface would need to be higher there to make up for this force. The equator would be farther from the center of the earth, and the poles would be closer to the center of the earth, but because the geoid is an equipotential surface, traveling from the equator to one of the poles would not involve going downhill. If the earth formed a rigid sphere, oceans would be much deeper at the equator than at the poles. However, the scale of the earth and the fact that it has existed for billions of years allowed it to deform much as if it were a fluid.

EVOLVING KNOWLEDGE OF THE GEOID

Suppose a model is allowed to assume the equilibrium shape of a fluid with the earth's mean density, rotating in space once a day. In 1686, Sir Isaac Newton determined that such a model would form an ellipsoid with a flattening of 1 part in 230. His solution piqued considerable interest. This much flattening would result in differences in the length of a degree of latitude between the equator and the poles, which should have been measurable using the techniques available in the early part of the eighteenth century. Expeditions were made to Lapland and Peru to do just this. The results showed a flattening, but of only about 1 part in 300. The current value is 1 part in 298.257, and many geoids are presented in terms of elevation above or below this reference ellipsoid.

As additional geodetic surveys, gravity surveys, and satellite orbit determinations were done, knowledge of the geoid evolved. It is now known that the Indian Ocean just off the southern tip of India is about 100 meters beneath the ellipsoid. Other ocean lows exist in the western North Atlantic Ocean (-50 meters), the eastern North Pacific Ocean (-50 meters), and the Ross Sea near Antarctica (-60 meters). On the continents, lows are present in central Asia (-60 meters) and northern Canada (-50 meters). High areas of the geoid occur over New Guinea (+75 meters), southeast of Africa halfway to Antarctica (+50 meters), in the North Atlantic Ocean (+60 meters), in western South

Geodesist Dr. John O'Keefe, with geodetic map of North America showing geoid contour. (Time & Life Pictures/Getty Images)

America (+40 meters), and in southern Alaska (+20 meters). These highs and lows dominate the geoid. Their existence and locations have been known since the 1970's.

UNDULATIONS AND GEOID ANOMALIES

The huge areas over which individual highs and lows extend require that they be produced by large, deep-seated density variations. Their existence suggests the earth is not in hydrostatic equilibrium, which in turn suggests that they result from density variations that do not persist for more than a few hundred million years. Therefore, mantle convection seems to be the most likely cause of these undulations, and many geophysical studies of the long-wavelength undulations of the geoid have concentrated on determining what they tell us about this convection. In general, lows on the geoid are above areas of rapid spreading, and highs are above subducted slabs. Although still evolving and hence subject to change, most of these investigations seem to suggest that mantle convection is driven by descending, not ascending, plumes; that convection involves the whole mantle, not just the seismically active (less than 670 kilometers deep) mantle; and that viscosity increases by a factor of about ten at some depth, probably 670 kilometers.

Other research involves the smaller-wavelength geoid anomalies, particularly over ocean areas. The data are filtered to remove the larger effects, and what remains is usually an excellent indicator of sea-bottom topography. The additional mass produced by a mountain on the sea floor attracts extra water, which piles up and causes a high on the geoid. The geoid is depressed above trenches because trench areas have less mass than the normal sea floor, so they attract less water. This small-wavelength low, with an amplitude of ten meters or so, is usually superimposed on a much larger wavelength high produced by the huge mass excess of the cold slab descending into the mantle nearby.

Undulations of the geoid give some of the best evidence for lateral density variations in the mantle. Seismic data also reveal lateral variations within the mantle, but these are variations in seismic wave velocities, which may or may not correspond directly to density variations. Eventually, the two lines of investigation promise to reveal the inner workings of the planet.

Laplace's Equation

Because gravity depends on the shape of the geoid, if there were enough accurate determinations of gravity at sea level, mathematical manipulations could be performed to find the shape of the geoid. However, gravity measurements taken above sea level can be adjusted in ways that convert them to equivalent values for sea-level readings. By the middle of the twentieth century, great progress had been made in mapping the gravity field over much of North America and Europe. This gave some indication of how the geoid undulated locally, but accurate determinations of geoid heights actually require considerable knowledge of gravity from around the entire earth. When the first satellite was launched in 1957, a new technique suddenly became available that permitted global gravity—and geoid height—to be calculated.

A satellite's orbit is influenced by the distribution of mass beneath it. The motion of the satellite in its orbit and the precession (gyration) and nutation (wobble) of the orbit are all influenced by the gravitational field it experiences. If the orbit is carefully tracked, it reveals much about the earth's shape and gravity. This requires a considerable mathematical effort, seeking solutions to a partial differential equation called Laplace's equation. These solutions take the form of coefficients of Legendre polynomials and associated polynomials. To see how they can describe the geoid, consider a surface suspended in space above a chessboard.

To describe the topography of this surface, each square could be designated by its row and column number and its elevation listed. However, there is another way to do this. The description can begin with the average elevation for the whole surface. Once that is determined, the surface is divided into quarters labeled "1" through "4"; the difference between the average elevation for each quarter and the average for the whole surface is then listed. Each quarter is then divided into four parts labeled "A" through "D," and the difference between the average elevation for each part and the average for the whole quarter is again listed. If this is done once more, labeling the divisions "a" through "d," elevation data will exist for all sixty-four squares of the chessboard. If the quarters are numbered clockwise from the upper left, then the square in row 3, column 7 would be designated 2Ca, and its elevation would equal the total surface average plus the difference for quarter 2, plus the difference for sixteenth 2C, plus the difference for sixty-fourth 2Ca.

One advantage this system has is that the earlier results do not change as the description gets more and more detailed. The geoid uses a similar system; however, it divides a spherical grid rather than a square one, so instead of repeatedly quartering, it uses Legendre polynomials. Instead of dividing the board horizontally, the geoid goes to the next degree; instead of dividing the board vertically, the geoid goes to the next order. A geoid determined to degree and order 12 will have much more detail than one determined to degree and order 6, but the coefficients for the first six degrees and orders should be the same.

Satellite Data

As satellites have been tracked through more orbits by better technology, the geoid has become known to increasingly better precision. It was known to degree and order 8 by 1961, 16 by 1971, and 20 by 1985. Current models use supercomputers to work through 10,000 parameters, yielding some results to degree and order 360. There are differences between them; however, typically they vary on the order of one-half meter at degree and order 50 or so. Much of the difficulty lies in correctly merging data from more than thirty satellites and tens of thousands of individual gravity measurements made on land.

Satellite data are influenced by factors such as atmospheric drag and refraction effects, and considerable thought and engineering have gone into finding ways around such problems. Geodetic satellites must be in near-Earth orbits to be influenced significantly by the undulations in the geoid, and in such orbits atmospheric drag is significant. One approach has been to design satellites out of uranium, which is denser than lead, giving them a great mass for a small cross-sectional area. Another solution has been to suspend a massive core inside a spherical shell. Atmospheric drag will act on the shell, pushing it closer to the core on the side experiencing the drag. Appropriate use of thrusters built into the shell can then ensure that these effects are perfectly canceled. Refraction effects also result from the atmosphere and can be minimized by determining the satellite's position relative to other satellites in higher orbit. Lasers are used for this, reflecting off corner cubes embedded in the outer shell of the satellite.

Time variations in the geoid have been measured by the GRACE (Gravity Recovery and Climate Experiment) satellites launched in 2002 and the GOCE (Gravity Field and Steady-State Ocean Circulation Explorer) satellite launched in 2009. The GOCE satellite uses a highly sensitive gradiometer to measure variations in density, and can measure the geoid to an accuracy of 1-2 cm. Time variations of the geoid are used in studies of hydrological cycles and of glacial melting.

Ground-based gravity data are influenced by instrument design, operator expertise, and the care with which topographic effects have been removed. Accurate elevations are essential, and much of the earth's land surface has not been mapped with sufficient topographic precision to provide this control. This, combined with political and economic realities, has resulted in a data set that includes very little information from Asia and not enough from South America. The global positioning system (GPS) constellation of satellites promises great improvement in the data set by providing better satellite tracking as well as better elevation control in poorly mapped areas.

Predictive Value

Better knowledge of the geoid will lead to better understanding of plate motions and convection in the mantle, permit more accurate placement of satellites in orbit, and provide an enhanced base for surveys on Earth. One of the outstanding questions in geophysics concerns the scale of convection in the mantle. No one knows if the convection cells extend throughout the entire mantle or if there are two convection regimes—one above a depth of 670 kilometers, the other below it. Unequivocal answers have yet to be found, but the study of the geoid offers great promise. Models for most of the highs of the geoid use the mass of the descending slabs to generate them. It appears that a slab reaching only to depths of 670 kilometers may not have sufficient mass. This somewhat tentative result favors deep mantle convection. As knowledge of the geoid improves, particularly in continental areas for which only sparse gravity data exist, the validity of this result may be established. An improved grasp of convection in the mantle will increase understanding of the plate-tectonic theory. As that develops, it may well produce a capability to predict earthquakes and volcanoes and to mitigate the death and destruction they cause.

As communications technology develops, the need for many more satellites to relay the ever-increasing traffic will grow. Just as minor perturbations in satellite orbits help us to define the geoid, a better understanding of the geoid will permit much better predictions of just how a satellite will behave in orbit. This is true even for the very high communications satellites. As the population of satellites grows, the need for such refinements in orbit calculations will become necessary to avoid collisions.

Ocean currents such as the Gulf Stream are powered by differences in elevation of the sea surface. These differences, on the order of a meter or so, occur because huge, warm lenses of water float on denser water below. Water at the surface, which in this case is not an equipotential surface, tries to flow down the slope, is affected by the earth's rotation, and ends up going around the lens of warmer water instead. Measuring such tiny variations in the level of the sea is difficult. As our knowledge of the geoid improves, however, we should be able to observe these discrepancies. This should provide important data from which changes in ocean currents can be predicted. Finally, because all surveys on land measure elevations with respect to sea level, improvements in charting sea level will make the elevation measures on land much better.

Otto H. Muller

Further Reading

Davidson, Jon P., Walter E. Reed, and Paul M. Davis. *Exploring Earth: An Introduction to Physical Geology.* 2d ed. Upper Saddle River, N.J.: Prentice Hall, 2001. An excellent introduction to physical geology, this book explains the composition of the earth, its history, and its state of constant change. Intended for high-school-level readers, it is filled with colorful illustrations and maps.

Fowler, C. M. R. *The Solid Earth.* 2d ed. Cambridge, England: Cambridge University Press, 2004. This college textbook contains a brief but easily understood treatment of the geoid and examples of geoid height-anomaly calculations.

Garland, G. D. *The Earth's Shape and Gravity.* Oxford, England: Pergamon Press, 1977. An excellent, easy-to-read general treatment that describes the geoid and the techniques and data manipulations needed to obtain it. Although somewhat dated (the geoid map included is from 1961), the treatment is thorough. Potential theory is discussed in an appendix, and the book tries to avoid higher math as much as possible, making this the best source for a quantitative introduction to the subject.

Greenberg, John L. *The Problem of the Earth's Shape from Newton to Clairaut.* New York, Cambridge University Press. 2010. This provides a detailed guide through the historical knowledge of earth's shape. Contains a high degree of mathematics and a strong history of science. An excellent text for a graduate-level student.

Hamblin, William K., and Eric H. Christiansen. *Earth's Dynamic Systems.* 10th ed. Upper Saddle River, N.J.: Prentice Hall, 2003. This geology textbook offers an integrated view of the earth's interior not common in books of this type. This text is well organized into four easily accessible parts. The illustrations, diagrams, and charts are superb. Includes a glossary and laboratory guide. Suitable for high school readers.

Heiskanen, Weikko A., and Helmut Moritz. *Physical Geodesy.* San Francisco: W. H. Freeman, 1967. The standard text on this subject for many years, this book presents a thorough treatment of the techniques and theoretical bases for classical geodesy. Somewhat heavy with equations.

King-Hele, Desmond. "The Shape of the Earth." *Scientific American* 192 (1976): 1293-1300. This easy-to-read paper captures the excitement and sense of discovery shared by scientists working with the early satellites. Although the data set has grown tremendously since this paper was written, most of the major features of the geoid can be seen on the map.

Lambeck, Kurt. *Geophysical Geodesy.* Oxford, England: Clarendon Press, 1988. The best book available on the subject, it includes a detailed treatment of the geoid and how it is obtained. Considerably better than other references in explaining satellite techniques and dynamic ocean topography. Although rich in equations, many containing Legendre polynomials, most of the relevant concepts and results are also described in words. Contains many maps, including one of global gravity and one of geoid heights from 1985.

Murthy, I. V. *Gravity and Magnetic Interpretation in Exploration Geophysics.* Bangaloree: Geological Society of India, 1998. This book is an excellent source of information about gravity and geomagnetism. In addition to useful illustrations, the book comes with a CD-ROM that complements the information in the chapters. Bibliography and index. Intended for the reader with some earth science knowledge.

Smith, James. *Introduction to Geodesy: The History and Concepts of Mode Geodesy.* New York: Wiley, 1997. Geared toward the college student, this book provides a nice introduction to the study of the earth's shape and rotation. Includes many illustrations and maps that add clarity to key concepts. Bibliography and index.

Tsuboi, Chuji. *Gravity.* London: George Allen & Unwin, 1983. A remarkable book by an accomplished Japanese scientist. The geoid is given unusual treatment, using Cartesian, cylindrical, and spherical approaches. Although quantitative and rigorous, many of the treatments are practical, drawn from work done by the author in the first half of the twentieth century. Almost entirely devoted to presatellite work.

See also: Continental Drift; Creep; Earthquakes; Earth Tides; Gravity Anomalies; Isostasy; Mantle Dynamics and Convection; Ocean-Floor Drilling Programs; Plate Motions; Plate Tectonics; Tectonic Plate Margins.

GEOLOGIC AND TOPOGRAPHIC MAPS

One of the earliest known geological maps is a geologic map of England constructed by William Smith in 1815. Topographic and geologic maps are basic tools used for the management and development of the earth's resources. Topographic maps represent on paper the earth's surface and its various landforms. Geologic maps show the distribution of the rocks that underlie the landforms and provide a view of the earth's surface to a depth of several thousand feet.

PRINCIPAL TERMS

- **contour lines:** on a topographic map, lines of equal elevation that portray the shape and elevation of the terrain
- **geologic map:** a representation of the distribution of mappable units (formations)
- **map scale:** the scale that defines the relationship between measurements of features shown on the map and the actual features on the earth's surface
- **topographic map:** a line-and-symbol representation of natural and selected human-made features of a part of the earth's surface, plotted to a definite scale

TOPOGRAPHIC MAPS

A topographic map is a line-and-symbol representation of natural and selected human-made features of a part of the earth's surface, plotted to a definite scale. Topographic maps portray the shape and elevation of the terrain by contour lines, or lines of equal elevation. The physical and cultural characteristics of the terrain are recorded on the map. Topographic maps thus show the locations and shapes of mountains, valleys, prairies, rivers, and the principal works of humans.

In the past, topographic maps were constructed by labor-intensive field methods, which involved detailed field measurements made with telescopic-type instruments. These data were translated in the field to actual distances and plotted by hand on field sheets for eventual office compilation and printing. More recently, however, most maps have been prepared using adjacent pairs of aerial photographs. Highly accurate, these photographs are further checked for accuracy by reference to global positioning satellites. With the advent of the global positioning system (GPS) network, location in respect to longitude and latitude or to Universal Metric Coordinates and elevations may be readily determined without extensive baselines and triangulation networks. In some instances, laser-beam surveys provide extremely precise control in areas that are subject to earthquakes, such as California, to monitor stress buildup. Complex stereoscopic plotting instruments are used by a trained observer to delineate contour lines and various features on a base map. Field verification of place names and features is required before the map is printed. The maps are then compiled on a stable base (a type of plastic that does not change dimensions during temperature and humidity fluctuations). Such a base helps to ensure the map's accuracy by preventing distortions. Modern photographic and photochemical techniques are used to prepare the map for printing.

All maps must meet accuracy standards established by the government. Special standardized symbols, each with its own meaning, are used to convey a wide variety of information. Also, colors are frequently used to show the more common features. Generally, blue indicates water bodies, brown indicates contour lines, red indicates map features with special emphasis (chiefly land boundaries), pink indicates built-up urban areas, and purple indicates revisions based on new photographic information since the original map was made.

GEOLOGIC MAPS

Geologic maps are a representation of the distribution of mappable units (formations). These maps provide the data for an accurate compilation of the rock units at the surface or in the subsurface. A geologist makes hundreds of observations each day in the field. Many of these observations are recorded in a notebook or on the field sheet that eventually becomes a geologic map. Some geologic maps are prepared from aerial photographs or from remotely sensed images created through satellites. New detailed geologic mapping frequently reveals information that may require reevaluation of previously mapped areas. Large-scale (1:24,000) geologic maps require detailed examination of the area being mapped. Field investigation describes outcrops as close as a spacing of several hundred feet using a topographic map as a base. Outcrop descriptions include determinations of fossils, rock type, and mineralogy, along with descriptions of rock properties, such as color, thickness, and type of bedding units, and attitudes of the rocks. Some studies are supplemented by geophysical surveys. Drilling (cores or cuttings) is integral to many studies, including oil and gas exploration, mining, and engineering. Samples are examined later in the laboratory. Some of the laboratory work includes microscopic study of thin slices of rock to establish mineral relations, binocular study and identification of fossils, or detailed chemical analyses of the whole rock or of separated minerals, and radiometric age dating.

Compilations of geologic maps require regular reviews of field notes and observations, laboratory data, and information from the scientific literature. Data are transferred to a topographic map database and hardcopies are stored in plastic to ensure stability during temperature or humidity variations. Finalized contacts are drawn that divide rocks of one unit from those of another. The degree of certainty of the contacts is shown by a standard set of line symbols. The orientation of the various rock units is indicated by uniform symbols. When a geologic map is complete, it is prepared for publication by conventional drafting methods or by digitization and computer plotting methods. Orientations of surfaces are given as "strike" and "dip," and orientations of linear features are given as "trend" and "plunge," giving the directions of maximum slope and their angles from the horizontal.

Geologic and topographic maps are added to large computer databases known as geographic information systems (GISs). Through computer manipulation of layers of information such as geology and topography, informed decisions can be reached by combining these with other data layers. Individual geologic maps are issued in a numbered series by the U.S. Geological Survey and related agencies, and a large number of these maps are available online.

Map Scales and Series

Map scale defines the relationships between measurements of features shown on the map and measurements of features on the earth's surface. These comparisons are numerically expressed as a ratio—for example, 1:24,000, 1:125,000, 1:250,000, 1:500,000, and 1:1,000,000 scale. Large-scale maps (1:24,000 or larger) are used when highly detailed information is required. Examples include proposed projects (roads, large construction projects, and so on) in highly developed or populated areas. Intermediate-scale maps are quite useful in land- and water-management planning projects and in resource management. The 1:100,000 scale (metric) maps have become popular for a growing number of applications, particularly environmental protection and planning. Small-scale maps (1:250,000 to 1:1,000,000) cover very large areas. They are useful mainly in regional planning.

The topographic map series of the National Mapping Program includes quadrangle and other map series published by the U.S. Geological Survey. A map series is a family of maps conforming to the same specification or having common characteristics, such as scale. Adjacent maps of the sample quadrangle series can be combined to form a single large map manually, photographically, or by computer methods. Geologic maps are prepared using existing topographic and/or planimetric base maps (maps showing boundaries but no indications of relief). Thus, the scales of geologic maps generally correspond to those of the common topographic and/or planimetric maps. In special cases, such as a major engineering project (for example, a dam or nuclear power plant), preparation of a site-specific large-scale topographic map may include detailed geologic mapping.

Practical Applications

Topographic and geologic maps help to reveal the structure and resources of the surrounding environment. These maps are basic tools for resource management and planning and for major construction projects. They are used in the planning of roads, railroads, airports, dams, pipelines, industrial and nuclear plants, and basic construction. Both types of maps are also used in environmental protection and management, water quality and quantity studies, flood control, soil conservation, and reforestation planning. In addition, topographic maps receive wide use in recreational activities such as hunting, fishing, boating, rock climbing, camping, and orienteering.

Geologic maps provide baseline data for the identification and orderly development of the earth materials required for modern civilization. Examples include sand and gravel, crushed stone and aggregate, clay, metal deposits, and hydrocarbon fields.

Geologic maps are also used extensively in environmental monitoring and protection, in local regional planning, and in scientific studies. They help to identify areas prone to landslides or earthquakes, groundwater recharge areas, and potential sand and gravel resources. They are therefore used in land-use and planning studies to determine technically suitable and environmentally safe locations for subsurface solid, hazardous, or low-level and high-level nuclear waste repositories and excavations, waste disposal, water resources investigations, and military applications.

Jeffrey C. Reid

FURTHER READING

Bohne, Rolf, and Roger Anson, eds. *Inventory of World Topographical Mapping.* 3 vols. New York: Pergamon Press, 1989-1993. This three-volume set, published under the auspices of the International Cartographic Association, provides an analysis of mapping worldwide, a guide to national mapping activities, and an inventory of topographical maps. Volume 1 covers Western Europe, North America, and Australia; volume 2 covers South and Central America and Africa; and volume 3 covers Eastern Europe, Asia, Oceania, and Antarctica.

Compton, Robert R. *Manual of Field Geology.* New York: John Wiley & Sons, 1962. This somewhat dated publication provides an excellent idea of the practical aspects of conducting fieldwork and the steps involved in making a geologic map. Functionally useful in the field, as it is a compact single-volume text, it is still used as reference material by many geologists.

Easterbrook, Don J., and Dori J. Kovanen. *Interpretation of Landforms from Topographical Maps and Air Photographs.* Upper Saddle River, N.J.: Prentice Hall, 2000. This book is a laboratory manual containing exercises that focus on developing problem-solving skills regarding the use and design of topographical maps. Assumes the reader has a basic knowledge of topographical maps and map symbols.

Illinois State Geological Survey. *How to Read Illinois Topographic Maps.* Champaigne, Ill. Illinois Department of Natural Resources. 2005. Written specifically for Illinois, yet applicable to any topographic map. This document provides an excellent overview of features common to topographical maps. The numerous images, complete with labels, enhance the utility of this reference material.

Longley, Paul A., et al. *Geographic Information Systems and Science,* 3rd ed. Hoboken: John Wiley & Sons, 2010. Well organized into five units—foundations, principles, techniques, analysis, and management and policy. This text contains the latest developments in the field of GIS and vital theoretical material to provide background for the practical uses of maps.

National Research Council. *Geologic Mapping: Future Needs.* Washington, D.C.: National Academy Press, 1988. This publication presents the results of a national survey on geologic maps. The survey was designed to identify current usage of geologic maps as well as future needs. Most important, the survey identified the relative needs for geologic maps by map scale, style of presentation, and type of user (for example, exploration, basic research, engineering, and hazard assessment).

Ormsby, Tim, et al. *Getting to Know ArcGIS Desktop.* 2d ed. New York: ESRI Press, 2010. This text was developed to explain the ArcGIS software, but is useful as a guide to the world of geographic information science. It is very useful for learning how topographic maps are read, built, and analyzed.

Steger, T. D. *Topographic Maps.* Denver, Colo.: U.S. Geological Survey, n.d. This free brochure provides a concise overview of topographic maps, their production, and their use.

U.S. Geological Survey. *COGEOMAP: A New Era in Cooperative Geological Mapping.* Circular No. 1003. Denver, Colo.: Author, 1987. Single copies of this circular are free upon application to the U.S. Geological Survey. Provides an overview of how cooperative geologic mapping between state and federal geological surveys is attempting to meet the need for large- and intermediate-scale geologic maps and other types of earth science maps.

_____. *Digital Line Graphics from 1:24,000-Scale Maps: Data Users Guide.* Denver, Colo.: Author, 1986. This free publication references how digital data were used to make topographic maps. This publication will be of value to students in understanding the foundation of computer applications.

_____. *Finding Your Way with Map and Compass.* Denver, Colo.: Author, n.d. This free brochure shows the hiker how to use a topographic map and describes the various map scales used on maps in the national topographic map series.

_____. *Large-Scale Mapping Guidelines.* Denver, Colo.: Author, 1986. This free publication provides basic information and aids in preparing specifications and acquiring large-scale maps for a variety of uses. Contains a large number of practical maps and an extensive applied glossary.

_____. *National Geographic Mapping Program: Goals, Objectives, and Long-Range Plans.* Denver, Colo.: Author, 1987. This free publication provides a nontechnical overview of the use and importance of geologic maps in the United States.

_____. *Topographic Map Symbols.* Denver, Colo.: Author, n.d. This free brochure summarizes all the symbols used on large-scale maps that are in the National Mapping Program. It indicates where and how to order topographic maps.

Van Burgh, Dana. *How to Teach with Topographical Maps.* Washington, D.C.: International Science Teachers Association, 1994. This volume provides basic information about the use and development of topographical maps, as well as ideas about how they can be used in the classroom to teach map-reading and problem-solving skills.

See also: Deep-Earth Drilling Projects; Earthquake Locating; Earth's Mantle; Electron Microprobes; Electron Microscopy; Elemental Distribution; Engineering Geophysics; Environmental Chemistry; Experimental Petrology; The Geoid; Infrared Spectra; Mass Spectrometry; Metamorphosis and Crustal Thickening; Mountain Building; Neutron Activation Analysis; Ocean-Floor Drilling Programs; Petrographic Microscopes; Relative Dating of Strata; X-ray Fluorescence; X-ray Powder Diffraction.

GEOTHERMOMETRY AND GEOBAROMETRY

Geothermometry and geobarometry use the difference in chemistry of minerals that exist together to estimate the temperature and pressure at which some geological processes occur. The estimation of temperatures and pressures is crucial to the understanding of such complex phenomena as the melting of the rocks that ultimately result in volcanic eruptions on the earth's surface and the mechanical properties of rocks that are involved in ruptures that result in earthquakes.

PRINCIPAL TERMS

- **component:** a chemical entity used to describe compositional variation within a phase
- **igneous rock:** any rock that forms by the solidification of molten material, usually a silicate liquid
- **metamorphic rock:** any rock whose mineralogy, mineral chemistry, or texture has been altered by heat, pressure, or changes in composition; metamorphic rocks may have igneous, sedimentary, or other, older metamorphic rocks as their precursors
- **mineral:** a naturally occurring solid compound that has a specific chemical formula or range of composition; minerals normally have regular crystal structures such that their internal arrangement of atoms is predictable
- **phase:** a chemical entity that is generally homogeneous and distinct from others in the system under investigation
- **phase equilibria:** the properties of chemical systems described in terms of classical thermodynamics; systems of specified composition are generally investigated as a function of temperature and pressure
- **thermodynamics:** the area of science that deals with the transformation of energy and the laws that govern these changes; equilibrium thermodynamics is especially concerned with the reversible conversion of heat into other forms of energy
- **volcanic rock:** a type of igneous rock that is erupted at the surface of the earth; volcanic rocks are usually composed of larger crystals inside a fine-grained matrix of very small crystals and glass

BASIC PRINCIPLES OF CHEMICAL THERMODYNAMICS

Geothermometry and geobarometry are methodologies used by geologists to determine the temperature and pressure attending igneous and metamorphic processes. The temperatures below the earth's surface at which magma (molten rock) is generated and subsequently crystallizes (solidifies) are an example of one type of information that may be obtained through geothermometry and geobarometry. Another is the temperature at which the constituent mineral phases of large masses of sediments are metamorphosed to different compositions or structures. To understand how geologists have developed the methodologies of geothermometry and geobarometry, one needs to understand some of the basic principles of chemical thermodynamics and how these principles are applied to rocks.

Temperature, along with pressure and bulk composition, largely controls the macroscopic physical and chemical properties of most materials. Based on experience, scientists know that certain substances behave in predictable ways under certain conditions. The substance may be as simple as a glass of water or a single grain of homogeneous iron metal, or it may be as complex as an igneous or metamorphic rock. For example, at atmospheric pressure (the pressure of the atmosphere at sea level, or 14.69 pounds per square inch), pure water will become solid ice at a temperature of 0 degrees Celsius (its freezing point) and gaseous steam at a temperature of 100 degrees Celsius (its boiling point). The behavior outlined above, however, may not be commonly observed, because pure water is rarely encountered in communities or homes. In the colder regions of the United States and Canada, many communities spread sodium chloride (common household salt) on the streets during the winter months. The introduction of salt to a pure water system changes its properties such that at atmospheric pressure, the water freezes at a temperature below 0 degrees Celsius. The new freezing point is a function of the salted water's composition and pressure. The addition of small amounts of salt to pure liquid water also raises its boiling point to a temperature above 100 degrees Celsius that, again, depends on the amount of salt and pressure. Many recipes in which food is cooked in boiling water call for small amounts of salt, allowing the liquid to attain a higher temperature, and hence heat content, before reaching its boiling point.

Phase Equilibria

To discuss these concepts in greater detail, it helps to define which system is under investigation and where it exists. In the first example described earlier, the system is the ice and liquid water on a road surface, where water existed in one of two distinct phases, either solid or liquid, depending on the temperature. In the ice-water system, salt spread on the road will primarily dissolve into the water phase, changing its composition from pure water to a saltwater solution. The composition is fixed by the amount of salt relative to the amount of water present, which is defined as the salt concentration in solution and is generally expressed in units of grams per liter. When the system obtains its minimum energy configuration for the specified set of conditions, it is at equilibrium.

For fixed temperature and pressure, equilibrium is characterized by a minimum in the Gibbs free energy function, which balances the heat content of the system against energy unavailable to the system to perform work on its surroundings. (The Gibbs free energy function takes its name from American theoretical physicist Josiah Willard Gibbs.) The type of work may be mechanical—for example, pushing a piston as a result of the rapid expansion of a gas—or electrochemical—for example, oxidation and reduction of chemical species at the positive and negative electrodes of a battery to produce an electric current. Gibbs formalism relies on the principle of chemical equilibrium, which allows scientists to relate changes in a chemical system, such as the ice-water-steam examples outlined previously, to specific thermodynamic properties. The principle of chemical equilibrium states that the chemical potential of each component in every phase in the system must be equal. Chemical potentials are used by scientists to relate the reactivity of individual chemical species at some specified temperature and pressure. At equilibrium, then, the component chemical potentials are equal and the total Gibbs free energy of the system under consideration is at a minimum. As a consequence, one would expect to observe no net change in the status of the system. A description of a well-defined chemical system in terms of classical thermodynamics is referred to as the study of the phase equilibria of that system.

When scientists analyze phase equilibria in terms of thermodynamics, they develop relationships between the composition of constituent phases and the temperature and pressure conditions under which they formed. In the salt-water system described earlier, if one could experimentally determine the relationship between the amount of freezing point depression and the concentration of salt in solution, one could then easily predict the temperature at which the road surface would freeze, given the amount of salt added to the system. The appropriate relationships applied to systems of geologic interest are commonly referred to as geothermometers, if they are used to determine temperature, or geobarometers, if they are used to determine pressure.

Application of Thermodynamics to Petrology

For the last 250 years, scientists from a variety of disciplines have carefully gathered experimental data to characterize the behavior of many materials of variable composition under a wide range of temperature and pressure conditions. These experimental data may be analyzed using a formalism called classical thermodynamics, which was largely developed by J. Willard Gibbs in the 1870's, based on the pioneering experiments and ideas of S. Carnot and E. Clapeyron. In the 1910's, N. L. Bowen, a geologist, pioneered the application of classical thermodynamics to the study of rocks. Bowen and his coworkers began a long and productive career of careful experimental characterization of geological systems. During this period, they developed experimental apparatus and techniques that are still in use presently with only moderate modification. Bowen mainly focused on examining the phase equilibria of silicate systems comprising two to three oxide components. Many of his original experimental investigations form the basis of modern igneous petrology.

Although Bowen and his coworkers did extensive experimental work to characterize the phase equilibria of many important geologic systems, the development and application of the methodologies of geothermometry and geobarometry did not gain wide acceptance until the 1960's. During the 1960's and 1970's, four common types of chemical reactions were developed and applied to rocks as both geothermometers and geobarometers: solid-solid exchange reactions, solid-liquid exchange reactions, solid-gas buffer reactions, and stable isotope exchange reactions. Each type of reaction is based on slightly different thermodynamic data and therefore has a somewhat different application to geological systems.

SOLID-SOLID GEOTHERMOMETRY AND GEOBAROMETRY

Solid-solid geothermometry or geobarometry is based on the exchange of one or more chemical components in the formation of coexisting solid phases that depends on temperature and pressure. In most igneous and metamorphic rocks, silicate phases (solids whose structure and properties are dominantly controlled by the presence of silicon, usually coordinated by four oxygens at the apexes of a tetrahedron) represent more than 90 percent of the rock's weight. The other 10 percent is composed of oxide phases (solids whose structure and properties are dominantly controlled by oxygen, usually in a close-packed arrangement). The exchange of a magnesium component between two similar silicate phases, clinopyroxene and orthopyroxene, is the basis of a commonly applied geothermometer. The exchange of an aluminum component between orthopyroxene and garnet (another common silicate mineral) is the basis of a commonly applied geobarometer.

There are several prerequisites for the application of either a geothermometer or a geobarometer to rocks. First, both phases must be present in the system of interest. Second, the two phases must be in demonstrable chemical equilibrium. Often, that is quite difficult to prove. However, evidence of disequilibrium is usually discernible in the form of compositionally zoned phases. Third, the thermodynamic properties of each of the phases must be known. Fourth, the composition of the two phases must be determined by chemical analysis. Last, to apply the geothermometer, one needs an independent determination of the pressure at which the two pyroxene phases formed, because the thermodynamic relations for the two pyroxenes are a function of pressure. Often, a geologist would use a geobarometer to obtain an estimate of the pressure. Concomitantly, to apply the geobarometer, one needs an independent determination of the temperature. Armed with these data, one may easily calculate the temperature at which the pyroxenes formed.

GEOTHERMS

Only the uppermost 10-20 kilometers of the earth's continental crust have been directly sampled by drilling; direct observation of the oceanic crust is even less extensive because of the extreme difficulty and expense of drilling from the ocean's surface through 5 kilometers of water prior to reaching the oceanic crust. There often are pieces of rock called xenoliths, however, which are accidentally entrained in magmas that originate from depths of 5 kilometers to as great as 200 kilometers below the earth's surface. Many of these xenoliths contain several mineral phases that are in equilibrium. The compositions of the constituent minerals are determined either by chemical analysis using traditional wet chemical methods, which require separation of individual phases, or by electron microprobe microanalysis, which allows direct determination of the composition of a spot with a diameter on the order of 1-10 microns. By examining these xenoliths in detail and applying a geothermometer and a geobarometer in concert, geologists have been able to determine temperature versus depth profiles for parts of the earth that are inaccessible to direct observation. Temperature-depth profiles are called geotherms, and they provide the link between the region of the earth where its composition is reasonably well known and the much deeper region where its composition and temperature are poorly constrained and subject to debate.

SOLID-LIQUID GEOTHERMOMETRY

The principles of solid-liquid geothermometry are much the same as those outlined previously for solid-solid geothermometry. In contrast to the simple example offered earlier, in which salt dissolved only in the liquid water phase, the chemical components in silicate systems tend to be soluble in several solid phases and the liquid phase simultaneously. That complicates the thermodynamic analysis considerably, and significantly more experimental data are needed to fully characterize the system. In addition, unlike the application of the two-pyroxene solid-solid geothermometer, where both phases are examined directly to obtain the temperature at which they formed, application of a solid-liquid geothermometer requires extensive assumptions about the composition and state of the liquid in equilibrium with the solid, as that liquid is no longer present in the rock. Plagioclase is a common silicate mineral found in rapidly frozen lavas that span a wide compositional spectrum. Because the plagioclase phase incorporates both sodium oxide and calcium oxide, which are both components in the liquid phase as well, the composition of plagioclase may be used to infer the composition of the coexisting liquid and the temperature at which the plagioclase formed. This method has been applied extensively to lavas that range from basaltic (silica poor) to andesitic (silica rich) in composition.

SOLID-GAS BUFFER REACTIONS

The reaction between one or more solid phases and a gas phase also yields information about temperature and the partial pressure of some gas species. One example of a solid-gas buffer reaction defines the coexistence of magnetite and hematite (oxide phases composed of iron). If both phases are pure—that is, they contain only iron and oxygen in the appropriate ratio that defines the phase—then their equilibrium with each other is directly related to their theoretical coexistence with an oxygen-bearing gas phase, which is defined by the partial pressure of oxygen (the amount of oxygen present, expressed as a percentage of the total pressure of the system).

The concept of a buffer arises because if oxygen is added to or subtracted from the system, the relative proportions of magnetite and hematite in the system will change to maintain internal equilibrium. This type of reaction does not specify the temperature explicitly, since magnetite and hematite may coexist over a range of temperatures and oxygen contents. Thus, if both pure solid phases are present and the temperature is varied, then the oxygen partial pressure will change sympathetically. Similarly, if the oxygen partial pressure is varied, then the temperature must also change. The observed coexistence of magnetite and hematite in many silica-rich igneous systems has been used to infer either the temperature or oxygen partial pressure attending formation, depending on which parameter may be fixed by independent information.

STABLE ISOTOPE EXCHANGE GEOTHERMOMETRY

The exchange of different stable isotopes of oxygen between coexisting phases forms the basis of the last common type of reaction used as a geothermometer. The isotopes of any chemical element—for example, oxygen—have the same number of protons but a different number of neutrons in their nuclei. Oxygen has two stable isotopes; one contains 8 protons and 8 neutrons, and one contains 8 protons and 10 neutrons. These two isotopes are referred to as oxygen-16 and oxygen-18. Their exchange between two coexisting phases in general is a function of temperature. At present, very few solid-solid, solid-liquid, or solid-gas exchange equilibria have been experimentally calibrated as a function of temperature and isotopic concentration of oxygen, however, and in the absence of such experimental calibrations, stable isotope exchange geothermometry has seen little application to igneous or metamorphic systems.

UNDERSTANDING OF VOLCANOES AND EARTHQUAKES

Many geologic processes can affect people's daily lives. Examples include potentially devastating earthquakes and the rapid and oftentimes catastrophic eruptions of volcanoes at the surface of the earth. By applying geothermometry and geobarometry methods to rocks, geologists have developed a much better understanding of how some aspects of these complex phenomena are initiated and how they evolve.

By determining temperature-depth profiles for the upper 200 kilometers of the earth, geologists have gained valuable information on the composition and state of regions of the earth that are not accessible to direct observation. This information is crucial to the understanding of how and why rocks deform during earthquakes. In fact, for rocks of fixed composition, ambient temperature is the primary variable that determines whether rocks will be able to deform in such a way as to produce an earthquake. In addition, temperature may be important in controlling the magnitude of some earthquakes.

The second and perhaps more direct application of geothermometry is in an effort to gain a greater understanding of the causes and warning signs of potentially catastrophic igneous eruptions, such as the one that occurred in 1980 at Mount St. Helens, Washington. Studies of volcanic rocks, for example, which generally integrate geothermometry, geobarometry, geochemistry, and field geology, have revealed that the processes that governed the eruption of Mount St. Helens continue to operate there as well as at other sites worldwide where oceanic crust is subducted below continental crust. In the Cascade province of the western United States, for example, Mount Shasta, Mount Bachelor, and Mount Rainier share many characteristics with Mount St. Helens, so scientists suspect that these volcanoes should erupt in a similar manner. Mount Rainier and Mount Bachelor represent a potential danger to the large population centers of Seattle and Portland, which are in close proximity to these volcanoes.

Glen S. Mattioli

FURTHER READING

Anderson, Greg M., and David A. Crerar. *Thermodynamics in Geochemistry: The Equilibrium Mode.* New York: Oxford University Press, 1993. An exploration of geochemistry and its relationship to thermodynamics and geothermometry. A thorough resource, but it can be somewhat technical at times. Recommended for the person with some chemistry and earth sciences background.

Atsuyuki, Inoue, et al. "Applications of Chemical Geothermometry to Low-Temperature Trioctahedral Chlorites." *Clays and Clay Minerals.* 57 (2009): 371-382. Provides an alternative method of geothermometry and comparison of estimates to earlier methods proposed by Walshe (1986) and Vidal et al. (2001).

Blatt, Harvey, and Robert J. Tracy. *Petrology: Igneous, Sedimentary, and Metamorphic.* 3rd ed. New York: W. H. Freeman, 2005. Undergraduate text in elementary petrology for readers with some familiarity with minerals and chemistry. Thorough, readable discussion of most aspects of geothermometry and geobarometry. Abundant illustrations and diagrams, good bibliography, and thorough indices.

Carmichael, Ian S. E., Francis J. Turner, and John Verhoogen. *Igneous Petrology.* New York: McGraw-Hill, 1974. A classic text used by most colleges for a first course in igneous petrology. Includes extensive discussions on all aspects of the formation of igneous rocks. The text is highly technical but quite readable. An extensive discussion of geothermometry and geobarometry and their application to solving geologic problems. Suitable for college-level students.

Correns, C. W. "Fluid Inclusions with Gas Bubbles as Geothermometers." In *Milestones in Geosciences: Selected Benchmark Papers Published in The Journal "Geologische Rundschau,"* edited by Wolf-Christian Dullo and Geologische Vereinigung e.V. New York, Springer-Verlag, 2010. The article was originally written in 1952, though it has been recognized multiple times since then as a significant paper in Geoscience. A background in geology is helpful in understanding this article.

Ernst, W. G. *Petrologic Phase Equilibria.* San Francisco: W. H. Freeman, 1976. This book outlines the elements of classical thermodynamics. Also discusses experimental approaches to acquiring the thermodynamic data necessary for geothermometry and geobarometry. Recommended for readers with a college-level background in chemistry.

Gregory, Snyder A., Clive R. Neal, and W. Gary Ernst, eds. *Planetary Petrology and Geochemistry.* Columbia, Md.: Geological Society of North America, 1999. A compilation of essays written by scientific experts, this book provides an excellent overview of the field of geochemistry and its principles and applications. The essays can get technical at times and are intended for college students.

Grotzinger, John, et al. *Understanding Earth.* 5th ed. New York: W. H. Freeman, 2006. An excellent general text on all aspects of geology, including the formation of igneous and metamorphic rocks. Contains some discussion of the structure and composition of the common rock-forming minerals. The relationship of igneous and metamorphic petrology to the general principles that form the basis of modern plate tectonic theory is discussed. Suitable for advanced high school and college students.

Klotz, Irving M., and R. Rosenberg. *Chemical Thermodynamics: Basic Theory and Methods.* 6th ed. New York: John Wiley & Sons, 2000. This book is designed to accompany a first course in classical thermodynamics at the college level. Knowledge of basic chemistry is desirable. Each chapter contains worked examples as well as problems to be solved by the reader. The answers are provided.

Mortimer, Charles E. *Chemistry: A Conceptual Approach.* 3rd ed. New York: D. Van Nostrand, 1975. An excellent basic chemistry text designed to accompany a first course in general chemistry, this book is suitable for advanced high school and beginning college students. It contains extensive descriptions of all basic chemical phenomena. Problems follow each chapter; a separate answer book is available. Several appendices and tables of data that the reader might find useful.

Oerter, Erik J. *Geothermometry of Thermal Springs in the Rico, Dunton, and West Fork Dolores River Areas, Dolores County, Colorado.* Colorado Geological Survey, Department of Natural Resources. 2011. Describes different types of geothermometers and geothermometry methods. Highly technical, written for professional geologists and graduate-level students.

Uyeda, Seiya. *The New View of the Earth: Moving Continents and Moving Oceans.* San Francisco: W. H. Freeman, 1971. This college-level text outlines the modern theory of plate tectonics in detail. Because the relevant observations are discussed in their historical context, the reader can learn about the personalities involved in the development of this central paradigm of the earth sciences. Many illustrations. The text is quite easy to read.

Winter, J. D. *Principles of Igneous and Metamorphic Petrology.* 2d ed. Pearson Education, 2010. A good undergraduate text, covers techniques of modern petrology. Recommended to have a geological dictionary on hand, as this has some parts with very technical writing.

Wood, B. J., and D. G. Fraser. *Elementary Thermodynamics for Geologists.* Oxford, England: Oxford University Press, 1976. A basic review of the thermodynamics necessary for understanding the chemistry of geologic processes. Outlines in great detail specific examples of geothermometers and geobarometers and their application to igneous and metamorphic petrology. Recommended for readers who have a college-level understanding of chemistry.

Wu, C.-M., and G. C. Zhao. "The Applicability of the GRIPS Geobarometry in Metapelitic Assemblages." *Journal of Metamorphic Geology* 24 (2006): 297-307. This article discusses the application of the GRIPS geobarometer to metapelitic rocks, detailing the calibration methods. Also compares GRIPS with the GASP barometer. A strong mathematics background needed to understand equations.

See also: Earthquake Distribution; Earthquakes; Earth's Core; Earth's Mantle; Elemental Distribution; Environmental Chemistry; Fluid Inclusions; Freshwater Chemistry; Geochemical Cycle; Heat Sources and Heat Flow; Mantle Dynamics and Convection; Nucleosynthesis; Oxygen, Hydrogen, and Carbon Ratios; Phase Changes; Phase Equilibria; Plate Tectonics; Volcanism; Water-Rock Interactions.

GLACIATION AND AZOLLA EVENT

More than 45 million years ago, the presence of stagnant, fresh water atop the Arctic Ocean fostered the growth of the fern Azolla. Over the course of hundreds of thousands of years, this plant's growth caused a climate shift that moved the warm Earth into the period known as the Ice Age. This period, which continues to this day, created massive ice sheets (glaciers) that at one point covered much of what is now Asia, northern Europe, and North America. Glaciation has shaped much of the earth's topography, leaving some areas flat and others rocky and mountainous.

PRINCIPAL TERMS

- *Azolla*: species of prehistoric fern that can reproduce rapidly; believed to be the initial cause of the climate change that led to the Ice Age
- **carbon sequestration**: a natural and artificial process whereby carbon dioxide is withheld from the atmosphere and instead retained in an idle location
- **geodesy**: study of the earth's shape, topography, and physical features and forces
- **greenhouse effect**: environmental phenomenon in which carbon dioxide and other gases are released into the atmosphere; these gases filter the sun's ultraviolet rays and redistribute warmth around the planet
- **last glacial maximum**: prehistoric period, approximately 30,000 years ago, in which the glaciers that covered the earth were at their thickest
- **plate tectonics**: theory stating that beneath the earth's outer crust is a series of plates in constant motion and through which magma flows
- **postglacial rebound**: process by which Earth's crust slowly returns to its original position after a glacier dissipates or moves away from the subduction zone
- **subduction**: geodynamic process whereby an extremely heavy object located on the outer crust pushes down on the crust and the tectonic plates beneath it

Basic Principles

About 49 million years ago, during the middle of what is known as the Eocene epoch, the Ice Age began with the introduction of a new species of fern, *Azolla*, to the Arctic Ocean. The warm and landlocked Arctic Ocean began to receive fresh rainwater, which set off a process in which the fern's rapid growth caused a chain reaction that reversed the earth's greenhouse effect and ushered in a cold climate highlighted by glaciation.

Glaciers have significantly affected the geodesy (the shape and topography) of the earth, causing the formation of mountain ranges and other features. Glaciers also are the primary source of the planet's fresh water and contribute to the stability of the planet's climate. Glaciers play an important role in the earth's interior processes, triggering volcanic activity and causing the movement of the tectonic plates beneath the earth's crust.

Glaciers were at their thickest and most numerous between about 27,000 and 19,000 years ago, during a period known as the last glacial maximum. However, glaciers still exist at high altitudes, and at the earth's polar regions, indicating that Earth remains in a waning period of the Ice Age.

Background and History

One of the most vexing questions paleontologists and other scientists have attempted to answer concerns the causes of the Ice Age. During this period, which began about 50 million years ago, the earth's climate shifted dramatically from warm to cold, giving rise to glaciation (the creation of massive sheets of ice). At the beginning of the twentieth century, this mystery seemed to be based on scientific curiosity. Scientists now are concerned that a similar climate shift could occur in the near future.

As early as the mid-nineteenth century, scientists began to explore the idea that the regions humans now live on were previously buried under ice sheets (glaciers) several kilometers thick. Soon afterwards, scientists agreed that not only did such glaciations occur, but other periods of major climate change also occurred before the Ice Age. The apparently cyclical patterns of ice ages led scientists to wonder about the causes of such events. Some speculated that these climate shifts could be attributed to extraterrestrial elements, such as the gravitational effects of other planets on the earth or changes in solar activity. Others argued that the climate shifts were caused by Earth's solar orbit: During cycles of about ten thousand years, areas in the Northern Hemisphere would see less sunlight, a phenomenon that caused snow and ice buildup.

The introduction of new technologies greatly aided the pursuit of scientific information about the Ice Age. One critical scientific practice, radiocarbon dating, was introduced in the 1950's, enabling scientists to date paleontological samples to their approximate dates of origin. In time, radiocarbon dating practices continued to improve, with core samples being successfully extracted from higher elevations and the ocean floor, offering much information about these climate shifts.

THE AZOLLA EVENT

Approximately 49 million years ago, during the early Eocene epoch, the Arctic Ocean was considerably warmer than it is today and was almost completely landlocked. Rain left a great amount of fresh surface water, and the rains also brought with them the nutrients that sustain plant life. Subsequently, a freshwater species of prehistoric fern, *Azolla*, began to develop in this thin layer of fresh water atop the ocean. Scientists believe that the *Azolla*, when exposed to optimal environmental conditions, could reproduce rapidly.

During this time in Earth's history, the rains over the Arctic Ocean were cyclical. Paleontologists and botanists believe that the ferns flourished during periods of high rains, such as those during the Eocene summers, and retreated when conditions changed. With the freshwater rainfall came Cyanobacteria, which produces organic nitrogen—a staple for plant life. Fossilized remains of Cyanobacteria located within sediment samples revealed that Cyanobacteria was present in great volumes beneath the Arctic Ocean.

Scientists believe that dead *Azolla* sank beneath the ocean's surface, and because oxygen at the bottom of the ocean was scarce (a condition known as anoxia), the ferns were immediately fossilized rather than consumed by bacteria. The fossilization of the dead *Azolla* also resulted in the draw-down (or sequestration) of massive amounts of carbon dioxide from Earth's atmosphere. This led scientists to believe that the fern played a direct role in shifting Earth's climate from a greenhouse environment to an "icehouse" environment.

GLACIATION

In time, the Azolla event led to a gradual cooling of the planet. Scientists have been able to uncover evidence of this transition by observing high sea levels that date to the late Paleocene epoch (which preceded the Eocene). Evidence also shows a gradual reduction in sea levels associated with global cooling starting to become manifest in the late Eocene and subsequent early Oligocene epochs. This drop in sea levels was caused by the development of enormous bodies of ice (the glaciers). The largest of these are called ice sheets, massive bodies of ice that could cover continent-sized regions and which began to develop in the eastern regions of the Antarctic during this period.

Throughout the Ice Age, glaciers spread and retreated as temperature cycles fluctuated. At one point, approximately 30,000 years ago, glacial thickness was at its highest for this era (and sea levels were at their lowest), marking the last glacial maximum.

The glaciers that formed during this period of global cooling began to move slowly, pushed by gravity and aided both by the deformation of internal ice crystals within them and by the lubricating effects of melting ice at the surface. When in motion, however, glaciers also have major effects on the surface over which they move, picking up and placing minerals great distances from their points of origin. Additionally, because of the enormous weight of these ice sheets, Earth's crust is pushed downward as they pass. After the glacier moves onward, the previously depressed crust slowly returns to its original shape—a process called postglacial rebound. The modification of the earth's surface caused by these bodies of ice is called glaciation.

THE HOLOCENE EPOCH

The full effects of the Azolla event and the Ice Age it created have not yet been fully realized. After all, many of the glaciers that were created during the Eocene epoch (most notably the ice sheets of Antarctica and Greenland, as well as glaciers found in higher elevations around the world) still exist. Put simply, the earth remains in an ice age and glaciations is ongoing.

Scientists, however, can study the effects of glaciation from other epochs on the timeline. For example, scientists have learned that, during the Miocene epoch (which followed the Oligocene), a period of relative warming and glacial retreat existed. Studying the North American area, researchers saw evidence of the postglacial rebound that caused the flattening of a number of regions. When the climate began to return to a cooler temperature in the subsequent Pliocene epoch, the glaciation caused during the Miocene enabled the Pliocene glaciers to be much larger and thicker. Glaciologists, climatologists, and geologists use the data obtained from such samples to create models of glacial development and climate change and to develop scenarios for how the current glacial period (the Holocene epoch) will end.

GEOLOGICAL AND MINERAL ANALYSIS

Evidence of both the Azolla event and the effects of glaciation have been found all over the world. In many cases, such evidence is carried to other parts of the world on icebergs, which broke away from polar ice sheets and carried scientific evidence across the ocean. In such situations, it is often difficult to separate the ice of today from the ice of prehistory. Scientist have been able to do so by looking past sea ice that has become embedded in icebergs and focusing on quartz and other minerals found within the body of the ice. In one study, scientists working on the Lomonosov Ridge (a suboceanic ridge of continental crust in the Artic Ocean) located the presence of quartz in a sample, giving researchers the ability to focus on the segments of ice that came from the middle Eocene epoch rather than sea ice found attached to the iceberg. In another study, scientists analyzed sediment found within the body of an ice raft. Based on a geological and mineralogical analysis of the sediment (which originated in the late Eocene/early Oligocene epochs), the researchers determined that the block had originated in a glacier found in the area of Norway, supporting the notion that the Azolla event caused glacier development in not only what is now the Antarctic but also elsewhere in the world.

SATELLITE TECHNOLOGY

Because of the sheer size and volume of glaciers, the effects of glaciation are often difficult to assess from a ground-level perspective. For this reason, satellite technology has become an invaluable tool for glaciologists. Satellites can collect an enormous amount of data on a single area as they fly over. Scientists may use satellite data to study the effects of multiple periods of glaciation on a particular region. For example, in the early twenty-first century, researchers utilized the *Landsat-7* satellite to observe the effects of glaciation on an area of Siberia, Russia. The *Landsat-7* data revealed a number of different eras in which glaciation had occurred, forging riverbeds in some areas and blocking waterways in others. In some regions the terrain had been flattened, while in others steep mountains were formed. These satellite data revealed multiple examples of glaciation from different epochs in Siberia's prehistory.

The Landsat satellite system also has proved useful in the study of the deterioration of glaciers during a period of climate change. In one case, scientists used the satellite's infrared and thermal sensors to study a glacier atop Mount Kenya in Africa. The satellite, compiling enormous volumes of data, provided a thirteen-year analysis of the glacier's slow reduction in volume and of the changes in the region's topography caused by the glacier.

COMPUTER MODELS

Compiling available data on the Azolla event and glaciation is a difficult process in light of the global distribution of evidence. It is in this area of research that computer models are invaluable tools in the study of these two connected concepts. For example, scientists studying glaciation of the Weichselian ice sheet, which during the last glacial maximum covered most of Scandinavia, used computer models to study the directions of ice flow since that event.

Computer models also can help scientists compare the Azolla event's influence on the Eocene climate with climate change occurring in the present. In one case, researchers created a computer model using chemical and biological data (such as marine fossils and oxygen isotopes) from the Eocene epoch to create a simulation of how the Eocene climate changed during and following the Azolla event. Such a simulation can help scientists look for clues regarding the current climate and changes that may occur in the future.

RELEVANT ORGANIZATIONS AND NETWORKS

The Azolla event and the glaciation that occurred as a result have strong implications for human civilization. Arguably, the most prominent of these implications is the ongoing scientific and political debate surrounding global warming and climate change. A number of organizations and institutions, including the following, play roles in the study of these two concepts.

Governments play important roles in terms of both funding scientific research on glaciation and climate change, and conducting their own research. For example, the U.S. Geological Survey includes a climate and land-use-change program, designed to study (among other issues) the effects of carbon sequestration, a natural and artificial process whereby carbon dioxide is withheld from the atmosphere and instead retained in an idle location, such as within the soil or at the bottom of the ocean. Carbon sequestration was a major contributor to the climate change caused by the Azolla event. By studying that event, governments hope to better understand carbon sequestration's potential affects on the environment.

Universities play a critical part in the study of climate change and glaciation. It is in such institutions that field research data are compiled, computer models generated, and theories about the effects of the Azolla event and subsequent glaciation are presented. Universities also serve as vehicles for public awareness of the significance of these phenomena, educating students and the general public on how to apply what has been learned about the last 50 million years to today's natural world.

The Internet has made possible the widespread work of many scientific societies and networks focused on glaciation and climate change. The International Glaciological Society, for example, provides symposia and conferences in which its members share theories and data. The society also features a publication for scientists to submit their own research.

Implications and Future Prospects

Radiocarbon dating has greatly benefited the study of the Eocene climate shift and the glaciation that resulted from the Azolla event. Technologies such as the Internet, computer modeling, and satellite systems have further enhanced the study of these events and processes. These advanced systems have enabled scientists to research both the prehistoric past and the present and to place any discoveries in this arena within the context of future climate change.

One of the most pressing aspects of the study of the Azolla event and glaciation is that scientists believe there is strong evidence that climate change is again occurring. Scientists look to the carbon sequestration that occurred during the Azolla event as a model for understanding examples of what is believed to be human-made carbon sequestration today. Furthermore, recently proven diminutions in the earth's glaciers and rising sea levels show a similarity to events that followed the early Eocene. Therefore, the Azolla event and the effects of glaciation in the last 50 million years are relevant to understanding what could happen to Earth should the planet's climate continue to warm.

Michael P. Auerbach

Further Reading

Benn, Douglas I., and David J. A. Evans. *Glaciers and Glaciation.* 2d ed. London: Hodder Education, 2010. This book provides a comprehensive review of the origins, composition, and nature of glaciers. The authors also discuss the paleogeological history and mechanics of glaciation and the role glaciers play in Earth's environment.

Florindo, Fabio, and Martin Siegert, eds. *Antarctic Climate Evolution.* Vol. 8. Miamisburg, Ohio: Elsevier Science, 2008. This book provides a review of the Antarctic ice sheet's 34 million year history. The editors discuss the Antarctic's past and present climate changes, glaciation, and geological features.

Peters, Shanan E., et al. "Large-Scale Glaciation and Deglaciation of Antarctica During the Late Eocene." *Geology* 38, no. 8 (2010): 723-726. This article discusses the changes that occurred to Earth's climate after the Eocene epoch and after the Azolla event. The authors review the sea level changes and ice sheet formations as manifest from evidence unearthed in Egypt.

Ruddiman, William F. *Earth's Climate: Past and Future.* 2d ed. Gordonsville, Va.: W. H. Freeman, 2007. In this book, the author reviews the Earth's history of climate change. Ruddiman discusses the breakthroughs in climatological and paleoclimatological research that have been introduced in the early twenty-first century. Also discusses past and present examples of climate change.

See also: Climate Change: Causes; Environmental Chemistry; Freshwater Chemistry; Geologic and Topographic Maps; Milankovitch Hypothesis.

GRAVITY ANOMALIES

Gravity anomalies are variations in expected values of measured gravity at specific locations and elevations on the earth. Gravity anomalies reveal changes in the density of rocks in the subsurface and were the first evidence for the existence of mountain roots and the differences between oceanic and continental crust.

PRINCIPAL TERMS

- **amplitude:** the positive or negative value (intensity) of an anomaly as measured against the background values in the region
- **Bouguer gravity:** a residual value for the gravity at a point, corrected for latitude and elevation effects and for the average density of the rocks above sea level
- **density:** the mass of a specific volume of a given material
- **free-air gravity:** a residual value for the gravity at a point, corrected for latitude and elevation effects; this value allows the scientist to determine differences in the densities of subsurface rocks
- **half-width:** the distance over which the amplitude of an anomaly falls from its maximum value to half the maximum amplitude
- **isostasy:** the concept of balance by which continental and oceanic crust are "floating" on the denser substrate of the mantle
- **milligal:** the basic unit of the acceleration of gravity, used by geophysicists in measurement of gravity anomalies equal to 0.001 centimeter per second squared
- **wavelength:** the distance over which an anomaly rises to its maximum amplitude and falls again to background values

GRAVITY VARIATION FACTORS

A gravity anomaly is a departure from the expected value of the acceleration of gravity at any point on the earth's surface. In general, such departures are small compared to the total gravity of the earth, which averages 980 centimeters per second squared. The actual value varies as a function of latitude and elevation. This variation occurs because the earth is not a perfect sphere but a spheroid of revolution. The equatorial radius is 6,378 kilometers, 21 kilometers longer than the polar radius of 6,357 kilometers. Since gravity decreases over distance, the gravity at the equator is less than that at the pole. Added to this is the effect of the earth's rotation. Together, these effects lead to gravity values of 978.0490 centimeters per second squared at the equator and 983.2213 centimeters per second squared at the pole, with a value for any latitude between predicted by a simple formula. Latitude explains the largest variation in gravity values of the earth. A second major effect results from elevation, which brings about a decrease in gravity of approximately 0.094 centimeter per second squared for every thousand feet of elevation above sea level.

The gravity at two points on the earth's surface will depend on latitude and elevation effects and on the densities of the rocks beneath the two points. The densities are of particular interest to the geophysicist. To evaluate these densities, the gravity values measured at two points on the earth's surface must be corrected for the latitude and elevation effects.

FREE-AIR AND BOUGUER ANOMALIES

The correction for the shape and rotation of the earth is made with the simple formula mentioned previously and reduces the effect of the latitude difference to zero. After this correction has been made, differences in gravity measured at two points may be attributed to differences in elevation and variations in the density of the underlying rocks. The correction for elevation is made relative to sea level and may be divided into two parts. The first involves the effect of being farther from the center of the earth as a result of the elevation. This correction is called the free-air correction, because it corrects the effect of distance as if there were only air between the point on the surface of the earth and sea level. The second part of the elevation correction involves the subtraction of the effect of the slab or rock between the surface and sea level. This latter correction is termed the Bouguer correction.

Gravity values corrected for latitude and incorporating the free-air correction are termed the free-air gravity or free-air anomaly. Gravity values corrected by latitude and by free-air and Bouguer corrections are called the Bouguer gravity values or Bouguer anomaly. Bouguer gravity values may also involve corrections for irregular topography and the curvature of the earth.

ROCK DENSITY

After the measured gravity values have been corrected, all differences caused by latitude, elevation, and topography have been removed mathematically, and the residual gravity anomalies reflect the lateral changes in rock density at depth. These gravity anomalies are typically small, usually less than 0.2 centimeter per second squared or 200 milligals—representing the differences between measured and ideal gravity values.

The value of the anomaly is related directly to a surplus or deficiency in mass in the subsurface. The effect of this mass difference is related to the acceleration of gravity by Isaac Newton's equation $\delta g = G\delta M/r^2$, where δg is the difference in gravity values over what is expected, G is the universal gravitational constant, r is the distance from the anomalous mass to the point on the surface of the earth where the gravity is measured, and δM is the increased or decreased mass.

This mass is usually expressed as the product of the change in density ($\delta\sigma$) times volume, or $\delta M = \delta\sigma \times V$. The density difference is the physical property of the rocks in the subsurface that causes the gravity anomaly. The amplitude and wavelength of the anomaly are caused by the size of the density contrast, the size and shape of the body, and the depth of the body.

This relationship can be illustrated by use of the simplest shape that may cause a gravity anomaly: a sphere. The figure shows spheres of different densities and sizes that are buried at different depths and the gravity anomalies associated with them. The anomalies have a maximum amplitude directly above the centers of the buried spheres, and the amplitudes decrease away from the bodies. In the figure, the horizontal axis marks distance in kilometers right and left of the center of the spheres, and the vertical axis marks the anomaly amplitude in milligals. In the first panel, three spheres are buried with their centers at the same depth of 20 kilometers. Most of the rocks in the area have a density of 2.7 grams per cubic

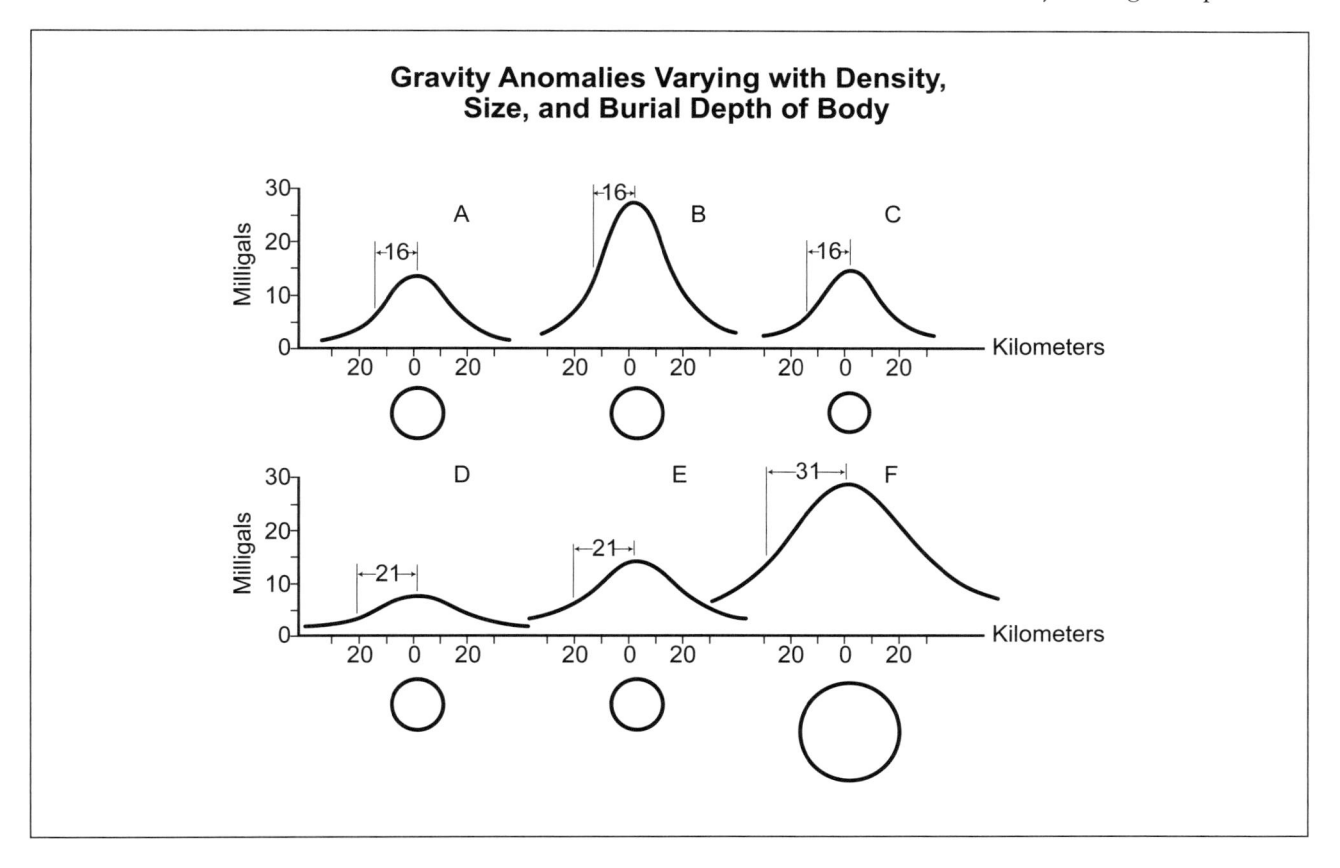

Gravity Anomalies Varying with Density, Size, and Burial Depth of Body

centimeter; this value represents the background value of density in the example. Sphere A has a radius of 10 kilometers and a density of 2.9 grams per cubic centimeter, giving a positive density contrast, $\delta\sigma$, of 0.2 gram per cubic centimeter. The excess mass causes a positive gravity anomaly with a maximum amplitude of 13.97 milligals. The second sphere, B, has a density of 3.12 grams per cubic centimeter, a density contrast of 0.42 gram per cubic centimeter, and a 10-kilometer radius, causing a maximum anomaly value of 27.24 milligals. The third sphere, C, has a density of 3.12 grams per cubic centimeter but is smaller, with a radius of only 8 kilometers. This sphere causes an anomaly of 13.95 milligals. Even though sphere C has the same density as sphere B, its size, and therefore its mass, is less, and the resulting anomaly is less.

Comparison of anomalies A and C shows them to be nearly identical. This occurs even for bodies of different densities when the product of volume and density yields the same mass. Note that the half-width of anomalies A, B, and C are all the same, with a value of 16 kilometers. This occurs because the center of mass of all three spheres is buried at 20 kilometers.

In the figure's second panel, spheres D and E are the same sizes and densities as A and B, respectively, but are buried at 28 kilometers depth. The amplitudes of each anomaly are diminished as a result of the greater distances to the bodies. Furthermore, the wavelengths and the half-widths of the anomalies are increased. Here, the half-widths are approximately 21 kilometers. This example shows that a pair of anomalies such as A and E with approximately the same amplitude will have different wavelengths or half-widths if the bodies causing the anomalies are at different depths.

The last sphere is buried at a depth of 40 kilometers with a radius of 20 kilometers and a density contrast of 0.2 gram per cubic centimeter. The maximum amplitude of the anomaly is 27.94 milligals, close to the amplitude of sphere B, but the half-width of the anomaly is 31 kilometers, nearly twice that of anomaly B.

While most rock bodies do not approximate spherical shapes, the above relationships show several general principles of gravity interpretation. The anomaly occurs because of an excess (or deficit) of mass in the subsurface. The amplitude is controlled by the density, size, and depth of occurrence of the rock body. The wavelength and half-width are related to the depth of the body. The longer the wavelength and half-width, the greater the depth of the source of the anomaly.

More complexly shaped geologic bodies cause more complex anomalies, and more complex equations are used to describe them. Shapes often used to simulate geologic bodies include cylinders, slabs, and three-dimensional prisms. The power of the gravity method involves the use of equations to calculate an anomaly that matches as closely as possible the gravity values measured in the field. The match allows the scientist to infer much about the character of rocks at depth.

DETERMINING ROCK DENSITY

Gravity anomalies show the density variations in the subsurface related to the occurrence of specific rock types. Practically, these variations must involve a lateral change in density from one place to another. Geologists and geophysicists try to understand the variation in terms of specific rock types. The densities of rocks vary as a function of composition, mineralogy, the occurrence of open spaces, and physical conditions such as temperature and pressure. Among sedimentary rocks, sandstone has a density range of 2.35-2.55 grams per cubic centimeter; shale, 2.25-2.45; limestone, 2.45-2.65; and loose sand, 1.90-2.00. Igneous rocks include granite, with a density range of 2.60-2.80 grams per cubic centimeter; gabbro, 2.85-3.10; and peridotite, 3.15-3.25. Metamorphic rocks include the following density ranges: granite gneiss, 2.60-2.70; schist, 2.70-2.90; amphibolite, 2.80-3.10; eclogite, 3.30-3.45; and marble, 2.70-2.75.

Using these data, one can get an idea of the rocks that are likely to cause a positive or negative gravity anomaly. If one were to rely only on these figures in interpreting rock types from gravity data, however, one would find a great many possibilities because of the overlapping of density ranges for different rock types. Thus, a limestone may have a density similar to that of a granite or a gneiss. For this reason, the geophysicist must be guided by what is known about the geology in the area being studied and by careful measurements of the actual densities for rocks from outcrop or drill holes in the region.

GRAVITY MODELING

The boundaries of the shapes of the rock bodies, which are mathematically determined by gravity modeling, are interpreted as having a geologic significance. Thus, these boundaries, representing the contact between rocks of different densities, may be interpreted as faults, intrusive contacts, unconformities, or normal depositional contacts, depending on what else is known about the geology.

By tracking changes in acceleration as a satellite orbits a planetary body, space scientists are able to map variations in the gravity field. These observations led to the discovery of lunar mascons (apparent mass concentrations in the lunar near-surface rocks). Mascons were eventually determined by gravity modeling to be caused by relatively thin layers of dense basalt pooled in large lunar basins mirrored by upward migration of dense mantle material. Gravity maps have been constructed for Mars and Venus by satellite measurements as well.

Gravity Anomaly Patterns

The largest differences in gravity occur as gravity anomalies between the ocean basins and the large continental masses. Ocean basins have positive gravity values, related to the dense rocks that underlie the oceans. Continents, by comparison, have negative Bouguer gravity values, which reveal that thick sections of low-density rock occur beneath the continents. The twin GRACE (Gravity Recovery and Climate Experiment) satellites launched in 2002 have produced large-scale anomaly maps from slight changes in their relative orbits.

The relationship between the density of underlying rock and elevation is a very general one that applies to most areas of the earth. As early as 1850, measurements had shown that mountains were underlain by rocks less dense than those underlying the surrounding lowlands. The relationship was explained by two models. The first was that continents and mountains were high because they were underlain by thicker sections of low-density rocks. Seismic data on the depths to the base of the continental crust have confirmed this model in most places. The second model suggested that mountains were high relative to lowlands because mountains are less dense than are the rocks under the lowlands. This relationship has also been verified in a number of areas and does explain the lower elevations found in some rift valleys.

Scientists have found that the patterns of gravity anomalies in a large area may give a distinctive grain or "fabric" to regions on the continents. The patterns in a particular region may involve a series of positive and negative anomalies of a certain amplitude aligned in a particular direction. Adjacent regions may have groups of anomalies of different amplitudes, wavelengths, or orientations that contrast with one another in the same way that the different sections of a quilt stand out against one another. The distinctive regional character of gravity in many places allows geologists to divide the crust into provinces. Other geologic observations and age determinations have shown these provinces to be pieces of crust that were assembled over time to make the continents.

Geophysical Applications

Gravity anomalies are a major source of geophysical understanding of the earth, but there are uncertainties in the interpretation and modeling of gravity data. These uncertainties are lessened by the use of geologic information and other geophysical survey techniques. Magnetic, electrical, or seismic data can provide additional information about the depth, size, and shape of the body in the subsurface. Knowledge of these variables improves the scientist's ability to define rock density and the rock type.

Gravity anomalies have helped to define the compositional and structural differences between oceanic and continental crust and the crustal thickening that occurs under mountain ranges. The patterns of gravity anomalies also reveal the internal structure of continents and the stages of continental development. The occurrence of gravity anomalies associated with the oceanic trench systems and island areas likewise attest to the dynamic character of these features and constitute evidence for the theory of plate tectonics.

Economic Applications

Of major importance is the application of gravity anomalies to natural resource discovery and evaluation. Many economically important features are related to changes in rock density and are detectable using gravity measurements. Lateral changes in density may occur in areas where anticlines or faulting have formed traps for oil and gas. These features may cause either positive or negative anomalies. One of the most successful of these applications is in the energy industry, where a large number of producing fields are related to thick, intrusive masses of salt called salt domes. The low-density salt moves upward in the sedimentary section, creating folds and faults that are excellent traps for oil and gas. The negative gravity anomalies associated with the salt domes have been used to locate pools of oil and gas in this geologic environment.

A second application of economic importance is in the location and evaluation of groundwater aquifers in certain parts of the country. Aquifers are often found where Quaternary (about 2 million years to the present) sand and gravel deposits are especially thick. Since the density of sand and gravel is much less than that of rock, thick sections cause negative gravity anomalies proportional to aquifer thickness. Information from gravity surveys may thus be useful in land-use planning and development.

Donald F. Palmer

FURTHER READING

Bott, M. H. P. *The Interior of the Earth: Its Structure, Constitution, and Evolution.* London: Edward Arnold, 1982. This text emphasizes the use of geophysics in the interpretation of the major compositional and structural components of the earth. The treatment is generally nonmathematical. The text integrates different geophysical methods pertaining to the different parts of the crust, mantle, and core. Discussion of gravity anomalies is found in sections dealing with general crustal structure, mountain ranges, rift systems, the mid-ocean ridges, continental margins, island arcs, and global gravity variations resulting from mantle inhomogeneities.

Brown, G. C., and A. E. Mussett. *The Inaccessible Earth: An Intergrated View to Its Structure and Composition.* London: Chapman and Hall, 1993. This book deals well with geophysics topics as they relate to the structure, chemical composition, and evolution of the earth. Its excellent line drawings helpfully illustrate complex ideas.

Chapis, D. A. "The Theory of the Bouguer Gravity Anomaly: A Tutorial." *The Leading Edge* (May 1996) 361-363. A classic article filling the gap of gravity anomalies found in basic texts.

Dobrin, M. B., and C. H. Savit. *Introduction to Geophysical Prospecting.* 4th ed. New York: McGraw-Hill, 1988. The newest edition of one of the most popular books in the application of geophysics to natural resource investigations. The majority of the book deals with seismic exploration. Three chapters on gravity prospecting treat the subject quite adequately. The mathematical treatment is good. Equations for gravity modeling of simple shapes are given.

Hamblin, William K., and Eric H. Christiansen. *Earth's Dynamic Systems.* 10th ed. Upper Saddle River, N.J.: Prentice Hall, 2003. This geology textbook offers an integrated view of the earth's interior not common in books of this type. The text is well organized into four easily accessible parts. The illustrations, diagrams, and charts are superb. Includes a glossary and laboratory guide. Suitable for high school readers.

Jacobs, John A. *Deep Interior of the Earth.* London: Chapman and Hall, 1992. An informative, mathematical treatment of the global variations of gravity as a function of latitude, elevation, and topography. Also provides an overview of the earth's structure and composition.

Robinson, Edwin S., and Cahit Coruh. *Basic Exploration Geophysics.* New York: John Wiley & Sons, 1988. The text includes three chapters on gravity, with clear descriptions and excellent diagrams that deal with the geologic interpretations that can be made based on gravity modeling.

Smith, P. J. *Topics in Geophysics.* Cambridge, Mass.: MIT Press, 1973. Contains an excellent, mathematically simple explanation of the free-air and Bouguer corrections, the principles of gravity modeling, and continental structures. This text is especially good for students at the high school level.

Stacey, F. D., and Paul M. Davis. *Physics of the Earth.* 4th ed. New York: Cambridge University Press, 2008. A mathematically rigorous treatment of the gravity of the earth and isostasy. Reformatted into units to make the topics more accessible. This text is best used by those who have a background in calculus and basic physics.

Tarbuck, Edward J., Frederick K. Lutgens, and Dennis Tasa. *Earth: An Introduction to Physical Geology.* 10th ed. Upper Saddle River, N.J.: Prentice Hall, 2010. This college text provides a clear picture of the earth's systems and processes that is suitable for the high school or college reader. It has excellent illustrations and graphics. Bibliography and index.

See also: Earth Tides; The Geoid; Isostasy; Metamorphism and Crustal Thickening; Mountain Building; Plate Tectonics; Relative Dating of Strata.

H

HEAT SOURCES AND HEAT FLOW

Several significant sources contribute to the internal heat of the earth. Radioactive decay of isotopes of thorium, uranium, and potassium in the crust produces most of the heat observed at the surface of the continents. Terrestrial heat flow is readily observed in deep wells, mines, and tunnels which penetrate below the narrow zone on the surface which is heated by daily and seasonal radiation changes. Another major source is heat convected outward from the earth's core and through the mantle. This heat produces the majority of the heat flow measured in the oceans, especially at the mid-oceanic ridges.

PRINCIPAL TERMS

- **asthenosphere:** the semi-molten portion of the outer mantle (ranging to a depth of 250 kilometers) which lies at the base of the lithosphere
- **basalt:** a fine-grained, dark extrusive igneous rock
- **conduction:** the transfer of heat caused by temperature differences
- **convection:** the transfer of heat by the movement or circulation of the heated parts of a liquid or gas
- **core:** the center portion of the earth which is divided into a liquid outer portion and a solid, denser inner section
- **crust:** the outermost layer of the earth, ranging in thickness from 5 to 60 kilometers; it consists of rocky material which is less dense than the mantle
- **lithosphere:** the outer, rigid portion of the earth which extends to a depth of 100 kilometers; it includes the crust and uppermost portion of the mantle
- **mantle:** the portion of the earth's interior extending from about 60 kilometers in depth to 2,900 kilometers; it consists of relatively high-density minerals which consist primarily of silicates

TERRESTRIAL HEAT FLOW

The surface of the earth receives heat from several different sources. Solar radiation provides the largest amount of heat—an amount that is approximately five thousand times greater than that moving outward from the subsurface. However, solar radiation is almost all reradiated into space, so it has very little effect on the earth's temperature deeper than a few meters. The amount of terrestrial heat flow is so small that it would take several months for a dish of water placed on the surface to heat up by an additional 1 degree Celsius. During that period of time, the surface temperature on the earth could easily fluctuate 20 to 30 degrees Celsius, depending on the time of year. Because of the large variation in surface temperatures, it is necessary to measure terrestrial heat flow at depths that lie well below those that are affected by these relatively short-term changes.

The fact that subsurface temperatures increase with depth has been known from deep mines and wells which penetrate several kilometers into the earth's crust. The temperature at a given depth cannot be determined with the degree of accuracy that pressures at depth can be calculated. Variations in temperature are attributable to several important variables, particularly radioactive heat production and the coefficient of heat transfer. Our inability to measure these parameters at great depths within the crust makes it difficult to establish a well-defined temperature versus depth graph.

In general, the amount of outward heat flow on the surface is small. The mean heat flow for the earth ranges between 60 and 70 milliwatts per square meter. Average values for oceanic and continental areas are essentially equal. Significant differences for values measured in the two general areas depend on the specific geologic setting within the oceanic or continental area. Observed values also differ significantly for measurements taken at specific locations in either the oceans or the continents. Heat-flow measurements in continental regions vary from 41 milliwatts per square meter in Precambrian shields (large masses of igneous and/or metamorphic rock) to 74 milliwatts per square meter for more active, younger Mesozoic and Cenozoic areas. Within the ocean basins, heat-flow values range from 49 to 80 milliwatts per square meter or more. Many researchers provide a multitude of different values for the various geologic regions.

OCEANIC HEAT FLOW

Molten basaltic rock formed by extrusion at the mid-oceanic ridges cools as it moves out from this primary heat source. This heat is being transferred by both convection from the mantle and conduction of heat produced by radioactive isotopes within the rocks in the oceanic layer. As the molten basalt cools, it contracts and undergoes a small density increase, which allows the newly formed layer to sink deeper into the asthenosphere. The depth to which the cooler basaltic layer sinks has been shown to be roughly proportional to the square root of its age. Oceanic crust that is 2 million years old is covered by about 1.5 kilometers of water, whereas crust that is 20 million years old would be located at a considerably greater depth below the ocean surface.

Subduction trenches have the lowest heat-flow values among measurement sites in the oceans. These trenches are deep depressions on the ocean floor usually adjacent to the boundary of continental-oceanic or oceanic-oceanic plate collisions. The lower values in those regions arise for several reasons. As discussed earlier, most of the heat observed in the oceans results from volcanic activity associated with the mid-oceanic ridges. This heat is lost once the hotter volcanic material moves away from the ridge. Trenches are very distant from the ridges, so the oceanic floor in the trench areas has long since been removed from the heat source as the floor has spread outward.

Trenches associated with continental-oceanic plate collisions are also areas that receive large volumes of sediment that are shed from the continents through normal erosion processes. In addition, the deposition of organic and fine-grained clastic material from seawater fills the trenches. All these sediments produce an insulating blanket that can be as thick as 1 kilometer overlying the basaltic oceanic floor. Any heat produced by the basalts is held in by this sedimentary layer. Heat-flow measurements taken in the oceans are made in this soft sediment layer, not in the underlying basalt.

RADIOACTIVE DECAY OF ISOTOPES

Heat is produced by the natural radioactive decay of isotopes of thorium, uranium, and potassium. The long-lived radioactive isotopes which contribute most to the overall heat production are thorium-232, uranium-238, potassium-40, and uranium-235 (listed in order of decreasing importance). All these isotopes have half-lives roughly the same as the age of the earth. These isotopes are not readily incorporated into the internal structure of the most common minerals. They do tend to become concentrated in minerals that have lower melting points, such as those occurring in granites. A granitic crust 20-25 kilometers thick can produce the amount of terrestrial heat flow observed over these areas. If a more intermediate composition for the continental crust is used, it is still possible to account for at least two-thirds of the heat flux of the continents as being generated from the crust. The remaining heat flow (about 20 milliwatts per square meter) is assumed to come from the mantle.

Oceanic crust, which has an average thickness of 5 kilometers, produces less than 3 percent of the heat flow observed in oceanic measurements. As previously mentioned, the oceans and continents exhibit near-equality in their total heat-flow measurements. The implication is that the mantle underlying oceanic rocks has a higher temperature and greater amounts of radioactive elements than continental mantle and thus serves as the primary heat source in the oceans. This higher heat flow is conducted through the basaltic layer and also convected upward at the mid-oceanic ridges. It must be remembered, however, that, as a result of the mobility of the lithosphere, this higher-temperature mantle material underneath the oceans eventually moves under the continents and helps compound the interpretation of actual heat flow being produced by the mantle. Almost 75 percent of the earth's total heat loss occurs through the ocean floors, a percentage similar to the amount of surface covered by the oceans (70 percent). The vast percentage of this oceanic heat loss is related to the formation of new oceanic crust at the mid-oceanic ridges.

HEAT FLOW VARIABLES

The amount of heat flow observed in the oceans and on the continents is dependent on a number of variables. The age of the rock is of primary importance in that older rocks produce less heat, because the radioactive isotopes in the rocks have had a longer time to break down and hence less of the original heat-producing parent isotope is present. Older basalts on the ocean floor are also farther removed from their primary heat source associated with volcanism along the mid-oceanic ridge, where they were first extruded.

Continental heat-flow values are dependent on this same concept of age of the parent rock. On the continents, the lowest heat-flow values are associated with Precambrian shield areas, those portions of the continents which represent the most stable and oldest central core of the landmasses. These areas have also been subjected to the most erosion, which has removed a significant portion of the rocks and minerals containing the radioactive constituents. Some portions of the continents have undergone extensional stresses, which have pulled the landmasses apart and thus thinned the continents. In these places, such as the Basin and Range area of Nevada, the crust has been thinned just as a rubber band becomes thinner when it is extended. The result is that the mantle is much closer to the surface and its influence is enhanced with respect to the amount of heat being moved upward toward the surface. Areas on the continents which have experienced Mesozoic and Cenozoic mountain-building activity also show high levels of heat flow, yet display the greatest amount of variation in observed values. However, these areas of tensional tectonics and mountain-building activity make up a small percentage of the continents and thus do not contribute much to the average values observed on land.

Surface temperatures can also be affected by groundwater flow, soil moisture, slope orientation, vegetative cover, topography, and sun angle. These contributing factors must be removed to obtain heat-flow readings that are representative of rocks at depth.

HEAT FLOW CONDUCTION AND TRANSFER

Scientists have found that the earth is losing heat. The heat that is being lost is produced partly by higher rates of radioactive heat production in the past and partly by heat generated by the formation of the earth. The inner core of the earth has a temperature estimated to be between 4,000 and 5,000 degrees Celsius. If the earth were a perfect conductor of heat, the rate of heat loss would offset the rate of heat generation. The mantle is a poor conductor of heat, however, so it stores heat, which is slowly released as the core and lower mantle cool. Calculations have shown that heat in the lower mantle is not completely transferred to the upper mantle and later to the surface. If large-scale convection cells existed within the mantle, there would be a more efficient transfer of this deep-seated heat; this is not observed, however. The best models of the internal transfer of heat seem to point to two levels of convection cells, one lying in the lower mantle and a second, separate level in the upper mantle. This latter level serves as an additional insulator from the deeper heat, which is trying to rise to the surface. The lateral motion of these convection cells in the upper mantle also serves as the primary mechanism to move the continental masses around. The continents are of a lower density than the underlying material; hence they "float" on the lithosphere and asthenosphere. These convection cells raft the continents around on the earth's surface, albeit at a slow rate (several centimeters per year). It must be remembered that heat-transfer processes in the mantle and core are not directly observable. Therefore, geophysicists do not totally agree on the mechanisms which explain the production and transfer of heat at great depth.

Heat flow itself is controlled by the second law of thermodynamics. This law states that for thermal equilibrium to be attained, heat must move from warmer to colder material. This means that in the earth heat flows from the warmer interior to the colder lithosphere and crustal-atmosphere boundary. Heat produced within the earth is transferred in two ways. Deep-seated sources in the mantle and core transfer heat to the upper mantle and crust by convection. Heat rises slowly up through the mantle until it encounters the base of the lithosphere. Some heat is conducted into the relatively cooler lithosphere, while the remainder moves laterally along the boundary of the lithosphere and mantle. As it does so, the temperature of the upper mantle lessens and eventually the cooler rock material sinks back into the mantle, where it is reheated and returned into the convective cycle. Within the oceanic and continental masses, heat is also conducted by the rock. Although rocks are generally poor conductors of heat, the vast quantities being generated by radioactive decay and moved by convection are conducted toward the surface.

Hot Spots

Heat sources mentioned so far are large, often extending hundreds or thousands of kilometers across and through the earth and displaying a broad horizontal and vertical expanse within the crust and mantle. Mantle plumes or hot spots represent much more localized heat sources. Only several dozen of these features have been recognized to date. Their spatial distribution is widespread and generally dispersed, with a slight concentration located along portions of the mid-oceanic ridges. Hot spots are usually only several tens of kilometers in diameter. They are thought to be conduits of heat rising from the mantle and intersecting the earth's surface. Several well-known examples include Iceland, the Hawaiian Islands, and Yellowstone National Park in Wyoming.

Iceland is the result of a hot spot which lies directly on the Mid-Atlantic Ridge. Its volcanic nature is direct evidence of the heat and type of rock produced by the upward movement of heat from beneath the ocean floor. Geothermal energy produced by the volcanism is used as the primary heat source for the island. An obvious hazard of the geologic setting of the island is that volcanic eruptions can adversely affect everyday life in the area.

The Hawaiian Islands are volcanic mountains which rise from the deep ocean floor to elevations of more than 4,200 meters above sea level. When considered in total, they are the highest mountains on earth. The entire string of islands in the Hawaiian Islands chain formed as the result of the Pacific plate having moved in two separate stages in a northwesterly direction over a hot spot. The present position of the Pacific plate has the hot spot centered on the southeastern corner of the island of Hawaii. This area has experienced very active volcanism in the past few centuries and has been the site of numerous eruptions since 1983. Ocean-floor reconnaissance has

Steam rises from vents of Porcelain Basin in Yellowstone National Park's Norris Geyser Basin. (AP Photo)

detected the formation of a new island to the southeast of the island of Hawaii. This submarine volcanic feature has been named Loihi, and it will continue to grow until it breaks through the ocean surface to produce another major island in the chain.

The Yellowstone caldera, located in and around Yellowstone National Park in the western United States, is an excellent example of a hot spot that has risen through the continental lithosphere. A topographic high is centered on the caldera. Elevations decrease as the cooler lithosphere moves out from the center of the hot spot. Evidence for the heat exists in geologically recent volcanic activity and the present-day hot springs and geysers found in the park.

HEAT-FLOW MEASUREMENT

Heat-flow determinations depend on two separate measurements: the rate of increase of temperature with depth (which is termed the vertical temperature gradient, r) and the thermal conductivity (K) of the rocks in which the temperatures are being measured. The flux, or rate, of heat flow (q) is calculated using the formula $q = GKr$. The units of q are watts per square meter, those of K are watts per meter per degree, and those of r are per degree per meter (deg $C^{-1}m^{-1}$). Absolute temperatures (in kelvins) can also be used to express these parameters. The minus sign denotes the fact that heat flows down the temperature gradient, from the warmer spots to the colder ones. It must be noted that it is standard practice to consider heat-flow values as positive numbers even though the values obtained from the equation are indeed negative.

The temperature gradient is fairly linear within certain depth ranges in the earth. David G. Smith, in *The Cambridge Encyclopedia of Earth Sciences* (1981), provides a figure which roughly defines these different gradients for varying depths in the crust and upper mantle. The rate of increase of the temperature gradient, which is also referred to as the geothermal gradient, is greatest in the outer 1,500 kilometers. The rate of increase decreases significantly at a depth of 2,900 kilometers, where the core-mantle boundary exists. This is attributable to a change in state of the minerals present at that depth.

From the above equation, it is clear that two variables must be measured to determine the rate of heat flow. The thermal conductivity (K) is usually measured at discrete points along a core sample taken from the borehole. Errors can be introduced because of temperature contamination as the cores are brought to the surface. Average values are obtained by taking a series of measurements along the retrieved core. Measurements of the temperature gradient (r) are obtained by placing a probe into the borehole (or driving it into the ocean sediment). The probe has a series of thermal sensors attached to it which record the temperatures at various distances along the probe. The gradient is calculated by dividing the temperature differences by the known distance of separation between the respective probes.

Observed heat-flow values of ocean bottoms are much less variable than those measured on the continents. In water depths exceeding several hundred meters, it is necessary only to measure temperatures in the upper few meters of the sediments and to establish the thermal conductivity over this same interval. These shallower probe depths in the oceans are permitted because of the more stable heat regime at the boundary of the cold seawater and the ocean sediments. On the continents, however, seasonal variations resulting from solar radiation can affect heat-flow measurements to varying depths. For an average continental rock, the daily change in temperature affects rocks only to a depth of about 15 centimeters; annual variations extend down about 3 meters, while longer-term variations can reach to more than 8 meters. In addition, the flow of groundwater in the upper 50-100 meters alters the heat regime in the subsurface. In some areas, such as highly fractured rocks, groundwater effects can penetrate to depths of as much as 1 kilometer. Therefore, heat-flow measurements must be taken at depths great enough to remove these effects.

David M. Best

FURTHER READING

Cook, A. H. *Physics of the Earth and Planets*. New York: Halsted Press, 1973. This book provides succinct presentations describing the physical properties of the earth observed both on the surface and in the interior. A relatively advanced text, suitable for college-level readers.

Diller, T. E. "Advances in Heat Flux Measurements." In *Advances in Heat Transfer*, Vol. 23, edited by J. P. Hartnett et al. Boston: Academic Press, 1993: 279-368. *Advances in Heat Transfer*, is one of a series of volumes developed to bridge the gap between academic journals and textbooks. This article provides a detailed description of measurement techniques and advances that were taking place during the early 1990's.

Duffield, Wendell, and John Sas. *Geothermal Energy; Clean Power from the Earth's Heat.* USGS Circular 1249. 2003. The report begins with a description of historical and current geothermal energy use. Content covered in this report includes global geothermal applications, mining of geothermal energy, hydrothermal systems and dry geothermal systems, and the environmental impact of geothermal energy use. This report includes numerous color diagrams and images.

Francheteau, Jean. "The Oceanic Crust." *Scientific American* 249 (September 1983): 114. A clearly written article in an issue which addresses all facets of the geology of the earth. This article has good diagrams showing the role the oceanic crust plays in the overall geologic setting of the earth. Suitable for high-school-level readers.

Jacobs, J. A. *The Earth's Core.* London: Academic Press, 1975. Deals in detail with all aspects of the earth's inner and outer core. The origin of the core, its constitution, and its thermal and magnetic properties are discussed in detail. Well suited to the serious science student.

Jacobs, John. *Deep Interior of the Earth.* London: Chapman and Hall, 1992. This introductory geophysics textbook is formidable for the average student because there is considerable mathematics in some chapters, but it does cover many useful topics. It contains a minimum of equations but many figures and graphs.

Lowell, Lindsay, et al., eds. *Geology and Geothermal Resources of the Imperial and Mexicali Valleys.* San Diego: San Diego Association of Geologists, 1998. This detailed account of the geothermal resources and processes of the Imperial and Mexicali Valleys in California contains many helpful illustrations and maps to help clarify the subject. Includes a bibliography.

Plate, Erich J., et al., eds. *Buoyant Convection in Geophysical Flows.* Boston: Kluwer Academic Publishers, 1998. Although highly technical at times, this collection of papers provides good descriptions and discussions about heat convection theories, geophysics, and heat flows. The collection also offers illustrations and a bibliography.

Skinner, Brian J., and Stephen C. Porter. *Physical Geology.* New York: John Wiley & Sons, 1988. A well-written, clearly illustrated text which provides all the basic concepts of geology at a level which advanced high school and college science students can understand.

Smith, David G., ed. *The Cambridge Encyclopedia of Earth Sciences.* New York: Crown Publishers, 1981. Organized as a compilation of high-quality and authoritative scientific articles rather than a typical encyclopedia. Chapter 9, "The Energy Budget of the Earth," offers a good summary of the external and internal energy sources affecting the earth, including a table showing the contributions that geologic regions and isotopes make to the total heat-flow regime on Earth. An excellent glossary and good graphics augment the text, which is aimed at college-level readers.

Stacey, Frank D., and Paul M. Davis. *Physics of the Earth.* 4th ed. New York: Cambridge University Press, 2008. This advanced text provides the mathematical basis for understanding many global geophysical processes. It also provides thorough explanations of the processes which involve the physics and geology both on and in the earth. Reformatted into units to make topics more accessible. Contains an excellent bibliography listing several hundred references on specific topics.

See also: Earth's Core; Earth's Differentiation; Earth's Lithosphere; Earth's Mantle; Elemental Distribution; Isotope Geochemistry; Lithospheric Plates; Mantle Dynamics and Convection; Plate Tectonics; Plumes and Megaplumes; Potassium-Argon Dating; Radioactive Decay; Subduction and Orogeny; Uranium-Thorium-Lead Dating.

IMPORTANCE OF THE MOON FOR EARTH LIFE

The giant-impact theory of the moon can account for the origin of the moon and many of Earth's unusual features, such as the right rotation rate, axial tilt, atmosphere, iron core, magnetic field, plate tectonics, and tidal history, all of which have created a unique environment for the existence of complex life forms.

PRINCIPAL TERMS

- **accretion:** the accumulation of gas and dust in the solar disc, forming celestial objects such as asteroids, planetesimals, planets, and moons
- **dynamo theory:** an attempt to explain the magnetic field of some celestial objects in terms of rotation, heat convection, and electrical conduction in a fluid and metallic inner core
- **ecliptic plane:** the plane of Earth's orbit around the sun
- **escape velocity:** the minimum speed required for an object to escape from the gravity of the earth or other celestial object
- **giant-impact theory:** a theory that describes a glancing collision with Earth of a Mars-size planetesimal that blasted enough debris into orbit to form the moon
- **greenhouse gases:** atmospheric gases such as water vapor, carbon dioxide, and methane that trap heat by absorption of solar radiation and re-emission of longer wavelengths that cannot escape from the atmosphere
- **magma ocean:** a deep layer of molten rocks, volatiles, and solids that may have covered large portions of Earth and the moon in their early stages of formation
- **planetesimals:** celestial objects that form by the accretion of dust and gravitational attraction that eventually combine to form planets
- **plate tectonics:** a geological theory describing the large-scale motions of the earth's lithosphere that account for continental drift and mountain building
- **solar wind:** the stream of high-energy charged particles flowing from the upper atmosphere of the sun, consisting mostly of protons and electrons
- **volatiles:** chemical elements and compounds, such as oxygen, hydrogen, nitrogen, carbon dioxide, and methane, that are found in Earth's crust and atmosphere and which have low boiling points

ORIGIN OF THE MOON

The giant-impact theory of how the moon was formed not only explains the origin and unusual features of the moon but also accounts for many unique properties of the earth that make it the only habitable planet in the solar system. Several historical theories for the origin of Earth's moon have been proposed, but none has been completely successful.

Immanuel Kant's co-accretion theory suggested that Earth and the moon formed from the same dust cloud. British astronomer and mathematician George Darwin's fission theory assumed that the moon split from Earth. American astronomer Thomas See's capture theory tried to explain how Earth's gravity could pull a planetesimal into its orbit. However, none of these theories could account for all of the moon's unusual features, including its large size relative to Earth, its orbital properties, its low density, its small iron core, and evidence from the Apollo space missions that the moon lacks volatiles and once had a magma ocean.

The origin of the moon and its unusual features were finally explained by the giant-impact theory, which succeeded by combining various aspects of earlier theories. The giant-impact theory was first proposed by Canadian geologist Reginald Daly at Harvard University in 1946. Daly suggested that a glancing collision of a large planetesimal with the Earth early in its history would blast vaporized debris into orbit, which would eventually coalesce to form the moon. This theory was ignored until after the Apollo missions, when American planetary scientists William K. Hartmann and Donald R. Davis began to apply computer programs to the problem of planetesimal formation in the early solar system and revived Daly's giant-impact idea in 1974.

About the same time, astrophysicist Alastair Cameron (another Canadian, also based at Harvard) and his student William Ward began to develop computer simulations of a glancing collision, finding that formation of the moon required an impactor about ten times larger than the moon itself. According to this model, one-half of the debris that had blasted into space would remain in orbit, and within a few weeks some of it would coalesce to form the moon. These simulations showed that the collision would increase Earth's mass by about 10 percent, increase its rate of rotation to about five hours, and produce a moon deficient in iron and volatiles.

The giant-impact theory made little progress until 1984, when a post-Apollo conference was held in Kona, Hawaii, about the origin of the moon. Several conferees presented papers on the giant-impact theory, leading to a growing consensus in favor of the theory. More improved simulations followed the Kona consensus, especially by Cameron, who had retired to Arizona, and by American astrophysicist Robin M. Canup at the Southwest Research Institute in Boulder, Colorado. The two scientists began using a method developed for modeling bomb explosions called the smooth particle hydrodynamics (SPH) method, in which their simulations differentiated between rock and iron particles (several thousand of each). These new simulations showed the melted iron core of the impactor blasting into space and then falling back to Earth and sinking into its core.

Planetesimal accretion models suggest that the giant impact occurred about forty million years after the formation of the solar system as determined from the oldest meteorites. An interesting extension to the theory was proposed in 2011 based on a computer simulation, suggesting that the ejected material from the giant impact formed two bodies, which eventually collided to form the moon. This proposal explains why the far side of the moon has a thicker crust and more highlands, resulting from the smaller body striking the larger on what is now the far side.

It has long been recognized that Earth's moon provides several benefits for life on Earth. These benefits include keeping time based on the phases of the moon, illumination of the night sky, and lunar tides for cleansing and oxygenating the oceans. The giant-impact theory suggests a number of additional benefits that are critical to providing the conditions that are needed for life on Earth. Ten such essential factors appear to be related to the formation of a large moon, assuming that complex life requires liquid water. The first five of these beneficial results relate to the giant impact itself, and the last five result from the moon's subsequent influence on Earth. The absence of any of these benefits might have prevented the development of life on Earth.

IMMEDIATE BENEFITS OF THE GIANT IMPACT

In the giant-impact theory, the glancing collision itself produced a minimum of five effects on the Earth that helped prepare it to support life. The first of these was an increase in Earth's spin to an initial five-hour rotation rate, much faster than any other rate in the solar system. This initial rate was fast enough so that over the time for life on Earth to develop, the rate could be slowed to the current twenty-four-hour day by the moon's tidal action on Earth's oceans. Earth's current rotation rate makes photosynthesis possible and moderates temperatures between the freezing and boiling points of water over most of the planet. By comparison, Mercury's rotation rate of fifty-nine days produces long 100 kelvin (-173 degrees Celsius [C]) nights and 700 kelvin (427 degrees C) days.

A second benefit of the giant impact was to provide Earth with a favorable axial tilt, leading to the current inclination of Earth's equatorial plane of about 23 degrees relative to its orbital plane (ecliptic). This axial tilt gives the earth its relatively mild seasonal variations; the tilt is large enough, however, to stimulate evolutionary processes. By contrast, Mercury's axis has no tilt and thus no seasonal variations. The tilt of Earth appears to have remained fairly steady through time, as evidenced by the growth pattern of an 850-million-year-old stromatolite. Analysis based on the assumption of growth toward the noontime sun suggests a 26.5 degree tilt at that time.

A third life-supporting benefit of the giant impact was the apparent removal of greenhouse gases. Several investigators have suggested that a glancing collision would have stripped Earth of its primordial atmosphere. By comparison, the atmospheric pressure of Venus without a glancing collision is ninety times that of Earth, producing greenhouse temperatures of about 700 kelvins, which has boiled away all surface water. On Earth, surface water absorbs much of the excess carbon dioxide.

After the giant impact and the resulting magma ocean on Earth, a new atmosphere formed from outgassing and comet collisions. Eventually, a new crust formed and water condensed and formed oceans to absorb carbon dioxide. The reformulated atmosphere was then thin enough to prevent a runaway greenhouse effect and to eventually allow photosynthesis to occur with its associated production of oxygen.

There are two further benefits from the giant impact: first, the action of the molten iron core of the impactor falling back to Earth and sinking into its core; and second, the transfer of the impactor's mass to Earth. Both of these benefits have been revealed by modern computer simulations.

The dynamo theory of Earth's magnetism shows that the enlargement of the liquid iron core, together with a much faster rotation rate, increased Earth's magnetic field to about one hundred times that of any other rocky planet. A strong magnetic field deflects charged particles in the solar wind. Without the magnetic field, this wind would have stripped away much of Earth's atmosphere and threatened its emerging life.

Computer simulations show that most of the mass of the Mars-size impactor transferred to Earth, increasing its mass by about 10 percent. This increase in mass was critical for life because it provided sufficient gravity to keep Earth's water vapor from escaping to space before it could condense to form the oceans.

Later Benefits for Life from a Large Moon

Five further benefits for life on Earth emerged after the giant impact, but resulted from the impact and the large moon it produced. The first of these benefits was to contribute to the conditions required for plate tectonics, which provides strong support for life and occurs on no other known planet. The glancing collision removed up to 70 percent of Earth's crust, added significantly to its core and mantle heat, and it increased radioactive isotopes to sustain this heat.

When a thinner crust re-formed on Earth after the collision, Earth was more susceptible to cracking and the driving forces of heat convection, enhanced by increased internal heat. Continuing plate tectonics built the mountains and continents of Earth, without which it would be mostly covered with water and have little chance for developing land-based life. Tectonic activity also recycles the crust, bringing minerals to the surface and controlling long-term climate by the carbon-rock weathering cycle, which helps to balance atmospheric carbon dioxide and prevent temperature extremes.

A second benefit that followed the giant impact came from huge tides early in Earth's history, which eroded the land and enriched the oceans with the minerals needed for life. Calculations show that the moon formed about fifteen times closer than it is today, when Earth had its initial five-hour day. In the late nineteenth century, George Darwin showed how tidal action slows Earth's rotation and increases the orbital distance of the moon. When the moon was about ten times closer than it is now, and when the day had slowed to perhaps ten hours, the tidal forces would have been one thousand times larger and hundreds of times higher than today's tides. Huge tides from the early moon would have eroded minerals from far inland about every five hours, enriching the oceans with life-sustaining minerals.

A third benefit after the giant impact resulted from the slowing effect on Earth's rotation caused by the tidal action from a large moon, also shown by Darwin. Early rapid rotation produced super-hurricane winds, similar to those produced on Jupiter by its rapid ten-hour rate of rotation, which would have posed severe threats to most life forms. Geological evidence for the slowing of Earth's rotation involves alternating layers of silt and sand offshore from tidal estuaries, showing that the earth's rotation had slowed to about an eighteen-hour day by 900 million years ago. A slower rotation rate optimized wind circulation and surface temperatures for the development of life.

Another benefit from early tides was their role in forming intertidal pools, which have long been recognized as ideal locations for concentrating nutrients by evaporation for emerging life forms. Rapid tidal cycling occurred when the day was shorter and the moon was closer, so that the tides would have been larger and tidal pools would have covered larger areas. Longer cycles between spring and neap tides might have given several days for drying. Some investigators have suggested that cycles of tides and evaporation along the shorelines of the early oceans might have provided the kind of environment in which protonucleic acid fragments could begin to associate and assemble molecular strands, leading to the origin of life.

A final beneficial result of a large moon is its stabilizing effect on the tilt of Earth's axis. In the early 1990's, scientists showed that gravitational forces from Earth's large moon keeps Earth's axis tilted in a narrow range between 22 and 25 degrees, stabilizing annual climate variations in a favorable range for living organisms and producing the regular seasons that occur on Earth. Thus, Earth's large moon prevents the kind of large and chaotic changes in tilt that have been shown to occur on Mars, which has two small moons.

IMPLICATIONS FOR PLANETARY LIFE

All the results described in computer simulations of the giant-impact theory and of the large moon it produced are apparent contributions to making life on Earth possible. It appears also that the lack of any one of these results might have prevented the development of complex life forms, including human life. Not only is it remarkable that Earth has all of these life-sustaining features but these features also appear to be the result of just the right kind of glancing collision to form a large enough moon. Beyond these lunar features, Earth has many other properties that are needed for life, such as a favorable location in the galaxy, the right size sun, the right distance from the sun, a sufficient amount of water, and an ozone layer to protect from ultraviolet radiation. These conditions greatly restrict the possibilities of life elsewhere in the galaxy, even without taking into account the probable requirement of a large moon. Optimistic estimates, however, claim that more than 10,000 planets with intelligent life should exist among the more than 100 billion stars in the Milky Way galaxy alone.

The unlikely possibility of a giant-impact formation of a large moon is consistent with an infrared survey by the National Aeronautic and Space Administration's Spitzer telescope of more than four hundred young stars just a few million years past their planet-forming age. The survey revealed only one dust cloud signature large enough to be a possible moon-forming collision. Computer studies have shown that any accreting planet has some chance of being hit by a planetesimal object about one-tenth its size (as in the apparent formation of Earth's moon).

However, it is also evident that the right kind of glancing collision has a very low probability. Any estimate of this probability should take into account a minimum of five independent parameters, each with its own estimated probability of less than one-tenth: the right size impactor, the right time for the impact to occur, the right direction for an effective glancing collision, the right point of impact on the proto-earth, and the right speed to place enough debris in orbit for a large moon. The product of these factors gives an estimated probability of about one-millionth for such an event, which combined with other requirements, makes it surprising that intelligent life exists on even one planet.

Joseph L. Spradley

FURTHER READING

Asphaug, E., and M. Jutzi. "Forming the Lunar Farside Highlands by Accretion of a Companion Moon." *Nature* 476 (August 3, 2011): 69-72. This article proposes a new extension of the giant-impact theory based on computer simulations that show two moons forming from the impact debris and later colliding to form Earth's moon.

Belbruno, E., and J. Richard Gott, III. "Where Did the Moon Come from?" *Astronomical Journal* 129, no. 3 (March 2005): 1724-1745. The authors analyze the behavior of a planetesimal in an orbit with Earth and the conditions that would lead to a collision and the formation of a large moon.

Benn, Chris R. "The Moon and the Origin of Life." *Earth, Moon, and Planets* 85/86, no. 6 (2001): 61-67. The author provides a concise discussion of the giant-impact theory of the formation of a large moon and its implications for the origin and evolution of life on Earth.

Comins, Neil F. *What If the Moon Didn't Exist? Voyages to Earths That Might Have Been.* New York: HarperCollins, 1993. This is one of the first popular books to dedicate substantial focus to the implications of a large moon for the existence of life on Earth.

Galimov, E. M., and A. M. Krivtsov. "Origin of the Earth-Moon System." *Journal of Earth Systems Science* 114, no. 6 (December 2005): 593-600. The authors establish certain geochemical constraints on the formation of the moon that are not completely met by the giant-impact theory in terms of similarities and differences in lunar and terrestrial materials.

Mackenzie, Dana. *The Big Splat: Or, How Our Moon Came to Be.* Hoboken, N.J.: John Wiley & Sons,

2003. This popular presentation discusses efforts to understand the origin of the moon and the development of the giant-impact theory. Also examines the implications for existence of life on Earth.

Spradley, Joseph. "Ten Lunar Legacies: Importance of the Moon for Life on Earth." *Perspectives on Science and Christian Faith* 62, no. 4 (December 2010): 267-275. This article describes the giant-impact theory leading to the formation of a large moon and gives ten life-supporting results of such an impact, including extensive documentation of original sources.

Ward, Peter, and Donald Brownlee. *Rare Earth: Why Complex Life Is Uncommon in the Universe*. New York: Copernicus, 2000. This groundbreaking book describes many unusual life-sustaining features of the earth, including a discussion in Chapter 10 of the giant-impact theory and the apparent necessity of a large moon for life on Earth.

See also: Earth-Moon Interactions; Earth's Differentiation; Earth Tides; Gravity Anomalies; Jupiter's Effect on the Earth; Lunar Origin Theories; Plate Tectonics; Solar Wind Interactions.

INFRARED SPECTRA

The infrared spectrum is part of the electromagnetic spectrum that lies beyond the red color that human eyes perceive as visible light. Every body emits some energy in the infrared or near infrared. Detection of infrared radiation by special instruments is used in numerous fields, including medicine, mapping, defense, communication, and astronomy.

PRINCIPAL TERMS

- **angstrom:** a unit of wavelength of light equal to one ten-billionth of a meter
- **bolometer:** a detector of radiant energy that works by determining the change in resistance of an electrical conductor due to temperature change
- **macroscopic:** large enough to be observed by the naked eye
- **photon:** a quantum of radiant energy
- **spectrograph:** an instrument for dispersing radiation (as electromagnetic radiation) into a spectrum and photographing or mapping the spectrum
- **spectrometer:** an instrument used in determining the index of refraction; a spectroscope fitted for measurements of the observed spectra
- **spectrophotometer:** a photometer for measuring the relative intensities of the light in different parts of a spectrum
- **spectroscopy:** the subdiscipline of physics that deals with the theory and interpretation of the interactions of matter and radiation (as electromagnetic radiation)
- **spectrum:** an array of the components of an emission or wave separated and arranged in the order of some varying characteristic (such as wavelength, mass, or energy)
- **spectrum analysis:** the determination of the constitution of bodies and substances by means of the spectra they produce

ELECTROMAGNETIC SPECTRUM

The small bands of infrared radiation seeping through the atmosphere were accidentally discovered in 1800 by German-born British astronomer William Herschel. When measuring the temperatures of the visible light spectra, he found a source of greater heat and wavelength radiation beyond the color red. However, it was not until 1881, when American physicist Samuel Pierpont Langley developed the bolometer, that the first in-depth studies of the infrared were possible. German physicist Max Planck's development of the quantum theory in 1900 and Einstein's discovery of photons in 1905 led to the development of quantum detectors, which further advanced the study of the infrared. These early detectors brought forth modern spectroscopes, spectrometers, and spectrophotometers.

The electromagnetic spectrum comprises visible light and six forms of invisible radiation: radio, microwave, infrared, ultraviolet, X, and gamma rays. All spectra travel at the speed of light in waves of energy bundles called photons and can be reflected, refracted, transmitted, absorbed, and emitted.

Infrared spectrometry encompasses the study of wavelengths in the electromagnetic spectrum that range between 0.7 micron in the near-infrared photographic region and 500 microns in the far-infrared rotation region. Wavelengths of the infrared spectra are most useful for detecting certain atoms and molecules visible in the infrared, such as hydrogen, the most abundant element in the universe. Infrared rays differ from the other components of the electromagnetic spectrum. Radio waves are propagated through the atmosphere and have wavelengths between 30,000 meters and 1 millimeter. Microwaves have wavelengths between 1 meter and 1 millimeter. X-rays have extremely short wavelengths, approximately 1 angstrom, and are generated by a sudden change in the velocity of an electrical charge. Gamma rays are similar to X-rays but are of a higher frequency and penetrating power. Ultraviolet rays are beyond violet in the spectrum and have wavelengths shorter than 400 angstroms.

INFRARED RADIATION

Infrared radiation is emitted in some amount by every macroscopic body in the universe that has a temperature above absolute zero (or -273 degrees Celsius). In each macroscopic body, molecules not only are moving in all directions but also are rotating, and at the same time the individual atoms within the molecule are vibrating with respect to one another. It is the interaction of molecules with radiation that is the essence of the study of infrared.

For a molecule to absorb radiation, it must have a vibrational or rotational frequency the same as that of the electromagnetic radiation. In addition, a change in the magnitude and/or direction of the dipole moment must take place. The dipole moment is a vector that is oriented from the center of gravity of the positive charges to that of the negative charges, and it is defined as the product of the size and the distance between these charges. Corresponding frequencies between radiation and molecules are possible because radiation has an electrical component, in addition to having a magnetic component. In contrast, a molecule has an electrical field. When the electrical field of the molecule is rotating or vibrating at the same frequency as is the incoming radiation, then it is possible for a transfer of energy to take place.

The second requirement for the study of infrared, the dipole moment change, must have something to couple the energy from the radiation to the molecule. If atoms differ in their electronegativity and they combine to form a molecule, the centers of the positive and negative electrical charges may not coincide, producing a permanent dipole moment. The energy to produce this work can come from the absorption of the incoming radiation by the molecule. A permanent dipole is also necessary for inducing rotation. Atoms rotate because the electric fields are not the same on each side, thereby allowing for a transfer of energy; when energy is transferred in this manner, a rotator will rotate faster under certain rotational frequencies, while a vibrator will not change its frequency but will increase its amplitude of vibration. Because vibrational frequencies are of the order of 10^{14} cycles per second and rotational frequencies are 10^{11} cycles, they fall within the infrared region. Absorption bands for rotational spectra are quite sharp, but the bands for vibrational spectra tend to become broader because of the rotational levels associated with each vibrational level.

INFRARED DETECTORS

Detectors, either thermal or quantum, are commonly used to study the infrared. Each type uses a different property of electromagnetic radiation to convert the infrared to an electrical signal with intensity equivalent to the amount of infrared striking the detector. A thermal detector measures heat-induced changes in a property of a material, usually electrical resistance. A quantum detector also measures change, although it uses a photon—not heat—to create successive events when it strikes a material. There are three types of quantum detectors, each of which uses a separation, or diffusion, of different types of electrons as a catalyst for an event. In brief, the photoconductive effect uses incidental radiation to increase electrical conductivity, the photovoltaic effect uses a special junction for diffusion which creates voltage from charge separation, and the photoelectromagnetic effect uses radiation falling on a semiconductor with a magnetic field. In addition to detectors, various other instruments are used in the study and application of the infrared, such as the radiometer, the comparator, the collimeter, and modulators, all of which perform unique and valuable tasks.

SPECTROMETERS

A device basic to understanding how the infrared spectra works is the spectrometer. All spectrometers use certain elementary components. These include a source of radiation, a condensing source for focusing energy onto the monochromatic (pertaining to one color or one wavelength) slit, a monochromator to isolate a narrow spectral range, a radiation detector, and some form of amplifying system and output recorder. Single-beam spectrometers record energy versus wavelength, whereas a double-beam spectrometer measures the ratio between energy transmitted by the sample and energy incident on the sample, and plot transmittance or a related quantity as a function of wavelength or wavenumber. One micron is equal to one-millionth of a meter, and wavenumber is obtained by dividing 10,000 by the wavelength in microns.

EMISSION SPECTROGRAPHY

An application of the method can be illustrated by emission spectrography, which allows the determination of major, minor, and trace elements in many materials. Approximately seventy elements can be determined in rocks and other geologic materials. When a sample of material is correctly excited by an electric arc or a spark, each element in the sample emits light of a characteristic wavelength. The light enters the spectrograph via a narrow opening and falls on a diffraction grating, which is a band of equidistant parallel lines (from 10,000 to 30,000 or more lines per inch) ruled on a surface of glass or polished metal used for obtaining optical spectra. The grating separates the reflected light of each wavelength by a different angle. The dispersed light is focused and registered on a photographic plate in the form of lines of the spectrum.

The comparator-densitometer is used to measure the intensity, or darkness, of the spectrum lines registered on the spectrograph photographic plate. Using standard films or plates for each element, scientists change the spectrograph markings to indicate the percent concentration of each element. For a visual estimate, a special screen permits a comparison of the spectrum of the sample with the spectra of standards containing known element concentrations. A direct-reading emission spectrometer is one that is tied in with a computer in which are stored electronic signals from specific parts of the spectrum during the burn of a sample. The stored signals are emitted in sequence to an electronic system that measures the intensities of the spectral lines.

Thermal infrared analytical techniques are being employed to detect the mineralogical composition of the surface of Mars. The Mars Global Surveyor sent into orbit in 1999 employed a thermal emission spectrograph (TES) to determine the composition of the Mars surface. Much of the surface varies from basaltic to andesitic-basaltic rock compositions (e.g., similar to lava flows from volcanoes of the west coast of the United States) and soils derived from these rocks. TES detected two areas of hematite-rich material and failed to detect large surface exposure of carbonates. Mini-TES instruments aboard rovers, such as the 1997 Pathfinder Mission and future rovers, will allow close-up analysis of rocks on the surface. The TES instruments offer a chance to identify mineralogy of the surface of Mars rather than elemental composition.

Use in Geochemical Research

Infrared is related to geophysics, geomorphology, structural geology, and exploration as well as to geochemistry. Specific areas in which the infrared spectra have been used in geochemical research include the study of the bonds between atoms in minerals and the gaining of unique information on features of the structure, including the family of minerals to which the specimen belongs, the mixture of isomorphic substituents, the distinction of molecular water from the constitutional hydroxyl, the degree of regularity in the structure, and the presence of both crystalline and noncrystalline impurities. For example, chalcedony, including flint, chert, and agate, has been shown by infrared spectroscopy and X-ray studies to contain hydroxyl in structural sites as well as in several types of nonstructural water that can be held by internal surfaces and pores. The content of the structural hydroxyl varies zonally in chalcedony fibers and in both natural and synthetic crystals of the same spectral type as chalcedony. The varieties of chalcedony, as well as rock crystal and amethyst formed at low temperatures and in association with chalcedony, together with crystals of synthetic quartz, show a distinctive infrared absorption spectrum in the region of 2.78-3.12 microns; natural quartz crystals formed at higher temperatures give a spectrum in this region. Structural hydroxyl is housed by different mechanisms in the two types of quartz. The fibrose nature of the low-temperature quartz may derive from the hydroxyl content and its effect on dislocations.

Use in Remote Sensing

The widest applications for the infrared spectra are in remote sensing, which is the process of detecting chemical and physical properties of an area by measuring its reflected and emitted radiation. Remote sensing has been a great aid to geologists in their study of the earth. Thermal infrared scanning has been used to monitor and update mine waste embankment data and to locate faults and fracture zones. Landsat thematic mapper and airborne thermal infrared multispectral scanner data have been used to do surface rock mapping in Nevada. Advanced visible and infrared imaging spectrometers and other remotely sensed data have been used to locate water-producing zones beneath the surface in parts of the Great Plains region. Infrared reflectance surveys have been used to locate an extinct hot spring system in the Idaho batholith. Infrared surveys have also included quantitative measurement of thermal radiation from localized heat flow in Long Valley, California, as well as surveys of the lava dome on Mount St. Helens, Washington. Reflectance variations related to petrographic texture and impurities of carbonate rocks have been analyzed by visible and near-infrared spectra.

In addition, infrared surveys and photography of volcanic zones around Mauna Loa and Kilauea were used to obtain information impossible to gather from the ground. The effects of the 1977 earthquake in Nicaragua were surveyed by infrared photography in an attempt to find access and evacuation routes as well as safe areas for temporary camps; in addition, the photographs provided geomorphological information for future study. Infrared surveying has been used to study California's San Andreas fault. In one section near the Indio Hills, the fault trace is not topographically distinguishable but can be located by a margin of vegetation on the northeast side of the fault. Infrared sensing verified its location by imaging a band of alluvium kept cool by the water dammed up by the displaced rock.

ADDITIONAL APPLICATIONS

On a large scale national defense system infrared is used in secret communications, night reconnaissance, missile guidance, and gun sitings and tracking. Weather and pollution control programs use infrared to detect levels of radiation and chemicals. In medicine, infrared is used to find hot areas on the surface of the human body that may indicate possible areas of disease. Studies have revealed positive applications in the detection of, for example, breast cancer, skin burns, frostbite, tumors, abscesses, and appendicitis. As an analyzer, infrared can pinpoint damage in semiconductors that results from overheating. It can also be used to study electrical circuit performance while in operation, detect underrated components in regular circuits, and predict component failure or shortened lifetime: These applications are only a few that demonstrate infrared's analytical ability. In space technology, infrared can be used in telescopes for locating new cosmic bodies, observing star formation, studying planet temperatures, and determining the chemical and/or physical nature of distant sources of infrared radiation. Other major uses of infrared include detecting forest infiltrations and spotting welding defects; other fields that utilize infrared include photography and organic chemistry.

Miscellaneous uses of infrared are as widely varied as the major applications already discussed. Infrared photography can reveal original charcoal sketches under oil paintings. Ecologists can study the thermodynamic world on the planet and observe the animals and plants as they adapt to the volatile thermal balance. They can also track schools of fish by mapping "warm" areas on the water's surface. Criminologists and police can use night vision to survey high crime areas in the dark without the use of a spotlight. Warm air masses associated with turbulence can be detected and avoided, resulting in smoother, safer airplane flights. Infrared studies are contributing to improved telecommunications as a result of the introduction of fiber optics, which uses glass as opposed to copper and other scarce metals. Other uses of the infrared are highly technical and are found in many fields, including geology, biology, agriculture, engineering, and defense.

Earl G. Hoover

FURTHER READING

Bernard, Burton. *ABC's of Infrared*. New York: Howard W. Sams, 1970. Bernard's book deals mostly with the theory and application of infrared but also includes segments on physics and optics. Written in the format of a college-level textbook, with questions at the ends of the chapters. While technical data are included, any reader can follow the carefully explained examples.

Brownlow, Arthur H. *Geochemistry*. 2d ed. Englewood Cliffs, N.J.: Prentice-Hall, 1995. This book is intended as an introductory text on geochemistry. Each chapter is devoted to a specific area of geochemistry. Pertinent chemical principles and concepts are reviewed. The basic data are summarized, and examples of applications to geological problems are given.

Campbell, James B., and Randolph H. Wynne. *Introduction to Remote Sensing*. 5th ed. Guilford Press, 2011. Provides an interdisciplinary introduction to the topic, with background information on the electromagnetic spectrum. Covers many aspects of digital imagery, from aerial photography and coverage, to image enhancement and interpretation.

Coates, John. "Interpretation of Infrared Spectra: A Practical Approach." In *Encyclopedia of Analytical Chemistry*, edited by R. A. Meyers. Chichester: John Wiley & Sons, 2000. This article begins with an overview of the infrared spectrum and follows with interpretation and methodology of infrared spectroscopy. It has detailed explanations that are not too technical.

Conn, George Keith. *Infrared Methods, Principles, and Applications*. New York: Academic Press, 1960. An introduction to infrared studies, this book is divided into two parts: principles of the chief components used in studying the infrared region and practical applications. Although the text is heavily weighted in mathematics and physics, it can be understood by the layperson.

Gibson, Henry Louis. *Photography by Infrared: Its Principles and Applications.* 3rd ed. New York: Wiley, 1978. This book's main topic is infrared photography. There are some aspects of indirect longwave infrared recording in which the image is formed by electronic or other means and copied onto ordinary film. Discusses how infrared photography can be used by geologists and geomorphologists.

Myers, Anne B., et al., eds. *Laser Techniques in Chemistry.* New York: Wiley, 1995. Several sections in this book cover the methods, theories, and applications of infrared and ultraviolet spectroscopy. Illustrations, diagrams, index, and bibliography included.

Roeges, Noel P. G. *A Guide to the Complete Interpretation of Infrared Spectra of Organic Structures.* Chichester, N.Y.: Wiley, 1994. An excellent handbook for exploring the field of infrared spectroscopy and its relationship to organic compounds. A good introduction for the reader without a scientific background.

Sabins, Floyd F. *Remote Sensing: Principles and Interpretation.* San Francisco: W. H. Freeman, 1978. This book is an overview of remote sensing for people with no previous training in that process. The presentation attempts to strike a balance between the physical principles that control remote sensing and the practical interpretation and use of the imagery for a variety of applications. The first chapter summarizes the important characteristics of electromagnetic radiation and the reactions with matter that are basic to all forms of remote sensing.

Salisbury, John W., et al., eds. *Infrared (2.1-25 µm) Spectra of Minerals.* Baltimore: John Hopkins University Press, 1992. A thorough look at infrared spectroscopy, this collection of essays examines the complete spectra of minerals and their properties. Includes a CD-ROM.

Siegel, Frederick R. *Applied Geochemistry.* New York: Wiley, 1975. A very good general text on applied geochemistry, with a brief discussion of emission spectrography. Written in an understandable format for the layperson, requiring a minimum background in mathematics and chemistry. Very good references.

Strahler, Alan H., and Arthur N. Strahler. *Environmental Geoscience: Interaction Between Natural Systems and Man.* Santa Barbara: Hamilton, 1973. A college text on a basic introductory level, this book contains good illustrations and general information on the earth sciences.

Watkins, Jim. "Use of Satellite Remote Sensing Tools for the Great Lakes." *Aquatic Ecosystem Health & Management* 13 (2010): 127-134. This article discusses the use of infrared spectra in measuring the temperature of the Great Lakes.

See also: Electron Microprobes; Electron Microscopy; Elemental Distribution; Experimental Petrology; Geologic and Topographic Maps; Geothermometry and Geobarometry; Heat Sources and Heat Flow; Mass Spectrometry; Neutron Activation Analysis; Petrographic Microscopes; X-ray Fluorescence; X-ray Powder Diffraction.

ISOSTASY

Isostasy is a principle that describes the vertical positioning of segments of the earth's lithosphere relative to the elevation of the land and depth to the top of the asthenosphere. It is, in effect, a restatement of Archimedes' principle or an application of that principle to the outer layers of the earth.

PRINCIPAL TERMS

- **Archimedes' principle:** the notion that a solid, floating body displaces a mass of fluid equal to its own mass
- **asthenosphere:** the layer immediately underneath the lithosphere, which acts geologically like a fluid
- **column:** a cylindrical segment of the earth oriented on a line from the center of the earth to any point on its surface, beginning somewhere in the asthenosphere and ending somewhere within the atmosphere
- **density:** the amount of mass per unit volume of a substance
- **lithosphere:** the outermost solid layer of the earth
- **sea level:** the position of the surface of the ocean relative to the surface of land
- **subsidence:** the sinking of the earth's surface or a decrease in the distance between the earth's surface and its center
- **uplift:** the rising of the earth's surface or the increase in distance between the earth's surface and its center
- **viscosity:** the ability of a fluid to flow

Archimedes' Principle

Isostasy, sometimes called the doctrine or principle of isostasy, is a fundamental principle of the earth sciences that describes the spacial positioning of lithospheric mass within the earth. Isostasy requires that the total mass of air, water, and rock within any vertical column extending from within the asthenosphere, through the lithosphere, to within the atmosphere is equal to the total mass of any other column in the same area of the earth, extending from the same depth in the asthenosphere to the same elevation in the atmosphere. The concept of isostasy is analogous to the concept of buoyancy in physics. Buoyancy was first explained by Archimedes, who, as legend tells it, lowered himself into bathwater, observed the level of the water rise against the wall of the bath pool, and thus realized that ships float because they displace a mass of water equal to the mass of the ship. This discovery came to be known as Archimedes' principle.

Many centuries later, scholars realized that Archimedes' principle could be used to explain why the earth has both high mountain ranges and deep ocean basins. The main obstacle to the acceptance of the principle was the belief of early scholars that the earth was a solid, rigid body. The idea that the ground on which they stood could be compared to a boat floating on the sea was totally beyond their comprehension. The knowledge needed to draw that analogy did not become available until the mid-nineteenth century, when British surveyors under the direction of Sir George Everest were engaged in the trigonometrical survey of India near the Himalaya. The surveyors noted that the distance between the towns of Kalianpur and Kaliana, when measured by triangulation methods, differed by 5.236 seconds of arc, or about 160 meters from the distance when measured by astronomical methods. Two British scholars, George Biddell Airy and John Henry Pratt, realized the cause of this apparent error, though each provided different interpretations of the geologic conditions that gave rise to the difference in distances. Their interpretations later came to be known as the Airy hypothesis and the Pratt hypothesis of isostasy.

Airy and Pratt Hypotheses

While both Airy and Pratt applied the Archimedes' principle to explain the elevation of the Himalayas and the discrepancy in distance between the two survey methods, their hypotheses differed in the way they explained how the mass is distributed below the mountains. Airy viewed that apparent mass deficiency below the mountains as a result of the mountains having a root of low-density rock that extends well into a lower-lying, denser, fluid layer upon which the mountains and all other surficial rock layers float. This lower-density mountain and mountain root combination were envisioned as being like a boat floating upon a denser fluid, which Airy thought was lava. The "boat" was thus made buoyant by the root's displacing a mass of the fluid equal in mass to the combined mountain and mountain root. To Airy, the higher the mountain, the deeper the root must extend to compensate for the elevated mass. By analogy, of two vessels of the same areal extent, one tall and the other of low profile, the tall vessel projects deeper into and rises higher out of the water.

Pratt saw the situation somewhat differently. He maintained that the position of the base of the solid crust must be the same everywhere. The differences in surficial elevations, Pratt thought, arise from some areas having experienced less "contraction" than other areas during the cooling of the earth. These areas of less contraction are also of less density and float higher in accordance with Archimedes' principle. Regional variations in surface elevation, according to the Airy hypothesis, result from variations in the thickness of the solid outer layer of the earth. Airy thought the density of the outer layer was the same everywhere, but the position of the base varies according to the magnitude of surface elevation. According to the Pratt hypothesis, the regional variations in surface elevation result from variations in density of the solid outer layer, with the base of that layer being of equal position everywhere. The Pratt model has a flat-bottomed crust.

Elevated terrains in both the Airy and the Pratt models have less mass near the surface than low-lying terrains. Therefore, according to these hypotheses, the plumb bob was pulled by gravity away from the area of the mountains and toward the lower-lying plains of India. Clarence Edward Dutton in 1889 recognized the significance of this variation in the amount of mass near the surface and concluded, "Where the lighter matter was accumulated there would be a tendency to bulge, and where the denser matter existed there would be a tendency to flatten or depress the surface." Dutton coined the term "isostasy" for this definition between land surface elevation and rock mass as mandated by Archimedes' principle. In the twentieth century, Airy's theory was refined by Finnish geodesist Veikko Heiskanen and Pratt's theory was developed by American geodesist John Hayford, giving the Airy-Heiskanen model for the thickness of the crust, and the Pratt-Hayford model for the density of the crust.

ISOSTATIC COMPENSATIONS

Earth scientists acknowledge the validity of both the Airy and the Pratt hypotheses. They consider large-scale or regional land surface elevation variation to result from variations in density and thickness of the lithosphere and also from variations in density of the asthenosphere. Furthermore, earth scientists recognize that the density and/or thickness of the lithosphere at any particular place can change through time and thus result in vertical movements of lithospheric plates to compensate for these changes. If one were to heat a solid object, such as a steel pipe, it would expand and thus decrease its density and increase its length. If a segment of the lithosphere of the earth were to be heated, the rock within that segment would also become less dense, the thickness of the lithospheric segment would increase, and the elevation of the land surface of that segment would rise in accordance with both the Airy and the Pratt hypotheses. This rising of the land surface is known as uplift. Similarly, if a segment of the lithosphere were to cool, the rock would increase in density, the thickness would be reduced, and the elevation of the surface would be reduced. This reduction of land surface elevation through time is referred to as subsidence. If cargo were added to the deck of a boat, it would be seen to ride lower in the water. The top of the cargo, however, would be at a greater distance above the water. Removing deck cargo has the opposite effect. If a segment of the lithosphere were to have sediment deposited on its surface, its base would project a greater distance into the asthenosphere, and its top would be described as being at a greater elevation. If material is removed from the lithosphere by erosion, the base of the lithosphere rises and the land surface elevation decreases.

The vertical adjustments in the position of the lithosphere to maintain equilibrium are referred to as isostatic compensations. To be in equilibrium, the total amount of mass within a column of the earth that extends from within the atmosphere, through the hydrosphere and lithosphere, and into the asthenosphere must be equal to the total mass of any other column of the same areal range that extends from the same elevation in the atmosphere to the same depth in the asthenosphere. A change in the lithosphere within a column in terms of mass or density will be compensated for by changes in the mass of the atmosphere, hydrosphere, and asthenosphere.

When sediment or rock is deposited or eroded from the top of the lithosphere, mass is added or subtracted from the lithosphere, and isostatic adjustments are made to compensate for this change. If sediment is deposited upon the surface of the lithosphere, the added load displaces some of the asthenosphere; thus, there is less asthenospheric mass in the column. If the top of the lithosphere were below sea level, then hydrospheric mass would also be displaced; if sediment accumulated until it were stacked above sea level, then mass within the atmosphere would be displaced as well. If sediment or rock is eroded from the top of the lithosphere, the base of the lithosphere rises, and mass is added to the asthenospheric portion of the column. If the top of the column were initially above sea level, then the mass of the atmospheric portion of the column would increase; if erosion cut below sea level, then hydrospheric mass would be added to the column. Depositional isostatic subsidence is seen along continental margins such as the Gulf coast of Texas, where there are great accumulations of sediment. Isostatic rebound is associated with erosion (melting) of the Pleistocene ice sheets. Such glacio-isostatic rebounds have been measured in eastern North America and Northern Europe.

THERMO-ISOSTATIC UPLIFT AND SUBSIDENCE

When the lithosphere is warmed or cooled, the situation becomes more complex. The warming or cooling of the lithosphere is geologically accomplished by changes in the temperature of the asthenosphere. Therefore, the density of both the lithosphere and the asthenosphere would be expected to vary with temperature changes. If temperature change were the only process operating, then the mass of the lithosphere would have remained constant regardless of its temperature; its thickness and density, however, would have changed. If the lithosphere were warmed, the mass in the hydrosphere and/or atmospheric portions of the column would have decreased in an amount equal to the increase in mass of the asthenospheric portion. The net effect would be an increase in the elevation of the land surface, or thermo-isostatic uplift. If the lithospheric portion of the column were to cool, there would be an increase in the mass of the hydrospheric and/or atmospheric portions of the column and a decrease in the asthenospheric portion. The net effect would be a decrease in land elevation, or thermo-isostatic subsidence.

Isostatic uplift is seen in the area of the Mid-Atlantic Ridge, the greatest mountain range on the surface of the earth. It also may explain why

Portrait of Greek mathematician and inventor Archimedes (287-212 B.C.). (Time & Life Pictures/Getty Images)

the continent of Africa has such a greater average elevation relative to sea level than do the other continents. Isostatic subsidence has been suggested to be the underlying mechanism for the formation of the thick sediment accumulations within continental areas. The Michigan Basin and the Williston Basin in North America are examples of these accumulations.

The processes that give rise to isostatic adjustments take millions of years. The resulting isostatic adjustments are also very slow to occur. When a person steps onto a boat, it instantly rides lower in the water, because the compensation of the boat for the additional load is immediate. The medium upon which the boat floats, water, has a very low viscosity. If the boat were afloat in a more viscous fluid, such as cold molasses, the adjustment to the added mass would be noticeably slower, perhaps taking a minute or more. The asthenosphere is very viscous. Consequently, isostatic adjustments to lithospheric changes may take tens of thousands of years.

EVALUATION OF GLACIO-ISOSTATIC REBOUND

Isostasy is a principle or law of the earth sciences, and, as such, it cannot be collected, observed, or quantified. What can be observed or quantified are the results of lithospheric segments satisfying or attempting to establish isostatic equilibrium. If a geologic process changes the mass or density of a segment of the lithosphere, vertical adjustments in the position of the lithosphere are necessary to reestablish the equilibrium. These vertical adjustments are slow; 1 centimeter per year would be considered fast. To measure the changes in lithosphere position caused by isostatic compensation, one needs a hypothetical measuring stick and a clock. The "stick" in nearly all cases measures the distance between the top of the lithospheric segment and sea level. Because the time over which the adjustment process occurs is quite long, the clock that is used is the decay of radioisotopes, such as carbon-14, potassium-40, and uranium-238.

The application of these tools to the study of isostatic compensation can be illustrated with the evaluation of the phenomenon of glacio-isostatic rebound. During the last glaciation, the Wisconsin, vast sheets of ice covered portions of Antarctica, North America, Europe, and the southern tip of South America. That ice constituted a load on the decks of several lithospheric boats. From 18,000 to 6,500 years ago, most of the ice sheets in North America, Europe, and South America melted. The meltwater increased the volume of water in the oceans. Consequently, the level of the oceans rose 100 meters relative to a fixed point on a landmass that was not glaciated, such as the island of Cuba. From 6,500 years ago to the present, little additional ice has melted. Thus, the amount of water in the oceans has been constant. Sea level, therefore, should have been constant worldwide.

During this period of time, however, sea level has not been constant in those areas where glacial ice had once loaded the lithosphere. In those areas, fixed points on the land surface are rising relative to sea level. Some areas are currently rising at the rate of 2 centimeters per year; other areas have already risen nearly 140 meters. Scientists can determine how far and how fast the lithosphere has rebounded or is rebounding by examining the locations of exposed shoreline sediments or marine terraces. The sediments would have been deposited and the terraces formed by waves on a beach when sea level was at that land point. Part of that sediment would have been the remains of plants and animals that were alive at the time of deposition of the sediment. By surveying the current difference in elevation between the ancient shoreline sediments and the present sea level, scientists can determine the amount of vertical uplift since the sediment was deposited. By determining the radiometric age of the remains of organic life using carbon-14 dating methods, scientists can calculate the length of the time over which that amount of rebound occurred. Several different shoreline deposits or terraces in the same region can reveal different land positions relative to sea level and how the rate of rebound has changed with time.

Geologists can therefore determine the viscosity of the asthenosphere, project how much rebound will occur in the future, and estimate how much rebound will have occurred when isostatic equilibrium is established. This estimate can be translated into how thick the ice was when the glaciers were present. Ice thickness equals the product of the total rebound times the ratio of the density of the asthenosphere to the density of the ice.

INTEREST TO GEOLOGISTS AND HISTORIANS

The relationship between isostasy and the surface of the land is analogous to the relationship between buoyancy and the deck of a ship. Humans can overload the deck of a ship and sink it into the sea, but they cannot overload the lithosphere and sink it into the asthenosphere. This area of nature is one of the few that is not heavily influenced by human activity. If all the engineers of the world used all the earth-moving equipment in the world to pile soil, sediment, and rock in one huge mound, they could not in their lifetimes cause a segment of the lithosphere to ride 1 millimeter lower in the asthenosphere. Nature, however, in a few hundred millennia can pile enough snow and ice on Antarctica to sink land surface so substantially that most of the subice rock surface (the preglaciation top of the lithosphere) now lies below sea level, several hundred meters below where it originally was. Besides geologists and geophysicists, isostasy touches the lives of very few people directly. The notable exceptions are those historians who ponder why certain Viking harbors in Scandinavia are now situated above sea level: The answer pertains to glacio-isostatic rebound.

James A. Dockal

FURTHER READING

Condie, Kent C. *Plate Tectonics and Crustal Evolution*. 4th ed. Oxford: Butterworth Heinemann, 1997. An excellent overview of modern plate tectonics theory that synthesizes data from geology, geochemistry, geophysics, and oceanography. A very helpful tectonic map of the world is enclosed. The book is nontechnical and suitable for a college-level reader. Useful "suggestions for further reading" follow each chapter.

Davidson, Jon P., Walter E. Reed, and Paul M. Davis. *Exploring Earth: An Introduction to Physical Geology*. 2d ed. Upper Saddle River, N.J.: Prentice Hall, 2001. An excellent introduction to physical geology, this book explains the composition of the earth, its history, and its state of constant change. Intended for high-school-level readers, it is filled with colorful illustrations and maps.

Dockal, J. A., R. A. Laws, and T. R. Worsley. "A General Mathematical Model for Balanced Global Isostasy." *Mathematical Geology* 21 (March 1989): 147. A comprehensive mathematical treatment of isostasy. Discusses the connection between isostatic adjustments and global sea-level changes. Suitable for college-level students with a working knowledge of algebra.

Hamblin, William K., and Eric H. Christiansen. *Earth's Dynamic Systems*. 10th ed. Upper Saddle River, N.J.: Prentice Hall, 2003. This geology textbook offers an integrated view of the earth's interior not common in books of this type. The text is well organized into four easily accessible parts. The illustrations, diagrams, and charts are superb. Includes a glossary and laboratory guide. Suitable for high school readers.

Hart, P. J., ed. *The Earth's Crust and Upper Mantle*. Washington, D.C.: American Geophysical Union, 1969. A somewhat technical book that gathers together many aspects of the crust and mantle or lithosphere and asthenosphere. A chapter by E. V. Artyushkov and Y. U. A. Mescherikov deals quite well with recent isostatic movements and provides a good bibliography of the foreign literature on isostasy. Suitable for college-level students.

Jordan, Thomas H. "The Deep Structure of the Continents." *Scientific American* 240 (January 1979): 92-107. Discusses new ideas on the makeup and nature of the lithosphere and asthenosphere. Provides considerable insight into how knowledge of the deep earth is obtained. Suitable for college-level students.

Mather, K. F., ed. *A Source Book in Geology, 1900-1950*. Cambridge, Mass.: Harvard University Press, 1967. A collection of major landmark geologic works dating from 1900 to 1950. Included in this collection is Joseph Barrell's "The Status of the Theory of Isostasy," which summarizes much of the early thinking on isostasy. Other relevant works include two studies of the properties of the asthenosphere, one by Felix Vening Meinesz and the other by Beno Gutenberg. Suitable for high school and college-level students.

Mather, K. F., and S. L. Mason, eds. *A Source Book in Geology, 1400-1900*. Cambridge, Mass.: Harvard University Press, 1970. A collection of landmark geologic works dating from 1400 to 1900. Each paper is condensed from its original length. The collection includes the works of G. B. Airy, J. H. Pratt, and C. E. Dutton. Brief biographic sketches are given for the authors. Suitable for high school and college-level students.

Plummer, Charles C., and Diane Carlson. *Physical Geology*. 12th ed. Boston: McGraw-Hill, 2007. A college-level introductory geology textbook that is clearly written and wonderfully illustrated. An excellent sourcebook of basic information on geologic terminology and fundamentals of geologic processes. An excellent glossary.

Tarbuck, Edward J., Frederick K. Lutgens, and Dennis Tasa. *Earth: An Introduction to Physical Geology*. 10th ed. Upper Saddle River, N.J.: Prentice Hall, 2010. This college text provides a clear picture of the earth's systems and processes that is suitable for the high school or college reader. It has excellent illustrations and graphics. Bibliography and index.

Walcott, R. I. "Late Quaternary Vertical Movements in Eastern North America: Quantitative Evidence of Glacio-isostatic Rebound." *Review of Geophysics and Space Physics* 10 (November 1972): 849-884. A review paper that collects and evaluates from published sources the evidence of vertical movements of the lithosphere that are attributed to glacio-isostatic rebound. Charts present accumulated data for sea-level changes in North America. Maps portray the magnitude of rebound. Contains an excellent bibliography on glacio-isostatic rebound. Suitable in part for advanced high school and college-level readers.

See also: Earth's Mantle; Earth Tides; The Geoid; Gravity Anomalies; Lithospheric Plates; Mantle Dynamics and Convection; Plate Motions; Plate Tectonics.

ISOTOPE GEOCHEMISTRY

The presence of an excessive or deficient number of neutrons in an atomic nucleus renders that nucleus unstable. The nucleus may then transmute into atoms of other elements through a number of nuclear fission processes. The transmutation is accompanied by the release of subatomic particles and energy, or radioactivity. The relative quantities of atomic isotopes, normally measured by mass spectrometry, can be used to determine the age of a sample or to identify the location and formation of a mineral source through forensic examination.

PRINCIPAL TERMS

- **alpha (α) particle:** a subatomic particle consisting of two protons and two neutrons, bearing two positive electrical charges, and emitted from an atomic nucleus through fission; the nucleus of a helium atom
- **beta (β) particle:** an electron or positron emitted from the nucleus of an unstable atom through the spontaneous decomposition of a neutron into a proton
- **exponential decay:** a process of decomposition or reaction, particularly in regard to nuclear fission, whose kinetics are described by the time-related function $A_t = A_o e^{-kt}$
- **fission:** splitting of a nucleus through the emission of nuclear particles to form atoms of different elements
- **forensic:** the examination of material clues to determine the cause and progression of an event that occurred in the past
- **fusion:** the combining of two separate nuclear entities to form a single entity having an identity different from either of the originals, as when a proton and an electron fuse to form a neutron
- **half-life:** the length of time required for one-half of an amount of material to decompose or be consumed through a process of exponential decay
- **irradiation:** the exposure of some material to radioactive emissions
- **radionuclides:** atoms of the same element, containing the same number of protons but different numbers of neutrons in the nucleus; a synonym for isotopes

Modern Theory of Atomic Structure

To understand the principles of isotope geochemistry, it is necessary to have a sound comprehension of the modern theory of atomic structure, realizing that it is no more and no less than a model. The most basic assumption of that theory is that all matter in the universe is composed of discrete particles called atoms. Based on and supported by experiment and observation, it is further supposed that each atom is composed of smaller particles in specific arrangements.

According to the strict mathematical principles of quantum mechanics, essentially all of the mass of each atom is located in a very small, dense nucleus that contains the two massive particles called protons and neutrons. Surrounding this nucleus is a much broader, but tenuous cloud containing the lightest subatomic particles, the electrons. Each proton carries a single electrical charge that has been designated as positive. Because like electrical charges repel each other, the presence of more than one proton in a nucleus requires the presence of roughly the same number of electrically neutral, equally massive particles called neutrons. The combined masses of the protons and neutrons account for no less than 99.98 percent of the mass of an atom, and the nuclear structure that they form places absolute restrictions on the properties of the surrounding electrons. Specifically, electrons can have only specific quantum energies and so are restricted to specifically defined regions of space around the nucleus.

At the miniscule size of atoms (it would require 6.023×10^{23} of the simplest atoms containing just one proton and one electron to achieve a total mass of one gram), the physics of the macroscopic world simply do not apply. Each electron, for example, can exist and behave both as a particle with mass or as a photon having a specific wavelength but no mass; each electron also can exist as both a particle and a photon, at the same time. All normal chemistry takes place at the level of the outermost electrons and does not involve the nuclei of the atoms.

Isotope chemistry, however, is a nuclear process. Naturally occurring atoms, up to and including the ninety-second element, uranium, occur primarily as stable atoms having a specific number of protons and an equally specific number of neutrons in the nucleus. It is a fundamental principal of atomic theory that each atom of a specific element must contain exactly the same number of protons in the nucleus; otherwise, the atoms are of different elements with different chemical identities. The number of neutrons can vary from atom to atom, however, and atoms that have the same number of protons but different numbers of neutrons in their nuclei are called isotopes of that particular element.

The simplest element, hydrogen, is known in three isotopic forms: protium (generally called hydrogen), deuterium, and tritium. These forms have zero, one, and two neutrons, respectively, in their nuclei. Their respective atomic masses are one, two, and three daltons (also called atomic mass units [amu]). Uranium, with ninety-two protons in its nucleus, is known to exist in several isotopic forms, the most common of which have atomic masses of 235 and 238 daltons.

RADIOACTIVITY

It is one of the supreme ironies of atomic structure that the neutron, when present in specific numbers in the nucleus, imparts extreme stability to the nucleus of an atom. When the neutron is present in different numbers, however, the atomic nucleus becomes extremely unstable and spontaneously decomposes by the process of nuclear fission.

In nuclear fission, an unstable atomic nucleus ejects some of its subatomic structure in a repetitive process until a stable nucleus forms. This can occur by different mechanisms. A neutron may decompose into an electron and a proton, producing an atom of an element with a higher atomic number. The electron that is ejected from the nucleus is called a β-particle (beta particle). In another process, the nucleus may emit an α-particle (alpha particle), which is the same as the nucleus of a helium atom, consisting of two protons and two neutrons carrying two positive electrical charges. More rarely, the unstable nucleus may split apart into the nuclei of two lighter elements. The emission of charged subatomic particles by unstable nuclei is termed "radioactivity" because the energy associated with those emissions was first detected by radio interference. The processes produced activity in the radio frequency range of the electromagnetic spectrum, hence they were radioactive. There are other modes of nuclear fission, but the foregoing three are the primary methods of relevance to isotope geochemistry.

It is important to remember that the mechanism of nuclear decomposition is very specific, in that discrete and well-defined subunits are removed or altered so that the product of each step is equally rigidly defined. Because of the specificity of nuclear fission mechanisms, and because of the precise mathematical description of the rate of nuclear decomposition, the relative amounts of specific isotopes in a material can be used to determine the age of the material, within experimental detection limits.

MASS SPECTROMETRY

The mass spectrometer is a device that measures the behavior of a charged particle in a magnetic field. The charge-to-mass ratio of the particle determines the radius of a circular path that the charged particle will travel within a uniform magnetic field. In the mass spectrometer, the strength of the magnetic field can be continuously varied in strength so that the number of particles of different masses in a sample can be measured by their flight time through the mass spectrometer. The technique itself is quite straightforward in concept.

A material sample is prepared containing only the material to be studied and is injected into the mass spectrometer. Once injected, it is subjected to an electrical discharge that fragments the sample and imparts a single electrical charge to each fragment. Because the interior of the mass spectrometer operates under high vacuum, the fragments or particles are present in the vapor state and are easily accelerated through a grid system into the magnetic field sector of the spectrometer. Each particle bears the same total charge and is accelerated by the same energy as it passes through the grid system, so that when it enters the magnetic field chamber, it has a specific velocity determined only by its mass. The mass and velocity of the particle interact with the magnetic field to determine the particular radius of its circular path through the magnetic field sector.

The operating program of the spectrometer varies the strength of the magnetic field at a closely defined rate, which in turn alters the radii of the particle trajectories so that particles in the magnetic field are brought to the detector in order of their mass. The detection pattern of the different masses is recorded as a mass spectrum. A typical mass spectrum is a bar graph displaying the relative numbers of particles of each mass that have been detected in a specific period of time, determined by the rate at which the magnetic field strength is varied during the analysis. The masses are displayed in incremental units because the structures of the different particles differ by the loss of discrete mass units. Accordingly, a sample in which a radioactive nucleus has ejected a single neutron by radioactive decomposition will exhibit two peaks that differ by one mass unit, while a sample that has emitted an α-particle will exhibit two peaks that differ by exactly four mass units, and possibly a peak at four mass units corresponding to the α-particle (a helium nucleus).

The preparation of the sample is of paramount importance in analyzing the isotope content of a material. The experimental error limits of this stage of the process ultimately determine the accuracy of the analysis. If the analysis is being carried out to determine the age of a material by radiometric dating, the more precisely the quantities of different isotopes can be determined, the narrower the range of dates that can be ascribed to the material will be.

Half-Life and Radiometric Dating

Radioactive elements undergo nuclear fission processes by exponential decay, a process that has a precise and strict mathematical description. The process is absolutely described by the mathematical equation $A_t = A_o e^{kt}$, where A_t is the amount of material A at time t, A_o is the original amount of material A, t is the elapsed time, and k is a constant value for the specific process being described. A special relationship exists when A_t is exactly one-half of the value of A_o; the time interval associated with that change in value is called the half-life of the process. It should be noted that it does not matter what was the original amount of the material; the same amount of time is required for any starting quantity to become halved in value. That is to say, it takes exactly the same amount of time for one kilogram of the specific material to decompose to one-half of a kilogram as it does for one milligram to decompose to one-half of a milligram. Thus, by determining the relative quantities of the starting material and the its daughter material or materials, the mathematical relationship allows the determination of the elapsed time, t. The relationship breaks down to the simple statement $A/A_o = (1/2)^n$, where n is the number of half-lives that have passed.

If the half-life is known, it then becomes a simple matter to convert the factor n to a specific number of years. It is at this point that the precision of the measurement becomes exceedingly important, because the error limit of the measurement determines the corresponding valid range of time. An error range of only ±0.1 percent in the measurement of quantities can translate to a range of as much as one million years for a process with a very long half-life.

In isotope geochemistry, the atoms of interest are those that naturally occur in rock structures and other consistent environmental structures and processes. Research in geochemistry (literally the chemistry of the earth) has determined many of the basic methods and quantities by which radioactive materials are formed and how their decay processes progress over time. This has provided consistent values for the initial quantities of specific radionuclides in a material and a reasonably sound basis for determination of the age of those materials within experimental error limits.

Principal Applications

Isotope geochemistry has its principal applications in three areas: radiometric dating, forensics, and isotopic signature identification. These three areas are discussed here.

The basic principle of radiometric dating is the mathematical relationship of exponential decay. The essential components for radiometric dating are knowledge of the specific mechanism of the fission reactions that are involved and the half-life of the specific process. Reference books such as the *CRC Handbook of Chemistry and Physics* (92d ed., 2011) list the known half-lives of all known radioactive isotopes and the specific type of particle that is ejected from the nucleus during fission. The half-lives listed range from 10^{-16} second to more than 10^9 years. This range clearly provides the opportunity to employ an appropriate radiometric dating methodology for different ages. Each methodology requires that the fission process ends with the formation of stable elements so that the proportion of starting radionuclide to final element can be determined; thus the utility of the methodology is limited to the identification of an appropriate, single and unambiguous process.

The most commonly used nuclear processes for radiometric dating are carbon-14 and nitrogen-14 (C^{14} and N^{14}), potassium-40 and argon-40 (K^{40} and Ar^{40}), and uranium-238 and lead-206 (U^{238} and Pb^{206}). The common key feature of these transitions is that the atoms of the elements that end the corresponding chain are entirely stable and do not decompose further. In the transition from C^{14} to N^{14}, it is believed that one of the extra neutrons in C^{14} decomposes into a proton by emitting a β-particle (an electron) to produce the more stable N^{14} atom. The half-life of C^{14} is 5,730 ± 30 years. The C^{14}-N^{14} transition is thus most useful for the radiocarbon dating of materials from organic sources and carbonaceous minerals. The fundamental premise of this application is that the influx of cosmic rays impinging on the upper atmosphere brings about a constant rate of conversion of normal carbon, C^{12}, to its heavier radionuclide, C^{14}, in atmospheric carbon dioxide that is then incorporated into the photosynthetic process in the usual way.

The K^{40}-Ar^{40} transition has a half-life of 1.28×10^9 years and presumably occurs by a similar but opposite process in which a proton and an electron combine to produce the extra neutron of the resulting highly stable argon atom. The U^{238}-Pb^{206} transition is a more complicated process, occurring by α-particle decay through rapid intermediate stages resulting in the formation of lead-206. These modes are used for dating rock structures and rely on all of the Pb^{206} and Ar^{40} that are present in the rock coming from the original U^{238} and K^{40} content. The helium that is produced by ejection of α-particles in the nuclear fission process typically remains trapped within the rock until something happens to bring about its release.

The forensic application of isotope geochemistry is related to its use in isotopic signature identification. In essence, the relative populations of isotopes in any material are unique and can be characterized in much the same way as a human's personal characteristics, though they are perhaps not as unique as fingerprints. By determining the isotopic content of a soil or mineral sample, that particular sample can be identified with locations that have the identical isotopic composition. In forensic examinations of crime scene evidence, for example, this can be used to place a suspect at a location. In archaeological investigations, which are forensic investigations, the use of carbon-14 dating is an invaluable tool in determining the absolute age of an artifact and the relative ages of surrounding materials.

For geological research, the isotopic signature of a mineral sample provides an unambiguous association with the location and formation of that mineral and is key to determining the geological age of the strata and structures from which it was taken. An isotopic signature also can be artificially induced in a material to provide a means of tracking and detection, as is sometimes used with gemstones. Irradiation of diamonds and other gemstones is often used to impart a specific isotopic signature that does not affect the appearance of the stone, but which can be readily detected and verified for identification purposes.

Richard M. Renneboog

FURTHER READING

Baskaran, Mark, ed. *Handbook of Environmental Isotope Geochemistry*. Vol. 1. New York: Springer, 2011. This book is designed to bring together in one location new and old applications of isotope geochemistry applicable to environmental problems, so as to be readily accessible.

Dalrymple, G. Brent. *Ancient Earth, Ancient Skies: The Age of Earth and Its Cosmic Surroundings*. Stanford, Calif.: Stanford University Press, 2004. The fourth chapter, "How Radiometric Dating Works," examines the principles of radiometric dating and also provides a list of isotopic materials used for dating rocks and minerals.

Hoefs, Jochen. *Stable Isotope Geochemistry*. 6th ed. New York: Springer, 2009. A college-level book suitable for reference, this work introduces and discusses the basic principles of stable isotopes, their isolation and analysis, and the application of the various analytical techniques to the solution of geochemical problems.

Holland, Heinrich D., and Karl K. Turekian, eds. *Isotope Geochemistry: From the Treatise on Geochemistry*. San Diego, Calif.: Academic Press/Elsevier, 2011. A collection of chapters from the larger work of the title. Describes the range of isotopic studies used in geochemical determinations and the use of mass spectrometry and other analytical techniques for the measurement of radioactive isotopes.

Van Kranendonk, Martin J., R. Hugh Smithies, and Vickie C. Bennett, eds. *Earth's Oldest Rocks*. Boston: Elsevier, 2007. The book is a specialist treatise on the age determination and properties of the oldest known rock structures of the earth, in which isotopic composition plays a central role.

See also: Biogeochemistry; Earth's Magnetic Field; Elemental Distribution; Environmental Chemistry; Freshwater Chemistry; Geobiomagnetism; Geochemical Cycle; Isotopic Fractionation; Magnetic Stratigraphy; Mass Spectrometry; Neutron Activation Analysis; Nucleosynthesis; Oxygen, Hydrogen, and Carbon Ratios; Potassium-Argon Dating; Radioactive Decay; Radiocarbon Dating; Rock Magnetism; Rubidium-Strontium Dating; Samarium-Neodymium Dating; Uranium-Thorium-Lead Dating.

ISOTOPIC FRACTIONATION

Processes as fundamental as evaporation, condensation, fluid movement through a rock, and many biological functions result in isotope fractionation. The record of these processes as they influenced the formation of various minerals is preserved in the distribution of stable isotopes within rock.

PRINCIPAL TERMS

- **chemical bond:** the force holding two chemical elements together as part of a molecule
- **depletion:** the process by which the light isotope is concentrated in either the reactants or the products of a chemical reaction
- **enrichment:** the process by which the heavy isotope is concentrated in either the reactants or the products of a chemical reaction
- **isotopes:** atoms containing the same number of protons but a different number of neutrons, giving the same chemical properties but different atomic weights
- **isotopic fractionation:** changes in the isotopic composition of natural substances, which result from small differences in the physical, chemical, and biological properties of isotopes
- **product:** the material that results from a reactant undergoing a chemical process
- **reactant:** the starting material or materials in any chemical reaction
- **standard:** a material of known isotopic composition; all enrichment and depletion is measured relative to the standard value

WEIGHT DIFFERENCE

Fractionation implies the breaking of a whole into its parts. In the case of a chemical element, the parts are the naturally occurring isotopes of that element. Some of the isotopes may be radioactive and may spontaneously decay to form another element. Most isotopes, however, are stable; they do not decay, and they differ from other isotopes only in their mass. Thus, stable isotope fractionation comprises several physical and chemical processes that can separate the stable isotopes of an element on the basis of weight difference.

The weight difference can be substantial. For example, deuterium is a stable isotope of hydrogen. The hydrogen nucleus contains one proton, whereas the deuterium nucleus contains one proton and one neutron. Because the proton and the neutron are about equal in mass (and the electron is so small it can be ignored), the atomic weight of hydrogen is 1 and the atomic weight of deuterium is 2; deuterium weighs 100 percent more than does hydrogen. In another example, the most common isotopes of carbon are carbon-12 (six protons and six neutrons in the nucleus) and carbon-13 (six protons and seven neutrons in the nucleus). Just as deuterium has one more neutron than does hydrogen, so carbon-13 has one more neutron than does carbon-12. The weight difference between carbon-13 and carbon-12, however, is less on a percentage basis—only 8 percent. It is apparent, then, that the addition of a neutron in an isotope has the greatest relative mass impact for the light elements (for example, hydrogen) and that the impact decreases as the elements become heavier. Thus, stable isotope fractionation processes are most obvious when light elements (those preceding sulfur in the periodic table) are involved.

EVAPORATION AND CONDENSATION

Fractionation includes those processes that separate light and heavy isotopes by physical means or through some chemical reaction. Evaporation and condensation are two physical processes that result in the separation of stable isotopes. For their impact on the earth's climate, the evaporation and condensation of water are arguably the most important physical processes.

Evaporation and condensation are mirror images of each other as far as isotope fractionation is concerned. In evaporation, energy is absorbed by water. This energy absorption is reflected by an increase in the temperature of the water. The individual water molecules absorb the energy and begin to move and to vibrate faster. Eventually, the individual molecules absorb so much energy that large quantities of them change from water molecules in a liquid to water molecules in a gas. The water molecules are said to have undergone a change of phase, from a liquid to a gas. If the process is reversed and energy is removed from the gas containing water molecules, the molecules slow down and begin to clump together. In the atmosphere, this condensation creates water droplets that may eventually produce a rain shower.

Water molecules are actually not all the same. With its two hydrogen atoms and one oxygen atom, the water molecule may have any of a range of molecular weights, depending upon the range of isotopic substitution. Stable isotopes of hydrogen (hydrogen and deuterium) and stable isotopes of oxygen (oxygen-16, oxygen-17, and oxygen-18) combine to yield water molecules of varying molecular weights. Normal water has a molecular weight of 18 (two hydrogens plus one oxygen-16), and heavier water has a molecular weight of 20 (two hydrogens and one oxygen-18). The heavier water molecule is 11 percent heavier than is light water.

If a mixture of light- and heavy-oxygen water is undergoing evaporation, it will take more energy input to "lift" the heavy water out of the liquid phase and into the gas phase. This process is analogous to a weightlifter's expending more energy to raise a 200-kilogram weight than a 180-kilogram weight. In the case of evaporation, the input energy is thermal (heat) rather than mechanical (physical lifting). The lighter oxygen-16 water will evaporate more readily, leaving the heavier oxygen-18 water behind. The gas phase above the evaporating water (for example, the atmosphere above the ocean) is dominated by light water, whereas the water remaining in the liquid phase is dominated by heavier water. This is not to say that heavy water molecules never evaporate; they do. Yet if one defines a number, R, as the ratio of heavy oxygen to light oxygen and measures R before the evaporation starts and then in the water vapor and the liquid water after some period of evaporation, the value of R will change. The ratio will be larger in the liquid (concentrating heavy oxygen) and smaller in the water vapor (dominated by light oxygen).

MAGNITUDE OF ISOTOPIC FRACTIONATION

Geochemists who study isotope fractionation use an equation which uses the R values calculated above to express the magnitude of isotopic fractionation. The equation $\delta = [(R_{sample} - R_{standard}) \div R_{standard}] \times 1,000$ defines a quantity called delta (δ), which represents the difference in the ratio of heavy oxygen to light oxygen between a sample (for example, the water vapor) and a standard (for example, seawater). This difference is multiplied by 1,000 and expressed as per mil (instead of percent).

The value of δ may be positive or negative, depending on the sign of the numerator. If the numerator is negative, R_{sample} is less than $R_{standard}$. In the evaporation example, that means that the sample has less heavy oxygen (or more light oxygen) than does the standard. The sample is depleted in the heavy isotope of oxygen and enriched in the light isotope of oxygen. In evaporation, the vapor phase is depleted in heavy oxygen and its δ value is negative. Conversely, the δ value for the remaining liquid phase will be positive (R_{sample} greater than $R_{standard}$), because the liquid phase is enriched in the heavy isotope of oxygen. When water vapor condenses to form raindrops, the first droplets to form are enriched in heavy-oxygen water (δ greater than 0), and the remaining vapor is enriched in light-oxygen water (δ less than 0).

SEAWATER AS STANDARD

Seawater makes up greater than 90 percent of the liquid water at the earth's surface and makes a good standard for identifying fractionation effects. It should be apparent from the previous example that rainwater will be isotopically lighter (in both hydrogen isotopes and oxygen isotopes) than seawater. Similarly, evaporative waters, such as the waters of the Red Sea, will be isotopically heavier than is normal seawater.

If an oxygen-containing mineral forms in contact with water, some of the oxygen in the mineral will probably come from the water. If the water is isotopically light (from a freshwater, terrestrial environment, for example), the mineral oxygen will be relatively light as well. If the mineral forms in contact with an evaporative brine, the mineral oxygen should be enriched in the heavy isotope of oxygen. Thus, the stable isotopic composition of a mineral can tell scientists about the composition of ancient waters.

BOND BREAKING

Chemical and biological reactions involve the breaking of chemical bonds. Just as it takes energy to evaporate a molecule of water, so it takes energy to break the bond between hydrogen and oxygen atoms in water or between carbon and oxygen atoms in carbon dioxide. The bond between atoms can be visualized as a spring. Inputs of energy cause the bond (spring) to vibrate. The atoms at the ends of the bond move apart as the bond stretches, and they move together as the bond restores the molecule to its original shape. If sufficient energy is applied, the atoms move so far apart (stretching the "spring") that there is no force remaining to pull them back together. At this point, the bond has been broken.

Bond breaking occurs in any chemical reaction, whether it involves living organisms (a biochemical reaction) or proceeds without biological intervention (an inorganic reaction). In the case of inorganic reactions, the energy to break bonds is usually supplied by the environment in the form of heat. At low temperatures, such as normal room temperature, bonds involving heavy isotopes are less likely to be broken than are those involving only light isotopes. For the same amount of energy input, a spring with heavy weights on each end vibrates and stretches more slowly than does the same spring with lighter weights on the ends. This analogy shows why it takes more energy input to break bonds involving heavy isotopes. As the environmental temperature increases, however, the energy necessary to break bonds becomes readily available, and the degree of isotopic fractionation decreases. The slight differences in bond strength are insignificant at higher temperatures.

When living organisms are involved, the fractionation can be exaggerated. In biochemical reactions, the organism is often the source of energy for bond breaking. When the isotopes are chemically identical, it does not make sense for an organism to expend the extra energy to break heavy-isotope bonds when the same reaction with light isotopes uses less energy. Even in the case of photosynthesis, where sunlight is the primary energy source used to break the carbon-oxygen bond in carbon dioxide, isotopically light carbon dioxide is more likely to be photosynthesized than is isotopically heavy carbon dioxide. The result is that light isotopes are concentrated in the reaction products of biochemical reaction. Thus, biochemical molecules, which contain elements such as hydrogen, carbon, oxygen, nitrogen, sulfur, and phosphorus, tend to concentrate the lighter isotopes of those elements. The surrounding environment, be it lakewater, seawater, or sediments, tends to accumulate the heavier isotopes.

MASS SPECTROMETRY

The mass spectrometer made it possible for scientists to identify and study the phenomenon of isotope fractionation. As its name implies, the instrument analyzes a spectrum based on mass, or weight. The rainbow is a common example of the spectrum of white light. The original light has been broken into its colorful components by passing through water droplets in the atmosphere that act like a glass prism. Each individual color represents light of a different wavelength. Water droplets and prisms, then, serve as simple wavelength spectrometers. In a mass spectrometer, a sample is analyzed for the range of masses of the elements it contains.

In an actual analysis of a geologic sample—a coal, for example—the sample is converted through chemical processes into gases suitable for analysis in the mass spectrometer. The gases produced by the sample—in this case probably carbon dioxide, water, and some nitrogen and sulfur gases—are separated and purified by passing them through a series of freezing and drying steps. When a pure gas sample is obtained, that portion of the sample is injected into the mass spectrometer. Inside the mass spectrometer, the sample first enters an ionization chamber, where one or two electrons are stripped from the molecule. The molecule is now a positively charged ion (it has lost one or more negative electrons). The ionized gas sample next approaches a negatively charged metal plate with a hole in the middle. Because the sample gas is positively charged and the metal plate is negatively charged, the sample molecules are attracted to the plate and accelerate toward it. Some of the sample molecules pass through the hole.

ISOTOPIC SEPARATION

At this point, isotopic separation becomes important. For example, in nitrogen gas (composed of two nitrogen atoms), the nitrogen molecule may contain either nitrogen-14 or nitrogen-15 isotopes. The molecule may weigh 28, 29, or 30 units. During the acceleration toward the metal plate, the lighter-isotope molecules will be moving faster than will the heavier molecules. Nitrogen gas molecules composed of two nitrogen 15 atoms are moving most slowly.

After passing through the hole, the nitrogen molecules enter a metal pipe with a slight bend. There is a vacuum in the pipe, and the pipe is surrounded by a strong magnet. As the accelerated, charged molecules enter the bent pipe, their flight path is bent by the presence of the magnetic field. The degree of bending depends on how fast the molecule is moving, which, in turn, depends on the mass of the molecule and the amount of charge on the metal accelerating plate. Thus, given one set of conditions (accelerating plate voltage and strength of the magnetic field), only the nitrogen gas molecules weighing 29 units will negotiate the bend in the pipe and reach the particle detector at the other end. On the one hand, the lighter nitrogen will be moving too fast to make the turn and will adhere to the outer wall of the pipe. On the other hand, the heavier nitrogen molecules will be moving too slowly and will be bent sharply into the inside wall of the pipe. By varying the accelerating voltage, the scientist can focus beams of all three types of nitrogen gas onto the detector and analyze the nitrogen isotopic composition of the sample. The same sort of analysis is done with hydrogen isotopes, using hydrogen gas and carbon, and with oxygen isotopes, using carbon dioxide gas.

Understanding Climate and Biochemical Processes

In general, a scientist analyzes a sample and seeks a measure of isotopic fractionation to determine either at what temperature a mineral formed or whether a mineral formed at lower temperatures is inorganic or biochemically influenced. The samples analyzed may occur naturally, or they may be minerals grown in carefully controlled laboratory systems where the precise isotopic compositions of the starting materials are known.

Stable isotope fractionation is a process that occurs continuously. The processes involved are as simple as evaporation and as fundamental to human survival as are photosynthesis and respiration. Many of the natural fractionation processes studied with stable isotopes are very simple, but others are extremely complex. Among the former, precipitation in the form of snow leaves a permanent, frozen record of climate in the polar ice caps. Studies of climatic change in these polar latitudes involve drilling ice cores out of the thick ice sheet and determining the stable isotope ratios in the water. Because human activities such as the burning of fossil fuel and forest clear-cutting are so extensive, some scientists believe that the earth's surface climate is being changed. Understanding the range of climate variability over the recent geologic past through the study of isotopic variations in ice cores may tell scientists whether the earth's climate can absorb and recover from such changes.

Biological processes would fit into the category of extremely complex processes. Scientists who study the details of biochemical processes must examine and understand each step in a series of many that make up the overall reaction. At any point in that series of reactions, if something goes wrong, the chance for development of a disease or abnormality enters. Isotopes in general, and stable isotopes in particular, can be used to trace the starting chemicals through the maze of reactions to the ultimate products. Stable isotopes act to identify certain elements throughout the reaction sequence, without introducing possible radiation effects (as when radioactive isotopes are used). Knowledge of fractionation processes allows the scientist to correct for or ignore certain amounts of enrichment or depletion in the reaction products. Identifying stable isotopes as parts of nonproductive side branches of the reaction series may provide a clue to the reaction process, which, in turn, may suggest a weakness in the process. Understanding important biochemical reactions on the molecular scale may help scientists to uncover cures for diseases and prevent birth defects.

Richard W. Arnseth

Further Reading

Albarede, Francis. *Geochemistry: An Introduction.* 2d ed. Boston: Cambridge University Press, 2009. A good introduction for students looking to gain some knowledge in geochemistry. Covers basic topics in physics and chemistry; isotopes, fractionation, geochemical cycles, and the geochemistry of select elements.

Bowen, Robert. *Isotopes and Climates.* London: Elsevier, 1991. Bowen examines the role of isotopes in geochemical phases and processes. This text does require some background in chemistry or the earth sciences but will provide some useful information about isotopes and geochemistry for someone without prior knowledge in those fields. Charts and diagrams help clarify difficult concepts.

Chacko, T., D. Cole, and J. Horita. "Equilibrium Oxygen, Hydrogen and Carbon Isotope Fractionation Factors Applicable to Geologic Systems." *Reviews in Mineralogy and Geochemistry* 43 (2001): 1-81. Provides an evaluation of the major methods used to determine fractionation factors.

Criss, Robert E. *Principles of Stable Isotope Distribution.* New York: Oxford University Press, 1999. Criss describes isotopes and their properties with clarity. In addition to well-written text, the book features diagrams and illustrations that present a clear picture of the different phases of isotopes and isotope distribution. Bibliography and index.

Drever, J. I. *The Geochemistry of Natural Waters.* 3rd ed. Englewood Cliffs, N.J.: Prentice-Hall, 1997. Nearly every general geochemistry text contains a chapter or two on isotopes, both radioactive and stable, but Drever's chapter is certainly one of the most accessible to the general college-level reader. His approach to explaining geochemistry in the text is very good, so the person without a strong chemistry background could use this text. Contains a reasonably helpful bibliography.

Faure, Gunter. *Isotopes: Principles and Applications.* 3rd ed. New York: John Wiley & Sons, 2004. Originally titled *Principles of Isotope Geology*, this is a standard text for undergraduate- to graduate-level courses in geochemistry and isotope geochemistry. The majority of the book is devoted to radioactive isotopes, but the last unit discusses the fractionation processes for stable isotopes of oxygen, hydrogen, nitrogen, carbon, and sulfur. In addition, the introductory chapter presents an interesting history of the development of isotope geology. Each chapter concludes with a summary section, followed by a few calculation problems and an extensive reference list.

Fritz, P., and J. C. Fontes, eds. *Handbook of Environmental Isotope Geochemistry.* Vol. 1, *Terrestrial Environment.* New York: Elsevier, 1981. Part of a five-volume set that covers the state of knowledge in isotope geochemistry. Later volumes in the series concentrate on the marine environment and high-temperature geologic environments. The discussions are highly technical and about evenly divided between stable and radioactive isotopes. Contains an extensive bibliography. The main strength of this set is that it is comprehensive and would serve as a good source for specific, detailed information. The technical nature of the presentation will limit its usefulness to college-level readers with good chemistry backgrounds.

Fry, Brian. *Stable Isotope Ecology.* New York: Springer Science, 2006. A good graduate-level text for those with a background in geology and looking to expand their knowledge of stable isotopes. Chapter 7, titled "Fractionation," covers fractionation in closed and open systems, equilibrium fractionation, and experiments. A CD-ROM is included.

Hoefs, J. *Stable Isotope Geochemistry.* 6th ed. New York: Springer-Verlag, 2009. A standard reference in the field of stable isotope geochemistry. The first chapter gives a thorough, though somewhat technical, introduction to the range of isotope fractionation processes and the basic principles of mass spectrometry. Chapter 2 discusses some of the fractionation processes observed for specific isotopes in geologically important processes. The remainder of the book systematically examines the range of geologic materials, from extraterrestrial dust to the composition of the mantle, and their isotopic signatures. A very comprehensive text. The reader should have some background in chemistry. An extensive bibliography makes this book especially valuable.

Rankama, K. *Progress in Isotope Geology.* New York: John Wiley & Sons, 1963. A sequel to the author's 1954 *Isotope Geology*, this book is basically a progress report on developments during the intervening decade. Though both the original and this text are old and highly technical, they do provide a wealth of information on isotopes, some of which is only rarely discussed in more recent texts. Suitable as a reference only. For college-level readers.

Schimel, David Steven. *Theory and Application of Tracers.* San Diego: Academic Press, 1993. Schimel examines the geochemistry of isotopes in plants, soils, and waters. He also describes techniques and methodology used in isotopic laboratories.

See also: Earth's Mantle; Elemental Distribution; Fluid Inclusions; Freshwater Chemistry; Geochemical Cycle; Geothermometry and Geobarometry; Isotope Geochemistry; Mass Spectrometry; Nucleosynthesis; Oxygen, Hydrogen, and Carbon Ratios; Phase Changes; Phase Equilibria; Rock Magnetism; Volcanism; Water-Rock Interactions.

J

JUPITER'S EFFECT ON EARTH

A number of theories propose that Jupiter has many affects upon Earth, some of which are being explored, some of which have long since been debunked. The relationship between Earth and Jupiter provides clues about the solar system's history and about how the planets will interact in the future.

PRINCIPAL TERMS

- **Centaur:** a large, icy planetoid orbiting between Jupiter and Neptune
- **Interplanetary magnetic field (IMF):** the sun's magnetic field that is carried from the sun toward the planets and other objects in the solar system
- **Jovian:** a gas giant planet; this classification includes Jupiter
- **Jupiter family comets:** a group of comets that orbit in the outer reaches of the solar system
- **Kepler's third law of planetary motion:** a theory introduced by Johannes Kepler that states that the solar orbit of a planet (or moon) increases rapidly as its radius increases
- **long-period comet:** a comet originating from the Oort cloud
- **near-Earth asteroid:** large rocks orbiting in the asteroid belt between Mars and Jupiter
- **oort cloud:** a massive region of comets believed to exist beyond Pluto

BACKGROUND AND HISTORY

Since the seventeenth century, scientists have worked to unlock the secrets of Jupiter, the largest planet in the solar system. In addition to attempting to study the planet itself, astronomers and astrophysicists are working to find evidence of how Jupiter's physical location, gravitational pull, and other elements affect Earth. In particular, researchers are investigating whether Jupiter's gravitational pull acts as a shield to block comets, asteroids, and even planets from approaching and impacting Earth.

Like Saturn, Uranus, and Neptune, Jupiter is a gas giant—a planet composed predominantly of gas. Jupiter, like the other planets in the solar system, moves around the sun in an elliptical orbit. The planet itself rotates quickly, which contributes to its powerful and expansive magnetosphere (the area in which the planet's magnetic fields, emanating from its polar regions, are contained). In 1979, the National Aeronautics and Space Administration (NASA) directed the Viking I spacecraft (which was launched to study Mars) to study the planet's rings. In 1995, the Galileo spacecraft began a multiyear analysis of Jupiter's magnetosphere and other aspects of the planet. During that orbit, Galileo joined with another spacecraft, Cassini-Huygens, which studied Jupiter's inner workings and phenomena in even greater detail.

The largest of the gas giants, which are also known as Jovian planets, Jupiter has long been an intriguing topic of study for astronomers. Ancient Romans named the planet for their highest god, even when they did not realize how large the planet was. Not until the early seventeenth century did the planet's size and complexity become apparent, when Italian mathematician and astronomer Galileo Galilei (1564-1642) used his invention, the telescope, to observe the colorful planet and several of its satellites. During the same period, German astronomer and mathematician Johannes Kepler (1571-1630) used four of Jupiter's moons (Io, Callisto, Ganymede, and Europa) to support his third law of planetary motion, which suggests that the solar orbit of a planet (or moon) increases rapidly as its radius increases.

Jupiter continued to be a major source of both study and speculation among scientific circles during the centuries that followed. In 1974, astronomers John Gribbin and Stephen Plagemann published *The Jupiter Effect*, which posited that, in 1982 (when all the planets were expected to align), the combined gravitational pulls of Jupiter and the other Jovian planets would cause untold devastation on Earth by triggering earthquakes and geomagnetic storms. Meanwhile, the United States launched a series of probes to Jupiter's system including Pioneer 10 and 11 (1972-1973), Voyager 1 and 2 (1977), and the Galileo probe and orbiter (1989). Each of these crafts gathered a significant amount of data from Jupiter and its moons, including data from the planet's atmosphere, magnetosphere, and gravitational field, and those of its largest moons. Two other spacecraft, Cassini (1997) and New Horizons (2006), also have flown near Jupiter on their way to other areas of the solar system.

THEORIES ABOUT JUPITER AND EARTH

Much of the speculation about Jupiter's perceived effects on Earth centers on Jupiter's gravitational and magnetic pull. Indeed, Jupiter has an enormous magnetic field—an estimated fourteen times the strength of the field found on Earth. This field is so significant that it is believed to have an effect on the interplanetary magnetic field (IMF), which is the part of the sun's magnetic field that is carried from the sun to the planets and other objects in the solar system. According to scientists, Jupiter's own magnetic field discharges enough energy particles to significantly alter the IMF, causing fluctuations in speed and intensity.

The Russian Federal Space Agency has planned a series of missions to Jupiter and its moon Europa. However, the high volume of charged particles and radiation emanating from Jupiter's

This image provided by NASA and taken by the New Horizons Long Range Reconnaissance Imager shows a 4-millisecond exposure of Jupiter and two of its moons on January 17, 2007. The spacecraft was 68.5 million kilometers (42.5 million miles) from Jupiter, closing in on the giant planet at 41,500 miles (66,790 kilometers) per hour. The volcanic moon Io is the closest planet to the right of Jupiter; the icy moon Ganymede is to Io's right. The shadows of each satellite are visible atop Jupiter's clouds; Ganymede's shadow is draped over Jupiter's northwestern limb. (AP Photo)

system could affect the trajectories and disrupt onboard systems of any spacecraft sent. Russian scientists are therefore carefully studying the Jovian magnetic field and radiation belts to avoid any program failures.

One of the most-discussed of Jupiter's perceived effects on Earth focuses on Jupiter's gravity. In *The Jupiter Effect*, Gribbin and Plagemann theorized that Jupiter's tremendous gravitational pull would contribute to complete devastation on Earth. This theory was largely discounted (the authors themselves would later admit that the theory was highly flawed), particularly given that Jupiter's gravity has only a miniscule influence on Earth. Indeed, Jupiter's gravitational pull on the earth, as evidenced by Earth's tides, is minimal; the strongest pull on the earth's tides is that of its own moon, followed by that of the sun (half of the moon's pull), and Venus (which is one thousand times weaker than the sun's pull). Jupiter's pull is far smaller than that of Venus.

Another theory continues to generate debate. Many scientists believe that Jupiter may be acting as a buffer for the earth against asteroids, meteors, and comets. The idea is that the Jovian gravitational field deflects such objects from an Earth-bound trajectory, and that life on Earth might not be possible without the presence of Jupiter. Furthermore, some scientists believe that this "protective" role played by Jupiter may be a key to finding other Earth-like planets in other solar systems. Astronomers are thus searching other solar systems for areas that feature significant gravitational signatures resembling those of Jupiter.

Recent studies, however, suggest that the "protective" theory is subject to debate. According to computer models in 2007, though Jupiter may block and deflect some centaurs (large, icy planetoids orbiting between Jupiter and Neptune) from entering the inner solar system, Jupiter also may be responsible for redirecting others toward Earth. Furthermore, the notion that Jupiter prevents so-called Jupiter family comets (a group of comets that orbit in the outer reaches of the solar system, including the famous Shoemaker-Levy 9 comet that crashed into Jupiter in 1994) may not be wholly true. The models suggest that Jupiter may deflect or pull some of these comets from harm but also may enable others to approach the inner solar system. Meanwhile, long-period comets, those comets from the Oort cloud (a massive group of comets located beyond Pluto's orbit), often travel through the solar system quickly and without warning, little influenced by Jupiter. Additionally, Jupiter does not appear to reduce the threat posed by near-Earth asteroids from the asteroid belt, the group of bodies orbiting between Mars and Jupiter. Such models add to a debate over Jupiter's protective role, which remains unsettled.

PROBES AND COMPUTER MODELS

Launching spacecraft toward Jupiter's system is one of the most effective ways to get a close-up look at the many complexities of the planet and its moons. In many cases, such spacecraft provide definitive answers to the mysteries surrounding the solar system's largest planet. The Galileo spacecraft, for example, largely disproved a popular theory that suggested Jupiter could one day turn into a star. As part of its mission, a probe plunged into the planet's atmosphere, recording high-speed winds and signs that Jupiter's core was generating heat. The data sent back from Galileo continue to provide a complex profile of Jupiter and its moons and how the planet interacts with the other planets in the solar system.

Because of the sheer distance between Earth and Jupiter, many of the theories surrounding how those two planets interact (if at all) are based on speculation. However, astronomers and other scientists are becoming increasingly able to explore this area through the use of computer modeling software. For example, scientists have for decades attempted to determine the composition of Jupiter's core. Unlike Earth and the other terrestrial planets (Mercury, Venus, and Mars), Jupiter is composed of mostly gaseous hydrogen and helium. Scientists do not know if the core, believed to be surrounded by metallic hydrogen, is solid or liquid. Telescopes and even deep space probes have yet to answer this mystery.

However, scientists have developed computer models based on chemical principles (such as molecular-to-metallic transition for hydrogen), and they now use data collected from Galileo and other missions. While no definitive conclusions have been made, the notion of Jupiter having either a solid, semisolid, or liquid core is gaining increased credence because of such computer models.

While examinations of Jupiter's core using computer models may help reveal information about how the solar system was formed, computer models also can help scientists study potential threats posed by Jupiter to its neighbors (including Earth). French and American scientists are using computer models to assess how Jupiter's gravitational pull may be affecting the erratic elliptical orbit of Mercury. Some models indicate that Jupiter could ultimately cause four scenarios: pushing Mercury into the sun, pushing Mercury out of the solar system altogether, pushing Mercury into Venus, or pushing Mercury into Earth. These models do indicate that chances of the latter scenario are extremely remote, however. Still, these computer models present a new way to study Jupiter's complexities and how they affect the other planets.

TELESCOPY

Just as Galileo used an early form of the telescope to examine Jupiter to support the theory that the planets revolve around the sun, scientists now continue to use much more advanced versions of the telescope to monitor Jupiter and its effects on the rest of the solar system. For example, one of the leading areas of discussion regarding Jupiter and Earth is the notion that Jupiter acts as a shield to either fling inbound comets and other objects away from Earth or to capture them and absorb their impacts.

In 1993, astronomers using ground- and space-based telescope systems monitored one of the best-known impacts on Jupiter's surface: The Shoemaker-Levy 9 comet (which was moving in a deteriorating orbit around the giant planet) steadily broke apart in Jupiter's gravitational field and fell into the planet's cloudy external membrane. Later, a number of amateur astronomers, with telescopes containing video recorders, caught small flashes of light on Jupiter, indicating a series of small-object impacts. As the telescope technology available to amateurs continues to improve, scientists believe more such images are likely to surface.

Telescopes are not limited to providing simple visual images, either. NASA's Infrared Telescope Facility in Hawaii captured thermal emissions from Jupiter that indicated a series of impacts. Such observations provide scientists with an understanding of how chemicals and dust released from Jupiter's atmosphere travel into other areas of the solar system. Meanwhile, the Chandra X-Ray Observatory, a satellite in an elliptical orbit around Earth, has been capturing X-ray emissions from Jupiter's northern polar region. X-rays are fostering increased attention to these emissions among astronomers, both in terms of their generation and in terms of how these emissions travel through the solar system, affecting other planets and celestial bodies.

RELEVANT GROUPS AND ORGANIZATIONS

The study of the Jovian influence on Earth is part of ongoing investigation of how the planets of the solar system formed and how they and other celestial bodies interact with one another. A number of organizations and agencies are dedicated to these questions. Among them are government agencies, universities, and private and amateur observatories.

Government agencies such as NASA take the most prominent role in the study of Jupiter. NASA has long been the main player in terms of spacecraft missions, having operated the Pioneer, Voyager, Galileo, Ulysses, and New Horizons missions unilaterally. The agency also has planned to launch the Juno probe to study Jupiter's northern pole. However, the European Space Agency and the Italian Space Agency partnered with NASA on the Cassini-Huygens program in the late 1990's. Furthermore, the Russian Space Agency has planned its own missions to Jupiter and Europa.

Universities also play a significant role in studying the effects of the Jovian system. Institutions such as the University of Hawaii, Massachusetts Institute of Technology, Cambridge University, and University of Tokyo have strong astronomy programs with state-of-the-art observatories on campus or in remote locations. Universities also are major sources of Jovian theories.

An increasing number of private observatories are focusing on Jupiter and its influences on the solar system. The impacts of large asteroids and other debris have been captured not by NASA or other major observatories but by amateurs with advanced telescopes and recording equipment. Amateur observatories are operating all over the world. In light of the ease by which information is transmitted and shared through the Internet and through satellite technologies, it is likely that private and amateur astronomers will continue to have an influence on the study of Jupiter.

IMPLICATIONS AND FUTURE PROSPECTS

In the 1970s, *The Jupiter Effect* caused considerable public debate, likely because of the book's sensational (if not dire) conclusion that Earth would be destroyed. A positive benefit of that theory, which scientists worked quickly to debunk, was that greater attention was cast toward Jupiter. The Galileo program may not have captured the imaginations of nearly all of humanity in the same way that the Apollo missions to the moon did, but the information it collected was nothing short of extraordinary and did much to pique the public's curiosity about Jupiter.

If the Galileo program helped humanity better understand how Jupiter formed and how the Jovian system formed, the probe Ulysses helped humanity understand how Jupiter interacts with the rest of the solar system. Ulysses was launched in the early 1990's to study the sun, solar winds, and the sun's magnetic field. However, the probe also made use of the gravity of Jupiter, which happened to be orbiting in the vicinity. During its approach to Jupiter, Ulysses studied how the Jovian magnetosphere influenced the sun's magnetosphere, and vice versa. This probe helped lay the groundwork for continued study of the Jovian gravitational and magnetic fields—research that will continue in greater depth with future missions and evolving technologies.

It is likely that humanity will build on the information Galileo and other missions have gathered (and the data collected by the Hubble Space Telescope and ground-based technologies) by returning to Jupiter. In addition to the planned Russian mission to Jupiter, NASA and the European Space Agency have a planned series of joint missions to Jupiter and its moons, beginning in 2020. Among the purposes of these missions is to gain a better understanding of how the Jovian system formed and how it has evolved in relation to the other planets (including Earth). As technologies continue to develop and more probes are launched into space, scientists hope that Jupiter's effects on Earth and the rest of the solar system will become more apparent and quantifiable.

Michael P. Auerbach

FURTHER READING

Bagenal, Fran, Timothy E. Dowling, and William B. McKinnon, eds. *Jupiter: The Planet, the Satellites, and Magnetosphere.* New York: Cambridge University Press, 2004. This book provides an extensive review of what is known about Jupiter and its system. The editors include discussions of many of the theories that have been formulated about the planet and an overview of the many technologies used to study it.

Biryukov, E. "Capture of Comets from the Oort Cloud into Halley-Type and Jupiter-Family Orbits." *Solar System Research* 41, no. 3 (2007): 211-219. This article discusses how comets from the Oort cloud are sometimes captured in the Jovian system. Some of these comets move toward the inner solar system, while others are retained in the Jupiter-family orbital system.

Glasby, Frank. *Planets, Sunspots, and Earthquakes: Effects on the Sun, Earth, and Its Inhabitants.* Bloomington, Ind.: iUniverse, 2002. In this book the author explores how the sun and the planets influence one another through gravitational pulls and magnetic field interactions. The author examines the effects of these forces on sunspot, volcanic, and seismic activities.

Harland, David M. *Jupiter Odyssey: The Story of NASA's Galileo Mission.* New York: Springer-Praxis, 2011. This book focuses on the five-year Galileo mission to Jupiter. Includes a summary of the mission's planning stages and launch, the images recorded, technical and scientific papers and collections of other data, and materials generated from that program.

Johnson, Torrence V. "The Galileo Mission to Jupiter and Its Moons." *Scientific American* 13, no. 3 (2003): 54-63. In this article, the author describes the work that went into the Galileo mission. Includes a review of the data from the orbiting spacecraft. The Galileo mission has proved enormously beneficial to the study of the Jovian system, including the planet's atmosphere, its gravity and magnetosphere, and its moons.

Timofeev, V. E., et al. "Variations of the Interplanetary Magnetic Field and the Electron and Cosmic-Ray Intensities under the Influence of Jupiter." *Astronomy Letters* 33, no. 1 (2007): 63-66. This article describes the interactions the enormous magnetic field emanating from Jupiter has with the interplanetary magnetic field. Discusses how Jupiter has the strongest influence of all the planets on the intensity of the electrons contained in cosmic rays.

See also: Earth-Moon Interactions; Earth's Magnetic Field; Gravity Anomalies; Importance of the Moon for Earth Life; Remote-Sensing Satellites; Solar Wind Interactions.

L

LITHOSPHERIC PLATES

Lithospheric plates are large, distinct, plate-like segments of brittle rock. They are composed of upper mantle material and oceanic or continental crust. The seven major and numerous minor plates fit together to form the outer crust of the earth. The seven major plates include the Eurasian plate, the North American plate, the South American plate, the Pacific plate, the African plate, the Australian plate, and the Antarctic plate.

PRINCIPAL TERMS

- **asthenosphere:** a layer of the mantle in which temperature and pressure have increased to the point that rocks have very little strength and flow readily
- **density:** the mass of a given volume of material as compared to an equal volume of water
- **felsic:** rocks composed of the lighter-colored feldspars, such as granite
- **isostasy:** the balance of all large portions of the earth's surface when floating on a denser material
- **lithosphere:** the outermost portion of the globe, including the mantle above the asthenosphere
- **mafic:** rocks composed of dark, heavy, iron-bearing minerals such as olivine and pyroxene
- **mantle:** the layer of the earth between the crust and the outer core

PLATE STRUCTURE AND COMPOSITION

The lithosphere is the sphere of stone or outer crust of the earth. It is composed of seven major plate-like segments and numerous smaller ones. These lithospheric plates fit together in jigsaw-puzzle fashion. The recognition of the existence of these plates and their distinct boundaries has led to the theories of plate tectonics and seafloor spreading.

Lithospheric plates are layered. The bottom layer is the rigid upper portion of the mantle. The upper mantle is composed of dense, grayish green, iron-rich rock. Some plates have another solid layer of oceanic crust; this crustal rock is composed primarily of basalt. Some plates consist of only upper mantle and a thin covering of oceanic crust, while other plates have mantle material, oceanic crust, and continental crust. The continental crust is primarily granitic and is less dense than the basalt of the oceanic crust. Until recently, it was assumed that all plates had a continuous layer of oceanic crust and that continental crust was an additional layer, riding on the top. That no longer appears to be the case. The continental crust may be underlain by areas of oceanic crust in a discontinuous fashion, but the two crustal types are actually complexly intermingled.

The upper crustal rocks range from 12 kilometers thick over the ocean plains to more than 30 kilometers thick on the continental masses. The *Mohorovičić* discontinuity defines the boundary between the crust and the upper mantle. This boundary is recognized because seismic waves suddenly accelerate at it. The lithospheric plates, including the rigid upper mantle, are 75 to 150 kilometers thick. They float on the asthenosphere, which is a deeper portion of the mantle. The rock of the asthenosphere is under such pressure and increased temperature that it has little strength and can readily flow in much the same fashion as warm candle wax. The contact between the plates and the asthenosphere is marked by a sudden decrease in the speed of seismic waves.

The ability of the lithospheric plates to float on the asthenosphere is a key to understanding them. In much the same way that ice floats on water, the plates float on the material below them. Ice is able to float because it is less dense than water. As ice forms, it crystallizes and expands to fill more space. A given volume of ice has less density than the same volume of water. The density of the different layers of the earth increases toward the solid iron and nickel core. The lower mantle floats on the outer core, the asthenosphere floats on the lower mantle, and the lithospheric plates float on the asthenosphere.

PLATE MARGINS

The plates fit together along margins. There are generally considered to be only three types of plate

302

margin: ridges, trenches, and transform faults. Ridges, such as the Mid-Atlantic Ridge, are characterized by rifts or spreading centers. Trenches are margins where one plate is being forced below another and are the deepest areas of the ocean floor. Transform faults, such as the San Andreas fault in California, are areas where two plates are sliding alongside each other. The complex interactions of the lithospheric plates have led to the formation of the continents as they now exist. The plate margins do not necessarily follow the continental outlines. Continents may be composed of more than one plate. All the rocks and minerals that are on or near the surface are located on these plates. Geologic processes such as mountain building, earthquakes, and volcanism can be observed at or near the plate margins.

FORMATION OF CRUSTAL MATERIAL

Both types of crustal material—oceanic and continental—form through crystallization. This process is dependent on time, temperature, and pressure. As a molten material cools, a complex series of reactions occurs. The denser minerals crystallize early in the cooling of a molten material. If there is sufficient time in the cooling process, these early-formed dense minerals will gradually react with the remaining molten materials to form less dense minerals.

At the divergent plate margin, melting of the upper mantle gives rise to a silicate magma rich in iron and magnesium. This magma intrudes along fractures to be emplaced in the ocean floor as dikes and erupts to the surface as lava flows. The mafic material thus formed is called "basalt" and makes up the ocean floor. As new material is added by injection into the basalt of the ocean floor, it must push the existing material out of the way. Thus, a new sea floor is added at the spreading centers, which is made up of progressively older material away from the spreading center. In contrast, continents are composed of mostly granitic material which is formed from silicate magmas that are low in iron and magnesium and high in alkali-elements such as sodium and potassium. These granites are less dense than basalt.

If a basaltic oceanic plate collides with lighter continental crustal material, the continental crust will ride up over the oceanic plate, and the oceanic plate will be pushed down into the hotter mantle, where it will be assimilated back into the mantle. This convergent margin is marked by a deep oceanic trench on the ocean side of the collision zone. The descending oceanic plate is known as a subduction zone.

When two plates of continental material collide, neither can be subducted. If they do not begin to slide alongside each other, the compressive forces will form mountains. These mountains cannot rise higher than their isostatic balance. They must either be eroded by wind and water or sink back into the asthenosphere. The eroded pieces of rock, called sediment, are transported to lower areas called basins. As the sediment becomes more deeply buried, the pressure of overlying sediments causes them to lithify or become sedimentary rock. These sedimentary rocks have considerable pore space between the individual grains of sediment or silt and, therefore, are not very dense. They become additional continental crust material.

If the sedimentary rocks are buried deep enough, the increased temperature and pressure will begin a process known as metamorphism. During metamorphism, the original minerals in the sedimentary rock react with each other to form new minerals that are stable in the new environment. As temperature and pressure increase with depth, the rock becomes more and more like granite. If the temperature reaches high enough, the rock may melt and become magma.

MULTIDISCIPLINARY STUDY OF LITHOSPHERIC PLATES

Lithospheric plates fit together to form the crust or rock surfaces of the earth. The study of the surface of the earth and its composition is an extremely broad subject, including many of the subdisciplines of geology and oceanography. The study of the earth's surface and extraction of economic minerals have been undertaken since humankind's earliest times. Flint and obsidian used in tool making were early trade items; mining geology and mineralogy are almost as old. The early Greeks and Romans wrote books on geology.

Humans have used minerals and the metallic minerals since prehistory. Much knowledge of the earth is essentially a by-product of what was learned during the search for minerals, mineral ores, and gems. Something as simple as the formation of a nail requires iron ore and carbon. Mining geologists assay ores looking for economic deposits, and mineralogists study minerals.

Geophysicists bounce sound waves through the earth to determine subsurface structures. With the use of seismographs, they listen to earthquakes to pinpoint their locations. They also measure the gravity and magnetic field of specific areas of the earth. Petroleum geologists search for oil and gas by drilling into the earth's surface. Their interpretation of drill cuttings and core samples provides information about ancient environments. Volcanologists study volcanoes. They employ lasers to measure any minute movements on the surface of a volcano. They also use seismographs to detect the earthquakes that may signal an onset of volcanic activity. Because of the potential devastation of volcanoes, prediction has become increasingly important. Petrologists examine rocks to understand the earth processes that formed them. Their primary tools are the scanning electron microscope and X-ray diffraction machines.

Geochemists analyze the chemical composition of rocks and minerals and the reactions that may have caused their formation and dissolution. Paleontologists study fossilized life-forms, while paleoecologists study ancient environments. Planetary scientists investigate meteorites and moon rocks to increase understanding of the earth and its lithospheric plates. Much of what is known about the mantle material is a result of the study of meteorites.

STUDY OF OCEANIC CRUST

Much early geologic work was done on the more readily accessible continental crust. Recently, scientists have made considerable progress in the study of the oceanic crust. Early exploration of the ocean floor was through simple depth measurements from ships. Sailors lowered a weighted line over the side of a ship and physically measured the depth to the sea floor. The echo sounders developed in the early 1900's allowed for more rapid measurements of the ocean depths. In time, continuous profiles of the sea floor were made. Instead of the featureless plain that was expected, oceanic ridges, deep trenches, and numerous submerged volcanoes appeared.

Dredging is an old but ongoing method of sampling the surface of the ocean floor. The deep-diving bathysphere paved the way for bathyscaphes and other high-technology submersibles. Much recent work has been done with television cameras. A major find was made by a geologist in the late 1970's. A seafloor volcanic vent actually had life-forms subsisting on the chemically rich waters near it. Until this time, it had been assumed that all life on the earth was dependent on photosynthesis. This initial television discovery of chemosynthetic life-forms shocked the scientific community.

Drilling on the ocean floor has been accomplished by drill ships such as the *Glomar Challenger*. The cores of the deep ocean floor indicated a much younger oceanic crust than had been expected. Much that was learned about the oceanic crust simply did not fit with the scientific theories of the day. Serious rethinking had to be done, and many theories had to be radically changed.

SIGNIFICANCE OF PLATE INTERACTIONS

Lithospheric plates are the brittle rocks that float on the hot, plastic asthenosphere. They fit together to form the crust of the earth. Each step people take is either on the surface of a lithospheric plate or on something that is directly or indirectly made from one. Weathered surface rock provides the soil in which plants grow. The plants provide a breathable atmosphere and sustain animal life. The interaction of lithospheric plates leads to earthquakes, volcanic activity, and tidal waves. These impressive geologic displays have caught the human imagination since earliest times.

Since the lithospheric plates form the solid surface of the earth, in a real sense everything humans touch is related to them. Even something as unlikely as plastic is made from petroleum products extracted from the earth's crustal rocks. Coal, oil, and gas are burned to provide heat and electricity. Minerals extracted from lithospheric plates become the gold that makes jewelry, crowns teeth, and is part of circuit boards and computer chips. The minerals and compounds extracted from the lithospheric plates provide the iron for skyscrapers, cars, and car fuel. Coal, oil, and gas are formed by complex interactions of ancient plant life during the rock-forming processes that have occurred during the formation of upper portions of the lithospheric plates. The surface of the earth and its ongoing geologic processes also affect the weather.

The study of the composition and motion of lithospheric plates has created nearly all the body of knowledge in the field of geology. Numerous subdisciplines have arisen to study specific areas of geology. Study of the oceanic crust is relatively new, and recent discoveries are changing commonly accepted views of the earth. As views change, more discoveries seem to become possible. As more is learned about the earth's surface, views must be altered and upgraded to explain the phenomena observed.

Raymond U. Roberts

FURTHER READING

Condie, Kent C. *Plate Tectonics and Crustal Evolution.* 4th ed. Oxford: Butterworth Heinemann, 1997. An excellent overview of modern plate tectonics theory that synthesizes data from geology, geochemistry, geophysics, and oceanography. A very helpful tectonic map of the world is enclosed. The book is nontechnical and suitable for a college-level reader. Useful "suggestions for further reading" follow each chapter.

Davies, Thomas A. *Glaciated Continental Margins: An Atlas of Acoustic Images.* London: Chapman and Hall, 1997. Written for the college student, this book explores glacial landforms and their relationship to sedimentation, plate tectonics, continental drift, and the lithosphere. Filled with helpful maps and illustrations.

Glen, William. *Continental Drift and Plate Tectonics.* Columbus, Ohio: Charles E. Merrill, 1975. A college-level introductory text, this volume covers the concepts of lithospheric plates and their formation and motion. Although technical and outdated, the relevant material is introduced in a fashion that does not require a background in geology and is therefore beneficial to the nonscientific reader. Includes a very good index and an extensive supplementary-reading reference section.

Gross, M. Grant. *Oceanography.* 7th ed. Columbus, Ohio: Charles E. Merrill, 1996. Designed as an introductory-course text in oceanography for the college student. The text discusses oceanic plates and the sea floor, or oceanic crust. The historical section on sea floor study and current methodology is valuable, as are the index and the extensive supplementary reading list.

Marvin, Ursula B. *Continental Drift.* Washington, D.C.: Smithsonian Institution Press, 1973. Taking a historical approach, Marvin provides considerable discussion of plates and plate theory, covering old theories and explaining their progression toward new ones; the book includes discussion of the views that disagree with current theory. Index and extensive bibliography. For college-level readers.

Miller, Russell. *Planet Earth: Continents in Collision.* Alexandria, Va.: Time-Life Books, 1990. A clear and excellently illustrated introduction to lithospheric plates and plate tectonics. Extensive historical background is provided, and the concepts are introduced in a logical fashion. Good index and extensive bibliography. For advanced high school and college readers.

Ogawa, Yujiro, Ryo Anma, and Yildirim Dilek. *Accretionary Prisms and Convergent Margin Tectonics in the Northwest Pacific Basin.* New York: Springer Science+Business Media, 2011. Discusses new techniques in plate tectonic studies. One volume of the series *Modern Approaches in Solid Earth Sciences.* Covers accretionary prisms, tectonics, and Pacific Ocean events.

Olsen, Kenneth H., ed. *Continental Rifts: Evolution, Structure, Tectonics.* Amsterdam: Elsevier, 1995. The various essays provide good explanations of plate tectonics and continental rifts. Slightly technical but suitable for the careful reader. Illustrated.

Reynolds, John M. *An Introduction to Applied and Environmental Geophysics.* 2d ed. New York: John Wiley, 2011. An excellent introduction to seismology, geophysics, tectonics, and the lithosphere. Appropriate for those with minimal scientific background. Includes maps, illustrations, and bibliography.

_____. *Planet Earth: Earthquake.* Alexandria, Va.: Time-Life Books, 1982. Although dated, this text offers some exceptional illustrations of lithospheric plates. Chapter 5, "Dreams of Knowing When and Where," contains a good discussion of the equipment and methodology used to determine plate movement. Well indexed, with a good bibliography for additional reading. For high school and introductory college students.

Walker, Bryce. *Geology Today.* 10th ed. Del Mar, Calif.: Ziff-Davis, 1974. An excellent introductory text for the study of lithospheric plates, crustal rocks, plate movement, and geologic processes. The progression of concepts is clear and logical, and the volume is exceptionally well illustrated and indexed. The bibliography includes listings of other technical reference books.

Wegener, Alfred. *The Origin of Continents and Oceans.* Dover Publications, 1966. A classic text in the field of geology, this text provides the genesis of the plate tectonics and continental drift theories. Although outdated, it provides a historical foundation of these concepts upon which many current topics in geology are built.

See also: Continental Drift; Creep; Earthquakes; Earth's Core; Earth's Differentiation; Earth's Lithosphere; Earth's Mantle; Heat Sources and Heat Flow; Mantle Dynamics and Convection; Plate Motions; Plate Tectonics; Plumes and Megaplumes; Subduction and Orogeny; Tectonic Plate Margins; Volcanism.

LUNAR ORIGIN THEORIES

For centuries, humans have had a scientific interest in the moon and how it relates to the earth and the rest of the solar system. Scientists have sought to understand the moon, as well as the origins of the earth itself and the planet's place in the universe. The scientific study of the moon's origins is a relatively new field, and research has shed light on the moon's composition, orbit, and surface. Some lunar origin theories have been discounted, while others continue to be explored.

PRINCIPAL TERMS

- **Apollo:** the National Aeronautics and Space Administration (NASA) lunar expedition program active from 1963 to 1972, which included the 1969 Apollo 11 mission that placed humans on the moon for the first time
- **capture theory:** a lunar origin theory suggesting that the moon was created elsewhere in the solar system and was drawn into Earth's orbit
- **Chandrayaan-1:** an Indian lunar probe that landed on the moon in 2008
- **Chang'e 1:** a Chinese lunar orbiter launched in 2007 to map the moon's surface in preparation for a piloted mission to the moon in 2012
- **condensation theory:** a lunar origin theory suggesting that the earth and its moon were created from the same nebula materials that formed the rest of the solar system
- **fission theory:** a lunar origin theory suggesting that the moon and the spinning earth were separated from each other as the solar system formed
- **giant impactor theory:** a lunar origin theory suggesting that the moon was formed from fragments after another planetoid collided with the still-forming earth
- **Luna:** a Soviet lunar orbit expedition that culminated in the first impact on the moon's surface in 1959
- **Lunar Reconnaissance Orbiter:** a 2009 craft that found evidence of large amounts of ice at the moon's south pole
- **SMART-1:** a satellite from the European Space Agency's Small Missions for Advanced Research in Technology program, which launched in 2003 and was deliberately crashed into the moon in 2006, putting a European spacecraft on the moon for the first time
- **volatile elements:** basic chemical elements and compounds, such as nitrogen, helium, water, and methane, which have low boiling temperatures

BASIC PRINCIPLES

The study of the moon's origins is an important, yet challenging pursuit within the field of astronomy. Scientists have been studying the moon for decades, but even after this long period of time, very little about the moon's composition, history, or origins has been definitively revealed. Americans and Russians have sent missions to and around the moon (including the famous Apollo missions to the moon's surface), and have obtained hundreds of pounds of moon samples. However, analysis continues without conclusive evidence to support a number of lunar origin theories.

The fission theory advocates argue that the moon, during the early formation of the earth, was composed of a large amount of stray material from Earth and that it spun itself into the spherically shaped satellite seen today. According to the capture theory, the moon was created elsewhere in the solar system and then caught in Earth's gravitational pull as it moved through the system to its present location. In the condensation theory, adherents argue that both Earth and the moon were created from the nebula that formed the rest of the solar system. According to the giant impactor theory, a small planet struck Earth while it was still forming, and the superheated fragments of both the planet and the object spun together to form the moon. Although this theory has been widely accepted, efforts to locate concrete proof for it are ongoing.

BACKGROUND AND HISTORY

The study of the moon's origins and its relationship with the earth dates back to the early seventeenth century. In 1610, Galileo Galilei observed that the moon, like Earth, was covered with mountains, plains, and valleys. This was a considerable departure from previous notions that the moon's environment was completely dissimilar to that of Earth's. For his beliefs, Galileo was faced the Spanish Inquisition of the Roman Catholic Church. (His theories were inspired by Copernicus, whose ideas caused great controversy and were therefore considered taboo.) The Inquisition convinced Galileo to recant his ideas or risk torture. Several decades later, mathematician and philosopher René Descartes, inspired by Galileo's revolutionary viewpoints, introduced his own theory of the moon's origins, suggesting that it had been captured in Earth's gravity.

In 1878, George Darwin (son of naturalist Charles Darwin) suggested that the moon was a part of the earth, arguing that during its earliest stages of development, Earth had spun so fast that pieces of it had been "pulled" off the planet and ultimately became the moon. This "fission" idea was furthered by English geologist and physicist Osmond Fisher, who claimed that the material that formed the moon left a massive scar on Earth's surface that became the Pacific Ocean.

The fission theory would prevail until the early twentieth century, when American astronomer Thomas See speculated that the moon was created somewhere else in the solar system (hence its considerable differences in appearance from that of Earth). See theorized that the moon was captured in Earth's gravitational field and has remained there since. Shortly thereafter, French astronomer and mathematician Édouard Roche speculated that the moon was formed at the same time and from the same materials as the earth, and that over time they simply evolved in a different way, but still in tandem. This condensation theory would be added to the number of theories already on hand, but like the others, it lacked complete evidence.

In the late 1950s, the Soviet spacecraft Luna performed the first close-range examinations of the moon and took samples after striking the surface. In the late 1960s and early 1970s, NASA's Apollo missions returned with more than 850 pounds of moon rocks and other lunar samples. After analyzing the samples, scientists began to offer a new theory, which came to be known as the giant impactor theory (or the "big whack"). In this framework, a rogue planetoid of unknown origin collided with the newly forming Earth. The impact obliterated the planetoid and sent fragments of both itself and Earth into space, where they ultimately formed Earth's moon. As the samples from the moon's surface continued to reveal clues, the big whack remained the prevailing theory on the moon's origins.

GIANT IMPACTOR THEORY

The study of the moon's origins is part of an ongoing effort to understand the history of both Earth and the entire solar system. Although it has not been conclusively verified, the giant impactor theory has been embraced by scientists as the most reasonable hypothesis on which to pursue this study.

The Apollo missions of the late 1960s and early 1970s provided the most supporting evidence. Before these flights, it had been assumed that the moon's surface was dense, that its core was large, and that its composition included volatile elements (basic chemical elements and compounds, such as nitrogen, helium, water, and methane, which have low boiling temperatures). When the Apollo teams landed on the moon, they immediately began searching for evidence of these three characteristics. However, they (and scientists) were surprised to learn that the moon's surface was not very dense. When the astronauts surveyed the moon's core, they were surprised again to discover that the core was considerably smaller than theorized. This unexpected set of discoveries seemed to support the capture theory, until further examinations of the samples uncovered many elements that are found on Earth.

These revelations about the moon's and the earth's similarities and dissimilarities suggest a framework that is consistent with the big whack. Meanwhile, another discovery concerning other planets in the solar system provided even more convincing evidence: Planets that have a tilt in their axes, such as Mars, Pluto, and Neptune, experienced such shifts after a major event, such as a collision with an asteroid, moon, or another planet. Earth is among those planets with such a tilt. In light of these new discoveries, scientists have largely accepted the giant impactor theory as explaining the moon's origins.

PHOTO ANALYSIS

Much in the same way that Galileo and other early lunar scientists developed their theories by simply looking skyward, scientists have learned much about the moon simply by studying photographs taken of it. Since Apollo, spacecraft from NASA, the European Space Agency, the Japan Aerospace Exploration Agency (JAXA), and the Russian Federal Space Agency have traveled to and photographed areas of interest on the lunar surface. Various photographic technologies, such as spectral and thermographic equipment, have enabled scientists to capture different types of minerals and find evidence of water, ice, and certain chemical compounds on the lunar surface.

For example, a French study focused on the moon's Aristarchus crater (or plateau). Using ultraviolet and spectral-imaging photographic systems aboard the Clementine orbiting spacecraft and the Lunar Prospector craft (both of which circled the moon in the 1990s), astronomers were able to present a framework of the speed at which the moon's crust develops, how it develops, and how it transforms over time. This analysis of the plateau's crust provided more clues as to how the moon has developed over time. Another set of photographs helped scientists piece together a profile of the moon's topography, surface composition, ice and water volumes, and magnetic fields, all of which provide clues to the moon's history.

The moon was full after the total eclipse in the sky in Beijing on December 10, 2011. People across China were able to observe a total eclipse of the moon. (AFP/Getty Images)

MINERALOGICAL ANALYSIS

The collection of hundreds of pounds of moon rocks during the Luna and Apollo missions has revealed invaluable information that continues to be analyzed. For example, a 2010 study of lunar rocks focused on a particular mineral, apatite. Apatite accepts a number of volatile elements into its crystal structure and is prevalent in lunar rocks. The study also revealed that the concentrations of these elements are similar to those found in Earth-based apatite samples. The study also revealed the presence of sulfur in those crystal structures. On Earth, sulfur plays a major role in volcanic activity. This mineralogical revelation could cast light on how the moon's magma, core, and crust could have been formed. Then again, the presence of such large quantities of volatile elements raises questions about how, when the giant impact took place, these elements were able to survive in the moon's core fragments.

DATING TECHNIQUES

An important part of the pursuit of the moon's origins is determining the age of this celestial body. The rocks collected during the Apollo missions have provided a great deal of information to this end. Using radioactive isotopes and other dating techniques, scientists have determined the age of some of the rocks collected to be between 4.4 billion and 4.6 billion years. The 200-million-year difference is a vexing disparity.

Decades ago, the rocks that were brought back from the moon were believed to be formed at the moon's core and should have traveled outward to the moon's crust through a theoretical "molten rock ocean" that was the lunar core before it cooled. The rocks crust to the moon's surface. Over time, dating techniques have improved, revealing that the rocks' age is considerably younger. Although the moon's age has yet to be definitively established, the dating technology used to this end may soon solve this mystery. If such dating practices reveal that the rocks are indeed 4.4 billion years old and not 4.6 billion years old, then the widely accepted theory of the molten rock ocean could be called into question, as the particular rocks that were analyzed could not have floated to the moon's crust after the molten core cooled. Using dating practices may answer some questions, but also may raise others about the moon's origins.

Searching for Water

An important area that researchers explore when studying lunar origins is the probability of water on the moon. The presence and amount of water on the moon is an important clue about how the moon formed. Studying the moon rocks brought back by Apollo missions, scientists uncovered evidence of water. How much water exists on the moon has been an area of considerable debate. However, infrared photographic evidence (first uncovered in the late 1990's) of the moon's south pole suggested the presence of water. In 2009, the NASA Lunar Reconnaissance Orbiter (LRO) probe confirmed the assertion using radar, thermographic imaging, and other sensory equipment. LRO revealed that water was prevalent—an important discovery because, for the giant impactor theory to be viable, there should not be much water on the moon. A giant impact would have vaporized the water.

Meanwhile, a 2010 chemical analysis of the moon's relatively high chlorine content posited that there may not be as much water under the moon's surface as previously thought. During intense heat conditions (such as volcanic eruptions), chlorine and water cancel each other out: Higher levels of chlorine in the moon's mantle indicate that there is not much hydrogen at the moon's interior, and the possibility exists that the water may have come from elsewhere. Clearly, the presence of water casts doubt on the giant impactor theory. As such, this area of lunar origin studies continues to be a source of more mystery than evidence.

Relevant Groups and Organizations

National governments play perhaps the most integral role in the exploration of the moon's origins. NASA is not the only government agency with activity in this arena. For example, the European Space Agency (ESA) is an intergovernmental organization dedicated to the exploration of space. In 2003, ESA launched the SMART-1 probe (the first of the Small Missions for Advanced Research in Technology program), which was the first European spacecraft to orbit the moon. The craft cataloged ice and used innovative X-ray cameras to capture the reflections of iron, calcium, and other elements in the sunlight. In 2008, the Indian Space Research Organization (ISRO) launched Chadrayaan-1, a probe that detected water molecules when it touched down on the moon's surface. In 2007, China launched a probe, Chang'e 1, which mapped landing sites for a 2012 program to return humans to the moon's surface. As the world prepares for this return, governments will play the most significant role in bringing such plans to fruition.

Although NASA and other government space agencies maintain research laboratories to examine lunar and data collected from moon missions, universities also play important roles in the analysis of this data. Many major public and private universities around the world have gained access to these small souvenirs from the Apollo missions. Massachusetts Institute of Technology, Cambridge University, University of Tokyo, and other major universities have all gained access to these rocks and to the data gathered by orbiting probes; the universities conduct experiments and uncover a wide range of information about the moon's origins.

Most of the equipment used in the pursuit of information about the moon's origins is manufactured by private technology and engineering companies. From the onboard equipment used in the spacecraft (and the crafts themselves) to the analytical technologies used in the laboratory, such machinery is produced by government contractors, built to the specification (and maintained on site) by these private (or semiprivate) organizations. Boeing, Teledyne Brown Engineering, PerkinElmer, and Astrium are among the largest private contractors constructing propulsion systems, landing systems, mirrors, and sensory equipment for government space agencies.

Implications and Future Prospects

Since 1972, scientists have continued to analyze the samples that the Apollo and Luna missions brought home. When water molecules were discovered in those samples, the public's interest was again piqued. When it was learned that a great deal more ice and water exists on the moon's surface (and that more may lie beneath it), that interest turned into a growing desire to return to the moon.

In the fall of 2011, NASA launched the Gravity Recovery and Interior Laboratory (GRAIL) mission. The mission's pair of probes is mapping the moon's gravity and topography with more complexity than ever before. Additionally, the probes will carry a system called MoonKam (Moon Knowledge Acquired by Middle school students). With MoonKam, young people will be able to take images and video themselves on the moon's surface, enabling nonscientists to study the moon using the same technologies that scientists have used for decades. With a Chinese mission to the lunar surface (set to explore how to use the moon one day as a launching point to deep space), it is likely that humanity's interest in unraveling the moon's mysteries will continue.

Michael P. Auerbach

FURTHER READING

Canup, Robin M., and Kevin Righter, eds. *Origin of the Earth and Moon.* Tucson: University of Arizona Press, 2000. This book discusses the challenges facing scientists in proving the giant impactor theory, but expresses optimism that the complex and mysterious relationship between the earth and the moon will be further explained as technologies advance and as scientists continue to utilize a wide range of research tools.

Elkins-Tantan, Linda T., Seth Burgess, and Qing-Zhu Yin. "The Lunar Magma Ocean: Reconciling the Solidification Process with Lunar Petrology and Geochronology." *Earth and Planetary Science Letters,* 304, nos. 3/4 (April 2011): 326-336. This article explores the process by which the magma center of the moon cooled to form the moon's mantle and crust.

Goswami, J. N., and M. Annadurai. "Chandrayaan-1: India's First Planetary Science Mission to the Moon." *Current Science* 96, no. 4 (February 25, 2009): 486-491. This article discusses the parameters of the first Indian lunar mission. The authors discuss the cutting-edge technologies onboard the spacecraft, which was used to study and map the moon's surface from the lunar orbit.

Harland, David M. *Exploring the Moon: The Apollo Expeditions.* 2d ed. New York: Springer-Praxis, 2008. This book provides an extensive review of the Apollo missions, examining the research, the probes that crashed into the moon's surface, the orbiters, and the first steps on the moon. Much of the analysis focuses on the Apollo program's investigations and the findings from that mission.

Lee, Der-Chuen, and Alex N. Halliday. "Age and Origin of the Moon." *Science* 278 (November 7, 1997). This article provides an overview of the research that contributed to the giant impactor theory. The author discusses isotopic studies of twenty-one lunar samples, including the use of a tungsten-based chronometer to calculate the age of those rock and soil samples.

MacKenzie, Dana. *The Big Splat: Or, How Our Moon Came to Be.* Hoboken, N.J.: Wiley, 2003. This book first examines the history of lunar observation, including ancient civilizations' concepts of the moon before it began to be studied in a scientific light. The book also discusses the giant impactor theory and other hypotheses about the origins of the moon, including how these theories came to fruition.

Yan, Bokun, et al. "Minerals Mapping of the Lunar Surface with Clementine UVVIS/NIR Data Based on Spectra Unmixing Method and Hapke Model." *Icarus* 208, no. 1 (July 2010): 11-19. This article discusses the study of minerals recorded by the Clementine probe and their relevance to the pursuit of the moon's origins. The authors suggest a new course of action, focusing on the search for important minerals that can provide clues as to how the moon was formed.

See also: Asteroid Impact Craters; Earth-Moon Interactions; Earth's Differentiation; Earth's Magnetic Field; Earth Tides; Gravity Anomalies; Importance of the Moon for Earth Life; Rare Earth Hypothesis.